Lecture Notes
in Computational Science
and Engineering

71

Editors

Timothy J. Barth
Michael Griebel
David E. Keyes
Risto M. Nieminen
Dirk Roose
Tamar Schlick

Barry Koren • Kees Vuik
Editors

Advanced Computational Methods in Science and Engineering

With 109 Figures and 4 Tables

Editors
Barry Koren
Centrum Wiskunde & Informatica
Group Computing and Control
Science Park 123
1098 XG Amsterdam
The Netherlands
barry.koren@cwi.nl

Kees Vuik
Delft University of Technology
Institute of Applied Mathematics
Mekelweg 4
2628 CD Delft
The Netherlands
c.vuik@tudelft.nl

ISSN 1439-7358
ISBN 978-3-642-03343-8   e-ISBN 978-3-642-03344-5
DOI 10.1007/978-3-642-03344-5
Springer Heidelberg Dordrecht London New York

Library of Congress Control Number: 2009932722

Mathematics Subject Classification (2000): 65-XX, 70-XX, 74-XX, 76-XX, 78-XX

© Springer-Verlag Berlin Heidelberg 2010
This work is subject to copyright. All rights are reserved, whether the whole or part of the material is concerned, specifically the rights of translation, reprinting, reuse of illustrations, recitation, broadcasting, reproduction on microfilm or in any other way, and storage in data banks. Duplication of this publication or parts thereof is permitted only under the provisions of the German Copyright Law of September 9, 1965, in its current version, and permission for use must always be obtained from Springer. Violations are liable to prosecution under the German Copyright Law.
The use of general descriptive names, registered names, trademarks, etc. in this publication does not imply, even in the absence of a specific statement, that such names are exempt from the relevant protective laws and regulations and therefore free for general use.

Cover illustration: © TU Delft, Delft Center for Computational Science and Engineering

Cover design: deblik, Berlin

Printed on acid-free paper

Springer is part of Springer Science+Business Media (www.springer.com)

Foreword

James Clerk Maxwell wrote in the introduction to his Treatise on Electricity and Magnetism, in 1873:

"As I proceeded with the study of Faraday, I perceived that his method of conceiving the phenomena was also a mathematical one, though not exhibited in conventional form of mathematical symbols. I also found that these methods were capable of being expressed in the ordinary mathematical forms, and thus compared with those of the professional mathematicians. For instance, Faraday, in his minds eye, saw lines of force traversing all space where the mathematicians saw centres of force attracting at a distance." (sic!)

I consider this as a fine example how the model of the practitioner of science stimulates the mathematical mind. This fruitful interaction was and still is the objective of the Delft Centre for Computational Science and Engineering, an interfaculty institution, founded in 2003 by Professor Piet Wesseling. In its present form the centre acts as a forum for the computational aspects of different branches of science and engineering, taught and studied at Delft University of Technology. The centre proofs its value for the university in a twofold way: scientists and engineers discuss their problems with mathematicians, and mathematicians find new applications for their theories and methods. This is a clear case of cross-fertilization. Certainly in a world with growing interaction between different disciplines, there is a growing need for a common, albeit more complex, model representation. Mathematics furnishes this framework because its language is commensurable over the fields of application.

The articles in this book are a fine collection of examples that show the strength and progress of applied mathematics in the realm of technology. Just like a museum director chooses paintings with care to express to the public the power of a certain art theme; the editors of this book have carefully chosen topics to demonstrate the power of computational science and engineering. I hope this book stimulates the reader to paint his/her own specimen.

Jacob Fokkema
Rector Magnificus
Delft University of Technology

Preface by the editors

Computational Science and Engineering (CSE) is of vital importance to today's and tomorrow's society. It enables the simulation of processes, phenomena and systems that can not be studied by real experiments because these are too dangerous, too expensive, unethical, or just technically impossible. Moreover, as opposed to experiments, CSE allows for automatic design and optimization. Every major discipline in science and engineering has its own computational branch now. Engineers, medical doctors, policy makers, et cetera rely more and more on CSE for decision support. With the longstanding, continuing growth in speed, memory and cost-effectiveness of computers, and with similar improvements in numerical algorithms, the existing and future benefits of CSE are enormous. CSE is and will be a crucial enabling technology.

Although CSE has spread over many different disciplines, it is to be regarded as a discipline in its own right, because of the specialized skills involved, the long learning curve required, and the rapid pace of innovation, which is impossible to keep track of by non-experts and casual users. The challenge of CSE is the development and exploitation of ever more realistic computational models. Perfect realism by direct (brute-force) simulation is impossible in most instances, and will remain to be so for a long time, if not forever. With increasing realism, in general, multi-scaledness and multi-disciplinarity also increase, two fundamental inhibitors for direct simulation. Significant disparity in scales motivates multi-scale modeling instead. Although multi-scale modeling has always been part and parcel of science (think of replacing a body by a point mass, or of the continuum approximation in fluid dynamics), CSE has led to a renewed interest in it, and has opened new perspectives and challenges. Concerning multi-disciplinarity, this motivates more and more the intelligent coupling of distinct, existing models: fluid-dynamics plus structural-mechanics models, thermal plus electrical models, et cetera.

Although CSE depends heavily on computer architectures, numerical mathematics is at its heart. A challenge is to ensure that numerical methods remain capable of optimally profiting from the growing speed and memory of tomorrow's computer architectures. With the increasing realism and hence complexity in CSE, numerical robustness and numerical efficiency are properties of growing importance. Baby-sitting a complex CSE simulation because of the use of non-robust numerical techniques in it, is unwanted. And so is quick filling of fast and large computers by the use of inefficient numerical methods. Numerical robustness is to be obtained by developing and applying numerical methods that are stable and well-conditioned for the specific problems at hand. Numerical efficiency has to be striven for by extracting the maximum amount of correct information from the minimum amount of grid points, cells, and alike, and by aiming at a linear proportionality between the computing time required and the number of unknowns. Problems with N unknowns are preferably to be solved in $\mathcal{O}(N)$ operations. With insufficient attention for the numerics, this may easily be $\mathcal{O}(N^2)$ operations, implying quick filling of the fastest and largest computers.

Numerical mathematics is found throughout the entire book, like oxygen in the earth's atmosphere; from low to high density, often imperceptible but always indispensable.

CSE profits from its strategic position along two axes: the methodology axis and the discipline axis. It finds its coherence and added value along the methodology axis; in the numerical methods in common use by the various disciplines. Along the discipline axis, CSE finds its research focus and relevance. CSE is fruitful breeding ground; its multi-methodological – multi-disciplinary character allows for very many cross-fertilizations.

In 2003, at TU Delft, the Delft Centre for Computational Science and Engineering (www.cse.tudelft.nl) was founded. Computational groups from various disciplines participate in the centre. In this book, we follow the discipline axis. Most of the disciplinary groups in the Delft Centre for CSE contributed with a chapter on one or more of their own favorite research topics. The book is intended for readers, not necessarily numerical specialists, who are involved in CSE somehow. Our aim is to provide the reader with an impression of CSE's challenges and opportunities.

This book is the work of many people, whom we owe thanks. We thank Piet Wesseling for his visionary work in founding the Delft Centre for Computational Science and Engineering. Rector Magnificus Jacob Fokkema is thanked for his stimulus in preparing this book. Last but not least, we thank the authors for their efforts put into the preparation of their chapters.

December 2008   Barry Koren and Kees Vuik

Contents

A Model-Order Reduction Approach to Parametric Electromagnetic Inversion   1
R.F. Remis and N.V. Budko

Shifted-Laplacian Preconditioners for Heterogeneous Helmholtz Problems   21
C.W. Oosterlee, C. Vuik, W.A. Mulder, and R.-E. Plessix

On Numerical Issues in Time Accurate Laminar Reacting Gas Flow Solvers   47
S. van Veldhuizen, C. Vuik, and C.R. Kleijn

Parallel Scientific Computing on Loosely Coupled Networks of Computers   79
T.P. Collignon and M.B. van Gijzen

Data Assimilation Algorithms for Numerical Models   107
A.W. Heemink, R.G. Hanea, J. Sumihar, M. Roest, N. Velzen, and M. Verlaan

Radial Basis Functions for Interface Interpolation and Mesh Deformation   143
A. de Boer, A.H. van Zuijlen, and H. Bijl

Least-Squares Spectral Element Methods in Computational Fluid Dynamics   179
M. Gerritsma and B. De Maerschalck

Finite-Volume Discretizations and Immersed Boundaries   229
Y. Hassen and B. Koren

Large Eddy Simulation of Turbulent Non-Premixed Jet Flames with a High Order Numerical Method   269
S. van der Hoeven, B.J. Boersma, and D.J.E.M. Roekaerts

A Suite of Mathematical Models for Bone Ingrowth, Bone Fracture Healing and Intra-Osseous Wound Healing   289
F.J. Vermolen, A. Andreykiv, E.M. van Aken, J.C. van der Linden, E. Javierre, and A. van Keulen

Numerical Modeling of the Electromechanical Interaction in MEMS 315
S.D.A. Hannot and D.J. Rixen

Simulation of Progressive Failure in Composite Laminates 343
F.P. van der Meer and L.J. Sluys

Numerical Modeling of Wave Propagation, Breaking and Run-Up on a Beach 373
G.S. Stelling and M. Zijlema

Hybrid Navier-Stokes/DSMC Simulations of Gas Flows with Rarefied-Continuum Transitions 403
G. Abbate, B.J. Thijsse, and C.R. Kleijn

Multi-Scale PDE-Based Design of Hierarchically Structured Porous Catalysts 437
G. Wang, C.R. Kleijn, and M.-O. Coppens

From Molecular Dynamics and Particle Simulations towards Constitutive Relations for Continuum theory 453
S. Luding

A Model-Order Reduction Approach to Parametric Electromagnetic Inversion

Rob F. Remis and Neil V. Budko

Abstract Inverse scattering is a systematic approach to nondestructive testing, geophysical prospecting, remote sensing, and medical imaging. Being both nonlinear and ill-posed, inverse scattering problems are among the most challenging in computational science. In this chapter we expose the intimate connection between these problems and reduced-order modeling techniques, which allow significant reduction in computational complexity and efforts. Exploiting the natural symmetries of the electromagnetic field equations we develop and analyze reduced-order models for the parametric inversion problems arising in effective medium theory, metamaterials, and photonics. These models are Padé approximations of the measured scattered field. The achieved significant computational gain allows us to solve these large numerical problems by simple inspection of a two-dimensional objective functional.

1 Introduction

In many applications we want to look inside an object without breaking it (nondestructive testing), or digging it up (geophysical prospecting), or cutting it open (medical imaging). Electromagnetic waves are often used for this purpose due to their relatively large penetration depth. The basic idea is to illuminate the object and to reconstruct its interior from the scattered field measurements made somewhere outside the object. This reconstruction problem is called an *inverse scattering problem*. The electromagnetic constitutive parameters one usually wants to reconstruct are the permittivity ε and the conductivity σ. If the object is inhomogeneous these parameters vary with position.

Rob F. Remis and Neil V. Budko
Faculty of Electrical Engineering, Mathematics and Computer Science, Laboratory of Electromagnetic Research, Mekelweg 4, 2628 CD Delft, The Netherlands, e-mail: R.F.Remis@TUDelft.NL and N.V.Budko@TUDelft.NL

The inverse scattering problem is nonlinear, ill-posed, and computationally intense. Many different solution methods exist, ranging from linearized methods to full Newton-type minimization schemes (see the topical IOP journal *Inverse Problems*). In this chapter we describe a reduced-order approach to electromagnetic inverse scattering. Just as many other linearized and fully nonlinear inverse solution strategies, our technique is iterative in nature. We exploit the reduced-order parametric models naturally emerging from Krylov subspace iterative schemes. Appearing in different contexts, such as many-mass computations in quantum chromodynamics and optimal regularization parameter selection, similar approaches are used by Frommer and Glässner [10] and Frommer and Maas [11].

The need for reduced-order methods stems from the nonlinearity of the inverse scattering problem. Since the electromagnetic field is measured only outside the object, the field inside is, in fact, another implicit unknown of the problem. Although, the internal field is never found explicitly, it is introduced as a constraint in a nonlinear minimization problem. Hence, even if our ultimate goal is the permittivity and/or conductivity, we need to deal with the internal field as well. For example, constitution of a homogeneous object of known shape is completely determined by two parameters only. However, the field inside this object is a nontrivial function of position. Even upon discretization, this leads to a very large number of additional discrete degrees of freedom, especially for electrically large objects (objects large compared with the wavelength of the incident field). In principle, it is possible to take the internal field constraint into account by formally solving a linear forward scattering problem. This results in a single nonlinear equation relating the unknown constitutive parameters to the measured scattered field data. However, this equation then contains an inverse of a very large matrix. For practical large-scale problems, repeatedly evaluating the action of this inverse on a given vector is out of the question. This is where the reduced-order modeling comes into play. Given the nature of our problem, it turns out that we can reuse a single Krylov subspace for many different values of the unknown constitutive parameters. This property is known as the shift-invariance property of the Krylov subspace. In practice it means that we have to generate only one Krylov subspace leading to a significant reduction of computational efforts.

Initially we successfully implemented this technique using the Arnoldi algorithm [4]. Here we exploit the symmetry of our equations and employ a much more economic Lanczos-type algorithm [17]. We also show that the approximations constructed by this algorithm are actually Padé approximations of a certain type, which is well known in the control and optimization communities. Finally, we mention that presently research on model-order reduction techniques using rational interpolants in inverse scattering problems is ongoing and an extension of the method presented here (in combination with a Gauss-Newton minimization algorithm) was recently proposed by Druskin and Zaslavsky [8].

2 Integral representations and their discretized counterparts

We consider a two-dimensional configuration that is invariant in the z-direction. The configuration is sketched in Figure 1 and consists of an inhomogeneous object embedded in a homogeneous and lossless background medium. The medium parameters of the background are the constant permittivity ε_b and the constant permeability μ_b, while the object is characterized by a permittivity $\varepsilon(\mathbf{x})$, a conductivity $\sigma(\mathbf{x})$, and a permeability μ_b. The object occupies a bounded domain $\mathbb{D}$ and shows no contrast in its permeability.

Fig. 1 To probe the inside of an inhomogeneous penetrable object, we illuminate it by electromagnetic waves. These waves are generated by an electric line source (black triangle) and the total electric field strength is measured by a receiver (white triangle). Both the source and the receiver are located outside the object.

An electric line source is located outside the object domain $\mathbb{D}$ and generates electromagnetic waves to probe the inside of the object. Specifically, the line source is described by the current density

$$J_z^{\text{ext}}(\mathbf{x}, \omega) = f(\omega)\delta(\mathbf{x} - \mathbf{x}^{\text{src}}),$$

where $f(\omega)$ is the source signature and the delta function is the Dirac distribution operative at $\mathbf{x} = \mathbf{x}^{\text{src}} \notin \mathbb{D}$.

For the configuration described above, we have for the scattered electric field strength $\tilde{u}^{\text{sc}}$ the integral representation

$$\tilde{u}^{\text{sc}}(\mathbf{x}, \omega) = \int_{\mathbf{x}' \in \mathbb{D}} G(\mathbf{x} - \mathbf{x}', \omega)\chi(\mathbf{x}', \omega)\tilde{u}(\mathbf{x}', \omega)\,d\mathbf{x}', \tag{1}$$

which holds for any $\mathbf{x} \in \mathbb{R}^2$. In Eq. (1), the scattered field is the difference between the total field $\tilde{u}$ and the incident field u^{inc}. The latter field is the field that is present if the object is absent. Moreover, the Green's function is given by

$$G(\mathbf{x}, \omega) = \frac{ik_b^2}{4} H_0^{(1)}(k_b |\mathbf{x}|), \qquad (2)$$

where $H_0^{(1)}$ is the zero-order Hankel function of the first kind, and k_b is the wave number of the background medium. Finally, the contrast function is given by

$$\chi(\mathbf{x}, \omega) = \frac{\varepsilon(\mathbf{x})}{\varepsilon_b} - 1 + i\frac{\sigma(\mathbf{x})}{\omega\varepsilon_b}. \qquad (3)$$

Notice that the permittivity determines the real part of the contrast function, while the conductivity determines its imaginary part.

Even though the object may be inhomogeneous, we try to find an effective constant permittivity ε_{eff} and conductivity σ_{eff} such that an object characterized by these medium parameters, and having the same support as the true scatterer, matches the measured total (or scattered) field of the true object as good as possible. We refer to such a scatterer as an *effective scatterer* and its contrast function is given by the complex scalar

$$\zeta = \frac{\varepsilon_{\text{eff}}}{\varepsilon_b} - 1 + i\frac{\sigma_{\text{eff}}}{\omega\varepsilon_b}. \qquad (4)$$

For the scattered electric field due to the effective scatterer we have the integral representation

$$u^{\text{sc}}(\mathbf{x}, \omega) = \zeta \int_{\mathbf{x}' \in \mathbb{D}} G(\mathbf{x} - \mathbf{x}', \omega) u(\mathbf{x}', \omega) \, d\mathbf{x}', \qquad (5)$$

and at a receiver location $\mathbf{x} = \mathbf{x}^{\text{rec}} \notin \mathbb{D}$ this becomes

$$u^{\text{sc}}(\mathbf{x}^{\text{rec}}, \omega) = \zeta \int_{\mathbf{x}' \in \mathbb{D}} G(\mathbf{x}^{\text{rec}} - \mathbf{x}', \omega) u(\mathbf{x}', \omega) \, d\mathbf{x}'. \qquad (6)$$

This equation is known as the *data equation*. Taking $\mathbf{x} \in \mathbb{D}$ in Eq. (5) and using $u^{\text{sc}} = u - u^{\text{inc}}$, where u is the total field in presence of the effective scatterer, we have

$$u(\mathbf{x}, \omega) - \zeta \int_{\mathbf{x}' \in \mathbb{D}} G(\mathbf{x} - \mathbf{x}', \omega) u(\mathbf{x}', \omega) \, d\mathbf{x}' = u^{\text{inc}}(\mathbf{x}, \omega), \qquad \mathbf{x} \in \mathbb{D}. \qquad (7)$$

This equation is known as the *object equation*. It is a Fredholm integral equation of the second kind for the total field u if the contrast ζ is known.

The data and object equation are the basic equations of our inverse scattering problem. The data equation describes the connection between the total field inside the object and the scattered field at the receiver location, while the object equation acts as a kind of constraint on the total field in the sense that although we do not know this field, we do know that it has to satisfy Eq. (7).

We discretize the data and object equation using a standard discretization procedure which is well documented in the literature. We therefore only indicate the basic steps here. Details can be found in Balanis [2] and Chew et al. [6], for example.

The first step consists of introducing a uniform grid consisting of N square cells with side lengths $\delta > 0$. The contrast is assumed to be constant within each cell. Second, the total electric field strength is approximated by a rectangular pulse expansion and, finally, point matching is used to arrive at the discretized data and object equations. We note that the singularity of the Green's function requires special treatment, but this step is also well documented in the literature and we do not repeat it here (see, again, Balanis [2], for example).

After the discretization procedure briefly outlined above, we arrive at the discretized data equation

$$u^{sc}(\mathbf{x}^{rec}, \omega) \approx \gamma_r \zeta \mathbf{r}^T \mathbf{u}, \tag{8}$$

and the discretized object equation

$$(\mathbf{I} - \zeta \mathbf{G})\mathbf{u} = \gamma_s \mathbf{s}. \tag{9}$$

In the above equations, we have $\gamma_r = i(k_b \delta)^2/4$, $\gamma_s = i\omega \mu_b f(\omega)/4$, $\mathbf{u}$ contains the expansion coefficients of the electric field strength pulse expansion,

$$\mathbf{r} = \text{vec}\left[H_0^{(1)}(k_b|\mathbf{x}^{rec} - \mathbf{x}_n|)\right] \quad \text{and} \quad \mathbf{s} = \text{vec}\left[H_0^{(1)}(k_b|\mathbf{x}^{rec} - \mathbf{x}_n|)\right], \tag{10}$$

where $\mathbf{x}_n$ is the position vector of the midpoint of the nth cell. The right-hand side of Eq. (8) defines what we call modeled scattered field data. This expression can actually be computed and we write it as

$$v^{sc} = \gamma_r \zeta \mathbf{r}^T \mathbf{u}. \tag{11}$$

Matrix $\mathbf{G}$ is a discrete representation of the continuous convolution operator in Eq. (7). We mention two properties of this matrix that will be exploited later on. The first one is that matrix $\mathbf{G}$ is complex-symmetric, that is, it has complex entries and satisfies $\mathbf{G}^T = \mathbf{G}$ [1]. The second property is that since the discretized object equation is obtained by discretizing a convolution operator on a uniform grid, its action on a vector can be computed at "FFT-speed." We take advantage of this property in the Lanczos algorithm discussed below. Moreover, from now on we assume that the object operator $\mathbf{I} - \zeta \mathbf{G}$ is not ill-conditioned for all ζ-values of interest. This assumption amounts to assuming that the forward problem of determining $\mathbf{u}$ for a given ζ is not ill-conditioned. Solving now the discretized object equation for the field vector $\mathbf{u}$ and substituting the result in the expression for the modeled scattered field data, we obtain

$$v^{sc} = \gamma \zeta \mathbf{r}^T (\mathbf{I} - \zeta \mathbf{G})^{-1} \mathbf{s}, \tag{12}$$

where $\gamma = \gamma_r \gamma_s$. Equation (12) clearly shows that the scattered field data depends nonlinearly on the contrast coefficient ζ.

Finally, we mention that if the source and receiver location do not coincide, we have $\mathbf{s} \neq \mathbf{r}$ and this is referred to as a *bistatic* source-receiver setup. For a setup where the source and the receiver are located at the same position we have $\mathbf{s} = \mathbf{r}$ and

---

[1] Matrix $\mathbf{G}$ is symmetric but not Hermitian, that is, $\mathbf{G}^H \neq \mathbf{G}$.

Eq. (12) becomes
$$v^{sc} = \gamma\zeta \mathbf{r}^T (\mathbf{I} - \zeta\mathbf{G})^{-1} \mathbf{r}. \tag{13}$$
This is referred to as a *monostatic* source-receiver setup.

3 Reduced-order models for the scattered field

The expression for the scattered field data as given by Eq. (12) is very similar in form as the Laplace domain expression for a certain output variable $y(s)$ of a single-input, single-output, linear, and time-invariant (SISO-LTI) system. More precisely, given a SISO-LTI system described by an input vector $\mathbf{s}$, an output vector $\mathbf{r}$, and a system matrix $\mathbf{A}$, we have for the Laplace-domain output variable of interest the expression (see Antsaklis and Michel [1])
$$y(s) = \mathbf{r}^T (\mathbf{A} + s\mathbf{I})^{-1} \mathbf{s}.$$
Comparing this with Eq. (12), we observe that ζ^{-1} plays the role of the Laplace domain parameter s (assuming that $\zeta \neq 0$, of course). Now if one is interested in the output variable $y(s)$ for a complete range of Laplace parameter values, then for each new s a system of equations needs to be solved if the above expression for $y(s)$ is used. Such an approach may be computationally very expensive and is avoided in the Padé Via Lanczos process (PVL process) as introduced by Feldmann and Freund [9]. In this PVL approach, one first constructs a certain low-degree Padé approximation of $y(s)$ and then uses this approximation to obtain accurate values for $y(s)$ on the desired range of s-values. The crux of the matter is that evaluating the Padé approximation for different values of s is much more economical than solving systems of equations for each new Laplace parameter. The Padé approximations themselves are called reduced-order models and can be constructed very efficiently via the Lanczos algorithm provided that matrix-vector multiplications with the system matrix $\mathbf{A}$ are efficient.

Now before we apply the PVL process to our inverse scattering problem, we first rewrite Eq. (12) in such a way that the construction of the reduced-order models requires less memory usage than if we base our construction on Eq. (12) directly. Specifically, we follow a similar approach as Golub and Strakos [12] and show that a bistatic source-receiver setup can be written in terms of two monostatic source-receiver setups by exploiting the symmetry of matrix $\mathbf{G}$. To show this, we first introduce the vectors
$$\mathbf{x} = \mathbf{r} + \mathbf{s} \quad \text{and} \quad \mathbf{y} = \mathbf{r} - \mathbf{s},$$
and compute
$$\mathbf{x}^T (\mathbf{I} - \zeta\mathbf{G})^{-1} \mathbf{x} - \mathbf{y}^T (\mathbf{I} - \zeta\mathbf{G})^{-1} \mathbf{y} = 2\mathbf{r}^T (\mathbf{I} - \zeta\mathbf{G})^{-1} \mathbf{s} + 2\mathbf{s}^T (\mathbf{I} - \zeta\mathbf{G})^{-1} \mathbf{r}.$$
Since matrix $\mathbf{G}$ is symmetric, we have

$$\mathbf{s}^T(\mathbf{I}-\zeta\mathbf{G})^{-1}\mathbf{r} = \mathbf{r}^T(\mathbf{I}-\zeta\mathbf{G})^{-1}\mathbf{s},$$

and the above reduces to

$$\mathbf{x}^T(\mathbf{I}-\zeta\mathbf{G})^{-1}\mathbf{x} - \mathbf{y}^T(\mathbf{I}-\zeta\mathbf{G})^{-1}\mathbf{y} = 4\mathbf{r}^T(\mathbf{I}-\zeta\mathbf{G})^{-1}\mathbf{s}.$$

Consequently, the expression for the modeled scattered field data can be written as

$$v^{sc} = \frac{\gamma\zeta}{4}\left[\mathbf{x}^T(\mathbf{I}-\zeta\mathbf{G})^{-1}\mathbf{x} - \mathbf{y}^T(\mathbf{I}-\zeta\mathbf{G})^{-1}\mathbf{y}\right], \tag{14}$$

and this expression will serve as a starting point for the construction of our reduced-order models.

We start by considering the first term between the square brackets in Eq. (14). This term is actually a rational function in ζ, denoted by $z(\zeta)$, in which the degree of the polynomial in the numerator is $N-1$, while the degree of the polynomial in the denominator is N. The idea is now to approximate z by a low-degree Padé approximation of a similar type as z. Specifically, we approximate z by the Padé approximation

$$z_k(\zeta) = \frac{n_{k-1}\zeta^{k-1} + \ldots + n_1\zeta + n_0}{d_k\zeta^k + \ldots + d_1\zeta + 1},$$

where the coefficients $n_0, n_1, \ldots, n_{k-1}$ and $d_1, d_2, \ldots, d_k$ follow from the requirement that the first $2k$ Taylor coefficients of z around $\zeta = 0$ agree with the Taylor series of z_k around the same expansion point. Notice that z_k is of the same type as z, that is, the degree of the polynomial in the numerator is one less than the degree of the polynomial in the denominator. Computing the Taylor series of z, we find

$$z(\zeta) = \sum_{j=0}^{\infty} m_j \zeta^j,$$

where the coefficients

$$m_j = \mathbf{x}^T \mathbf{G}^j \mathbf{x}, \quad j = 0, 1, \ldots, \tag{15}$$

are called the moments of z. The desired Padé approximation that agrees with this Taylor series in the first $2k$ moments can be computed via a Lanczos-type algorithm. In this algorithm we exploit the symmetry of matrix $\mathbf{G}$ and the fact that matrix-vector products with matrix $\mathbf{G}$ can be computed at "FFT-speed." The details are as follows. We first introduce the indefinite inner product

$$\langle \mathbf{u}, \mathbf{v} \rangle = \mathbf{v}^T \mathbf{u}.$$

This inner product is indefinite, since it is defined over the complex vector space $\mathbb{C}^N$. Using such an inner product has consequences for the Lanczos algorithm (it may terminate prematurely) as will be discussed later on. Since matrix $\mathbf{G}$ is symmetric with respect to the indefinite inner product, we can reduce it to a sequence of tridiagonal matrices using a single three-term recurrence relation. This is the Lanczos-type algorithm that we are after. Specifically, the reduction is computed as follows:

Lanczos-type algorithm

1. Set $a_x = \langle \mathbf{x}, \mathbf{x} \rangle$, $\beta_1 = a_x^{1/2}$, $\mathbf{q}_0 = \mathbf{0}$, and $\mathbf{q}_1 = \beta_1^{-1} \mathbf{x}$;
2. Give k, the maximum number of iterations;
3. For $j = 1, 2, ..., k$ compute

$$\mathbf{w}_j = \mathbf{G}\mathbf{q}_j - \beta_j \mathbf{q}_{j-1},$$
$$\alpha_j = \langle \mathbf{w}_j, \mathbf{q}_j \rangle,$$
$$\beta_{j+1} = \langle \mathbf{w}_j - \alpha_j \mathbf{q}_j, \mathbf{w}_j - \alpha_j \mathbf{q}_j \rangle,$$
$$\mathbf{q}_{j+1} = \beta_{j+1}^{-1}(\mathbf{w}_j - \alpha_j \mathbf{q}_j).$$

Notice that matrix $\mathbf{G}$ is only required to form matrix-vector products in this algorithm. Furthermore, it may happen that $\beta_{j+1} = 0$. If this is because $\mathbf{w}_j - \alpha_j \mathbf{q}_j = \mathbf{0}$ then we have computed an invariant subspace for matrix $\mathbf{G}$ and we are done. This is called a regular termination of the algorithm, but unfortunately such terminations hardly occur in practice. If $\beta_{j+1} = 0$ and $\mathbf{w}_j - \alpha_j \mathbf{q}_j \neq \mathbf{0}$, the algorithm cannot continue since division by β_{j+1} is required in the next step. This is called a breakdown of the algorithm. Nearly as bad are near breakdown for which $|\beta_{j+1}| \approx 0$. Breakdowns of the Lanczos algorithm can be cured by so-called look-ahead techniques. We do not discuss these techniques here, since we have never detected a (near) breakdown in our numerical work. We emphasize, however, that we cannot guarantee that no breakdowns will occur even if we assume exact arithmetic.

Suppose now that no breakdown occurred during the first k iterations of the algorithm. These k iterations can be summarized into a single equation:

$$\mathbf{G}\mathbf{Q}_k = \mathbf{Q}_k \mathbf{T}_{\mathrm{x};k} + \beta_{k+1} \mathbf{q}_{k+1} \mathbf{e}_k^T, \tag{16}$$

where $\mathbf{e}_k$ is the kth column of the k-by-k identity matrix $\mathbf{I}_k$, and $\mathbf{Q}_k$ is an N-by-k matrix with a column partitioning

$$\mathbf{Q}_k = (\mathbf{q}_1, \mathbf{q}_2, ..., \mathbf{q}_k).$$

Matrix $\mathbf{Q}_k$ is complex orthogonal, that is, it satisfies $\mathbf{Q}_k^T \mathbf{Q}_k = \mathbf{I}_k$. Furthermore, $\mathbf{T}_{\mathrm{x};k}$ is a complex, symmetric, and tridiagonal matrix of order k given by

$$\mathbf{T}_{\mathrm{x};k} = \mathrm{tridiag}(\beta_j, \alpha_j, \beta_{j+1}).$$

The subscript x is added to the tridiagonal matrix to indicate that it was generated with vector $\mathbf{x}$ as a starting vector. We are interested in cases where k, the order of the tridiagonal matrix, is much smaller than N, the order of matrix $\mathbf{G}$.

Using the connection between matrix $\mathbf{G}$ and matrix $\mathbf{T}_{\mathrm{x};k}$ as given by Eq. (16) it is possible to show that

$$\mathbf{G}^j \mathbf{x} = \beta_1 \mathbf{Q}_k \mathbf{T}_{\mathrm{x};k}^j \mathbf{e}_1 \quad \text{for } j = 0, 1, ..., k-1. \tag{17}$$

The proof of this relation is by induction and can be found in Druskin and Knizhnerman [7], for example. With the help of Eq. (17) and using the complex orthogonality of matrix $\mathbf{Q}_k$, we can write the first $2k$ moments of z as

$$m_j = a_x \mathbf{e}_1^T \mathbf{T}_{x;k}^j \mathbf{e}_1 \quad j = 0, 1, \ldots, 2k-1.$$

Now the moments as given by the above equation are also the first $2k$ moments of

$$a_x \mathbf{e}_1^T (\mathbf{I}_k - \zeta \mathbf{T}_{x;k})^{-1} \mathbf{e}_1 \tag{18}$$

and this shows that Eq. (18) is in fact the desired Padé approximant of z around $\zeta = 0$.

To construct a Padé approximant for the second term in Eq. (14), we apply the Lanczos algorithm on matrix $\mathbf{G}$ and use $\mathbf{y}$ as a starting vector. Following similar steps as above, we arrive at the Padé approximant

$$a_y \mathbf{e}_1^T (\mathbf{I}_k - \zeta \mathbf{T}_{y;k})^{-1} \mathbf{e}_1, \tag{19}$$

where $a_y = \langle \mathbf{y}, \mathbf{y} \rangle$ and $\mathbf{T}_{y;k}$ is the tridiagonal matrix generated by the Lanczos algorithm with vector $\mathbf{y}$ as a starting vector.

Replacing now the two terms between the square brackets in Eq. (14) by their Padé approximants, we arrive at the reduced-order model

$$v_k^{sc} := \frac{\gamma \zeta}{4} \left[a_x \mathbf{e}_1^T (\mathbf{I}_k - \zeta \mathbf{T}_{x;k})^{-1} \mathbf{e}_1 - a_y \mathbf{e}_1^T (\mathbf{I}_k - \zeta \mathbf{T}_{y;k})^{-1} \mathbf{e}_1 \right]. \tag{20}$$

To compute these models, only tridiagonal systems of order k need to be inverted for each new value of ζ.

For a monostatic source-receiver setup ($\mathbf{s} = \mathbf{r}$) we have $\mathbf{y} = \mathbf{0}$ and $\mathbf{x} = 2\mathbf{r}$ so that we have to apply the Lanczos algorithm only once (with starting vector $\mathbf{r}$) to obtain the reduced-order model

$$v_k^{sc} = \gamma \zeta a_r \mathbf{e}_1^T (\mathbf{I}_k - \zeta \mathbf{T}_{r;k})^{-1} \mathbf{e}_1.$$

Finally, we mention that it is not necessary to use the expansion point $\zeta = 0$. Any other (physically acceptable) nonzero expansion point can be chosen as well, but then matrix factorization is required to construct the Padé approximation (see Feldmann and Freund [9]). Although such a factorization has to be computed only once, it is often too expensive to compute and we therefore expand z and its Padé approximation around $\zeta = 0$ only.

4 The reduced-order model objective function

In the previous section we constructed reduced-order models for the scattered field data. These models should be accurate for all permittivity and conductivity values of interest. More precisely, with ζ given by

$$\zeta = \frac{\varepsilon_{\text{eff}}}{\varepsilon_b} - 1 + i\frac{\sigma_{\text{eff}}}{\omega \varepsilon_b},$$

we require that our models should be accurate for all permittivity and conductivity values satisfying

$$\varepsilon_{\min} \leq \varepsilon_{\text{eff}} \leq \varepsilon_{\max} \quad \text{and} \quad \sigma_{\min} \leq \sigma_{\text{eff}} \leq \sigma_{\max},$$

respectively. This defines our domain of interest $\mathbb{A}$ in the complex ζ-plane. The maximum and minimum permittivity and conductivity values obviously satisfy $\varepsilon_{\max} \geq \varepsilon_{\min}$ and $\sigma_{\max} \geq \sigma_{\min}$ and are given a priori. Note that these minimum and maximum values should be chosen such that the unknown effective permittivity and conductivity values belong to the intervals of interest. This constitutes a priori information about the true scatterer, of course.

Now let $\tilde{u}^{\text{sc}}$ denote the measured scattered data at the receiver location. If $\tilde{u}^{\text{sc}} = 0$ then we take $\zeta = 0$ (no contrast) as a solution of our inverse scattering problem. If $\tilde{u}^{\text{sc}}$ does not vanish, consider minimizing the objective function

$$F(\zeta) = \frac{|\tilde{u}^{\text{sc}} - v^{\text{sc}}(\zeta)|^2}{|\tilde{u}^{\text{sc}}|^2}$$

on $\mathbb{A}$, our domain of interest in the complex ζ-plane. The problem with the objective function F is that it requires the computation of the scattered field v^{sc} for each new value of the contrast coefficient. This amounts to solving a forward problem for each new ζ. However, we have a kth-order model v_k^{sc} available that approximates v^{sc} on $\mathbb{A}$. If this approximation is accurate for all ζ-values belonging to $\mathbb{A}$, then it makes sense to minimize the reduced-order model objective function

$$F_k(\zeta) = \frac{|\tilde{u}^{\text{sc}} - v_k^{\text{sc}}(\zeta)|^2}{|\tilde{u}^{\text{sc}}|^2} \tag{21}$$

on $\mathbb{A}$. The problem is, of course, how to determine the order of the reduced model. Our solution to this problem is very practical. First, we determine the scattered field data due to the effective scatterer with the largest contrast coefficient (in magnitude) of interest. This scattered field is computed by solving the discretized object equation for this particular contrast value and substituting the result in the discretized data equation. We use Bi-CGSTAB (see van der Vorst [20]) to solve the object equation, but any other suitable iterative solver may be used as well, of course. Having found the scattered field data v^{sc}, we run the Lanczos algorithm and construct the reduced-order model v_k^{sc} for the largest contrast coefficient. As soon as k, the order of the model, is such that

$$|v^{\text{sc}} - v_k^{\text{sc}}| \leq \varepsilon_{\text{rom}},$$

where $\varepsilon_{\text{rom}} > 0$ is a given tolerance, we stop the Lanczos algorithm and we have determined the order of the reduced-order model. Excessive numerical testing indicates that the model determined in this way satisfies

$$|v^{\text{sc}}(\zeta) - v_k^{\text{sc}}(\zeta)| \leq \varepsilon_{\text{rom}},$$

for all $\zeta \in \mathbb{A}$. In other words, a reduced-order model which is accurate for the largest contrast coefficient is also accurate for all other contrast coefficients. The largest contrast coefficient is the worst case.

Different methods can be used to find a minimum of the reduced-order model objective function (Newton's method, for example, or one of its variants). However, since the reduced-order models can be computed very efficiently, we follow a much more straightforward approach and solve our inverse scattering problem simply by *inspection*. More precisely, we discretize our domain of interest on a uniform grid and look for a minimum of the objective function on this grid. A disadvantage of this approach is that the number of reduced-order model evaluations may be much larger compared with the number of evaluations required in a minimization procedure such as Newton's method. This is a small price to pay, however, since evaluating reduced-order models is very efficient. An advantage of our approach is that objective functions which are not differentiable are easily handled as well. As an example, in many applications the phase of the electric field strength cannot be determined through measurement and only its magnitude can be measured. In this case we cannot use the objective function as given by Eq. (21), since phase information is required to form this objective function. Having only the amplitude available, we can look for a minimizer of the objective function

$$F_k^{\text{am}}(\zeta) = \frac{||\tilde{u}| - |v_k^{\text{sc}}(\zeta) + u^{\text{inc}}||}{|\tilde{u}|} \tag{22}$$

instead. This function is not differentiable, but finding a minimum through inspection is straightforward (see Remis and Budko [18]). Notice that we have to include the incident field in the objective function defined above, since it is the total field that is measured by the receiver.

4.1 Multiple frequencies

Let us assume that the permittivity and conductivity of the object do not vary with frequency within the frequency band

$$\Omega = \{\omega \in \mathbb{R}^+; \omega_l \leq \omega \leq \omega_u \text{ with } \omega_u > \omega_l > 0\}.$$

We select $Q \geq 1$ different frequencies from this band and perform a scattering experiment at each frequency. In this way we obtain the measured electric field data

$$\tilde{u}(\omega_1), \tilde{u}(\omega_2), ..., \tilde{u}(\omega_Q),$$

with $\omega_q \in \Omega$ for $q = 1, 2, ..., Q$. Having this data available, we determine the effective permittivity and conductivity by looking for a minimum of the objective function

$$F_k^{\Omega}(\zeta) = \sum_{q=1}^{Q} w_q \frac{|\tilde{u}^{sc}(\omega_q) - v_k^{sc}(\zeta, \omega_q)|^2}{|\tilde{u}^{sc}(\omega_q)|^2}, \tag{23}$$

where the $w_q \geq 0$ are weighting coefficients satisfying the normalization condition

$$\sum_{q=1}^{Q} w_q = 1,$$

and $v_k^{sc}(\zeta, \omega_q)$ is the kth order reduced-order model for the scattered field at frequency ω_q. Notice that we have to apply the Lanczos algorithm Q times to obtain the reduced-order models for the scattered field at the Q different frequencies, since matrix $\mathbf{G}$ and the source and receiver vector are frequency dependent. Furthermore, we require that the reduced-order models for all frequencies are accurate on the largest domain of interest, which is determined by the smallest frequency of operation (see Eq. (4)).

Finally, for completeness we mention that if only the amplitude of the data available then we can look for a minimizer of the objective function

$$F_k^{am;\Omega} = \sum_{q=1}^{Q} w_q \frac{||\tilde{u}(\omega_q)| - |v_k^{sc}(\zeta, \omega_q) + u^{inc}(\omega_q)||}{|\tilde{u}(\omega_q)|}, \tag{24}$$

where, again, the w_q are weighting coefficients (see Remis and Budko [18]).

5 Numerical experiments

In our first set of experiments we consider the scattering setup shown in Figure 2 and investigate the difference between reflection and transmission data inversion results. The configuration consists of a homogeneous block and this block is surrounded by four antennas labeled N (North), W (West), E (East), and S (South). Antenna N acts as a source as well. The true medium parameters of the block are given by $\varepsilon_r = 4.5$ and $\sigma = 5$ mS/m. The minimum and maximum permittivity and conductivity values are taken as $\varepsilon_{r;min} = 1$, $\varepsilon_{r;max} = 9$, $\sigma_{min} = 0$, and $\sigma_{max} = 10$ mS/m. These values determine our domain of interest. This domain is discretized on a 50-by-50 grid and consequently we have to evaluate 2500 reduced-order models for each source-receiver pair. Furthermore, the source operates at a frequency of 36 MHz and the side length of the block is $d = \lambda$.

To obtain reflection data, we first consider a monostatic source-receiver setup, that is, antenna N acts as a source as well as a receiver. The objective function for this setup is shown in Figure 3 (top-left). The true medium parameters can be recovered from this function, but quite a number of false minima are present as well.

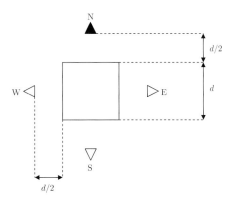

Fig. 2 Scattering configuration with one electric line source (black triangle) and four receivers.

To obtain partly reflected and partly transmitted data, we keep the source the same and measure the field with antenna W. The resulting objective function is shown in Figure 3 (top-right). Again, the true medium parameters can be recovered from this function and false minima are present. We do observe, however, that the number of false minima in this case is less than in the previous experiment. Measuring the field with antenna E produces the same results as with antenna W because of the symmetry of the configuration. We therefore do not show the corresponding objective function.

Finally, to obtain transmission data we measure the field with antenna S. The objective function obtained with this source-receiver setup is shown in Figure 3 (bottom). Identifying the true medium parameters from this objective function is straightforward and there are essentially no false minima present in our domain of interest. The results of this set of experiments indicate that transmission data provide better inversion results than reflection data. We have carried out many additional experiments and all these experiments confirm this conclusion.

In our second set of experiments we consider the scattering setup shown in Figure 4. As opposed to the previous set of experiments, we now keep the source-receiver unit the same throughout all experiments and use multi-frequency data instead. A square block with side length d is embedded in a vacuum domain and consists of two concentric square parts. The outer part is characterized by a relative permittivity $\varepsilon_{r;1}$ and a conductivity σ_1, the inner part has a relative permittivity $\varepsilon_{r;2}$ and a conductivity σ_2. The source and receiver are located 2 cm apart and are located a distance $d/2$ above the object (see Figure 4).

In our first experiment we operate at a frequency of 36 MHz. The wavelength in vacuum corresponding to this frequency is denoted by λ and for the side length d of the object we take $d = \lambda$. Initially, we consider a homogeneous block with medium parameters $\sigma_1 = \sigma_2 = 7.5$ mS/m and $\varepsilon_{r;1} = \varepsilon_{r;2} = 5$. The minimum and maximum relative permittivity and conductivity values are taken to be $\varepsilon_{r;min} = 1$, $\varepsilon_{r;max} = 6$, $\sigma_{min} = 0$ mS/m, and $\sigma_{max} = 10$ mS/m. Figure 5 (left) shows the base 10 logarithm

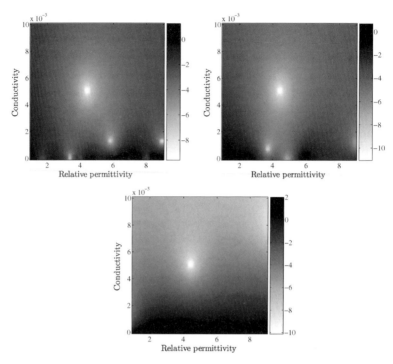

Fig. 3 Base 10 logarithm of the objective function on the domain of interest with antenna N acting as a source and antenna N acting as receiver (top-left), antenna W acting as a receiver (top-right), and antenna S acting as a receiver (bottom). The object is homogeneous and has a side length $d = \lambda$, where λ is the free-space wavelength corresponding to a frequency of 36 MHz.

of the reduced-order objective function. The true medium parameters of the object are easily recovered by inspecting this figure. However, just as in the previous set of experiments, we do observe a number of false minima. To see if the false minima can be removed by including multi-frequency data (instead of using different source-receiver setups as in the first set of experiments), we carry out two additional scattering experiments, namely, one at a frequency of 30 MHz and one at a frequency of 42 MHz. We now have scattered field data for three different frequencies and the corresponding multi-frequency objective function (with weights $w_q = 1/3$ for $q = 1, 2, 3$) is shown in Figure 5 (right). We observe that this objective function does not have any false minima on the domain of interest.

In our third set of experiments we consider an inhomogeneous block. Specifically, the medium parameters of the outer part are $\varepsilon_{r;1} = 3$ and $\sigma_1 = 3.0$ mS/m, while the medium parameters of the inner part are given by $\varepsilon_{r;2} = 5$ and $\sigma_2 = 5.0$ mS/m. Furthermore, the area of the inner block is 50 % of the total area of the block and we use the same three frequencies as in the previous experiment. For a block with a side length $d = \lambda/4$ we then obtain the objective function as shown in Figure 6 (left). In this figure the plus signs indicate the true medium parameters of the inner and outer

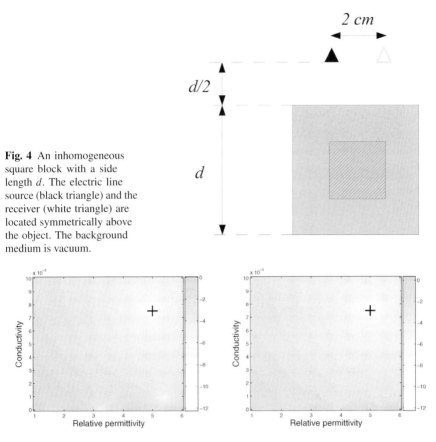

Fig. 4 An inhomogeneous square block with a side length d. The electric line source (black triangle) and the receiver (white triangle) are located symmetrically above the object. The background medium is vacuum.

Fig. 5 Base 10 logarithm of the single-frequency (left) and multi-frequency (right) objective function on the domain of interest. The object is homogeneous and has a side length $d = \lambda$, where λ is the free-space wavelength corresponding to a frequency of 36 MHz. The other two frequencies used in the multi-frequency case are $f = 30$ MHz and $f = 42$ MHz. The plus sign indicates the relative permittivity and conductivity values of the object.

part of the object. We observe that the minimum of the objective function is located somewhere in between the medium parameter values of the inner and outer part of the object. This can be expected, given the true medium parameters of the object. However, if we take a larger block with $d = \lambda$ we obtain the objective function as shown in Figure 6 (right). The effective medium parameters obtained by inspecting this objective function are smaller than the smallest medium parameters of the object itself. This is a surprising result and a common phenomenon in such areas as *photonic crystals* and *metamaterials*. These artificial media usually consist of a periodic lattice of relatively small dielectric or metal-dielectric objects [13, 3, 14]. Analysis of infinite periodic media involves expansion of the field in terms of the Bloch modes and leads to the *band-gap* phenomenon. Practically this means that

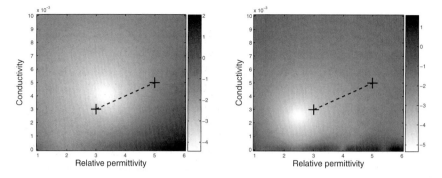

Fig. 6 Base 10 logarithm of the multi-frequency objective function on the domain of interest for the inhomogeneous block with side length $d = \lambda/4$ (left) and side length $d = \lambda$ (right). The frequencies used are $f = 30$ MHz, $f = 36$ MHz, $f = 42$ MHz, and λ is the free-space wavelength corresponding to a frequency of 36 MHz. The plus signs indicate the relative permittivity and conductivity values of the inner and outer parts of the object.

for certain frequencies, which usually form a continuous band, plane electromagnetic waves have an imaginary wavenumber and therefore cannot propagate in such a medium. Hence, it is convenient to describe periodic media in terms of dispersion, i.e., dependence of the wavenumber or permittivity/permeability on the angular frequency. Thus, we have two kinds of dispersion. One is due to atomic or molecular resonances, referred to as *microscopic dispersion*. Another is the *effective dispersion* due to periodicity and mesoscopic resonances. The latter are employed in metamaterials where each elementary scatterer in the periodic lattice is also a small resonator tuned to a specific frequency band.

Although infinite periodic media are easy to analyze, finite samples of such crystals cannot be studied analytically. Yet it is only natural to try to extend the notion of the effective dispersion and effective permittivity/permeability on finite crystals as well. This is how the so-called negative refraction materials were invented [16, 19]. To demonstrate that something like a band-gap exists even in finite photonic crystals let us consider one depicted in Figure 7. This crystal consists of a triangular lattice of lossless dielectric cylinders with relative permittivity $\varepsilon/\varepsilon_0 = 9.61$, suspended in and surrounded by air. The cylinders have a square cross-section, 3.2 mm on the side. The lattice period – the length of the edge of the equally-sided triangle formed by the centers of white squares – is approximately 4.79 mm. The line source is situated in the middle, 18.6 mm down from the lower interface of the crystal. Figure 8 illustrates the changes in the field intensity within the crystal as the frequency of the source varies. The third image corresponds to the frequency inside the photonic band-gap as calculated for an infinite crystal. Evidently, the strong decay of the field happens with this finite crystal as well. Hence it makes sense to try to determine the effective permittivity of this object, and our algorithm is particularly suited for this task. We place our receiver in the transmission regime 18.6 mm above the middle of the upper interface. The effective scatterer is a homogeneous rectangular object occupying the same area as the complete photonic crystal of Figure 7.

A Model-Order Reduction Approach to Parametric Electromagnetic Inversion 17

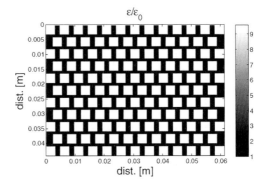

Fig. 7 Finite two-dimensional photonic crystal with triangular lattice - relative permittivity distribution (no losses).

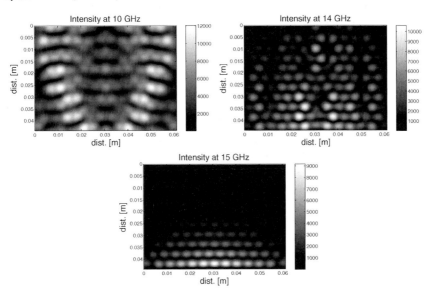

Fig. 8 Field intensity inside a finite photonic crystal at different frequencies. The bottom image demonstrates the photonic band-gap phenomenon – waves of particular frequencies cannot penetrate the crystal.

Since the effective permittivity is supposed to depend on the frequency in a nontrivial way, we shall not sum over frequencies, but instead visualize the objective function for each frequency separately. This time we shall work in terms of real and imaginary parts of the complex permittivity. Moreover, now we are interested in a wider domain of parameters as it is known that the effective permittivity of periodic crystals may take non-physical values as well [15]. In the present case this means looking at the negative values for the imaginary part of the relative permittivity, corresponding to media with negative losses. Mathematically, we shall be entering the domain where the forward scattering problem does not have a unique solution. Certain values of effective permittivity with negative imaginary part are in fact eigenvalues of the scattering operator (matrix) [5]. This will lead to a non-invertible matrix and thus a very high numerical value of the objective function for that particular effective permittivity and in its immediate neighborhood. This is indeed observed in all three images of Figure 9, where we see a tail of dark spots stretching to the right of the zero just below the real axis. Curiously, there are also white spots, i.e. minima

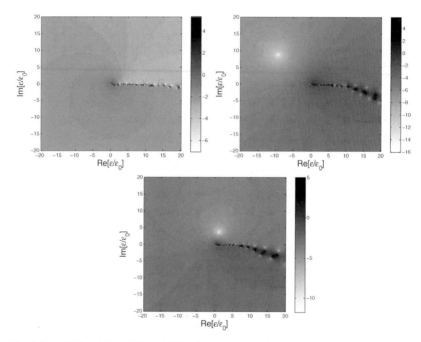

Fig. 9 Base 10 logarithm of the multi-frequency objective function on the domain of interest at 10, 14, and 15 GHz. The tail with many clustered maxima and minima corresponds to the eigenvalues of the scattering operator. Notice the negative real part of the effective permittivity at the lower edge of the photonic band-gap (top-right), and almost purely imaginary effective permittivity inside the band-gap (bottom).

of the objective function, surrounding the tail of eigenvalues. This means that on this large domain of interest, including non-physical media with negative losses, the effective permittivity of a finite photonic is not uniquely defined. Of course, one can exclude the lower half of the image from consideration and focus on the obvious distinct minima in the upper part only. However, in that case the effective permittivity at lower frequencies, far outside the photonic band-gap, is not defined at all, as can be seen from the left image of Figure 9. On the other hand, the location of the main distinct minimum in the middle image, shows significant negative real part of the effective permittivity as well as large effective losses (positive imaginary part). This happens at a frequency of 14 GHz, just before the band-gap. This frequency and the corresponding effective permittivity are important for the phenomenon of negative refraction on photonic crystal slabs. Inside the band-gap at 15 GHz we see that the effective permittivity is predominantly imaginary, indicating, as was expected, high losses.

References

1. P. J. Antsaklis, A. N. Michel, *Linear Systems*, (Birkhäuser, Boston, 2006).
2. C. Balanis, *Antenna Theory - Analysis and Design*, (John Wiley and Sons, New York, 1982).
3. H. Benisty, V. Berger, J.-M. Gerard, D. Maystre, A. Tchelnokov, *Photonic Crystals*, (Springer, 2005).
4. N. V. Budko, R. F. Remis, Inverse Problems **20**, 2004, pp. S17 – S26.
5. N. V. Budko, A. B. Samokhin, SIAM Journal on Scientific Computing **28**, 2006, pp. 682 – 700.
6. W. C. Chew, J. M. Jin, E. Michielssen, J. Song, *Fast and Efficient Algorithms in Computational Electromagnetics*, (Artech House, Norwood, 2001).
7. V. L. Druskin, L. A. Knizhnerman, USSR Computational Mathematics and Mathematical Physics **29**, 1989, pp. 112 – 121.
8. V. Druskin, M. Zaslavsky, Inverse Problems **23**, 2007, pp. 1599 – 1610.
9. P. Feldman, R. W. Freund, IEEE Transactions on Computer-Aided Design of Integrated Circuits and Systems **14**, 1995, pp. 639 – 649.
10. A. Frommer, U. Glässner, SIAM Journal on Scientific Computing **19**, 1998, pp. 15 – 26.
11. A. Frommer, P. Maas, SIAM Journal on Scientific Computing **20**, 1999, pp. 1831 – 1850.
12. G. H. Golub, Z. Strakos, Numerical Algorithms **8**, 1994, pp. 241 – 268.
13. J. D. Joannopoulos, R. D. Meade, J. N. Winn, *Photonic Crystals*, (Princeton Univ. Press, 1995).
14. C. M. Krowne, Y. Zhang (editors), *Physics of Negative Refraction and Negative Index Materials*, (Springer, 2007).
15. R. Liu, T. J. Cui, D. Huang, B. Zhao, D. R. Smith, Phys. Rev. E **76(2)**, 2007, 026606.
16. J. B. Pendry, Phys. Rev. Lett. **85**, 2000, pp. 3966 – 3969.
17. R. F. Remis, PIERS Online **2**, 2006, pp. 206 – 209.
18. R. F. Remis, N. V. Budko, Proceedings ICEAA05 and EESC05, 2005, pp. 409 – 412.
19. D. R. Smith, W. J. Padilla, D. C. Vier, S. C. Nemat-Nasser, S. Schultz, Phys. Rev. Lett. **84**, 2000, pp. 4184 – 4187.
20. H. A. van der Vorst, *Iterative Krylov Methods for Large Linear Systems*, (Cambridge University Press, Cambridge, 2003).

Shifted-Laplacian Preconditioners for Heterogeneous Helmholtz Problems

C.W. Oosterlee, C. Vuik, W.A. Mulder, and R.-E. Plessix

Abstract We present an iterative solution method for the discrete high wavenumber Helmholtz equation. The basic idea of the solution method, already presented in [18], is to develop a preconditioner which is based on a Helmholtz operator with a complex-valued shift, for a Krylov subspace iterative method. The preconditioner, which can be seen as a strongly damped wave equation in Fourier space, can be approximately inverted by a multigrid method.

1 Introduction

The efficient numerical solution of the Helmholtz equation with spatially-dependent high wavenumbers is a difficult task. The prescription of reasonable boundary conditions, on a domain which is truncated for computational reasons, further complicates the numerical treatment in a real-life setting. A thorough overview of the issues regarding the numerical solution of the Helmholtz equation was presented in [56].

In a series of papers, [16, 17, 18, 47, 48, 20, 44, 57], supported by Royal Dutch Shell and Philips via the Dutch Ministry of Economic Affairs, project BTS01044, we have systematically developed a robust and efficient numerical solution technique for the heterogeneous high wavenumber Helmholtz equation. We have subsequently looked into the method's performance for spatially-dependent wavenum-

---

C.W. Oosterlee
CWI, Center for Mathematics and Computer Science, Amsterdam, and Delft University of Technology, Delft, the Netherlands, e-mail: c.w.oosterlee@cwi.nl

C.Vuik
Delft University of Technology, Delft, e-mail: c.vuik@tudelft.nl

W.A. Mulder
Shell International Exploration and Production, Rijswijk, NL e-mail: Wim.Mulder@shell.com

R-E Plessix
Shell International Exploration and Production, NL, e-mail: ReneEdouard.Plessix@shell.com

bers, for high wavenumber problems, for the so-called absorbing boundary layer boundary conditions, for fourth-order discretizations, for 2D and 3D applications, academic and industrial applications. Former PhD student Yogi Erlangga, one of the driving forces behind this series of papers, has published an overview article on the topic, see [19], and a theoretical analysis has been presented in [20].

In [18], a preconditioned Bi-CGSTAB method has been introduced, in which the preconditioner is based on a second Helmholtz equation with an imaginary shift. This preconditioner is the basis of our work. It appears as a member of the family of shifted Laplacian operators, introduced in [34]. An interesting aspect is that its inverse can be efficiently approximated by means of a multigrid iteration, which is somewhat surprising as the original Helmholtz equation cannot be solved efficiently with off-the-shelf multigrid solvers. The particular preconditioner presented can be viewed as a generalization of the work by Bayliss, Goldstein, and Turkel [4] from the 1980s, where the Laplacian was proposed as a preconditioner for Helmholtz problems.

The idea of preconditioning the indefinite operator with a shifted Laplacian has now been considered also at other places in the research community. In [55] for example, the idea is highlighted. Further, a Finnish group at Jyväskylä has adopted the approach for their Helmholtz applications in [25, 26, 28, 1]. The preconditioner is also discussed and considered, for other Helmholtz applications in [12, 45, 7], and in different research areas that need to deal with indefinite problems, like electromagnetics or optics, in [6, 22, 59, 36, 42].

As an example of the solver, we discuss in this paper a version of the preconditioner for a fourth-order 2D finite-difference discretization of the Helmholtz operator. In the multigrid preconditioner we replace the point-wise Jacobi smoother, from [18], by a variant of the incomplete lower-upper factorization smoother, ILU(0). Furthermore, we show the performance of a prolongation scheme that originates from algebraic multigrid (AMG) [49]. We show that these enhancements to the iterative solver, proposed in [57], can reduce both the number of iterations and the total CPU time needed for convergence. Moreover, we aim to reduce the size of the imaginary shift parameter in the shifted Laplacian preconditioner, compared to the choice in [18], so that an even faster solution method is obtained. A fourth-order Helmholtz discretization enables us to use fewer grid points per wavelength compared to a second-order discretization.

Finally, the overall solution method with these algorithmic improvements is not limited to structured Cartesian grids, as it can be set up fully algebraically (a similar goal has been pursued in [1]). Although our method extends to solving problems on unstructured grids, we focus here on heterogeneous Helmholtz problems on Cartesian grids. We focus on the two-dimensional case; however, all of the method's ingredients can be easily generalized to three dimensions. Previously obtained 3D Helmholtz results, with the shifted Laplacian multigrid preconditioner, with a point-wise smoother, can be found in [48] (academic test problems) and [44] (industrial test problems).

This article is set up as follows. In Section 2, we discuss the Helmholtz equation, its field of application, and the discrete finite-difference formulations of second-

and fourth-order. The iterative solution method, including the preconditioner and its components, is presented in Section 3. Numerical results are presented in Section 4.

2 The Helmholtz equation and a seismic application

Accurately imaging the Earth is one of the major challenges in the hydrocarbon industry. Subsurface formations are mapped by measuring the time required for a seismic pulse to return to the surface after reflection from interfaces between formations with different physical properties. Variations in these reflection times, as recorded on the Earth's surface, usually indicate structural features in the strata below. Depths to reflecting interfaces can be determined from the times, using velocity information that can be obtained from the reflected signals themselves. At the same time the amplitudes of the signals provide valuable information.

In geophysics, numerical solutions for the wave equation are used in seismic imaging to map in depth the information recorded in time in the seismic data [13]. In the oil and gas industry, until approximately 1995, three-dimensional applications mainly relied on solutions of a high-frequency approximation of the wave equation, due to computer constraints. When the medium is very complex (containing heterogeneities that result in strong contrast) or in the case of crossing rays (multivaluedness), however, these so-called ray-based high-frequency migration techniques reach their limits.

With the increase of computer power and the need to image increasingly complex geological terrains, the so-called par-axial approximation of the wave equation [13] became popular. The accuracy of this approximation is however limited to certain angles of incidence which does not allow one to image steep reflectors accurately [40]. Nowadays, the three-dimensional (full) wave equation can be solved on a large cluster of computers with a finite-difference scheme and a time marching approach. This leads to the so-called reverse-time imaging algorithms. While still expensive, these imaging algorithms are now routinely applied in industry.

Typically an imaging algorithm requires the computation of thousands of wave equation solutions for given velocity and density fields. The complexity of a time-domain solution for a given source position is $O(n^4)$ with n the number of grid points in one direction (We can assume that the number of time steps is proportional to n). When the full time response is required, the time-domain solution has an optimal complexity if only one source is considered. However, it was noticed [39] that the matrix of the linear system associated with the discretization of the *frequency-domain wave equation*, i.e., the time-harmonic wave equation obtained by Fourier transformation, is independent of the source term. This implies that, when we can decompose this matrix, for instance by an LU decomposition, the frequency-domain approach can be more efficient than the time-domain approach when many solutions have to be computed with different source terms. This is the case in two dimensions [39]. Unfortunately, in three dimensions, the fill-in with an LU decomposition is very unfavorable, which makes the frequency-domain approach less attractive.

However, attempts to use massively parallel linear solvers are pursued in the context of full-wave form inversion when only a limited number of frequency responses are processed [43].

When the imaging algorithms require to localize the information in depth with a high resolution, as with the so-called migration algorithm, a full band-limited frequency response is needed, according to the Nyquist theorem. In this case, in three dimensions, it is more efficient to solve the time-domain wave equation than the Helmholtz equation. However, when only the solution at a limited number of frequencies is required, the choice of either the time or frequency domain remains open. This is the case for the full waveform inversion algorithm [46]. A competitive frequency-domain solver would need to be more efficient than a time-domain solver followed by a Fourier transform [44]. Since a direct solver is a-priori too expensive because of the filling of the LU decomposition, an iterative solver should be considered. For a given frequency and a given source, the optimal complexity of the iterative frequency-domain solver would be $O(n^3)$, which is better than the $O(n^4)$ complexity of the time-domain solver. Unfortunately, an optimal frequency-solver at seismic frequencies does not yet exist. The aim of the research presented here is to develop a robust and efficient solver of the Helmholtz equation at high wavenumbers. The Helmholtz equation corresponds to the frequency-domain acoustic wave-equation with constant density.

From the exploration-seismology point of view, the Earth is a heterogeneous semi-infinite medium. The wavenumber can be large, which implies that the discretized Helmholtz operator gives rise to both positive and negative eigenvalues and, therefore, the discretization matrix, A_h, is indefinite. For 2D problems, however, the computation can be performed efficiently by using, for example, direct methods combined with nested-dissection reordering [21]. Only one LU decomposition is needed to calculate the solutions at multiple source locations. The result can be used for the computation of all of the wavefields, for all shots and, also, for the back-propagated receiver wavefields [41]. However, for 3D problems, the matrix sizes and bandwidths rapidly become too large and one has to fall back on iterative methods. In that case, one no longer has the advantages in the frequency domain related to the LU decomposition.

For the Helmholtz equation, unfortunately, many iterative methods suffer from slow convergence, especially if high frequencies need to be resolved, due to the indefiniteness. The development of fast iterative methods for high-frequency Helmholtz problems remains a subject of active research. One approach to iteratively solving this equation is presented below. We focus on the 2D case, but provide a solution method which can easily be generalized to 3D.

2.1 Mathematical problem definition

We start with the description of the 2D Helmholtz problem of interest,

$$-\nabla^2 u(x,\omega) - k(x)^2(1-\alpha i)u(x,\omega) = g(x,\omega), \quad x \in \Omega. \tag{1}$$

Unknown $u(x,\omega)$ represents the pressure field in the frequency domain, ∇^2 is the Laplacian operator, $k(x) = \omega/c(x)$ is the wavenumber, with $c(x)$, the acoustic-wave velocity, which varies with position, and $\omega = 2\pi f$ denotes angular frequency, a scalar measure of rotation rate (f is the frequency in Hertz). Wavenumber k depends on x because of a spatially dependent speed of sound, $c(x)$. The source term is denoted by g. The medium is called barely attenuative if $0 \le \alpha \ll 1$, with α indicating the fraction of damping in the medium (and $i = \sqrt{-1}$, the imaginary unit). In geophysical applications, which are of our main interest, this damping can be up to 5% ($\alpha = 0.05$). While Equation (1) arises through the Fourier transform of a wave equation with a very simple model of damping, $(-\nabla^2 + (1-\alpha i)\partial_t^2)u = g$, it is closely related to the Fourier transform of the strongly damped wave equation, $(-\nabla^2 + \tau\partial_t\nabla^2 + \partial_t^2)u = g$ that yields, after scaling, $(-\nabla^2 - \frac{k^2}{1+\tau ik})u$. For small values of $\alpha = \tau k$, Equation (1) is an accurate approximation of the Fourier-domain strongly damped wave equation.

The semi-infinite physical domain needs to be truncated for a numerical treatment. A popular approach in geophysics in order to obtain a satisfactory near-boundary solution, without artificial reflections, is to use the absorbing boundary layer (ABL) approach; see, for example, [29] or [37]. This unphysical boundary layer is used to gradually damp the outgoing waves by adding dissipation in the equation outside the domain of interest. An efficient numerical solution technique should be robust with respect to this kind of feature. The absorption layers (denoted by Ω^e) are attached to the physical domain, Ω, (see Figure 1). In Ω^e, a damped Helmholtz equation (1) should be satisfied [51], with

$$\alpha = 0.25 \frac{\|x - x_d\|^2}{\|x_e - x_d\|^2}, \quad x \in \Omega^e, \tag{2}$$

where point x_d is a point at the boundary, Γ, and x_e a point at Γ^e (see Figure 1). The boundary conditions at the boundary Γ^e are in the form of first- or second-order absorbing boundary conditions. We use approximate radiation (or non-reflecting) boundary conditions at the artificial boundary. The well-known second-order radiation boundary condition [15], to avoid unphysical reflections at boundaries, reads

$$\mathscr{A}_\Gamma u := \frac{\partial u}{\partial \nu} - iku - \frac{i}{2k}\frac{\partial^2 u}{\partial \tau^2} = 0 \quad \text{on } \Gamma^e, \tag{3}$$

with ν the outward normal direction to the boundary and τ pointing in the tangential direction. At the corner-points the suggestions in [3] to avoid corner reflections have been adopted.

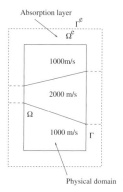

Fig. 1 A 2D domain with ABL in the case of a regular heterogeneous wedge medium.

2.2 Discretization

The equations are discretized here either by a second- or a fourth-order finite-difference scheme, resulting in the linear system:

$$A_h \phi_h = b_h, \qquad (4)$$

where ϕ_h and b_h represent the discrete frequency-domain pressure field and the source, respectively.

In a heterogeneous medium, the smallest velocity is usually selected based on the representative wavelength, λ_f. The number of wavelengths in a domain of size L equals L/λ_f. A dimensionless wavenumber, k, on a non-dimensional $[0,1]^2$ domain is defined by $k = 2\pi f L/c$, and a corresponding mesh size by $h = \lambda_f/(n_w L)$, with n_w the number of points per wavelength.

The usual 5-point stencil related to a second-order accurate discretization reads:

$$A_h^{2o} \triangleq \frac{1}{h^2} \begin{bmatrix} & -1 & \\ -1 & 4 - k^2 h^2 (1 - \alpha i) & -1 \\ & -1 & \end{bmatrix}_h . \qquad (5)$$

With domain size $L = 1$, an accuracy requirement, for second-order discretizations, is that $kh \leq \pi/5 (\approx 0.63)$ for $n_w = 10$ points per wavelength, and $kh \leq 0.53$ with $n_w = 12$ points per wavelength. The number of grid points used assumes a linear connection between k and h. In order to avoid a reduction of accuracy due to *pollution of the solution*, however, $k^2 h^3$ should be chosen constant, as stated in [5, 27]. For an iterative solution method, the requirement that kh should be constant is more severe and, so, this is the constraint that we consider here.

For the absorbing boundary conditions at Γ_e, we also apply central differences. Another discretization that we consider in this work is the $O(h^4)$ accurate discretization based on the Padé approximation. It is called the HO discretization in [50], with stencil

$$A_h^{HO} \triangleq \frac{1}{h^2} \begin{bmatrix} -\frac{1}{6} & -\frac{2}{3} - \frac{(kh)^2(1-\alpha i)}{12} & -\frac{1}{6} \\ -\frac{2}{3} - \frac{(kh)^2(1-\alpha i)}{12} & \frac{10}{3} - \frac{2(kh)^2(1-\alpha i)}{3} & -\frac{2}{3} - \frac{(kh)^2(1-\alpha i)}{12} \\ -\frac{1}{6} & -\frac{2}{3} - \frac{(kh)^2(1-\alpha i)}{12} & -\frac{1}{6} \end{bmatrix}. \tag{6}$$

An important reason for choosing a higher-order discretization method is that the number of grid points per wavelength can be reduced compared to a second-order discretization. Here, for example, we will show numerical experiments in which $kh = 0.8$ is set. This results in smaller matrices for the same level of accuracy and, thus, may lead to an algorithm that is more efficient overall, if the matrices associated with the higher-order discretization can be solved efficiently.

These matrices remain positive definite as long as k^2 is smaller than the first eigenvalue of the discrete Laplacian. The wavenumber in geophysical applications can, however, be large, which implies that the discretized Helmholtz equation gives rise to both positive and negative eigenvalues and, therefore, the discretization matrix, A_h, is indefinite. The size of the system of linear equations (4) gets very large for high frequencies. So, A_h in (4) is a large but sparse matrix, with complex-valued entries, because of the absorbing boundary conditions and the attenuative medium. It is symmetric but non-Hermitian.

2.2.1 Validation of the discretization

In order to validate the choice of boundary condition, ABL, and discretization, we first compute the solution for a constant wavenumber, and $\alpha = 0$, problem in a homogeneous medium with the source function, representative for a seismic pulse, chosen as

$$g_h = \frac{1}{h^2} \delta(x_1 - \frac{1}{2}, x_2 - \frac{1}{32}).$$

Here, $\delta(\cdot, \cdot)$ represents the Dirac delta function, which is 1 when its argument is $(0,0)$, and 0 elsewhere. The scaling by h^2 guarantees that the solutions on fine and coarse grids are of the same amplitude, giving a discrete approximation of a δ-function distribution. For this problem at constant wavenumber, the analytic solution is known, as the Green's function is available. The 2D solution reads:

$$u(r) = \frac{i}{4} H_0^{(1)}(k|r|), \quad r = \sqrt{(x-1/2)^2 + (y-1/32)^2}, \tag{7}$$

where $H_0^{(1)}$ is the Hankel function of the first kind of order 0. So, we can compare a numerical solution with this analytic solution.

Two formulations of the boundary discretization are compared here. In the first, we prescribe the second-order absorbing boundary conditions directly at the physical boundaries, whereas, in the second formulation, the boundary discretization is based on an extra absorbing boundary layer, placed along all physical domain boundaries. An ABL of $n/4$ points is added to each side.

In this first numerical experiment, we fix the wavenumber, $k = 40$, and use a model domain, $(0,1)^2$, covered by a fine grid consisting of 256^2 points ($kh = 0.156$). Figure 2 presents the two corresponding solutions with the second-order discretization. An unphysical damping of the solution without the ABL can be observed near the domain boundaries.

Turning to the fourth-order discretization we drastically reduce the number of grid points. Figure 3 compares, for wavenumber $k = 40$, by means of a *solution profile* at $x = 0.125$, the second- and fourth-order discretizations. Coarse grids consisting of $32^2, 48^2$ and 64^2 points are chosen. As shown, the profile lines generated by the fourth-order discretization converge very nicely towards the physical solution (7), whereas the profile with the second-order discretization on the 32^2-grid is too inaccurate to show, and the solutions on the 48^2- and 64^2-grids are shifted in phase. Thus, the fourth-order discretization with the ABL leads, in this case, to an accurate numerical solution, already on relatively coarse grids, although the accuracy requirement (for second-order discretizations) $kh = 0.625$ is not satisfied there.

3 Iterative solution method

Before we discuss the solution method of choice, we outline the general convergence problems when using the multigrid method directly for the discrete Helmholtz equation of interest. This serves as an illustration of the difficulties one can encounter when solving this equation in a robust and efficient way.

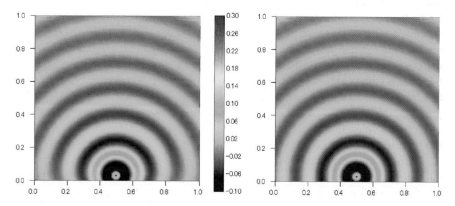

Fig. 2 Numerical solutions for $k = 40$ and $h = 1/256$, without (left-side) and with (right-side) ABL.

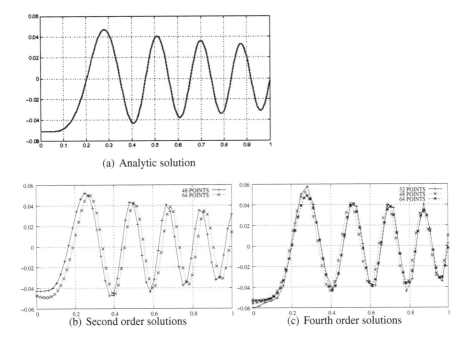

Fig. 3 Comparison of the vertical line solutions on coarse grids, at $x = 0.125$, $k = 40$ with the analytic solution, left side: second-order discretization, right side: fourth-order discretization. ABL is included.

3.1 Multigrid for the Helmholtz equation

Many authors, e.g. [23, 10, 14, 32], have contributed to the development of appropriate multigrid methods for the Helmholtz equation, but an efficient direct multigrid treatment of heterogeneous problems with high wavenumbers arising in engineering settings has not yet been proposed in literature. The multigrid method [8, 24] is known to be a highly efficient iterative method for discrete Poisson-type equations, even with fourth-order accurate discretizations [11, 54]. The Helmholtz equation, however, does not belong to the class of PDEs for which off-the-shelf multigrid methods perform efficiently. Convergence degradation and, consequently, loss of $O(N)$ complexity are caused by difficulties encountered in the smoothing and coarse-grid correction components.

Textbook multigrid methods are typically set up so that a smoothing method reduces high frequency components of an error, between the numerical approximation and the exact discrete solution, and a coarse-grid correction component handles the low frequency error components. Whereas such methods are easily defined for elliptic Poisson-like equations, this is not the case for the Helmholtz equation without any damping in (1), i.e., $\alpha = 0$. Depending on the particular value of the wavenumber, this equation gives rise to both smoothing and coarse grid correction difficulties. For analyzing multigrid algorithms quantitatively, Fourier smoothing, two-, and three-grid analysis [8, 54, 61] are the tools of choice. They have given the following insights. The matrix, resulting from a discretization of the Helmholtz equation, has eigenvalues in only the right half-plane as long as k^2 is less than the smallest eigenvalue of the Laplacian. For larger values, this matrix does not have positive eigenvalues only. Point-wise Jacobi iteration with under-relaxation does not *converge* in that case, but since its smoothing properties are satisfactory, the multigrid convergence will deteriorate only gradually for increasing k^2. By the time it approaches a certain eigenvalue (the 6th), the standard multigrid method diverges. The Jacobi relaxation now diverges also for some smooth eigenfrequencies. Consequently, the multigrid method will still converge as long as the coarsest level used is fine enough to represent these smooth eigenfrequencies sufficiently. So, the coarsest level chosen limits the convergence. When the wavenumber gets larger more variables need to be represented on the coarsest level for standard multigrid convergence. Eventually, this does not result in an $O(N)$ iterative method.

In addition to this feature, the Helmholtz equation also suffers from its coarse-grid correction components. Eigenvalues close to the origin may undergo a sign change after discretization on a coarser grid. If a sign change occurs, the coarse-grid solution does not give a convergence acceleration but gives a severe convergence degradation (or even divergence) instead. In [14] this phenomenon is analyzed and a remedy for the coarse-grid correction related to these problematic eigenvalues is proposed. The efficient treatment in [14] is that the multigrid method is combined with Krylov subspace iteration methods on both fine and coarse grids.

Standard multigrid will also fail for k^2-values very close to eigenvalues. In that case subspace correction techniques should be employed [9].

3.2 Shifted Laplacian preconditioned Krylov subspace method

Iterative solution methods for complex-valued indefinite systems based on Krylov subspace methods [53] are typically generalizations of the conjugate-gradient (CG) method. The Bi-conjugate gradient stabilized (Bi-CGSTAB) algorithm [58] is one of the better known Krylov subspace algorithms for non-Hermitian problems, which has been used for Helmholtz problems, for example, in [18, 1]. One of the advantages of Bi-CGSTAB, compared to full GMRES [52], is its limited memory requirements. Bi-CGSTAB requires only 7 vectors to be stored. Bi-CGSTAB is based on the idea of computing two mutually bi-orthogonal bases for the Krylov subspaces based on matrix, A_h, and its conjugate transpose, A_h^H and is easy to implement.

Without a preconditioner, however, the Krylov subspace methods converge very slowly, or not at all, for the problems of interest [17]. By preconditioning with a matrix, M_h^{-1}, we solve an equivalent linear system,

$$A_h M_h^{-1} \tilde{\phi}_h = b_h, \qquad \tilde{\phi}_h = M_h \phi_h. \tag{8}$$

The challenge, then, is to find a matrix, M_h, such that $A_h M_h^{-1}$ has a spectrum that is favorable for iterative solution with Krylov subspace methods, and whose inverse, M_h^{-1}, can be efficiently approximated.

In [18], a shifted-Laplacian operator was proposed as a preconditioner for the Helmholtz equation, with M_h defined as a discretization of

$$\mathcal{M} = -\nabla^2 - k^2(x)(\beta_1 - \beta_2 i). \tag{9}$$

Equation (9) looks like equation (1), but is much easier to solver with multigrid methods. It will serve as the preconditioner here. Boundary conditions were set identically to those for the original Helmholtz equation. The influence of parameters β_1 and β_2 was evaluated in [18], and the optimal values for the solver proposed there were $(\beta_1, \beta_2) = (1, 0.5)$. Here, we will also consider $\beta_2 = 0.4$, as in [57]. Smaller values of β_2 do not lead to a converging algorithm with the components to be introduced below. The matrix after discretization of (9), M_h, is obtained from either the 5-point, $O(h^2)$, or the 9-point, $O(h^4)$, finite-difference discretization.

3.3 Fourier analysis

The discrete Helmholtz matrix, A_h, as well as the preconditioner, M_h, allow us, assuming a constant wavenumber and Dirichlet boundary conditions, to apply Fourier analysis on the basis of discrete sine-eigenfunctions,

$$v_h^{p,q} = \sin(p\pi x)\sin(q\pi y), \tag{10}$$

to gain insight into the spectrum of $A_h M_h^{-1}$. With these discrete sine functions, $A_h M_h^{-1}$ is diagonalized, and the eigenvalues can easily be determined. As long as k^2 is not equal to any of the eigenvalues of the discrete Laplace operator, $A_h M_h^{-1}$ is nonsingular. Otherwise, the matrix is singular and its nullspace is spanned by the corresponding eigenfunctions (10). The assumption of homogeneous Dirichlet boundary conditions greatly simplifies the analysis (for radiation boundary conditions the Helmholtz operator is non-normal so that eigenvalue analysis alone would not be sufficient for analyzing preconditioned Krylov subspace methods).

We perform Fourier analysis here to visualize the effect of the choice of the parameters, (β_1, β_2), as well as the choice of discretization on the clustering of the eigenvalues of the preconditioned system. This analysis gives only a first indication of what we can expect from the solver. For both A_h and M_h, we use either the second-order discretizations or the fourth-order, HO stencils. Initially, we do not include damping in A_h in the analysis (we take $\alpha = 0$ in (1)).

First, we visualize the effect of the choice of (β_1, β_2) in the preconditioner on the clustering of the eigenvalues of the preconditioned system. For both A_h and M_h we choose the 5-point stencil. Figure 4 presents the spectra of $A_h M_h^{-1}$ for $(\beta_1, \beta_2) = (0, 0)$ (Laplacian preconditioner [4]), $(\beta_1, \beta_2) = (-1, 0)$ (Laird precondi-

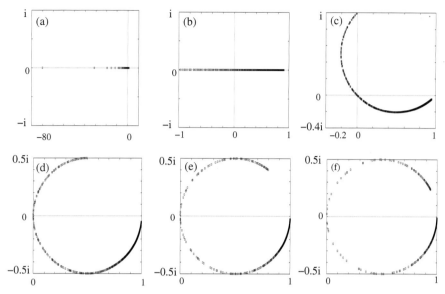

Fig. 4 Spectral pictures of $A_h M_h^{-1}$ with $\alpha = 0$ and different values of (β_1, β_2) in (8), see also [18]. (a) $(\beta_1, \beta_2) = (0,0)$, (b) $(-1,0)$, (c) $(0,1)$, (d) $(1,1)$, (e) $(1,0.5)$, and (f) $(1,0.3)$.

tioner [34]), $(\beta_1, \beta_2) = (0, 1)$ (preconditioner from [16]), $(\beta_1, \beta_2) = (1, 1)$ (basic parameter choice in [18]), $(\beta_1, \beta_2) = (1, 0.5)$, and $(\beta_1, \beta_2) = (1, 0.3)$ (more advanced parameters). The results are for $k = 40$ ($k^2 = 1600$) and $h = 1/64$.

From the spectra presented, the lower pictures of Figure 4 are favorable as their real parts vary between 0 and 1. The Laplacian preconditioner in Figure 4(a) exhibits large isolated eigenvalues; for the Laird preconditioner the eigenvalues in Figure 4(b) are distributed between -1 and 1 on the real axis. The preconditioners with complex Helmholtz terms give rise to a curved spectrum, see also [20]. Whereas the real part of the spectrum in Figure 4(c) still includes a part of the negative real axis, this is not the case for the (β_1, β_2)-preconditioners with $\beta_1 = 1$. The difference between Figures 4(d), 4(e), and 4(f) is that, with a smaller value of β_2, fewer outliers close to the origin are observed. This is favorable for the convergence of the preconditioned Krylov method. The approximate inversion of the preconditioner itself by multigrid, however, will be harder for smaller values of β_2. In Figure 5 the spectra for $k = 100 (k^2 = 10^4)$ are presented on a grid with $h = 1/160$ for $\beta_1 = 1$ and β_2 varying between 1 and 0.3. The spectra are very similar to those in Figure 4. More eigenvalues lie, however, in the vicinity of the origin due to the higher wavenumber and the correspondingly finer grid. Figure 6 then presents the distribution of eigenvalues for the case that 5% damping ($\alpha = 0.05$) is set in $\mathscr{A}$ from (1). Parameters in the preconditioner are $(\beta_1, \beta_2) = (1, 0.5)$. Again the 5-point stencil as in (5) is used for discretization. Figure 6(a) presents the spectrum for $k = 40, h = 1/64$, and Figure 6(b) presents the spectrum for $k = 100, h = 1/160$. An interesting observation is that now the eigenvalues move away from the origin into the right half-plane.

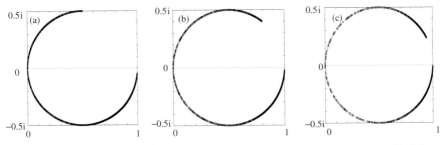

Fig. 5 Spectral pictures of $A_h M_h^{-1}$ for $k = 100$, $h = 1/160$, and $\alpha = 0$, see also [18]; (a) $(\beta_1, \beta_2) = (1,1)$, (b) $(\beta_1, \beta_2) = (1, 0.5)$, and (c) $(\beta_1, \beta_2) = (1, 0.3)$.

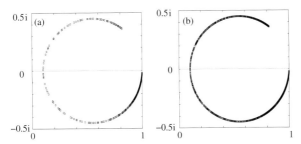

Fig. 6 Spectral pictures of AM^{-1} with 5 % damping in A and $(\beta_1, \beta_2) = (1, 0.5)$, see also [18]; (a) $k = 40, h = 1/64$ and (b) $k = 100, h = 1/160$.

This is beneficial for iterative solution methods. From the spectra in Figure 6 it is expected that the Bi-CGSTAB (and GMRES) convergence in the case of damping in the original equation will be considerably faster than in the undamped case. As the circles have moved away from the origin it is possible to apply the classical theory of the GMRES convergence [52, 53], for example.

Returning to the undamped case, $\alpha = 0$, we concentrate on the choice of discretization, and fix $k = 100$ ($k^2 = 10^4$), $h = 1/160$. Figure 7 presents the curved spectrum of $A_h M_h^{-1}$ for $(\beta_1, \beta_2) = (1, 0.4)$ in M_h, where both operators, A_h and M_h, are discretized by the fourth-order stencil. A very similar eigenvalue distribution is obtained as for the second-order discretization above.

The eigenvalues closest to the origin are the most problematic ones for the convergence of the Krylov subspace method. Figure 8a shows a zoom of the spectrum near the origin, comparing (for $\beta_2 = 0.5$) the location of the eigenvalues near the origin with the second- and the fourth-order discretizations. In Figure 8b, we also compare the location of the eigenvalues near the origin for $\beta_2 = 0.4$ and $\beta_2 = 0.5$, with the fourth-order discretization, keeping $k = 100$, $h = 1/160$. With the fourth-order discretization, the eigenvalues stay further from the origin as compared to the second-order discretization. This should have a positive effect on the convergence of the Krylov subspace method. The same is true when comparing the eigenvalues with $\beta = 0.4$ and $\beta_2 = 0.5$ where, as expected, the clustering with $\beta_2 = 0.4$ is more favorable for iterative solution.

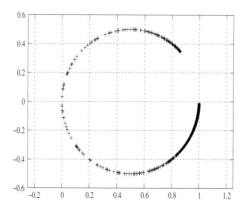

Fig. 7 Spectral picture of $M_h^{-1}A_h$ with $\alpha = 0$, $(\beta_1, \beta_2) = (1, 0.4)$, $k = 100, h = 1/160$. Both operators are discretized by the fourth-order, HO discretization.

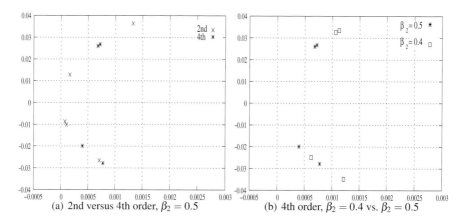

Fig. 8 Comparison of zoomed spectral pictures of $M_h^{-1}A_h$ with $\alpha = 0$, $k = 100, h = 1/160$. (a): Comparison of eigenvalues near the origin discretized with second-order and with fourth-order discretizations; (b) Comparison of eigenvalues for $\beta_2 = 0.4$ and $\beta_2 = 0.5$, fourth-order discretizations.

When discretized with second-order finite differences, M_h, with $(\beta_1, \beta_2) = (1, 0.5)$, can be relatively easily handled by a multigrid method, which is confirmed by Local Fourier Analysis, a quantitative multigrid analysis tool [54]. This is due to the imaginary term, $\beta_2 i$ in the shifted Laplacian. LFA also indicates that M_h based on the fourth-order discretization can be dealt with in multigrid as efficiently as the second-order discretization. With the multigrid components from [57], including an ILU(0) smoother within the preconditioner, we aim to decrease the value of β_2, to $\beta_2 = 0.4$, and obtain an efficient preconditioned Krylov subspace method.

3.4 Multigrid preconditioner

One multigrid cycle, based on standard grid coarsening and point-wise smoothing, can be used as an approximation to M_h^{-1} with $(\beta_1, \beta_2) = (1, 0.5)$. In [18], an F(1,1)-cycle [54], with one pre- and one post-smoothing iteration, with a point-wise Jacobi smoother with under-relaxation parameter $\omega = 0.5$ was chosen for the high wavenumber problems. The other multigrid components were:

- Restriction operators, I_h^H, based on 2D full weighting, whose stencil [54] reads:

$$I_h^H \triangleq \frac{1}{16} \begin{bmatrix} 1 & 2 & 1 \\ 2 & 4 & 2 \\ 1 & 2 & 1 \end{bmatrix}_h^H, \qquad (11)$$

  with h denoting the fine-, and H denoting the coarse-mesh size.
- Prolongation operators, I_H^h, were 2D matrix-dependent interpolations, based on de Zeeuw's interpolation weights [63]. The interpolation weights are especially tailored to the symmetric complex Helmholtz equation, i.e., the asymmetric components in [63] have been removed. As for symmetric problems with jumping coefficients, the prolongation operator by de Zeeuw [63] is very similar to the operator-dependent prolongation in [2]. For satisfactory convergence it is, however, important to consider the *moduli* of the complex-valued operator elements in the definition of the interpolation weights.
- Coarse-grid matrices were based on Galerkin coarse-grid discretizations, defined as $M_H = I_h^H M_h I_H^h$.

In [18], it was shown that the full-weighting restriction combined with the matrix-dependent prolongation resulted in a stable convergence for a variety of problems with irregular heterogeneities and strong contrasts. The inclusion of an ABL in the discretization does not lead to any multigrid convergence difficulties, as the multigrid components chosen are especially designed for problems with varying coefficients.

With a more powerful smoother, however, a robust multigrid method can be developed for approximately inverting matrices M_h that originate from a fourth-order discretization. As the smoother in the multigrid preconditioner, the point-wise Jacobi smoother was replaced by an ILU smoother in [57]. ILU smoothing is well-known in the multigrid literature [30, 31, 62, 60, 64]. We choose here the ILU(0) variant, meaning that we do not allow any additional fill-in in the lower- and upper-triangular factors outside of the nonzero pattern of matrix M_h. An ILU(0) smoother is known to be more powerful than a point-wise Jacobi smoother for a number of test problems [60]. Strictly speaking, ILU methods do not only have a smoothing effect on the errors. A lexicographical version may also reduce low-frequency errors, especially when the entries of the remainder matrix, R_h, in

$$M_h = \hat{L}_h \hat{U}_h - R_h,$$

are relatively small. Parallelization of an ILU smoother is, however, less trivial, but possible, compared to a point-wise Jacobi smoother.

3.5 AMG type interpolation

An efficient multigrid scheme relies on the effective complementarity of the chosen relaxation and interpolation procedures in reducing the error components in an approximate solution. The coarse-grid correction operator is designed to reduce errors that the chosen smoother is slow to attenuate. Such errors should lie in the range of interpolation, so that the coarse-grid correction may be effective. We consider a fixed choice of coarse grid, i.e., Cartesian (doubling the mesh size in each direction) as in geometric multigrid and in [18], but employ an interpolation operator that is chosen based on algebraic multigrid (AMG) principles, evaluated in [57]. The interpolation developed is largely based on the real-valued AMG interpolation from [49], and discussed for complex-valued equations in [33, 38].

Consider, then, an error, e_h, that is not quickly reduced by relaxation. For many standard problems and smoothers, these errors coincide with those vectors that yield small residuals. For the purpose of interpolation, AMG assumes that the error, e_h, is much larger than its residual when measured point-wise, $(A_h e_h)_j \ll (e_h)_j$, for each fine-grid index j. Based on this property, we have

$$(A_h e_h)_j \approx 0 \Rightarrow a_{jj}(e_h)_j \approx -\sum_{k \neq j} a_{jk}(e_h)_k, \qquad (12)$$

meaning that the value of the error at a fine-grid node, j, can be accurately approximated by the values from its neighboring nodes. If all neighboring nodes are also coarse-grid nodes, then (12) is easily turned into an interpolation formula.

With the fixed coarsening considered here, fine-grid node j will have both fine-grid and coarse-grid nodes as neighbors. Designing an interpolation procedure can, then, be thought of as modifying the balance in (12) in such a way as to remove connections to other fine-grid neighbors of j while preserving the overall balance. This is typically done by applying a partition to the neighboring nodes of j that identifies some nodes as important, or strong, connections and other nodes as unimportant, or weak connections. That is, we write the set, $\{k \neq j\} = C_j \cup F_j^s \cup F_j^w$, where C_j is the set of strongly connected coarse-grid neighbors of j, and the disjoint sets, F_j^s and F_j^w, denote the strong fine-grid and weak connections, respectively.

The matrix arising from the Helmholtz equation is complex and, typically, the sum of the moduli of the off-diagonal elements is larger than that of the diagonal element in each row. In this case, a different criterion should be considered as a measure of the strong connections. Here, we give two common criteria for defining the set, S_j, of strong connections for node j, defining

$$S_j = \left\{ k : |a_{jk}| \geq \theta \max_{l \neq j} |a_{jl}| \right\},$$

or
$$S_j = \left\{ k : -Re(a_{jk}) \geq \theta \max_{l \neq j} -Re(a_{jl}) \right\}.$$

Parameter θ allows some adjustment of the number of connections chosen as strong (relative to the strongest connection); for many problems, $\theta = 0.25$ is considered to be a standard choice. Numerical experiments with the discrete complex-valued shifted Laplacian have revealed that sometimes divergence is observed for high wavenumber problems if we use the measure based on the norm. The measure based on the real part of the matrix elements gave a satisfactory multigrid performance over a large range of wavenumbers and, thus, is used in the numerical results that follow.

It is expected that the weak connections of fine-grid node j can be discarded from the balance in (12). To remove these terms (in particular, the appearance of $(e_h)_k$ for $k \notin S_j$) without upsetting the balance, these terms are "lumped to the diagonal". In effect, this means that we make the approximation that $(e_h)_k \approx (e_h)_j$ for $k \in F_j^w$; while this approximation may not be very accurate, it is not harmful to make such a choice, since the connections involved are not important. Treating the strongly connected fine-grid neighbors of j is much more important, as these are connections that (by definition) cannot be easily dropped. In classical AMG methods, one assumes that these connections are well-represented on the coarse grid, by their values at neighboring points. Then, an approximation may be made by considering the weighted average of the values at common coarse-grid neighbors of node j and its fine-grid neighbor, node k, resulting in the expression,

$$(e_h)_k \approx \frac{\sum_{l \in C_j} a_{kl}(e_h)_l}{\sum_{l \in C_j} a_{kl}}.$$

If there is no point in C_j such that $a_{kl} \neq 0$ (or if $\sum_{l \in C_j} a_{kl} = 0$), then node k is neglected in the interpolation formula. Making these substitutions in (12) and choosing for equality, we then have

$$a_{jj}(e_h)_j = -\sum_{k \in C_j} a_{jk}(e_h)_k - \sum_{k \in F_i^s} a_{jk} \frac{\sum_{l \in C_j} a_{kl}(e_h)_l}{\sum_{l \in C_j} a_{kl}} - \sum_{k \in F_i^w} a_{jk}(e_h)_j,$$

or $(e_h)_j = \sum_{k \in C_j} w_{jk}(e_h)_k$, for

$$w_{jk} = -\frac{a_{jk} + \sum_{m \in F_i^s} \frac{a_{jm} a_{mk}}{\sum_{l \in C_j} a_{ml}}}{a_{jj} + \sum_{m \in F_i^w} a_{jm}}.$$

With these weights, we can form the coarse-to-fine transfer matrix, W, from which we can express the overall prolongation matrix, I_H^h, as

$$I_H^h = \begin{bmatrix} W \\ I \end{bmatrix}.$$

We stress that while we only investigate the use of this interpolation for structured grids in this work, the use of these multigrid components enable the solution of unstructured-grid Helmholtz problems.

4 Numerical experiments

In this section, we perform several numerical 2D experiments of increasing complexity. We start with the constant wavenumber problem, which serves as a benchmark for the algorithmic choices, after which we evaluate the method's performance for a Helmholtz problem with a wedge heterogeneity, and a model of the Sigsbee field.

The experiments have been performed with a C++ Helmholtz code on an Intel Core2 Duo 1.66GHz CPU, with 1.0GB RAM memory.

4.1 Homogeneous problem

The first numerical experiments are based on the *homogeneous* Helmholtz problem on the square domain, $(0,1)^2$, to gain insight into the overall performance of the solvers. The pulse source, g, is located near the surface, at $(\frac{1}{2}, \frac{1}{32})$ and is represented by the scaled delta function.

We set $\alpha = 0$ in equation (1) in all numerical experiments, unless stated otherwise.

We will evaluate the following two multigrid preconditioners:

1. A multigrid V(1,1)-cycle with de Zeeuw's prolongation operator, FW restriction and Jacobi smoothing with relaxation parameter $\omega = 0.5$. This is the solver from [18].
2. A multigrid V(0,1)-cycle with AMG's prolongation operator, FW restriction and ILU(0) post-smoothing. This is the solver from [57].

The quality of the multigrid preconditioner was assessed in [18] and [57], respectively. These preconditioners are combined with the Bi-CGSTAB Krylov subspace solver. The value of β_1 in the shifted Helmholtz preconditioner equals 1, β_2 is set to either 0.4 or 0.5.

4.1.1 Second- and fourth-order discretizations

We compare the convergence of the Krylov subspace solvers for the second- and fourth-order discretizations of both the original operator and the preconditioner. The ABL is not included in this experiment. We test each setting with a random initial guess and a point source as a right-hand side. The iteration is terminated as soon as

preconditioner	β_2	$h:$ 1/64	1/128	1/256	1/512
ω-Jacobi	0.4	39 (0.26)	75 (1.7)	139 (12.6)	266 (99)
Zeeuw-V(1,1)	0.5	36 (0.22)	69 (1.5)	125 (11.3)	236 (88)
ILU(0)	0.4	26 (0.17)	43 (0.94)	88 (7.9)	218 (83)
AMG-V(0,1)	0.5	27 (0.19)	42 (0.92)	83 (7.4)	162 (61)

Table 1 Bi-CGSTAB performance for the homogeneous model (*second-order discretization*) in terms of number of iterations and CPU time in seconds (in brackets).

multigrid preconditioner	β_2	$h:$ 1/64	1/128	1/256	1/512
ω-Jacobi	0.4	35 (0.30)	70 (2.0)	122 (14.1)	215 (103)
Zeeuw-V(1,1)	0.5	32 (0.25)	62 (1.8)	110 (13.0)	201 (96)
ILU(0)	0.4	16 (0.13)	26 (0.78)	45 (5.7)	84 (45)
AMG-V(0,1)	0.5	19 (0.16)	30 (0.91)	52 (6.6)	95 (51)

Table 2 Bi-CGSTAB performance for the homogeneous model (*fourth-order discretization*), $kh = 0.625$, in terms of number of iterations and CPU time in seconds (in brackets).

multigrid preconditioner	β_2	$h:$ 1/64	1/128	1/256	1/512
ω-Jacobi	0.4	48 (0.42)	111 (3.2)	267 (31.1)	> 500
Zeeuw-V(1,1)	0.5	46 (0.44)	107 (3.1)	242 (28.1)	> 500
ILU(0)	0.4	21 (0.17)	35 (1.06)	65 (8.3)	151 (81.1)
AMG-V(0,1)	0.5	23 (0.25)	41 (1.25)	73 (9.3)	132 (71.0)

Table 3 Bi-CGSTAB performance for the homogeneous model (*fourth-order discretization*), $kh = 0.8$, in terms of number of iterations and CPU time in seconds (in brackets).

the relative residual is reduced to a prescribed tolerance of 10^{-6},

$$\frac{||r_i||}{||r_0||} \leq 10^{-6}. \tag{13}$$

In the Tables 1 and 2 we present, for $kh = 0.625$, the Bi-CGSTAB performance on four meshes, with the two multigrid preconditioners, for the second- and fourth-order discretizations, respectively. Furthermore, in Table 3 we repeat the experiment with the fourth-order discretization, now setting $kh = 0.8$.

For all solvers, we observe a more-or-less linear increase in the number of iterations for increasing wavenumbers. The performance of the V(0,1) multigrid preconditioner with the ILU(0) smoothing appears to be the most robust among these choices. Its convergence for both values of β_2, and for both discretizations, is very satisfactory. An interesting observation, comparing the results in the Tables 1 and 2, is that the performance of the AMG interpolation (as in Table 2) is, especially on the finest grid, significantly better for the fourth-order discretization. Setting $kh = 0.8$ in Table 3 gives us an increase in the number of iterations, compared to the perfor-

number of points		discretization	
n	β_2	2nd order	4th order
32	0.4	109 (0.42)	56 (0.31)
	0.5	120 (0.48)	67 (0.31)
64	0.4	34 (0.45)	32 (0.33)
	0.5	37 (0.48)	39 (0.55)
128	0.4	29 (1.4)	28 (2.0)
	0.5	30 (1.5)	29 (2.1)
256	0.4	27 (5.8)	25 (7.6)
	0.5	29 (6.3)	27 (8.2)

Table 4 Bi-CGSTAB performance for second- and fourth-order discretizations for the homogeneous model in terms of number of iterations and CPU time in seconds (in brackets).

mance in Table 2. Note, however, that we deal with higher wavenumbers in Table 3, when we fix kh to 0.8 and fix h.

The CPU times per iteration reported for the fourth-order discretization are always somewhat higher than for the second-order problem, as we deal with 9-point discretization stencils on all grids.

4.1.2 Fixed wavenumber, increasing mesh sizes

We reconsider the 2D homogeneous model from Subsection 2.2.1, discretized on the unit square with the ABL. Wavenumber $k = 60$ is set in this experiment, and $kh = 0.8$. Note that this choice of the linear relation between k and h does not lead to accurate solutions with the second-order discretization on the coarse meshes.

The number of grid points increases in order to confirm the asymptotic grid independent convergence of the preconditioned Bi-CGSTAB solver for a fixed continuum problem. We terminate the iterations as soon as the relative residual is less than 10^{-6}.

The iterative solver in this experiment is based on the Bi-CGSTAB method with a V(0,1) multigrid preconditioner, with $\beta_1 = 1$, in which the ILU(0) smoother and the AMG prolongation are incorporated. Table 4 presents the number of Bi-CGSTAB iterations, plus the CPU time to reach the termination criterion, for two values of β_2, $\beta_2 = 0.4$ and $\beta_2 = 0.5$, with the second- and fourth-order discretizations. We observe the asymptotic h-independent convergence rate for the iterative solver on the finer grids; with fixed wavenumber and h decreasing, approximately the same number of iterations is needed to satisfy the termination criterion, for both discretizations. However, on the coarse grids, where we have $kh > 1$, we see a drastic increase in the number of iterations needed to converge, especially for the second-order discretization.

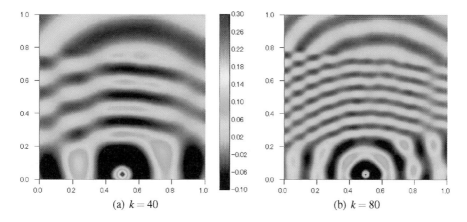

Fig. 9 Numerical solutions for the wedge problem with $k=40$ and $k=80$ with ABL.

4.2 The wedge problem

In this section, we present numerical results for the wedge problem. The domain, as in Figure 1, is a box, $(0,1)^2$, in which a wedge-shaped heterogeneity is placed, and the location of the source is $(1/2, 1/32)$. The wave number inside the wedge region is k, and outside the wedge it is set to $k/2$. We employ the fourth-order discretization with the ABL here, with $n/4$ points on both sides in the ABL.

Figure 9 presents the solutions of the wedge problem for $k=40$ and $k=80$.

We also examine the convergence of the preconditioned Bi-CGSTAB method, with the shifted Laplacian $V(0,1)$-multigrid preconditioner with the ILU(0) smoother and the AMG prolongation. We again set $kh=0.8$ here. We also present results with some damping included in the original Helmholtz equation. Parameter α in (1) varies between 0 and 0.05. The number of iterations and CPU time (in seconds) are presented in Table 5. We notice a significant improvement of the method's convergence, already when 1% damping is included in the original problem. With 5% damping, we even observe a more-or-less constant number of iterations, for varying k. Compared to the performance of the solution method with the damped Jacobi smoother (not shown here), the results in Table 5 are significantly improved, both in terms of the number of iterations and in terms of the CPU time.

4.3 The Sigsbee problem

The Sigsbee2A synthetic data set models the geologic setting found on the Sigsbee escarpment in the deep-water Gulf of Mexico. There is a substantial uniform layer of water at the top of the model. Here, we use a version of the original Sigsbee model to test our iterative Helmholtz solver, see Figure 10. The size of the domain

β_2	damping α	\multicolumn{5}{c}{number of points (kh constant)}				
		32	64	128	256	512
0.4	0.0%	18 (0.14)	27 (0.45)	48 (3.39)	88 (26.8)	175 (209)
	1.0%	17 (0.14)	24 (0.41)	38 (2.67)	54 (16.4)	86 (103)
	2.5%	14 (0.08)	21 (0.36)	29 (2.03)	38 (11.6)	53 (63.6)
	5.0%	12 (0.06)	16 (0.28)	20 (1.41)	24 (7.31)	28 (33.6)
0.5	0.0%	20 (0.14)	33 (0.58)	55 (3.88)	101 (30.7)	189 (227)
	1.0%	20 (0.16)	31 (0.53)	44 (3.09)	64 (19.4)	94 (113)
	2.5%	17 (0.09)	23 (0.39)	34 (2.39)	43 (13.1)	52 (62.5)
	5.0%	14 (0.08)	20 (0.34)	24 (1.69)	28 (8.52)	30 (26.0)

Table 5 Bi-CGSTAB performance for the fourth-order discretization of the wedge model, with α % damping, in terms of number of iterations and CPU time in seconds (in brackets).

is $15000^2 m$ and a source is placed at $(7500, 117)$, near the top wall. The frequency chosen for this computation is $5 Hz$.

The grid size consists of 512^2 points with an ABL of 128 points on each edge. The largest value of kh is 0.6135. Figure 11 presents the solution of this Helmholtz problem, where the fourth-order discretization is used. With the linear solver based on preconditioned Bi-CGSTAB with the V(0,1) multigrid preconditioner for the shifted Laplacian, using the ILU(0) smoother and AMG-based interpolation as essential components, we solve this problem in 61 iterations and 74.2 CPU seconds with $\beta_2 = 0.4$, and in 68 iterations and 85.5 CPU seconds for $\beta_2 = 0.5$. This convergence is highly satisfactory. As a comparison, the solver with $\beta_2 = 0.5$, the multigrid V(1,1)-cycle preconditioner, point-wise Jacobi smoothing and de Zeeuw's interpolation needed 216 iterations and 237 seconds CPU time.

Fig. 10 Domain for the scaled Sigsbee problem and the distribution of velocity, $c(x)$, and wavenumber, k.

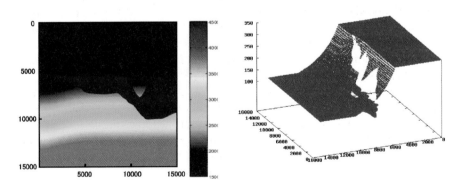

(a) Speed of sound, $c(x)$ for Sigsbee scaled domain

(b) Distribution of k

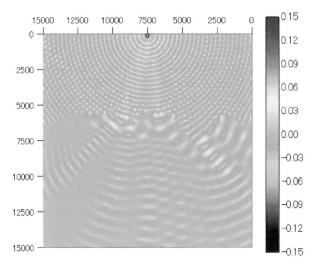

Fig. 11 Solution of the scaled Sigsbee problem with ABL, frequency $5Hz$.

5 Conclusion

In this paper, we have discussed the ingredients of a robust and efficient iterative solver for heterogeneous high wavenumber Helmholtz problems. A preconditioned Bi-CGSTAB solver has been developed in which the preconditioner is based on a shifted Laplacian with a complex-valued shift. We have shown that it is possible to work with fourth-order finite differences, both in the discrete original problem, as well as in the preconditioner. An absorbing boundary layer improves the quality of the solution significantly and does not pose difficulties to the solution method proposed. We have focused on fourth-order discretizations mostly obeying a linear relation between the wavenumber and the mesh size, $kh = 0.8$, here.

The fourth-order accurate shifted Laplacian preconditioner can be approximated by one V(0,1)-cycle of multigrid. In the multigrid preconditioner, we have compared a point-wise Jacobi smoother with an ILU(0) smoother and we included an AMG-based prolongation scheme. This enables us to choose a small imaginary shift parameter ($\beta_2 = 0.4$) in the preconditioner, which improves the solver's convergence (especially for high wavenumbers on fine meshes). The generalization to 3D is straightforward; the parallelization of this solver requires, however, some considerations as an ILU smoother is not easily parallelized.

Acknowledgment The authors would like to thank Shell International Exploration and Production and Philips for their financial support to this BTS project. Furthermore, they would like to thank Y.A. Erlangga, Ch. Dwi Riyanti, A. Kononov, M.B.

van Gijzen for their input in the project. In particular, we would also like to thank S.P. MacLachlan and N. Umetani for their contribution to this article, as the present article relies on paper [57] of which they are the co-authors. N. Umetani performed some additional numerical experiments, especially for this paper.

References

1. T. AIRAKSINEN, E. HEIKKOLA, A. PENNANEN, J. TOIVANEN, An algebraic multigrid based shifted-Laplacian preconditioner for the Helmholtz equation. *J. Comp. Physics,* 226: 1196–1210, 2007.
2. R. E. ALCOUFFE, A. BRANDT, J. E. DENDY JR., J. W. PAINTER, *The multi-grid method for the diffusion equation with strongly discontinuous coefficients*, SIAM J. Sci. Comput., 2 (1981), pp. 430–454.
3. A. BAMBERGER, P. JOLY, J.E. ROBERTS, Second-order absorbing boundary conditions for the wave equations: A solution for the corner problem. *SIAM J. Numer. Anal.,* 27: 323–352, 1990.
4. A. BAYLISS, C. I. GOLDSTEIN, E. TURKEL, *An iterative method for Helmholtz equation*, J. Comput. Phys., 49 (1983), pp. 443–457.
5. A. BAYLISS, C.I. GOLDSTEIN E. TURKEL, On accuracy conditions for the numerical computation of waves. *J. Comput. Phys.,* 59: 396–404, 1985.
6. M. BENZI, D. BERTACCINI, Block Preconditioning of Real-Valued Iterative Algorithms for Complex Linear Systems, October 2006, revised September 2007, 21 pages. To appear in IMA Journal of Numerical Analysis.
7. M. BOLLHÖFER, M. GROTE, O. SCHENK, Algebraic multilevel preconditioning for Helmholtz equation, *In:* Proc. Europ. Conf. on Comput. Fluid Dynamics (ECCOMAS CFD 2006), Egmond aan Zee, The Netherlands, Sept. 5-8, 2006.
8. A. BRANDT, Multi-level adaptive solutions to boundary-value problems. *Math. Comput.* 31: 333–390, 1977.
9. A. BRANDT, S. TA'ASAN, *Multigrid method for nearly singular and slightly indefinite problems*, in Proceedings EMG'85 Cologne, Multigrid Methods II, W. Hackbusch, U. Trottenberg, eds., Springer, Berlin, 1986, pp. 99–121.
10. A. BRANDT, I. LIVSHITS, Wave-ray multigrid methods for standing wave equations. *Elect. Trans. Numer. Anal.* 6: 162–181, 1997.
11. W.L. BRIGGS, V.E. HENSON, S.F. MCCORMICK, A multigrid tutorial: SIAM, Philadelphia, USA, 2000.
12. I. DUFF, S. GRATTON, X. PINEL, X. VASSEUR, Multigrid based preconditioners for the numerical solution of two-dimensional heterogeneous problems in geophysics. *Intern. J. Computer Math.,* 84(8):1167–1181, 2007.
13. J. CLAERBOUT, *Imaging the Earth's interior*, Blackwell Science Inc., 1985.
14. H. R. ELMAN, O. G. ERNST, D. P. O'LEARY, A multigrid method enhanced by Krylov subspace iteration for discrete Helmholtz equations, *SIAM J. Sci. Comput.*, 23: 1291–1315, 2001.
15. B. ENGQUIST, A. MAJDA, Absorbing boundary conditions for the numerical simulation of waves, *Math. Comput.*, 31: 629–651, 1977.
16. Y.A. ERLANGGA, C. VUIK, C.W. OOSTERLEE, *On a class of preconditioners for the Helmholtz equation*, Appl. Numer. Math., 50 (2004), pp. 409–425.
17. Y.A. ERLANGGA, C. VUIK, C.W. OOSTERLEE, Comparison of multigrid and incomplete LU shifted-Laplace preconditioners for the inhomogeneous Helmholtz equation. *Applied Num. Math.* 56: 648-666, 2006.
18. Y.A. ERLANGGA, C.W. OOSTERLEE, C. VUIK, A novel multigrid based preconditioner for heterogeneous Helmholtz problems. *SIAM J. Sci. Comput.* 27: 1471-1492, 2006.

19. Y.A. ERLANGGA, Advances in Iterative Methods and Preconditioners for the Helmholtz Equation *Archives Comput. Methods in Engin.*, 15: 37-66, 2008.
20. M. B. VAN GIJZEN, Y. A. ERLANGGA, C. VUIK, Spectral Analysis of the Discrete Helmholtz Operator Preconditioned with a Shifted Laplacian *SIAM J. Sci. Comput.*, 29: 1942-1958, 2007.
21. A. GEORGE, J.W. LIU, *Computer solution of large sparse positive definite systems*, Prentice-Hall, New Jersey, 1981.
22. S.GHEORGHE, *On Multigrid Methods for Solving Electromagnetic Scattering Problems*. PhD Thesis, Univ. Kiel, Germany, 2006.
23. J. GOZANI, A. NACHSHON, E. TURKEL, Conjugate gradient coupled with multigrid for an indefinite problem, in *Advances in Comput. Methods for PDEs V*, 425–427, 1984.
24. W. HACKBUSCH, *Multi-grid methods and applications*. Springer, Berlin, 1985.
25. E. HEIKKOLA, S. MÖNKÖLÄ, A. PENNANEN, T. ROSSI, Controllability method for acoustic scattering with spectral elements, *J. Comput. Appl. Math.*, 204(2): 344-355, 2007.
26. E. HEIKKOLA, S. MÖNKÖLÄ, A. PENNANEN, T. ROSSI, Controllability method for the Helmholtz equation with higher-order discretizations. *J. Comp. Phys.*, 225(2): 1553-1576, 2007.
27. F. IHLENBURG, I. BABUSKA, Finite element solution to the Helmholtz equation with high wave numbers. *Comput. Math. Appl.*, 30: 9-37, 1995.
28. K. ITO, J. TOIVANEN, A Fast Iterative Solver for Scattering by Elastic Objects in Layered Media. *Appl. Numerical Math.*, 57: 811-820, 2007.
29. C.-H. JO, C. SHIN, J.H. SUH, An optimal 9-point, finite-difference, frequency space, 2-D scalar wave extrapolator, *Geophysics* 61(2): 529–537, 1996.
30. R. KETTLER, Analysis and comparison of relaxation schemes in robust multigrid and preconditioned conjugate gradient methods. *In:* W. Hackbusch, U. Trottenberg (eds.), Multigrid methods, *Lecture Notes in Mathematics* 960: 502–534, Springer, Berlin, 1982.
31. M. KHALIL, *Analysis of linear multigrid methods for elliptic differential equations with discontinuous and anisotropic coefficients*. Ph.D. Thesis, Delft University of Technology, Delft, Netherlands, 1989.
32. S. KIM, S. KIM, Multigrid simulation for high-frequency solutions of the Helmholtz problem in heterogeneous media, *SIAM J. Sci. Comput.* 24: 684–701, 2002.
33. D. LAHAYE, H. DE GERSEM, S. VANDEWALLE, K. HAMEYER, *Algebraic multigrid for complex symmetric systems*, IEEE Trans. Magn., 36 (2000), pp. 1535–1538.
34. A. L. LAIRD, M. B. GILES, *Preconditioned iterative solution of the 2D Helmholtz equation*. Report NA 02-12, Comp. Lab., Oxford Univ., 2002.
35. B. LEE, T. A. MANTEUFFEL, S. F. MCCORMICK, J. RUGE, *First-order system least-squares for the Helmholtz equation*, SIAM J. Sci. Comput., 21 (2000), pp. 1927–1949.
36. S.S. LI, X.W. PING, R.S. CHEN, A Kind of Preconditioners Based on Shifted Operators to Solve Three-Dimensional TVFEM Equations *In:* Int. Symp. on Microwave, Antenna, Propagation and EMC Technologies for Wireless Communications, 842-844, 2007
37. Q. LIAO, G.A. MCMECHAN, Multifrequency viscoacoustic modeling and inversion. *Geophysics* 61(5): 1371–1378, 1996.
38. S.P. MACLACHLAN, C.W. OOSTERLEE, Algebraic multigrid solvers for complex-valued matrices, *SIAM J. Sci. Comput.*, 30:1548-1571, 2008.
39. K.J. MARFURT, Accuracy of finite-difference and finite-element modeling of the scalar and elastic wave-equations, *Geophysics*, 49: 533-549, 1984.
40. W.A. MULDER, R.-E. PLESSIX, A comparison between one-way and two-way wave-equation migration, *Geophysics*, 69: 1491–1504, 2004.
41. W.A. MULDER, R.-E. PLESSIX, How to choose a subset of frequencies in frequency-domain finite-difference migration, *Geophys. J. Int.* 158: 801–812, 2004.
42. O.V. NECHAEV, E.P. SHURINA, M.A. BOTCHEV, Multilevel iterative solvers for the edge finite element solution of the 3D Maxwell equation. *Comp. Mathem. Applications*. doi:10.1016/j.camwa.2007.11.003, to appear 2008.

43. S. OPERTO, J. VIRIEUX, P. AMESTOY, J.Y. L'EXCELLENT, L. GIRAUD, B. BEN HADJ ALI, 3D finite-difference frequency-domain modeling of visco-acoustic wave propagation using a massively parallel direct solver: A feasibility study, *Geophysics,* SM195-SM211, 2007.
44. R.-E. PLESSIX, A Helmholtz iterative solver for 3D seismic-imaging problems, *Geophysics,* 72: 185–194, 2007.
45. P. POULLET, A. BOAG, Incremental unknowns preconditioning for solving the Helmholtz equation. *Num. Methods Partial Diff. Equations,* 23(6): 1396-1410, 2007.
46. R.G. PRATT, Seismic waveform inversion in frequency domain. Part I: theory and verification in a physical scale domain, *Geophysics,* 64: 888–901, 1999.
47. C.D. RIYANTI, Y.A. ERLANGGA, R.-E. PLESSIX, W.A. MULDER, C.W. OOSTERLEE, C. VUIK, A new iterative solver for the time-harmonic wave equation, *Geophysics* 71: 57-63, 2006.
48. C.D. RIYANTI, A. KONONOV, Y.A. ERLANGGA, C. VUIK, C.W. OOSTERLEE, R-E PLESSIX W.A. MULDER, A parallel multigrid-based preconditioner for the 3D heterogeneous high-frequency Helmholtz equation. *J. Comp. Physics* 224: 431-448, 2007.
49. J.W. RUGE, K. STÜBEN, Algebraic Multigrid (AMG). *In:* S.F. McCormick (ed.), Multigrid Methods, Frontiers in Appl. Math., SIAM Philadelphia, 5: 73–130, 1987.
50. I. SINGER, E. TURKEL, High Order Finite Difference Methods for the Helmholtz Equation, *Comp. Meth. Appl. Mech. Eng.* 163:343-358, 1998.
51. I. SINGER, E. TURKEL, A Perfectly Matched Layer for the Helmholtz Equation in a Semi-infinite Strip. *J. Comput. Phys.* 201 (2004) 439-465.
52. Y. Saad, M.H. Schultz, GMRES: A generalized minimal residual algorithm for solving non-symmetric linear system, *SIAM J. Sci. Comput.,* 7: 856–869, 1986.
53. Y. SAAD, *Iterative Methods for Sparse Linear Systems.* SIAM, Philadelphia, 2003
54. U. TROTTENBERG, C.W. OOSTERLEE, A. SCHÜLLER, *Multigrid,* Academic Press, London, 2001.
55. E. TURKEL, Numerical Methods and Nature. *J. of Scientific Computing,* 28 2/3: 549-570, 2006.
56. E. TURKEL, *Numerical difficulties solving time harmonic equations,* in Multiscale Computational Methods in Chemistry and Physics, A. Brandt, et. al., eds., IOS Press, Tokyo, 2001, pp. 319–337.
57. N.UMETANI S.P.MACLACHLAN, C.W. OOSTERLEE, *A multigrid-based shifted-Laplacian preconditioner for a fourth-order Helmholtz discretization.* Working paper, Delft Univ. Techn. Delft, the Netherlands, 2008. Submitted for publication.
58. H.A. VAN DER VORST, Bi-CGSTAB: a fast and smoothly converging variant of Bi-CG for the solution of nonsymmetric linear systems, *SIAM J. Sci. Comput.,* 13: 631-645, 1992.
59. X. WEI, H. P. URBACH, A. J. WACHTERS, Finite-element model for three-dimensional optical scattering problems. *J. Opt. Soc. Am. A,* 24: 866-881, 2007.
60. P. WESSELING, *An introduction to multigrid methods.* John Wiley, Chichester, 1992.
61. R. WIENANDS, C. W. OOSTERLEE, *On three-grid Fourier analysis for multigrid,* SIAM J. Sci. Comput., 23 (2001), pp. 651–671.
62. G. WITTUM, On the robustness of ILU smoothing. *SIAM J. Sci. Comput.* 10: 699–717, 1989.
63. P. M. DE ZEEUW, Matrix-dependent prolongations and restrictions in a blackbox multigrid solver, *J. Comput. Appl. Math.,* 33: 1–27, 1990.
64. P.M. DE ZEEUW, Incomplete line LU as smoother and as preconditioner. *In:* W. Hackbusch, G. Wittum (eds.), Incomplete decompositions (ILU) - algorithms, theory, and applications. Vieweg, Braunschweig, 215–224, 1993. *Int. J. Num. Methods in Fluids* 20: 59–74, 1995.

On Numerical Issues in Time Accurate Laminar Reacting Gas Flow Solvers

S. van Veldhuizen, C. Vuik, and C.R. Kleijn

Abstract The numerical modeling of laminar reacting gas flows in thermal Chemical Vapor Deposition (CVD) processes commonly involves the solution of advection-diffusion-reaction equations for a large number of reactants and intermediate species. These equations are stiffly coupled through the reaction terms, which typically include dozens of finite rate elementary reaction steps with largely varying rate constants. The solution of such stiff sets of equations is difficult, especially when time-accurate transient solutions are required. In this study various numerical schemes for multidimensional transient simulations of laminar reacting gas flows with homogeneous and heterogeneous chemical reactions are compared in terms of efficiency, accuracy and robustness. One of the test cases is the CVD process of silicon from silane, modeled according to the classical 17 species, 26 reactions chemistry model for this process as published by Coltrin and coworkers [4]. It is concluded that, for time-accurate transient simulations the conservation of the non-negativity of the species concentrations is much more important, and much more restrictive towards the time step size, than stability. For this reason we restrict ourselves to the first order, unconditionally positive Euler Backward method. Since positivity of the solution is very important, the use of Newton methods to solve the nonlinear problems is only feasible in combination with direct solvers. When using iterative linear solvers, it appears that the approximate solutions may have small negative elements. To circumvent this, we introduce a projected Newton method. Choosing the best preconditioners, combined with our projected Newton method, enables us

---

S. van Veldhuizen
Delft University of Technology, Delft Institute of Applied Mathematics and J.M. Burgerscentrum, Mekelweg 4, 2628 CD Delft, The Netherlands, e-mail: s.vanveldhuizen@tudelft.nl

C. Vuik
Delft University of Technology, Delft Institute of Applied Mathematics and J.M. Burgerscentrum, Mekelweg 4, 2628 CD Delft, The Netherlands, e-mail: c.vuik@tudelft.nl

C.R. Kleijn
Delft University of Technology, Department of Multi Scale Physics and J.M. Burgerscentrum, Prins Bernardlaan 6, 2628 BW Delft, The Netherlands, e-mail: c.r.kleijn@tudelft.nl

to reduce the computational time of the time accurate simulation of the classical 17 species, 26 reactions chemistry model by a factor 20 on a single processor.

1 Introduction

Chemical Vapor Deposition (CVD) is a chemical process which transforms gaseous molecules into high-purity, high-performance solid materials in the form of, for instance, a thin film or a powder. The production of thin films via CVD is of considerable importance for the micro-electronics industry, but also in other technological areas applications of thin solid films via CVD can be found. For instance, in the glass industry protective and decorative layers may be deposited on glass via CVD. In a typical CVD process the material to be deposited is introduced into the reactor chamber in one or more gas(es), called precursors, which react and/or decompose on the substrate surface to produce the desired deposit. The volatile byproducts are removed by the gas flow through the reactor.

Numerical simulations are a widely used tool to design CVD reactors and to optimize the process itself [16]. In most cases, these are steady state simulations, since the total process time is large compared to the transients during start-up and shut-down. However, with the deposited films getting thinner and thinner, process times are reduced and transient times become of more importance. Also for the study of inherently transient CVD processes, such as Rapid Thermal CVD (RTCVD) and Atomic Layer Deposition (ALD), transient simulations are indispensable.

The aim of this study is to develop nonstationary solvers for CVD processes. The numerical modeling of realistic CVD processes based on detailed chemistry, involves the solution of multi-dimensional advection-diffusion-reaction equations for a large number of reactants and intermediate species. These equations are nonlinearly and stiffly coupled through the reaction terms, which typically include dozens of elementary finite rate reactions with largely varying rate constants. It is difficult to find the solution of such stiff systems, especially in the case one wants to find a time-accurate transient solution.

The numerical solvers present in most commercial CFD codes have great problems producing time-accurate, transient (and even steady state) results for laminar reacting flow simulations such as CVD. Although some commercial CFD codes claim to be able to handle stiff chemistry, no successful attempts to model multi-dimensional gas-flow with multi-species, multi-reaction CVD chemistry using commercial CFD codes have been reported in literature.

Alternative solution methods developed in the computational physics communities include ideas as the following. A first approximation would be to integrate the advection and diffusion terms explicitly, and the reaction terms implicitly. Then, per grid point the solution of a nonlinear systems of the size of the number of reactive species is needed. This could easily be done by means of a Newton method. A fully coupled approach, called the IMEX Runge-Kutta Chebyshev method, is discussed in Section 4.2. When using this approach, then it is important to integrate the discretized diffusion part stably, which is certainly not a straightforward task to accomplish. Secondly, one could also march through the grid from grid point to grid point and solve the nonlinear system of reaction terms. This strategy looks like a nonlinear Gauss-Seidel method and is used by Kleijn, see [15].

For strongly diluted laminar reacting flow problems, in which the flow and temperatures are not influenced by the reactions, the computation of the flow and temperature field is rather trivial in comparison with the solution of the system of advection-diffusion-reaction equations. Therefore, we mainly focus on efficient solution techniques for the stiff set of advection-diffusion-reaction equations. Besides stability requirements for stable integration in time, preservation of non-negativity of the species concentrations is required as well. In [24] we discussed this so-called positivity property, which puts severe restrictions on the time step size. In Section 4 we present further details on stability and positivity conditions for time integration methods.

It appears that the first order Euler Backward method is both unconditionally stable and unconditionally positive. Because of these properties, we restrict ourselves to the Euler Backward method for time integration. The species concentrations remain positive within Newton's method, used to solve the nonlinear systems within Euler Backward, when a direct solver is used to obtain the solution of the linear systems. However, when iterative linear solvers are used, the solutions of the nonlinear systems may have small negative components, despite the unconditional positivity of Euler Backward. In order to prevent this, we introduce a projected version of Newton's method. Further details are discussed in Section 5.1.

The linear systems which have to be solved in the Euler Backward method, are sparse and large. Preconditioned Krylov solvers are suitable to reduce computational costs. In Section 5.2 we report on the comparison of various preconditioners and their relationship to the positivity of the solution. We conclude with a description of the chemistry models in the test problems, and discuss the numerical results obtained. We want to emphasize that all results presented in this paper have been presented in previous work. To summarize, References [22, 24] are summarized in Section 4, whereas the the discussion on the simulation results can also be found in [21].

Finally, we remark that the computational methods presented in this paper are not only applicable in CVD, but also in applications such as laminar combustion [8] and Solid Oxide Fuel Cell modeling [27]. Under the restriction that the spatial discretization is structured, we conjecture that our methodology is also applicable in these applications.

2 Numerical modeling of chemical vapor deposition

Thin solid films are widely used in many technological areas such as microelectronics and the glass industry, with applications varying from insulating and (semi-)conducting layers in the micro-electronics, to optical, mechanical and/or decorative coatings on glass. The production of such thin layers can be done by various deposition processes, e.g. sputtering, evaporation and CVD. Involving chemical reactions clearly distinguishes CVD from the other production technologies, whereby the most important advantage is its capability of depositing films of uniform thickness on highly irregularly shaped surfaces [11].

Basically, a CVD system is a chemical reactor in which precursor gases containing the atoms to be deposited are introduced, usually diluted in an inert carrier gas. Furthermore, the reactor chamber contains substrates on which the deposition takes place. In this study it is assumed that the energy to drive the (gas phase and surface) reactions is thermal energy.

Basically, the following six steps occurring in every CVD process have to be mathematically modeled:

1. Convective and diffusive transport of reactants from the reactor inlet to the reaction zone within the reactor chamber
2. Chemical reactions in the gas phase leading to a multitude of new reactive species and byproducts,
3. Diffusive transport of the initial reactants and the reaction products from the homogeneous reactions to the susceptor surface, where they are adsorbed on the susceptor surface,
4. Surface diffusion of adsorbed species over the surface and heterogeneous surface reactions catalyzed by the surface, leading to the formation of a solid film,
5. Desorption of gaseous reaction products, and their diffusive transport away from the surface,
6. Convective and/or diffusive transport of reaction products away from the reaction zone to the outlet of the reactor.

For fully heterogeneous CVD processes the second step in the above enumeration does not take place. Steps one to six are illustrated in Figure 1

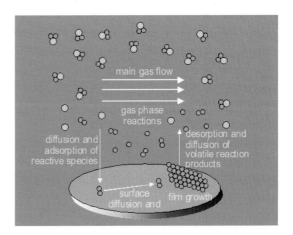

Fig. 1 Schematic representation of the six basic steps in CVD. This illustration is taken from [13].

2.1 Transport model for the gas species

To mathematically model a CVD process, the gas flow, the transport of thermal energy, the transport of species and the chemical reactions in the reactor have to be described. We assume that the gas mixture in the reactor behaves as a continuum, as an ideal gas and in accordance with Newton's law of viscosity. The gas flow in the reactor is assumed to be laminar.

The composition of the N component gas mixture is described in terms of the dimensionless mass fractions ω_i, which sum up to one. Transport of total mass, momentum and heat are described respectively by the continuity equation, the Navier-Stokes equations and the transport equation for thermal energy. Note that the consumption and production of heat due to the chemical reactions is also included in the energy equation. For most CVD systems, especially when the reactants are highly diluted in an inert carrier gas, the heat of reactions has a negligible influence on the gas temperature distribution. For such systems, the computation of the laminar flow and the temperature field is a relatively trivial task, which can be performed preceding to and independently of the calculation of the species concentrations. The difficulty, however, lies in solving the set of highly nonlinear and strongly coupled species equations.

Transport of mass fraction ω_i is described by the i-th species equation

$$\frac{\partial(\rho\omega_i)}{\partial t} = -\nabla\cdot(\rho\mathbf{v}\omega_i) + \nabla\cdot[(\rho\mathbb{D}_i\nabla\omega_i) + (\mathbb{D}_i^T\nabla(\ln T))] + m_i\sum_{k=1}^{K}v_{ik}R_k^g, \quad (1)$$

where diffusive mass fluxes are due to concentration diffusion and thermal diffusion. In (1) $\mathbb{D}_i$ is the effective ordinary diffusion coefficient and $\mathbb{D}_i^T$ is the effective thermal diffusion coefficient for species i [15], $\mathbf{v}$ the mass averaged velocity obtained from the Navier-Stokes equations, and ρ the density of the gas mixture. Suppose that K reversible gas-phase reactions of the form

$$\sum_{i=1}^{N}v'_{ik}\mathscr{A}_i \underset{k_{k,\text{backward}}^g}{\overset{k_{k,\text{forward}}^g}{\rightleftarrows}} \sum_{i=1}^{N}v''_{ik}\mathscr{A}_i \quad (2)$$

take place. The net molar reaction rate R_k^g for the kth reaction in the last term on the right-hand side of (1) is computed as

$$R_k^g = k_{k,\text{forward}}^g \prod_{i=1}^{N}\left(\frac{P\omega_i m}{RTm_i}\right)^{v'_{ik}} - k_{k,\text{backward}}^g \prod_{i=1}^{N}\left(\frac{P\omega_i m}{RTm_i}\right)^{v''_{ik}}. \quad (3)$$

In eq. (2), $\mathscr{A}_i$ are the species in the gas mixture, v'_{ik} the forward stoichiometric coefficient for species i in reaction k, v''_{ik} the backward stoichiometric coefficient for species i in reaction k. The net stoichiometric coefficient v_{ik} is then defined as $v_{ik} = v''_{ik} - v'_{ik}$. In eq. (3), P is the pressure in Pa, T the temperature, R the universal gas constant, m_i the molar mass of species i and m the average molar mass, computed

as
$$m = \frac{1}{\left(\sum_{i=1}^{N} \frac{\omega_i}{m_i}\right)}. \tag{4}$$

Usually, the forward reaction rate constant $k^g_{k,\text{forward}}$ is fitted according to a modified Arrhenius expression:
$$k^g_{k,\text{forward}}(T) = A_k T^{\beta_k} e^{\frac{-E_k}{RT}}, \tag{5}$$

where A_k, β_k and E_k are fit parameters. For the CVD process considered in the present paper, these fit parameters are available through the references presented in Section 3. The backward reaction rate constants $k^g_{k,\text{backward}}$ are computed self-consistently from the forward reaction rate constants and reaction thermo-chemistry, see [14]. The time constants of the fastest and slowest reactions can differ many (e.g. 25) orders of magnitude, introducing stiffness into the species equations (1). For a detailed description of the mathematical model for CVD, and the corresponding boundary conditions, we refer to [14, 15] and to Section 3.

2.2 Modeling of surface chemistry

We assume that at the wafer surface S irreversible surface reaction takes place. The s-th transformation of gaseous reactants into solid and gaseous reaction products is of the form
$$\sum_{i=1}^{N} \sigma'_{is} \mathscr{A}_i \xrightarrow{R^S_s} \sum_{i=1}^{N} \sigma''_{is} \mathscr{A}_i + \sum_{j=1}^{M} \chi''_{js} \mathscr{B}_j, \tag{6}$$

with $\mathscr{A}_i$ as before, $\mathscr{B}_j$ the solid reaction products, M the number of solid reaction products, σ'_{is} and σ''_{is} are the stoichiometric coefficients for gaseous species i in surface reaction s and χ_{is} the stoichiometric coefficient for the solid species. Again, the net stoichiometric coefficient σ_{is} is defined as $\sigma_{is} = \sigma''_{is} - \sigma'_{is}$.

Usually, heterogeneous surface reactions are characterized by complicated reaction mechanisms that consist of a number of steps. The surface reaction rate R^S_s will therefore depend on the partial pressures of gaseous species, the rate constants of the individual steps (as functions of local temperature), temperature and other surface properties. However, there is in general little or no information available on the individual reaction steps and rate constants. The most common way to model the surface reaction is to propose a mechanism and assume that one of the reaction steps is rate limiting, whereas all other reaction steps are in equilibrium. In this study we are not interested in the qualitative modeling of surface chemistry; we will make use of surface reaction mechanisms for which the surface reaction rates are known. For further details on surface reaction modeling we refer to [14].

If the surface reaction rate R^S_s can be computed, then the growth rate $\mathscr{G}_j$ in nm/s of solid species j is defined as

$$\mathcal{G}_j = 10^9 \frac{m_j}{\rho_j} \sum_{s=1}^{S} R_s^S \chi_{js}, \qquad (7)$$

with m_j the molecular mass of solid species j and ρ_j the density of solid species j.

3 Chemistry models

Numerical simulations presented in sections 4.2.1 and 6 are done for the CVD of silicon from silane according to one of the two reaction models presented in this section. For all simulations the reactor configuration is identical. Its geometry and the matching boundary conditions are shown in Figure 3. As computational domain one half of the $(r\text{-}z)$ plane is taken. From the top a gas-mixture, consisting of 0.1 mole% silane diluted in helium, enters the reactor with a uniform temperature $T_{in} = 300$ K and velocity $u_{in} = 0.1\ \frac{m}{s}$. In the hot region above the susceptor with temperature $T_s = 1000$ K the reactive gas silane decomposes into silylene and hydrogen. This first gas phase reaction initiates a chain of gas phase reactions, and the formation of solid species on the wafer surface. In Section 3.1 a chemistry model consisting of 7 species and 5 gas phase reactions without surface chemistry is presented, whereas in Section 3.2 we comprehensively describe the classical model as published by Coltrin and coworkers [4], which includes 17 gas species, 26 gas phase reactions and 14 surface reactions.

Further, in the processes considered in the present paper, silane and the formed reactive intermediates are highly diluted in the inert carrier gas helium. Therefore, it is justified to assume that the velocity, temperature, density and pressure fields are in steady state and not influenced by the transient chemistry. Since the focus of this study is on solving the species equations (1), we compute these fields using the software of Kleijn [15]. Buoyancy has not been accounted for in these simulations. In Figure 3 the streamlines and temperature field are shown.

3.1 Chemistry model I: 7 species and 5 gas phase reactions

The 7 species and 5 gas phase reactions chemistry model has also been used in [20]. The 5 reactions are listed in Table 1, in which all reactive species, except for the carrier gas helium He, can also be found. Note that for this model only 6 nonlinearly and stiffly coupled species equations (1) have to be solved, because the mass fraction of He can be computed via the property that all mass fraction should add up to one. The reaction terms in eq. (1) are constructed as in eqs. (3) and (5), of which the fit parameters can be found in Table 1. The backward rates are computed selfconsistently from

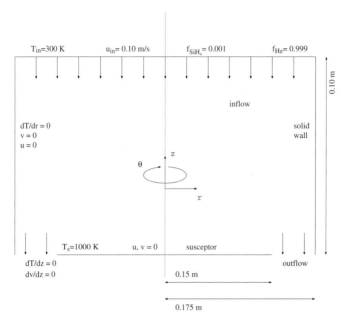

Fig. 2 Reactor geometry and boundary conditions.

$$k^g_{\text{backward}}(T) = \frac{k^g_{\text{forward}}(T)}{K^g(T)} \left(\frac{RT}{P^0}\right)^{\sum_{i=1}^N v_{ik}}, \tag{8}$$

with $K^g(T)$ the reaction equilibrium constants. To facilitate easy reproduction of the solutions presented in the present paper, the reaction equilibrium constants are fitted to a modified Arrhenius expression

$$K^g(T) = A_{\text{eq}} T^{\beta_{\text{eq}}} e^{\frac{-E_{\text{eq}}}{RT}}, \tag{9}$$

with $A_{\text{eq}}, \beta_{\text{eq}}$ and E_{eq} fit parameters which can be found in Table 1.

Table 1 Gas phase reaction mechanism and fit parameters for the 6 species/5 reactions model of Section 3.1. The parameters β_k and β_{eq} are dimensionless, while E_k and E_{eq} have unit $\frac{kJ}{mol}$. The units of A_k and A_{eq} depend on the order of the reaction, but are expressed in units mole, m^3 and s.

Reaction	A_k	β_k	E_k	$A_{k,eq}$	$\beta_{k,eq}$	$E_{k,eq}$
$SiH_4 \rightleftarrows SiH_2 + H_2$	1.09×10^{25}	-3.37	256	6.85×10^5	0.48	235
$Si_2H_6 \rightleftarrows SiH_4 + SiH_2$	3.24×10^{29}	-4.24	243	1.96×10^{12}	-1.68	229
$Si_2H_6 \rightleftarrows H_2SiSiH_2 + H_2$	7.94×10^{15}	0	236	3.70×10^7	0	187
$SiH_2 + Si_2H_6 \rightleftarrows Si_3H_8$	1.81×10^8	0	0	1.36×10^{-12}	1.64	-233
$2SiH_2 \rightleftarrows H_2SiSiH_2$	1.81×10^8	0	0	2.00×10^{-7}	0	-272

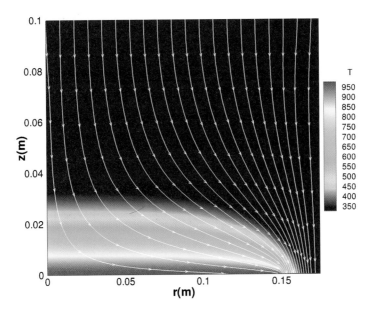

Fig. 3 Streamlines and temperature field in Kelvin for the right half part of the reactor illustrated in Figure 3.

3.2 Chemistry model II: 17 species and 26 gas phase reactions

The other test case in this study, is the same CVD process of silicon from silane, now modeled according to the classical 17 species and 26 reactions chemistry model for this process as published by Coltrin and coworkers [4]. In this model the decomposition of silane into silylene and hydrogen, initiates a chain of 25 homogeneous gas phase reactions leading to the (de)formation of 14 silicon containing gas phase species. Again, the reaction terms in eq. (1) are constructed as in eqs. (3) and (5). The backward rates are computed selfconsistently from eqs. (8) and (9). The 26 reactions and the fit parameters needed in eqs. (5) and (9) are listed Table 2.

Each of the silicon containing species may diffuse towards and react at the susceptor. In this model it is assumed that film growth is due to irreversible, unimolecular decomposition reactions of these species at the surface, leading to the deposition of solid silicon atoms and the desorption of gaseous hydrogen according to:

$$\mathrm{Si}_n\mathrm{H}_{2m} \xrightarrow{R^S_{\mathrm{Si}_n\mathrm{H}_{2m}}} n\,\mathrm{Si}\,(s) + m\,\mathrm{H}_2\,(g), \tag{10}$$

$$\mathrm{Si}_n\mathrm{H}_{2m+1} \xrightarrow{R^S_{\mathrm{Si}_n\mathrm{H}_{2m+1}}} n\,\mathrm{Si}\,(s) + m\,\mathrm{H}_2\,(g) + \mathrm{H}\,(g), \tag{11}$$

Table 2 Fit parameters for the forward reaction rates (5) and gas phase equilibria constants (9) for the benchmark problem. The parameters β_k and β_{eq} are dimensionless, while E_k and E_{eq} have unit $\frac{kJ}{mol}$. The units of A_k and A_{eq} depend on the order of the reaction, but are expressed in units mole, m^3 and s.

Reaction	A_k	β_k	E_k	$A_{k,eq}$	$\beta_{k,eq}$	$E_{k,eq}$
$SiH_4 \rightleftharpoons SiH_2 + H_2$	1.09×10^{25}	-3.37	256	6.85×10^5	0.48	235
$SiH_4 \rightleftharpoons SiH_3 + H$	3.69×10^{15}	0.0	390	1.45×10^4	0.90	382
$Si_2H_6 \rightleftharpoons SiH_4 + SiH_2$	3.24×10^{29}	-4.24	243	1.96×10^{12}	-1.68	229
$SiH_4 + H \rightleftharpoons SiH_3 + H_2$	1.46×10^7	0.0	10	1.75×10^3	-0.55	-50
$SiH_4 + SiH_3 \rightleftharpoons Si_2H_5 + H_2$	1.77×10^6	0.0	18	1.12×10^{-6}	2.09	-6
$SiH_4 + SiH \rightleftharpoons Si_2H_3 + H_2$	1.45×10^6	0.0	8	1.82×10^{-4}	1.65	21
$SiH_4 + SiH \rightleftharpoons Si_2H_5$	1.43×10^7	0.0	8	1.49×10^{-10}	1.56	-190
$SiH_2 \rightleftharpoons Si + H_2$	1.06×10^{14}	-0.88	189	1.23×10^2	0.97	180
$SiH_2 + H \rightleftharpoons SiH + H_2$	1.39×10^7	0.0	8	2.05×10^1	-0.51	-101
$SiH_2 + H \rightleftharpoons SiH_3$	3.81×10^7	0.0	8	2.56×10^{-3}	-1.03	-285
$SiH_2 + SiH_3 \rightleftharpoons Si_2H_5$	6.58×10^6	0.0	8	1.75×10^{-12}	1.60	-241
$SiH_2 + Si_2 \rightleftharpoons Si_3 + H_2$	3.55×10^5	0.0	8	5.95×10^{-6}	1.15	-225
$SiH_2 + Si_3 \rightleftharpoons Si_2H_2 + Si_2$	1.43×10^5	0.0	68	2.67×10^0	-0.18	59
$H_2SiSiH_2 \rightleftharpoons Si_2H_2 + H_2$	3.16×10^{14}	0.0	222	1.67×10^6	-0.37	112
$Si_2H_6 \rightleftharpoons H_3SiSiH + H_2$	7.94×10^{15}	0.0	236	1.17×10^9	-0.36	235
$H_2 + SiH \rightleftharpoons SiH_3$	3.45×10^7	0.0	8	1.42×10^{-4}	-0.52	-183
$H_2 + Si_2 \rightleftharpoons Si_2H_2$	1.54×10^7	0.0	8	7.47×10^{-6}	-0.37	-216
$H_2 + Si_2 \rightleftharpoons SiH + SiH$	1.54×10^7	0.0	168	1.65×10^3	-0.91	180
$H_2 + Si_3 \rightleftharpoons Si + Si_2H_2$	9.79×10^6	0.0	198	1.55×10^2	-0.55	189
$Si_2H_5 \rightleftharpoons Si_2H_3 + H_2$	3.16×10^{14}	0.0	222	1.14×10^6	0.08	210
$Si_2H_2 + H \rightleftharpoons Si_2H_3$	8.63×10^8	0.0	8	3.43×10^{-4}	-0.31	-149
$H + Si_2 \rightleftharpoons SiH + Si$	5.15×10^7	0.0	22	1.19×10^3	-0.88	29
$SiH_4 + H_3SiSiH \rightleftharpoons Si_3H_8$	6.02×10^7	0.0	0	7.97×10^{-16}	2.48	-233
$SiH_2 + Si_2H_6 \rightleftharpoons Si_3H_8$	1.81×10^8	0.0	0	1.36×10^{-12}	1.64	-233
$SiH_3 + Si_2H_5 \rightleftharpoons Si_3H_8$	3.31×10^7	0.0	0	1.06×10^{-14}	1.85	-318
$H_3SiSiH \rightleftharpoons H_2SiSiH_2$	1.15×10^{20}	-3.06	28	9.58×10^{-3}	0.50	-50

where $n = 1, 2,$ or 3, and $m = 0, 1, 2, 3,$ or 4. The molar reaction rate R_i^S for the decomposition of gas species i is given as

$$R_i^S = \frac{\gamma_i}{1 - \frac{\gamma_i}{2}} \frac{Pf_i}{(2\pi m_i R T_s)^{\frac{1}{2}}}, \qquad (12)$$

where T_s denotes the temperature of the wafer surface and f_i is the species mole fraction computed as

$$f_i = \frac{\omega_i m}{m_i}. \qquad (13)$$

The sticking coefficient γ_i is equal to one for all silicon containing species, except for

$$\gamma_{Si_3H_8} = 0, \quad \gamma_{Si_2H_6} = 0.537 \exp\left(\frac{-9400}{T_s}\right), \quad \text{and} \quad \gamma_{SiH_4} = \frac{1}{10} \gamma_{Si_2H_6}. \qquad (14)$$

There is some ambiguity as to which values of sticking coefficients were used in [4]. The values of the sticking coefficients used in this paper are the ones from [15].

4 Numerical difficulties

A lot of time dependent numerical solvers follow the popular Method of Lines (MOL) approach, in which space and time discretizations are considered separately. The popularity of this approach is based on its simple concept, flexibility, the fact that various discretizations can easily be combined and that nowadays many well developed ODE methods exist.

Here, the spatial discretization of the stiff system of species equations (1) is done in a Finite Volume (FV) setting, yielding a semi-discrete system

$$w'(t) = F(t, w(t)), \quad t \geq 0, \tag{15}$$

with $w(0)$ given. According to the MOL approach, fully discrete approximations are obtained by applying a suitable time integration method with time step size τ for the time levels $t_n = n\tau, n = 1, 2, \ldots$

Stiffness in the reaction terms of eq. (1) requires implicit time integration of these terms in order to maintain stability. In this section we summarize the results obtained in [24], in which various ODE methods for reacting flow problems are compared. It should be noted that if the computational costs of one time step is expensive, a time integration method that needs more but computationally cheaper time steps can be more efficient.

On the other hand, a natural property of species mass fractions and/or species mass concentrations is their natural non-negativity. Of course, we want this property to be conserved in both spatial and time discretization, as well as in the (time accurate) solution. It appears that for time integration methods this natural property is hard to be fulfilled for stiff problems. In Section 4.1 the conservation of non-negativity is comprehensively described.

4.1 Positivity conservation of mass fractions

Conserving the natural non-negativity property of species concentrations is not easily maintained when integrating the species equations (1). Generally speaking, positivity should hold for

1. the mathematical model,
2. spatial discretization,
3. time integration, and,
4. iterative solvers.

It can be shown that the mathematical model as presented in Section 2.1 and 2.2 preserves positivity [24]. Subsequently, we demand that neither spatial discretization nor time integration should introduce wiggles or negative components into the solution vector. Spatial discretization is done in a Finite Volume (FV) setting, in which positivity is preserved since we use a hybrid scheme in which the fluxes are approximated by central differences if possible, and by first order upwind if necessary, see [18]. For details see, for instance, [14, 22, 24]. For time integration, on the other hand, preserving positivity is certainly not easily guaranteed. Implicit time integration gives rise to huge nonlinear systems, which are solved by iterative solution methods, such as Newton method. In the next sections we discuss positivity for time integration, and on our strategies to maintain this property on the lower level of iterative solvers.

4.1.1 Positive time integration

Definition 1. An ODE system $w'(t) = F(t, w(t))$, $t \geq 0$, is called positive, or non-negativity preserving, if $w(0) \geq 0$ (component-wise) $\implies w(t) \geq 0$, for all $t > 0$.

The next theorem provides a simple criterion on $F(t, w(t))$ to test whether the system $w'(t) = F(t, w(t))$, $t \geq 0$, is positive. For a proof we refer to [12].

Theorem 1. *Suppose that $F(t, w)$ is continuous and satisfies a Lipschitz condition with respect to w. Then the system $w'(t) = F(t, w(t))$, $t \geq 0$, is positive if and only if for any vector $w \in \mathbb{R}^m$ and all $i = 1, \ldots, m$, and $t \geq 0$ yields*

$$w \geq 0 \quad (componentwise), \quad w_i = 0 \implies F_i(t, w) \geq 0. \tag{16}$$

It is interesting to investigate positivity for semi-discrete systems. Consider, for instance, the one dimensional linear advection-diffusion equation

$$\frac{\partial}{\partial t} u(x,t) + \frac{\partial}{\partial x}(a(x,t)u(x,t)) = \frac{\partial}{\partial x}\left(d(x,t)\frac{\partial}{\partial x}u(x,t)\right), \tag{17}$$

with periodic boundary conditions, and where $a(x,t)$ is the space and time dependent advection coefficient, and $d(x,t) > 0$ the space and time dependent diffusion coefficient. Application of Theorem 1 shows that finite difference discretization by means of central differences gives a positive semi-discretization if and only if the cell Péclet numbers, defined as ah/d, satisfy

$$\max_{x,t} \frac{|a(x,t)|h}{d(x,t)} \leq 2. \tag{18}$$

Discretizing the advection part by means of first order upwind, and second order central differences for the diffusive part, gives an unconditionally positive semi-discretization. The reaction terms (3) can be written in the production-loss form

$$\bar{R}_k^g(t, w) = p(t, w) - L(t, w)w, \tag{19}$$

where $p(t,w) \geq 0$ (componentwise) is a vector and $L(t,w) \geq 0$ (componentwise) a diagonal matrix, whose components $p_i(t,w)$ and $L_i(t,w)$ are of polynomial type with non-negative coefficients and can easily be found. Addition of reaction terms according to eq. (3), which can be written in the production-loss form (19), to the advection-diffusion eqn. (17) and applying Theorem 1 gives a positive semi-discretization for the one dimensional advection-diffusion-reaction equation if and only if $p(t,w) \geq 0$, see also [12, Section I.7].

The one-dimensional results above are easily generalized to higher dimensions and to FV schemes. Therefore, discretizing the species equations in space by means of a hybrid FV scheme as introduced in [14, 22, 24], which uses the central difference scheme if possible and the first order upwind scheme if necessary, maintains positivity. We remark that for higher order upwinding, such as, for example, third order upwinding, positivity is not ensured for all step-sizes, see [12, Section I.7].

Definition 2. A time integration method $w_{n+1} = \varphi(w_n)$ is called positive if for all $n \geq 0$ holds, $w_n \geq 0 \Longrightarrow w_{n+1} \geq 0$.

Positivity restricts the use of time integration methods. In this section we will present results for non-linear systems $w'(t) = F(t,w(t))$. First, we start exploring the positivity property for Euler Forward and Backward time integration.

4.1.2 Positivity for Euler Forward (EF) and Euler Backward (EB)

Suppose that the right hand side of the non-linear semi-discretization $w'(t) = F(t,w(t))$ satisfies:

Condition 2 *There is an $\alpha > 0$, depending on $F(t,w)$, such that for a time step τ holds: if $\alpha\tau \leq 1$, then $w + \tau F(t,w) \geq 0$ for all $t \geq 0$ and $w \geq 0$.*

Provided that $w_n \geq 0$, Condition 2 guarantees positivity for w_{n+1} computed via EF. For linear semi-discrete systems $w'(t) = Aw(t)$ with entries $A_{ij} \geq 0$ for $i \neq j$, $A_{ii} \geq -\zeta$ for all i and $\zeta > 0$ fixed, Condition 2 is easily illustrated. Application of Euler Forward to this systems gives a positive solution if $1 + \tau A_{ii} \geq 0$ for all i. This will hold if $\alpha\tau \leq 1$. To write down such an expression for α for eq. (1) is almost undoable, because of the complicated structure of the chemical source terms.

Furthermore, assume that $F(t,w(t))$ also satisfies:

Condition 3 *For any $v \geq 0, t \geq 0$ and $\tau > 0$ the equation $w = v + \tau F(t,w)$, has a unique solution w that depends continuously on τ and v.*

According to the following theorem we have unconditional positivity for EB. The proof is taken from [12].

Theorem 4. *Condition 2 and 3 imply positivity for EB for any step size τ.*

Proof. For given t, v and with a chosen τ, we consider the equation $w = v + \tau F(t,w)$ and we call its solution $w(\tau)$. We have to show that $v \geq 0$ implies $w(\tau) \geq 0$ for all positive τ. By continuity it is sufficient to show that $v > 0$ implies $w(\tau) \geq 0$. This is true because if we assume that $w(\tau) > 0$ for $\tau \leq \tau_0$, except for the ith component $w_i(\tau_0) = 0$, then $0 = w_i = v_i + \tau_0 F_i(t, w(\tau_0))$. According to Condition 2 we have $F_i(t, w(\tau_0)) \geq 0$ and thus $v_i + \tau_0 F_i(t, w(\tau_0)) > 0$, which is a contradiction.

Remark 1. Application of EB to the nonlinear semi-discretization $w'(t) = F(t, w(t))$ needs the solution of the nonlinear vector equation

$$w_{n+1} - \tau F(t_n, w_{n+1}) = w_n. \tag{20}$$

Theorem 4 ensures for every time step size τ positivity of the exact solution of (20). In practice, however, the solution of (20) is approximated iteratively, and therefore, it is not guaranteed to be positive. In Section 5.1 we present a class of projected Newton methods to prevent this undesired behavior.

4.1.3 Positive time integration continued: general remarks

For reacting flow simulations, time integration schemes can only be unconditionally positive if they are implicit, i.e., the severe stiffness in these problems rules out explicit integration of the reaction terms. One might hope to find accurate higher order methods with this unconditional positivity property. However, this hope is dashed by the following result, due to Bolley and Crouzeix [1].

Theorem 5. *Any unconditionally positive time integration method has order $p \leq 1$.*

For a proof we refer to [1]. The consequence is that the only well-known method having unconditional positivity is EB. Finally, we remark that for higher order methods the need to preserve positivity may necessitate the use of impractically small time steps.

To conclude, higher order methods are expected to integrate eq. (1) with small time steps, which is not feasible for simulations over relatively long time frames. The first order Euler Backward scheme is capable, from a theoretical point of view, of taking large time steps, which are only bounded by the required accuracy. With respect to accuracy, the first order consistency is a drawback.

Although higher order schemes suffer from severe positivity restrictions, we compare three classes of higher order ODE methods with EB, in terms of efficiency. The selection of these higher order ODE methods is based on properties like good performance for diffusion-reaction problems, or on efficiency of the solver itself. The results are presented in the next section.

4.2 Comparison of some stiff ODE methods

While on the one hand the intuitive meaning of stiffness is clear to computational scientists, on the other hand a mathematical definition is missing. Following the interpretation given in [12], we say that a PDE is stiff when implicit ODE methods perform considerably better than explicit ODE methods. Certainly, this property also depends on the eigenvalue distribution of the Jacobian matrix of F in eq. (15) and on the smoothness of the solution.

As said before, a lot of research has been done in the field of ODE methods. The result is that a huge amount of literature is available on time integration methods for stiff problems, see for instance [10, 12]. In [24] we compared the unconditionally stable and unconditionally positive Euler Backward method with the following second order methods:

- Rosenbrock methods are linearly implicit Runge-Kutta methods for stiff ODEs, which have proven to be effective for various stiff problems, see [10, 12]. The second order Rosenbrock scheme ROS2, which depends on the choice of the parameters b_2 and γ,

$$\begin{aligned} w_{n+1} &= w_n + b_1 k_1 + b_2 k_2 \\ k_1 &= \tau F(t_n, w_n) + \gamma \tau J_F k_1 \\ k_2 &= \tau F(t_n + \alpha_{21}\tau, w_n + \alpha_{21} k_1) + \gamma_{21} \tau J_F k_1 + \gamma \tau J_F k_2, \end{aligned} \quad (21)$$

with coefficients $b_1 = 1 - b_2$, $\alpha_{21} = \frac{1}{2b_2}$ and $\gamma_{21} = -\frac{\gamma}{b_2}$, has been implemented. In (21) J_F is the Jacobian of $F(t_n, w_n)$. ROS2 is second order consistent for arbitrary γ and $b_2 \neq 0$, A-stable for $\gamma \geq \frac{1}{4}$ and L-stable if $\gamma = 1 \pm \frac{1}{2}\sqrt{2}$. By selecting for γ the larger value $\gamma_+ = 1 + \frac{1}{2}\sqrt{2}$, we have the property that $\mathscr{R}(z) \geq 0$, for $z \in \mathbb{R}^-$, where $\mathscr{R}(z)$ is the stability function of ROS2. For diffusion-reaction problems, which have a Jacobian with negative real eigenvalues, this property ensures a positive solution. Adding advection introduces imaginary parts to the eigenvalues, such that positivity is no longer guaranteed. However, as has been experienced in [26], the ROS2 scheme performs quite well with respect to positivity for advection-diffusion-reaction problems. Although there is no explanation for this unexpected behavior, it is conjectured that the property that $\mathscr{R}(z) \geq 0$ for all $z \in \mathbb{R}^-$ plays a role [26].

- Backward Differentiation Formulas (BDF) belong to the most widely used methods to solve stiff chemical reaction equations, due to their favorable stability properties. The k-step BDF methods are implicit, of order k and defined as

$$\sum_{j=0}^{k} \alpha_j w_{n+j} = \tau F(t_{n+k}, w_{n+k}), \quad n = 0, 1, \ldots, \quad (22)$$

which uses the k past values $w_n, \ldots, w_{n+k-1}$ to compute w_{n+k}. Note that the most advanced level is t_{n+k} instead of t_{n+1}. The 1-step BDF method is EB, whereas the 2-step method is

$$\frac{3}{2} w_{n+2} - 2 w_{n+1} + \frac{1}{2} w_n = \tau F(t_{n+2}, w_{n+2}). \quad (23)$$

The BDF-1 and BDF-2 methods are A-stable, but for $k > 2$ they are $A(\alpha)$-stable and for $k > 6$ even unstable [9]. Remark that the first $(k-1)$ approximations cannot be computed with the k-step BDF scheme, and should thus be obtained by another scheme. Under Condition 2 and 3 we obtain positivity for BDF2 whenever $\alpha\tau \leq \frac{1}{2}$, provided that w_1 is computed positively from w_0. Remark that this positivity condition is a factor 2 tighter than the condition for EF.

- IMEX Runge-Kutta Chebyshev methods (IRKC) are the IMEX extension of the class of Runge-Kutta Chebyshev (RKC) methods, developed by Verwer et. al. [25]. The class of explicit RKC schemes was designed to integrate moderately stiff problems, for instance discretized diffusion, as computationally cheap as possible. In [25] this class of methods was extended such that much stiffer problems can also be integrated efficiently. The exact description of the scheme is rather spacious, and is therefore omitted here. In [24, 25] the scheme is comprehensively described, but it is worthwhile to remark that advection and diffusion are integrated explicitly and the reaction terms implicitly. Two versions of IRKC have been implemented; the IRKC(full) uses the CFL-condition to integrate advection and diffusion stably, see also [25]. The 'on the fly' version of IRKC, called IRKC(fly), uses only stability conditions for diffusion. To conclude, for this scheme no conditions are known to guarantee positivity, although we know that the scheme is not unconditionally positive.

Details on the exact nonlinear solvers used for the simulations can be found in [24]. Basically, a Newton method extended with an Armijo rule to enforce global convergence has been implemented. Further, there is also the option to update the Jacobian occasionally. In that case the Jacobian is only computed when it is necessary to maintain a sufficient decrease in the Newton residual. For further details we refer to [24]. The linear systems are solved by means of an LU factorization. It has to be remarked that direct linear solvers are only feasible when a suitable ordering of unknowns is used, see [24].

4.2.1 Numerical results

The ODE methods presented in the above section have been tested on the benchmark problem of Kleijn [15]. Recall that the chemistry model of this CVD process consists of 17 different species and 26 reactions. Furthermore, surface chemistry as described in Section 3.2 is included. The reactor configuration for all simulations has been illustrated in Figure 3.

The simulation runs from the the instant that the reactor is completely filled with helium carrier gas and a mixture of helium and silane starts to enter the reactor, until steady state. The spatial computational grid consists of 35 equidistant grid points in radial direction, and 32 non-equidistant grid points in axial direction. The grid spacing in axial direction gradually decreases towards the wafer surface. In our experiments steady state is assumed to be obtained when for a certain time step t_n the inequality

$$\frac{\|w_{n+1} - w_n\|_2}{\|w_n\|_2} \leq 10^{-6}, \tag{24}$$

holds, where w_n is the numerical solution of the semi-discretization $w'(t) = F(t, w)$ on time $t = t_n$. The validation and interpretation of the results is done in Section 6. Right now, we are only interested in the performance of the various ODE integrators.

In Table 3, 4 and 5 numerical results for the various time integration methods, with either the full or modified Newton iteration to solve the nonlinear systems, and the relative errors with respect to a time accurate ODE solution, on some fixed times in the L_2 norm, are given. We used relative errors, because the solution contains relatively small components. The user-specified quantity TOL to monitor the local truncation error is taken equal to 10^{-3}. For the time accurate ODE solution this value was set to 10^{-6}. We observe that for the global errors as presented in Tables 3, 4 and 5, the behavior is as expected.

For the unconditional positive EB time integration scheme it can be remarked that modified Newton (see above) influences the positivity of the solution, i.e., the number of rejected time steps due to negative species increases (compare Tables 3 and 4), and is in this case equal to 31. Rejected time steps due to negative entries in the solution vector should be redone with smaller time steps, resulting in a larger number of F evaluations (the number of Jacobian evaluations is approximately equal). Thus, as a result of an increasing number of Newton iterations, the total computational costs increase.

For the BDF2 scheme (compare Tables 3 and 4), application of modified Newton strategy, as explained above, gives more satisfying results. From Table 4 it can be concluded that for BDF2 an increasing number of cheaper Newton iterations is computationally cheaper than factorizing the Jacobian in every Newton iteration.

With respect to the other higher order time integration schemes (see Table 5), we note the following. ROS2 is the cheapest higher order time integrator for this CVD process. For the IRKC scheme we see that both versions perform equally well. Since there is no gain in efficiency by using 'on the fly' stability conditions for the explicit part, the more robust fully CFL-protected IRKC(full) is preferred.

Number of	EB	BDF-2
F	190	757
F'	94	417
Linesearch	11	0
Newton iters	94	417
Rej. time steps	1	10
Acc. time steps	38	138
CPU Time	6500	30500
Relative error ($t = 1.6$ s /$t = 3.2$ s)	$6.8 \cdot 10^{-3}$/$7.9 \cdot 10^{-4}$	$2.2 \cdot 10^{-3}$ / $1.4 \cdot 10^{-4}$

Table 3 Integration statistics for EB and BDF-2, with full Newton solver

With respect to positivity of the solution during transient simulations we note the following. Omission of the reacting surface and thermal diffusion in the reaction Jacobian gives very poor Newton convergence. We also observed that in this case the solution conserves positivity for very small time steps only, even for EB. We

Number of	EB	BDF-2
F	720	1786
F'	84	163
Linesearch	39	33
Newton iters	463	1441
Rej. time steps	31	33
Acc. time steps	88	121
CPU Time	10800	17000
Relative error ($t = 1.6$ s/$t = 3.2$ s)	$6.8 \cdot 10^{-3}/7.9 \cdot 10^{-4}$	$2.2 \cdot 10^{-3} / 1.4 \cdot 10^{-4}$

Table 4 Integration statistics for EB and BDF2, with modified Newton, as explained in Section 4.2

Number of	ROS2	IRKC(fly)	IRKC(full)
F	424	429662	427911
F'	142	2005	2008
Linesearch	0	50	30
Newton iters	0	17425	17331
Rej. time steps	2	729	728
Acc. time steps	140	1276	1284
CPU Time	8000	20000	19500
Relative error($t = 1.6$ s/$t = 3.2$ s)	$1.1 \cdot 10^{-3}/2.5 \cdot 10^{-4}$	$1.8 \cdot 10^{-3}/8.3 \cdot 10^{-5}$	

Table 5 Integration statistics for ROS2, IRKC(fly), where stability for the explicitly integrated part is tested for diffusion only, and IRKC(full), where stability conditions are forced for both advection and diffusion, schemes.

conclude that for this CVD problem it is required to use the exact Jacobian, in which also the derivatives of the reacting surface and thermal diffusion are included.

From the integration statistics presented in Tables 3-5 it is concluded that for long time steady state simulations Euler Backward is, in spite of its first order accuracy, the most efficient time integrator. In [24] we concluded that the unconditional positivity of Euler Backward is preferred over the conditional higher order methods present in this section.

Off the shelf stiff ODE solvers like ODEPACK and DASSL do not test for positivity. The ODE suite ODEPACK is a collection of Fortran solvers for ODE systems. This code and related codes, like LSODE and VODE, were the result of a long development which started with Gear, see [7, 3]. The DASSL code, which is also suitable for differential algebraic equations, employs the BDF formulas and is described in detail in [2]. The purpose of their design is to integrate (stiff) ODEs by means of BDF methods of the highest order possible in order to reduce computational costs. These codes automatically select their order of accuracy based upon the time step size and a user-defined tolerance of required accuracy.

If these codes would test for positivity, then they should all switch back to first order accuracy to maintain the positive solutions. For this reason, we designed a computationally efficient Euler Backward solver. The remaining sections in this chapter are devoted on the design of this solver. In particular we pay attention to the robustness, and thus positivity, of the solver, and, of course, the reduction of computational costs.

5 Design of the Euler Backward solver

The combination of the severe stiffness and the required positivity basically reduces the time integration method to Euler Backward, as we have discussed in the previous section. In practice, however, the use of Newton methods to iteratively solve the nonlinear systems, does not guarantee positivity of the solution. In Section 5.1 we propose a projected version of Newton's method to overcome this problem, and consequently, make the Euler Backward solver more robust.

Generally speaking, the large number of species results in huge nonlinear systems. The linear systems in Newton's method are, consequently, also large. Second, they are also sparse, which makes Krylov solvers suitable candidates to solve them. Newton's method combined with Krylov solvers are called Inexact Newton methods.

5.1 Globalized Inexact Projected Newton methods

In Inexact Newton solvers the Newton step, i.e. $s_k = -[F(x_k)]^{-1}F(x_k)$, is approximated by an iterative linear solver, which is in our case a preconditioned Krylov method. The approximated Newton step s_k has to satisfy the so-called Inexact Newton condition

$$\|F(x_k) + F'(x_k)s_k\| \leq \eta_k \|F(x_k)\|, \quad (25)$$

for a certain 'forcing term' $\eta_k \in [0,1)$. Note that the forcing term expresses a relative stopping criterion for the Krylov method.

In [23] it has been concluded that for laminar reacting gas flow simulations the best choice for the forcing term is to base it on residual norms as

$$\eta_k = \gamma \frac{\|F(x_k)\|^2}{\|F(x_{k-1})\|^2}, \quad (26)$$

with $\gamma \in [0,1)$ a parameter. The order of convergence of Inexact Newton with this forcing term is two. In our experiments we put $\gamma = 0.5$, which worked fine.

It is generally known that Newton's method converges towards a solution if and only if the initial guess is in a neighborhood of this solution, see for instance [17]. Global convergence is enforced by implementing a line-search algorithm in the Inexact Newton algorithm. This has been done as was proposed in [6].

In [23] we proposed an extension of the Globalized Inexact Newton method such that it preserves positivity of species concentrations. We call this Globalized Inexact Projected Newton (GIPN). As has been remarked in [23], repetitions of negative species concentrations can be observed in practice. The i-th entry of the projection $\mathscr{P}$ is defined as:

$$\mathscr{P}_i(x) = \begin{cases} x_i & \text{if } x_i \geq 0. \\ 0 & \text{if } x_i < 0. \end{cases} \tag{27}$$

In every Globalized Inexact Projected Newton step we test whether the projected solution satisfies the sufficient decrease condition, i.e.,

$$\|F(\mathscr{P}(x_k + s_k))\| > (1 - \alpha(1 - \eta_k))\|F(x_k)\|. \tag{28}$$

For a comprehensive description and background of the GIPN methods we refer to [23]. The present GIPN method, which is implemented in our code, is summarized in Algorithm 1.

Algorithm 1 Globalized Inexact Projected Newton

1: Let x_0, $\eta_{\max} \in [0,1)$, $\alpha \in (0,1)$ and $0 < \lambda_{\min} < \lambda_{\max} < 1$ be given.
2: **for** $k = 1, 2, \ldots$ until 'convergence' **do**
3:     Find some $\eta_k \in [0, \eta_{\max}]$ and s_k that satisfy
4:     $\|F(x_k) + F'(x_k)s_k\| \leq \eta_k \|F(x_k)\|$.
5:     **while** $\|F(\mathscr{P}(x_k + s_k))\| > (1 - \alpha(1 - \eta_k))\|F(x_k)\|$ **do**
6:         Choose $\lambda \in [\lambda_{\min}, \lambda_{\max}]$
7:         Set $s_k \leftarrow \lambda s_k$ and $\eta_k \leftarrow 1 - \lambda(1 - \eta_k)$
8:         If such λ cannot be found, terminate with failure.
9:     **end while**
10:    Set $x_{k+1} = \mathscr{P}(x_k + s_k)$.
11: **end for**

5.2 Preconditioned Krylov solver

Roughly speaking, there are two classes of Krylov solvers to solve large, sparse linear systems. On the one hand we have the class of GMRES solvers, and on the other hand we have the Bi-CGSTAB solvers. In the numerical linear algebra community it is well known that both methods have advantages and disadvantages. The linear systems in our code result from spatial discretization on a structured grid. Consequently, the matrix-vector multiplication is cheap to implement. Bi-CGSTAB needs two matrix-vector multiplications per linear iteration and needs seven vectors in memory. GMRES, however, needs only one matrix-vector multiplication, but needs more vectors in memory. Because the matrix-vector multiplication is cheap, we use from a memory usage point of view, the Bi-CGSTAB method.

The partial derivatives of the stiff and highly nonlinear reaction terms in eq. (1) cause ill-conditioned Jacobian matrices. In Figure 4 the typical order of magnitude of the condition-number of the Jacobian is shown as a function of (real) time (in seconds).

If the eigenvalues of a matrix A are not clustered, it probably has a large condition number. It is well known that solving ill-conditioned linear systems with Krylov solvers, leads to bad convergence or divergence of the Krylov method. Fast convergence can be achieved by multiplying the A with a preconditioner M^{-1}, such that the eigenvalues of AM^{-1} are efficiently clustered in comparison with those of A. For the simulations in the present paper, Krylov solvers without preconditioning are ruled out, because the condition numbers of the Jacobian matrices are far too large to expect Bi-CGSTAB convergence at all.

As a final remark before we discuss a set of suitable preconditioners, we want to emphasize on the following rule of thumb. The partial derivatives of the reaction terms, which contribute considerably to the large condition numbers, should be inverted in order to reduce the condition number of the preconditioned systems. Further, we mention that the ordering of unknowns and equations influences both the performance and construction of a preconditioner. In the present paper, we consider only the alternate blocking per grid point ordering, in which the unknown species concentrations are ordered per grid point. The corresponding non-zero pattern of the Jacobian is shown in Figure 5.

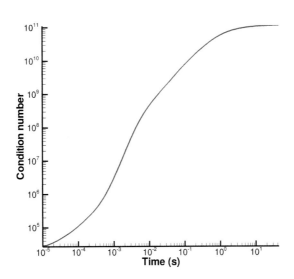

Fig. 4 Condition-number of the Jacobian as function of time (in seconds)

For the acceleration of the internal linear algebra problem in Algorithm 1, the following preconditioners are compared:

- Incomplete factorization without fill-in, denoted as ILU(0),

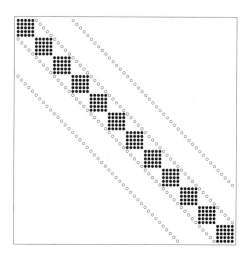

Fig. 5 Non-zero structure of the Jacobian for the alternate blocking per grid point ordering. The partial derivatives of the chemistry terms in Eq. (1) are represented by dots, and the super- and subdiagonals marked by circles are the off-diagonal discretized advection-diffusion terms in Eq.(1).

- Lumped Jacobian: a relatively good approximation of the Jacobian, which also is easy to invert, can be obtained by adding all off-diagonal elements related to the same species of a row to the main diagonal element, see [23],
- Block D-ILU: a block version of the D-ILU preconditioner, see for instance, [19],
- Block diagonal: omitting off-diagonal blocks gives an easy invertible block-diagonal structure that resembles the Jacobian quite well.

A comprehensive discussion on these preconditioners combined with the GIPN method can be found in [23]. In this paper we restrict ourselves to the conclusions drawn in [23]. In terms of efficiency the incomplete factorization based preconditioners perform excellent. Taking into account other arguments like the number of Jacobian evaluations, the total number of linear iterations, positivity and robustness, makes the block D-ILU preconditioner, combined with GIPN, the best preconditioner for this class of problems.

6 Numerical results

Numerical experiments are done with the Euler Backward solver as presented in the previous section. In all experiments we used block D-ILU as preconditioner. The numerical tests are done for the two models of the CVD process of silicon from silane, as presented in Section 3, on three different grid sizes. The simulations are done on a grid with $n_r = 35$ grid points in radial direction and $n_z = 32$ grid points in axial direction, an $n_r = 35$ by $n_z = 47$ grid and an $n_r = 70$ by $n_z = 82$ grid. For all grid holds that the grid points in radial direction are equidistant. The grid points in axial direction have a gradually decreasing grid spacing towards the wafer surface. For the finest grid, i.e., the 70×82 grid, the axial distance from the wafer to the first grid point equals $1 \cdot 10^{-6}$ m, and thereafter the grid spacing will gradually increase to $\Delta z = 5 \cdot 10^{-3}$ m for $z \geq 0.04$ m. For the coarsest grid the distance between the wafer and the first grid point is $2.5 \cdot 10^{-4}$ m. The 35×32 grid is illustrated in Figure 6.

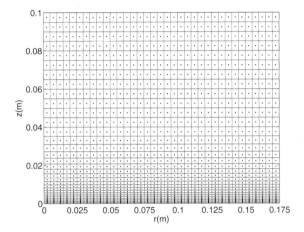

Fig. 6 Computational 35×32 grid. Species mass fractions are computed in the cell centers (represented as dots).

First, we present and discuss the integration statistics for the time accurate transient simulations of both chemistry models introduced in Section 3. Thereafter, some time accurate results and their validation are presented.

6.1 Discussion on the integration statistics

For all simulations in this section, of which we present integration statistics, holds that the simulations run from inflow conditions until the steady state solution is reached. In all these simulations we allow the maximum number of time steps to be 1000. With respect to the number of allowed Newton iterations per time step we remark the following. The strongly nonlinear reaction terms sometimes cause difficulties in finding the correct search direction. More specific, in the time frame right before steady state is reached, we experienced that to find the correct search direction might take a few extra Newton iterations. Therefore, the maximum number of Newton iterations is set to 80.

Further, it has to be mentioned that in this paper time accurate transient results are shown for different wafer temperatures varying from 900 up to 1100 K. Because of the large activation energies of some of the reactants (see Table 1 and 2), such temperature differences lead to large qualitative and quantitative differences in the solutions. The behavior and the integration statistics of the computational method is, however, not influenced by the wafer temperature. Therefore, we will restrict ourselves to present the integration statistics for one wafer temperature per computational grid.

In Table 6 relevant integration statistics are listed for the 7 species and 5 reactions model of Section 3.1. For this problem the grid size has no large influence on the total number of Newton iterations. The effect of different grid sizes is reflected in the number of linear iterations and CPU time. Due to the quality of the block D-ILU preconditioner, no rejected time steps are observed in these simulations. Weaker preconditioners can result in rejected time steps due to Newton divergence and/or negative species concentrations, see [23].

Due to its stronger nonlinearity, the number of Newton iterations increases for the 17 species and 26 gas phase reactions CVD model, see Table 7. Again, the block D-ILU preconditioner shows excellent performance with respect to positivity and fast Bi-CGSTAB convergence. However, for the finest grid, with more grid cells in the reaction zone, the semi-discrete problem is considerably stiffer than for the other two grids. This greater stiffness is especially reflected in the increasing condition numbers of the Jacobian, which are no longer easily cancelled by the preconditioner. This explains the relatively large number of linear iterations on the finest grid, see Table 7. When a 'weaker' preconditioner is used in this case, it is not possible to do a complete simulation from inflow conditions until steady state, due to many time step rejections caused by Newton divergence [23].

With respect to the total computational costs of these simulations the following has to be remarked. Since in almost any case the required accuracy of the approximated linear solutions is low, the Bi-CGSTAB algorithm converges in a small amount of iterations. However, when the stiff chemistry comes into play and an accurately approximated linear solution is needed, the Bi-CGSTAB algorithm needs to overcome a stagnation phase in order to obtain superlinear convergence. In the stagnation phase the Ritz values belonging to the 'bad' eigenvalues have not fully converged to the eigenvalues of the preconditioned system. If they have, then these 'bad' eigenvalues do not contribute to the effective condition number, and, hence, Bi-CGSTAB converges superlinearly to the solution.

Note that accurate linear approximations are only needed when the steepest descent direction is found in Newton's method. In the first few Newton iterations the required accuracy of the linear solutions is relatively low [23] and thus are the costs of constructing the Jacobian more important than the costs of the preconditioned Bi-CGSTAB solver to find an approximation of the linear solution.

Table 6 Number of operations for the 7 species and 5 reactions problem on three computational grids. The wafer temperature is different for each computational grid.

Grid size	35×32	35×47	70×82
Wafer temperature	1000 K	950 K	900 K
Newton iters	80	89	91
Line search	9	6	7
Total linear iters	430	1,137	1,003
CPU time (sec)	140	230	690

Table 7 Number of operations for the 17 species and 16 reactions problem on three computational grids. The wafer temperature is different for each computational grid.

Grid size	35×32	35×47	70×82
Wafer temperature	1000 K	950 K	900 K
Newton iters	93	148	306
Line search	4	28	96
Total linear iters	718	859	2,144
CPU time (sec)	320	470	4,175

6.2 Model validation and time accurate solutions

Correctness of our steady state solution, obtained after long time integration, is validated against the steady state solution obtained with the software of Kleijn [15]. The solver of Kleijn [15] computes the steady state solution of the system of species equations (1). The approach followed in [15] is to march through the computational domain from grid point to grid point and solving (1) by a time-relaxation algorithm. Kleijn's approach is robust, but computationally inefficient. All simulations presented in this paper are test cases where the wafer is *not* rotating.

In Figure 7 steady state mass fraction profiles are presented for some selected species, as well as the ones obtained by Kleijn [15], for a wafer temperature equal to 1000 K. In this case, the total steady state deposition rate of silicon at the symmetry axis as found by Kleijn [15] is 1.92 $\frac{nm}{s}$, whereas we found a deposition rate of 1.93 $\frac{nm}{s}$. Both values compare excellently to those obtained with the well-known 1-dimensional CVD simulation code SPIN within the Chemkin family [5]. In Figure 8 transient deposition rates are presented for some selected species, as well as the transient total deposition rate. It can be seen that the time dependent behavior of these deposition rates is monotonically increasing and stabilizes when the solution is in steady state. Also shown are the steady state deposition rates obtained with the software of Kleijn [15], which are in very good agreement with our current results.

In Figure 9 we present transient total deposition rates for simulations with wafer temperatures varying from 900 K up to 1100 K. The time dependent behavior of all deposition rates is monotonically increasing until the species concentrations are in steady state. Note that the relative contributions of the various silicon containing species to the total deposition rate is a function of the wafer temperatures, with the relative contribution of Si_2H_2 increasing with increasing temperature, and the relative contribution of H_2SiSiH_2 decreasing with increasing temperature. In Figure 10 the species concentrations of SiH_2 for wafer temperatures $T_s = 900$ K and $T_s = 1100$ K are shown. We see that for $T_s = 1100$ K the concentrations of SiH_2 are much higher along the reacting surface than for $T_s = 900$ K. Note that in Figure 10 the legends of both concentration fields differ two orders of magnitude. For the species concentrations of H_2SiSiH_2, which are shown in Figure 11, we see that the concentration of H_2SiSiH_2 for $T_s = 1100$ K is nearly zero along the wafer. This results in a relatively small contribution of H_2SiSiH_2 to the deposition rate.

In Figure 12 the transient behavior of the gas phase chemistry can be seen quite clearly. At time $t = 0.5$ s we see that reactive silane is entering the reactor from the top, but has not reached the reactive susceptor surface. At an inlet velocity of 0.1 m/s and a distance between the inlet and the susceptor of 0.1 m this actually takes 1 s. This is confirmed by Figure 8 and 9, in which it can be seen that deposition does not start until $t \sim 1$ s. A couple of seconds later at time $t = 5$ s, when the CVD process is almost in steady state, we see that along the reacting surface almost all silane molecules either have been decomposed into volatile reaction products, or have been adsorbed to the susceptor surface to form a solid silicon film (see Figure 12).

Figure 13 shows radial profiles of the total steady state deposition rates for both of Kleijn's steady state computations [15], and our steady state results obtained with the Euler Backward solver as discussed in Section 5, for wafer temperatures varied from 900 K up to 1100 K. Again, the agreement is excellent for all wafer temperatures. For all studied temperatures, the steady state growth rates obtained with the present transient solution method were found to differ less than 5% from those obtained with Kleijn's steady state code.

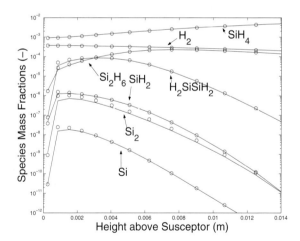

Fig. 7 Axial steady state concentration profiles along the symmetry axis for some selected species. Solid lines are Kleijn's solutions [15], circles are long time steady state results obtained with the present transient time integration methods.

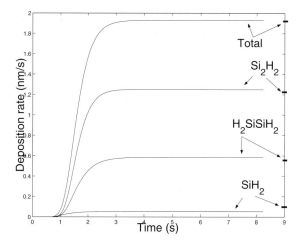

Fig. 8 Transient deposition rates due to some selected species on the symmetry axis for simulations with a non-rotating wafer at 1000 K. On the right vertical axis: steady state deposition rates obtained with Kleijn's steady state code [15].

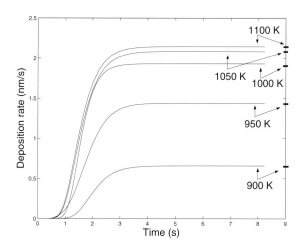

Fig. 9 Transient total deposition rates on the symmetry axis for wafer temperatures varying from 900 K up to 1100 K. On the right vertical axis: steady state total deposition rates obtained with Kleijn's steady state code [15].

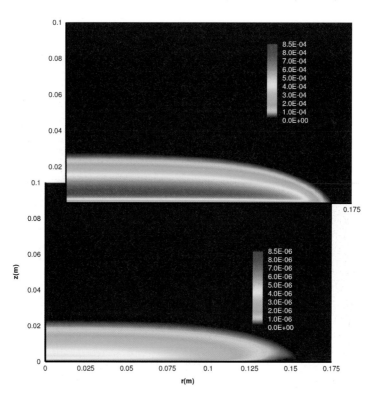

Fig. 10 Concentration profiles of Si_2H_2 for wafer temperature $T_s = 900$ K (bottom) and $T_s = 1100$ K (top). Note that the legends differ two orders of magnitude.

7 Conclusions

The computational methods proposed in this paper are suitable for laminar reacting gas flow simulations, which are typically found in applications such as Chemical Vapor Deposition (CVD), Solid Oxide Fuel Cells (SOFC), and laminar combustion. In the present paper we restrict ourselves to time accurate transient CVD simulations. In order to design time accurate simulation software for these kinds of problems, it is important to understand the numerical difficulties one might encounter. The two major difficulties, in our opinion, are the stiffness of the system of PDEs describing the species transport, and maintaining the solutions positive.

A lot of research has been done by the ODE community, which resulted in a huge amount of literature of the stable integration of stiff ODEs. However, the search for a stiffly stable time integration method that also preserves positivity of the solution can be restricted to first order accurate time integration methods, see [12, 24].

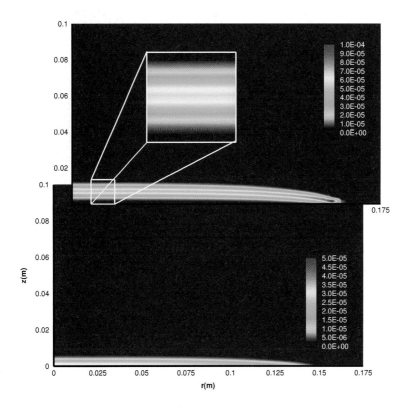

Fig. 11 Concentration profiles of H_2SiSiH_2 for wafer temperature $T_s = 900$ K (bottom) and $T_s = 1100$ K (top). Note that the legends are not identical.

Therefore, we concluded that time integration should be done by the first order unconditionally stable and unconditionally positive Euler Backward method.

In order to maintain the unconditional positivity of Euler Backward, we developed a Globalized Inexact Projected Newton solver, which returns, if it converges towards a solution, a positive solution of the nonlinear problem within Euler Backward. Further, the interior linear algebra problem in the Globalized Inexact Projected Newton method, is taken care of by means of a preconditioned Krylov solver.

The total amount of CPU time needed to compute a time accurate solution, from inflow conditions until steady state, on the 35×32 computational grid has been evolved from approximately 20,000 CPU seconds with a direct linear solver, to about 6500 CPU seconds with direct linear solver and an efficient reordering of unknowns. The introduction of preconditioned Krylov solvers within Newton's method roughly halved the total computational costs to 3000 CPU seconds. Chosing the best preconditioners combined with Globalized Inexact Projected Newton

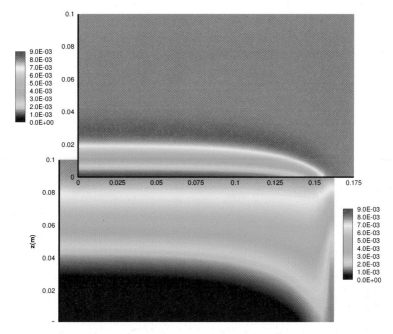

Fig. 12 Concentration profiles of silane on time $t = 0.5$ s (bottom) and $t = 5$ s (top).

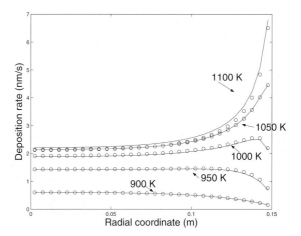

Fig. 13 Radial profiles of the total steady state deposition rate for wafer temperatures varied from 900 K up to 1100 K. Solid lines are Kleijn's steady state results, circles are long time steady state results obtained with the present transient time integration method.

methods enables us to reduce the required CPU time by an order of magnitude to 300 CPU seconds.

Acknowledgements The work of S. van Veldhuizen was financially supported by the Delft Center for Computational Science and Engineering.

References

1. C. Bolley and M. Crouzeix. Conservation de la positivité lors de la discrétisation des problèmes d'évolution paraboliques. *RAIRO Anal. Numer.*, 12:237–245, 1973.
2. K. Brenan, S. Campbell, and L. R. Petzold. *Numerical solution of initial value problems in differential-algebraic equations*. SIAM, Philadelphia, 1989. Second Edition.
3. P. N. Brown, G. D. Byrne, and A. C. Hindmarsch. VODE, a variable coefficient ODE solver. *SIAM J. Stat. Comput.*, 10:1038–1051, 1989.
4. M. E. Coltrin, R. J. Kee, and G. H. Evans. A mathematical model of the fluid mechanics and gas-phase chemistry in a rotating Chemical Vapor Deposition reactor. *J. Electrochem. Soc*, 136:819–829, 1989.
5. M. E. Coltrin, R. J. Kee, G. H. Evans, E. Meeks, F. M. Rupley, and J. F. Grcar. Spin (version 3.83): A FORTRAN program for modeling one-dimensional rotatingdisk/stagnation-flow Chemical Vapour Deposition reactors. Technical Report SAND91-80, Sandia National Laboratories, Albuquerque, NM/Livermore, CA, USA, 1993.
6. S. C. Eisenstat and H. F. Walker. Choosing the forcing terms in an inexact Newton method. *SIAM J. Sci. Comput.*, 17:16–32, 1996.
7. C. W. Gear. *Numerical Initial value problems in ordinary differential equations*. Prentice Hall, Englewood Cliffs, 1971.
8. M. Graziadei and J. H. M. ten Thije Boonkkamp. Local defect correction for laminar flame simulation. In A. Di Bucchianico, R.M.M. Mattheij, and M.A. Peletier, editors, *Progress in industrial mathematics at ECMI 2004*, volume 8 of *Math. Ind.*, pages 242–246, Berlin, 2006. Springer.
9. E. Hairer, S. P. Nørsett, and G. Wanner. *Solving ordinary differential equations I: nonstiff problems*. Number 8 in Springer Series in Computational Mathematics. Springer, 1987.
10. E. Hairer and G. Wanner. *Solving ordinary differential equations II: stiff and differential-algebraic problems*. Number 14 in Springer Series in Computational Mathematics. Springer, Berlin, 1996.
11. M. L. Hitchman and K. F. Jensen. *Chemical Vapor Deposition- Principles and Applications*. Academic Press, London, 1993.
12. W. Hundsdorfer and J. G. Verwer. *Numerical Solution of Time-Dependent Advection-Diffusion-Reaction Equations*. Number 33 in Springer Series in Computational Mathematics. Springer, Berlin, 2003.
13. K. F. Jensen. Modeling of chemical vapor deposition reactors. In *Modeling of Chemical Vapor Deposition reactors for semiconduction fabrication*. Berkeley, 1988. Course notes.
14. C. R. Kleijn. *Transport phenomena in Chemical Vapor Deposition reactors*. PhD thesis, Delft University of Technology, Delft, 1991.
15. C. R. Kleijn. Computational modeling of transport phenomena and detailed chemistry in Chemical Vapor Deposition- A benchmark solution. *Thin Solid Films*, 365:294–306, 2000.
16. C. R. Kleijn, R. Dorsman, K. J. Kuijlaars, M. Okkerse, and H. van Santen. Multi-scale modeling of chemical vapor deposition processes for thin film technology. *J. Cryst. Growth*, 303:362–380, 2007.
17. J. M. Ortega and W. C. Rheinboldt. *Iterative solution of nonlinear equations in several variables*. Number 30 in Classics in Applied Mathematics. SIAM, Philadelphia, 2000. Reprint of the 1970 original.

18. S. V. Patankar. *Numerical Heat Transfer and Fluid Flow*. Hemisphere Publishing Corp., Washington DC, 1980.
19. C. Pommerell. *Solution of large unsymmetric systems of linear equations*, volume 17 of *Series in Micro-electronics*. Hartung-Gorre Verlag, Konstanz, 1992.
20. S. van Veldhuizen, C. Vuik, and C. R. Kleijn. Numerical methods for reacting gas flow simulations. In V.N. Alexandrov, G.D. van Albada, P.M.A. Sloot, and J. Dongarra, editors, *Computational Science–ICCS 2006: 6th International Conference, Reading, UK, May 28-31, 2006. Proceedings, Part II*, pages 10–17, Berlin, 2006. Springer. Lecture Notes in Computer Science 3992.
21. S. van Veldhuizen, C. Vuik, and C. R. Kleijn. Comparison of numerical methods for transient CVD simulations. *Surf. Coat. Technol.*, 201:8859–8862, 2007.
22. S. van Veldhuizen, C. Vuik, and C. R. Kleijn. Numerical methods for reacting gas flow simulations. *Internat. J. Multiscale Eng.*, 5:1–10, 2007.
23. S. van Veldhuizen, C. Vuik, and C. R. Kleijn. A class of projected Newton methods to solve laminar reacting flow problems. Report 08-03, Delft University of Technology, Delft Institute of Applied Mathematics, Delft, 2008.
24. S. van Veldhuizen, C. Vuik, and C. R. Kleijn. Comparison of ODE methods for laminar reacting gas flow simulations. *Num. Meth. Part. Diff. Eq.*, 24:1037–1054, 2008.
25. J. G. Verwer, B. P. Sommeijer, and W. Hundsdorfer. RKC time-stepping for advection-diffusion-reaction problems. *J. of Comp. Physics*, 201:61–79, 2004.
26. J. G. Verwer, E. J. Spee, J. G. Blom, and W. Hundsdorfer. A second order Rosenbrock method applied to photochemical dispersion problems. *SIAM J. Sci. Comput.*, 20:1456–1480, 1999.
27. H. Zhu, R. J. Kee, V. M. Janardhanan, O. Deutschmann, and D. G. Goodwin. Modeling elementary heterogeneous chemistry and electrochemistry in solid-oxide fuel cells. *J. Electrochem. Soc.*, 152:A2427–A2440, 2005.

Parallel Scientific Computing on Loosely Coupled Networks of Computers

Tijmen P. Collignon and Martin B. van Gijzen

Abstract Efficiently solving large sparse linear systems on loosely coupled networks of computers is a rich and vibrant field of research. The induced heterogeneity and volatile nature of the aggregated computational resources present numerous algorithmic challenges. Designing efficient numerical algorithms for said systems is a complex process that brings together many different scientific disciplines. This book chapter is divided into two distinct parts. The purpose of the first half (Sect. 2–4) is to give a bird's view of the issues pertaining to designing efficient numerical algorithms for *Grid computing*. It kicks off by clearly stating the problem and exposing the various bottlenecks, subsequently followed by the presentation of potential solutions. Thus, the stage is set and Sect. 3 proceeds by detailing *classical* iterative solution methods, along with the concept of *asynchronism*, which is a highly favorable quality in the context of Grid computing. The first half is wrapped up by explaining how asynchronism can be introduced into faster but more complicated *subspace methods*. The general idea is that by using an asynchronous method as a *preconditioner*, the best of both worlds can be combined. The advantages and disadvantages of this approach are discussed in minute detail. The second half (Sect. 5) contains discussions on the various intricacies related to *implementing* the proposed algorithm on Grid computers. Section 6 gives some concluding remarks along with suggestions for further reading.

T. P. Collignon
Delft University of Technology, Delft Institute of Applied Mathematics and J. M. Burgerscentrum, Mekelweg 4, 2628 CD Delft, the Netherlands, e-mail: t.p.collignon@tudelft.nl

M. B. van Gijzen
Delft University of Technology, Delft Institute of Applied Mathematics and J. M. Burgerscentrum, Mekelweg 4, 2628 CD Delft, the Netherlands, e-mail: m.b.vangijzen@tudelft.nl

1 Introduction

Solving extremely large sparse linear systems of equations is the computational bottleneck in a wide range of scientific and engineering applications. Examples include simulation of air flow around wind turbine blades, weather prediction, option pricing, and Internet search engines. Although the computing power of a single processor continues to grow, fundamental physical laws place severe limitations on sequential processing. This fact accompanied by an ever increasing demand for more realistic simulations has intensely stimulated research in the field of *parallel and distributed computing*. By combining the power of multiple processors and sophisticated numerical algorithms, simulations can be performed that perfectly imitate physical reality.

Traditional parallel processing was and is currently performed using sophisticated supercomputers, which typically consist of thousands of identical processors linked by a high–speed network. They are often purpose–built and highly expensive to operate, maintain, and expand.

A poor man's alternative to massive supercomputing is to exploit existing non–dedicated hardware for performing parallel computations. With the use of cost–effective commodity components and freely available software, cheap and powerful parallel computers can be built. The *Beowulf* cluster technology is a good example of this approach [51]. A major advantage of such technology is that resources can easily be replaced and added. However, this introduces the problem of dealing with *heterogeneity*, both in machine architecture and in network capabilities. The problem of efficiently *partitioning* the computational work became an intense topic of research.

The nineties of the previous century ushered in the next stage of parallel computing. With the advent of the Internet, it became viable to connect geographically separate resources — such as individual desktop machines, local clusters, and storage space — to solve very large–scale computational problems. In the mid–1990s the SETI@home project was conceived, which has established itself as the prime example of a so–called *Grid computing* project. It currently combines the computational power of millions of personal computers from around the world to search for extraterrestrial intelligence by analysing massive quantities of radio telescope data [1].

In analogy to the Electric Grid, the driving philosophy behind Grid computing is to allow individual users and large organisations alike to access *casu quo* supply computational resources without effort by plugging into the Computational Grid. Much research has been done in Grid software and Grid hardware technologies, both by the scientific community and industry [31].

The fact that in Grid computing resources are geographically separated implies that *communication* is less efficient compared to dedicated parallel hardware. As a result, it is naturally suited for so–called *embarrassingly parallel* applications where the problem can be broken up easily and tasks require little or no interprocessor communication. An example of such an application is the aforementioned SETI@home project.

For the numerical solution of linear systems of equations, matters are far more complicated. One of the main reasons is that inter–task communication is both unavoidable and abundant. For this application, developing efficient parallel numerical algorithms for dedicated homogeneous systems is a difficult problem, but becomes even more challenging when applied to heterogeneous systems. In particular, the heterogeneity of the computational resources and the variability in network performance present numerous algorithmic challenges. This book chapter highlights the key difficulties in designing such algorithms and strives to present efficient solutions.

One of the latest trends in parallel processing is *Cell* or *GPU computing*. Modern gaming consoles and graphics cards employ dedicated high–performance processors for specialised tasks, such as rendering high–resolution graphics. In combination with their inherent parallel design and cheap manufacturing process, this makes them extremely appropriate for parallel numerical linear algebra [61]. The Folding@Home project is a striking example of an embarrassingly parallel application where the power of many gaming consoles is used to simulate protein folding and other molecular dynamics [30].

Nowadays, multi–core desktop computers with up to four computing cores are becoming increasingly mainstream. Many existing software products such as graphics editors and computer games cannot benefit from these additional resources effectively. Such software often needs to be rewritten from scratch and this has also become an intensive topic of research.

The book chapter is divided into two distinct parts. The purpose of the first half (Sect. 2–4) is to give a bird's view of the issues pertaining to designing efficient numerical algorithms for Grid computing and is aimed at a general audience. The second half (Sect. 5) deals with more advanced topics and contains detailed discussions on the issues related to implementing said algorithm on Grid computers. Section 6 gives some concluding remarks along with suggestions for further reading.

2 The problem

Large systems of linear equations arise in many different scientific disciplines, such as physics, computer science, chemistry, biology, and economics. Their efficient solution is a rich and vibrant field of research with a steady supply of important results. As the demand for more realistic simulations continues to grow, the use of direct methods for the solution of linear systems becomes increasingly infeasible. This leaves iterative methods as the only practical alternative.

The main characteristic of such methods is that at each iteration step, information from one or more previous iteration steps is used to find an increasingly accurate approximation to the solution. Although the design of new iterative algorithms is a very active field of research, physical restrictions such as memory capacity and computational power will always place limits on the type of problem that can be solved on a single processor.

Table 1 Parallel and distributed computing on cluster and Grid hardware.

Cluster computing	Grid computing
local–area–networks	wide–area networks
dedicated	non–dedicated
special–purpose hardware	aggregated resources
fast network	slow connections
synchronous communication	asynchronous communication
fine–grain	coarse–grain
homogeneous	heterogeneous
reliable resources	volatile resources
static environment	dynamic environment

The obvious solution is to combine the power of multiple processors in order to solve larger problems. This automatically implies that memory is also distributed. Combined with the fact that iterations may be based on previous iterations, this suggests that some form of *synchronisation* between the processors has to be performed.

Accumulating resources in a local manner is typically called cluster computing. Neglecting important issues such as heterogeneity, this approach ultimately has the same limitations as with sequential processing: memory capacity and computational power. The next logical step is to combine computational resources that are geographically separated, possible spanning entire continents. This idea gives birth to the concept of Grid computing. Ultimately, the price that needs to be paid is that of synchronisation.

Table 1 lists some of the classifications that may be associated with cluster and Grid computing, respectively. In real life, things are not as clear–cut as the Table might suggest. For example, a cluster of homogeneous and dedicated clusters connected by a network is considered a Grid computer. Vice versa, a local cluster may consist of computers that have varying workloads, making the annotations 'dedicated' and 'static environment' unwarranted.

The high cost of global synchronisation is not the only algorithmic hurdle in designing efficient numerical algorithms for Grid computing. In Tab. 2 the main problems are listed, along with possible solutions. Clearly there are many aspects that need to be addressed, requiring substantial expertise from a broad spectrum of mathematical disciplines.

When designing numerical algorithms for general applications, a proper balance should be struck between *robustness* (predictable performance using few parameters) and *efficiency* (optimal scalability, both algorithmic and parallel). At the risk of trivialising these two highly important issues, the ultimate numerical algorithm wishlist for Grid computing contains the following additional items: coarse–grain, asynchronous communications, minimal number of synchronisation points, resource–aware, dynamic, and fault tolerant. The ultimate challenge is to devise an algorithm that exhibits all of these eight features.

Table 2 Main difficulties and possible solutions associated with designing efficient numerical algorithms in Grid computing.

Difficulties and challenges	Possible solutions
– **Frequent synchronisation.** One of the reasons for synchronisation is global reduction. Compared to the overhead, the data that is being exchanged is relatively small, making this an extremely expensive operation in Grid environments. The most important example is the computation of an inner product.	– **Coarse–grained.** Communication is expensive, so the amount of computation should be large in comparison to the amount of communication. – **Asynchronous communication.** Tasks should not have to wait for specific information from other tasks to become available. That is, the algorithm should be able to incorporate any newly received information immediately. – **Minimising synchronisation points.** Many iterative algorithms can be modified in such a manner that the number of synchronisation points is reduced. These modifications include rearrangement of operations [16], truncation strategies [23], and the type of reorthogonalisation procedure [22].
– **Heterogeneity.** Resources from many different sources may be combined, potentially resulting in a highly heterogeneous environment. This can apply to machine architecture, network capabilities, and memory capacities.	– **Resource–aware.** When dividing the work, the diversity in computational hardware should be reflected in the partitioning process. Techniques from graph theory are extensively used here [53].
– **Volatility.** Large fluctuations can occur in things like processor workload, processor availability, and network bandwidth. A huge challenge is how to deal with failing network connections or computational resources.	– **Dynamic.** Changes in the computational environment should be detected and accounted for, either by repartitioning the work periodically or by using some type of diffusive partitioning algorithm [53]. – **Fault tolerant.** The algorithm should somehow be (partially) resistant to failing resources in the sense that the iteration process may stagnate in the worst case, but not break down.

3 The basics: iterative methods

The goal is to efficiently solve a large algebraic linear system of equations,

$$Ax = b, \qquad (1)$$

on large heterogeneous networks of computers. Here, A denotes the coefficient matrix, b represents the right–hand side vector, and x is the vector of unknowns.

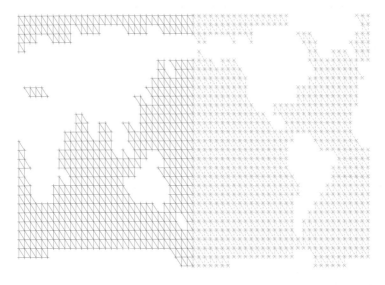

Fig. 1 Depiction of the oceans of the world, divided into two separate computational subdomains.

3.1 Simple iterations

Given an initial solution $x^{(0)}$, the classical iteration for solving the system (1) is

$$x^{(t+1)} = x^{(t)} + M^{-1}(b - Ax^{(t)}), \qquad t = 0, 1, \ldots, \qquad (2)$$

where M^{-1} serves as an approximation for A^{-1}. For practical reasons, inverting the matrix M should be cheap and this is reflected in the different choices for M. The simplest option would be to choose the identity matrix for M, which results in the *Richardson* iteration. Another variant is the *Jacobi* iteration, which is obtained by taking for M the diagonal matrix having entries from the diagonal of A. Choices that in some sense better approximate the matrix A naturally result in methods that converge to the solution in less iterations. However, inverting the matrix M will be more expensive and it is clear that some form of trade–off is necessary.

The iteration (2) can be generalised to a block version, which results in an algorithm closely related to *domain decomposition* techniques [48]. One of the earliest variants of this method was introduced as early as 1870 by the German mathematician Hermann Schwarz. The general idea is as follows. Most problems can be divided quite naturally into several smaller problems. For example, problems with complicated geometry may be divided into subdomains with a geometry that can be handled more easily, such as rectangles or triangles.

Consider the physical domain Ω shown in Fig. 1. The objective is to solve some given equation on this domain. For illustrative purposes, the domain is divided into two subdomains Ω_1 and Ω_2. The matrix, the solution vector, and the right–hand side are partitioned into blocks as follows:

Algorithm 1 Block Jacobi iteration for solving $Ax = b$.

OUTPUT: Approximation of $Ax = b$;
1: Initialize $x^{(0)}$;
2: **for** $t = 0, 1, \ldots$, until convergence **do**
3:     **for** $i = 1, 2, \ldots, p$ **do**
4:         Solve $A_{ii} x_i^{(t+1)} = b_i - \sum_{j=1, j \neq i}^{p} A_{ij} x_j^{(t)}$;
5:     **end for**
6: **end for**

$$A = \begin{bmatrix} A_{11} & A_{12} \\ A_{21} & A_{22} \end{bmatrix}, \quad x = \begin{bmatrix} x_1 \\ x_2 \end{bmatrix}, \quad b = \begin{bmatrix} b_1 \\ b_2 \end{bmatrix}. \quad (3)$$

The two matrices on the main diagonal of A symbolise the equation on the subdomains themselves, while the coupling between the subdomains is contained in the off–diagonal matrices A_{12} and A_{21}.

Block Jacobi generalises standard Jacobi by taking for M the block diagonal elements, giving

$$M = \begin{bmatrix} A_{11} & \varnothing \\ \varnothing & A_{22} \end{bmatrix}. \quad (4)$$

This results in the following two iterations for the first and second domain respectively,

$$\begin{cases} x_1^{(t+1)} = x_1^{(t)} + A_{11}^{-1} \left(b_1 - A_{11} x_1^{(t)} - A_{12} x_2^{(t)} \right); \\ x_2^{(t+1)} = x_2^{(t)} + A_{22}^{-1} \left(b_2 - A_{21} x_1^{(t)} - A_{22} x_2^{(t)} \right), \end{cases} \quad t = 0, 1, \ldots. \quad (5)$$

On a parallel computer, these iterations may be performed independently for each iteration step t. This is followed by a synchronisation point where information is exchanged between the processors. Algorithm 1 shows the general case for p processors and/or subdomains. An extra complication is that the block matrices located on the diagonal need to be inverted. In most cases these matrices have the same structure as the complete matrix. Therefore, systems involving these matrices are usually solved using some other iterative method, possibly block Jacobi. Another important issue is how accurately these systems should be solved.

3.2 Impatient processors: asynchronism

Parallel asynchronous algorithms can be considered as a generalisation of simple iterative methods such as the aforementioned block Jacobi method. Instead of exchanging the most recent information with other processes at each iteration step, an asynchronous algorithm performs their iterations based on information that is

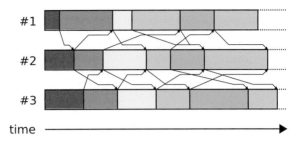

Fig. 2 Time line of a certain type of asynchronous algorithm, showing three (Jacobi) processes. Newly computed information is sent at the end of each iteration step and newly received information is not used until the start of the next iteration step. The graphic scheme is inspired by [3].

available at that particular time. Therefore, the iteration counter t loses its global meaning. The classification *asynchronous* pertains to the type of communication.

In Fig. 2 a schematic representation is given which illustrates some of the important features of a particular type of asynchronous algorithm. Time is progressing from left to right and communication between the three (Jacobi) iteration processes is denoted by arrows. The erratic communication is expressed by the varying length of the arrows. At the end of an iteration step of a particular process, locally updated information is sent to its neighbour(s). Vice versa, new information may be received multiple times during an iteration. However, only the most recent information is included at the start of the next iteration step. Other kinds of asynchronous communication are possible [4, 5, 20, 33, 38]. For example, asynchronous iterative methods exist where newly received information is immediately incorporated by the iteration processes.

Thus, the execution of the processes does not halt while waiting for new information to arrive from other processes. As a result, it may occur that a process does not receive updated information from one of its neighbours. Another possibility is that received information is outdated in some sense. Also, the duration of each iteration step may vary significantly, caused by heterogeneity in computer hardware and network capabilities, and fluctuations in things like processor workload and problem characteristics.

Some of the main advantages of parallel asynchronous algorithms are summarised in the following list.

- *Reduction of the synchronisation penalty.* No global synchronisation is performed, which may be extremely expensive in a heterogeneous environment.
- *Efficient overlap of communication with computation.* Erratic network behaviour may induce complicated communication patterns. Computation is not stalled while waiting for new information to arrive and more Jacobi iterations can be performed.
- *Coarse–graininess.* Techniques from domain decomposition can be used to effectively divide the computational work and the lack of synchronisation results in a highly attractive computation/communication ratio.

In extremely heterogeneous computing environments, these features can potentially result in improved parallel performance. However, no method is without disadvantages and asynchronous algorithms are no exception. The following list gives some idea on the various difficulties and possible bottlenecks.

- *Suboptimal convergence rates.* Block Jacobi–type methods exhibit slow convergence rates. Furthermore, if no synchronisation is performed whatsoever, processes perform their iterations based on potentially outdated information. Consequently, it is conceivable that important characteristics of the solution may propagate rather slowly throughout the domain.
- *Non–trivial convergence detection.* Although there are no synchronisation points, knowing when to stop may require a form of global communication at some point.
- *Partial fault tolerance.* If a particular Jacobi process is terminated, the complete iteration process will effectively break down. On the other hand, a process may become unavailable due to temporary network failure. Although this would delay convergence, the complete convergence process would eventually finish upon reinstatement of said process.
- *Importance of load balancing.* In the context of asynchronism, dividing the computational work efficiently may appear less important. However, significant desynchronisation of the iteration processes may negatively impact convergence rates. Therefore, some form of (resource–aware) load balancing could still be appropriate.

4 Acceleration: subspace methods

The major disadvantage of block Jacobi–type iterations — either synchronous or asynchronous — is that they suffer from slow convergence rates and that they only converge under certain strict conditions. These methods can be improved significantly as follows. Using a starting vector x_0 and the initial residual $r_0 = b - Ax_0$, iteration (2) may be rewritten as

$$Mu_k = r_k, \quad c_k = Au_k, \quad x_{k+1} = x_k + u_k, \quad r_{k+1} = r_k - c_k, \quad k = 0, 1, \ldots. \quad (6)$$

Instead of finding a new approximation x_{k+1} using information solely from the previous iteration, *subspace methods* operate by iteratively constructing some special subspace and extracting an approximate solution from this subspace. The key difference is that information is used from several previous iteration steps, resulting in more efficient methods. This is accomplished by performing (non–standard) projections, which suggests that inner products need to be computed. As mentioned before, in the context of Grid computing this is an expensive operation and should be avoided as much as possible.

Some popular subspace methods are: the Conjugate Gradient method, GCR, GMRES, Bi–CGSTAB, and IDR(s) [29, 35, 42, 49, 55]. Roughly speaking, these methods differ from each other in the way they exploit certain properties of the underlying linear system. Purely for illustrative purposes, the Conjugate Gradient method

Algorithm 2 The preconditioned Conjugate Gradient method.

INPUT: Choose x_0; Compute $r_0 = b - Ax_0$;
OUTPUT: Approximation of $Ax = b$;

1: **for** $k = 1, 2, \ldots$, until convergence **do**
2:     Solve $Mz_{k-1} = r_{k-1}$;
3:     Compute $\rho_{k-1} = (r_{k-1}, z_{k-1})$;
4:     **if** $k = 1$ **then**
5:         Set $p_1 = z_0$;
6:     **else**
7:         Compute $\beta_{k-1} = \rho_{k-1}/\rho_{k-2}$;
8:         Set $p_k = z_{k-1} + \beta_{k-1} p_{k-1}$;
9:     **end if**
10:    Compute $q_k = Ap_k$;
11:    Compute $\alpha_k = \rho_{k-1}/(p_k, q_k)$;
12:    Set $x_k = x_{k-1} + \alpha_k p_k$;
13:    Set $r_k = r_{k-1} - \alpha_k q_k$;
14: **end for**

is listed in Alg. 2, which is designed for symmetric systems. The four main building blocks of a subspace method can be identified as follows.

1. *Vector operations.* These include inner products and vector updates. Note that classical methods lack inner products.
2. *Matrix–vector multiplication.* This is generally speaking the most computationally intensive operation per iteration step. Therefore, the total number of iterations until convergence is a measure for the cost of a particular method.
3. *Preconditioning phase.* The matrix M in the iteration (6) is sometimes viewed as a *preconditioner*. The art of preconditioning is to find the optimal trade–off between the cost of solving systems involving M and the effectiveness of the newly obtained update u_k. That is, an effective but costly preconditioner will reduce the number of (outer) iterations, but the cost of solving said systems may be too large. Vice versa, applying some cheap preconditioner may be fast, but the resulting number of outer iterations may increase rapidly.
4. *Convergence detection.* Choosing an appropriate halting procedure is not entirely trivial. This has two main reasons: (i) the residual r_k that is computed does not need to resemble the actual residual $b - Ax_k$, and (ii) computing the norm of the residual requires an inner product.

For most applications, finding an efficient preconditioner is more important than the choice of subspace method and it may be advantageous to put much effort in the preconditioning step. A popular choice is to use so–called *incomplete factorisations* of the coefficient matrix as preconditioners, e.g., ILU and Incomplete Cholesky. Another well–known strategy is to approximate the solution to $A\varepsilon = r$ by performing one or more iteration steps of some iterative method, such as block Jacobi or IDR(s). Algorithms that use such a strategy are known as *inner–outer* methods.

A direct consequence of the latter approach is that the preconditioning step may be performed *inexactly*. Unfortunately, most subspace methods can potentially break down if a different preconditioning operator is used in each iteration step. An example is the aforementioned preconditioned Conjugate Gradient method. Methods that can handle a varying preconditioner are called *flexible*, e.g., GMRESR [56], FGMRES [41], and flexible Conjugate Gradients [2, 39, 46]. A major disadvantage of some flexible methods is that they can incur additional overhead in the form of inner products.

4.1 Hybrid methods: best of both worlds

The potentially large number of synchronisation points in subspace methods make them less suitable for Grid computing. On the other hand, the improved parallel performance of asynchronous algorithms make them perfect candidates.

To reap the benefits and awards of both techniques, the authors propose in [18] to use an asynchronous iterative method as a preconditioner in a flexible iterative method. By combining a slow but coarse–grain asynchronous preconditioning iteration with a fast but fine–grain outer iteration, it is believed that high convergence rates may be achieved on Grid computers.

For their particular application the flexible method GMRESR is used as the outer iteration and asynchronous block Jacobi as the preconditioning iteration. The proposed combined algorithm exhibits many of the features that are on the algorithmic wishlist given in Sect. 2. These include the following items.

- *Coarse–grained.* The asynchronous preconditioning iteration can be efficiently performed on Grid hardware with the help of domain decomposition techniques.
- *Minimal amount of synchronisation points.* When using this approach, a distinction has to be made between global and local synchronisation points. Global synchronisation occurs when information is exchanged between the preconditioning iteration and the outer iteration, whereas local synchronisation only takes place within the outer iteration process. By investing a large amount of time in the preconditioning iteration, the number of expensive global synchronisations can be reduced to a minimum. Subsequently, the number of outer iterations also diminishes, reducing the number of local synchronisation points.
- *Multiple instances of asynchronous communication.* Within the preconditioning iteration asynchronous communication is used, allowing for efficient overlap of communication with computation. Furthermore, the outer iteration process does not need to halt while waiting for a new update u to arrive. It may continue to iterate until a new complete update can be incorporated.
- *Resource–aware and dynamic.* A simple static partitioning scheme may be used for the preconditioner and repartitioning can be performed each outer iteration step. Any load imbalance that may have occurred during the preconditioning iteration will then automatically be resolved.
- *Increased fault tolerance.* In the preconditioning phase, each server iterates on a unique part of the vector u. In heterogeneous computing environments, servers

may become temporarily unavailable or completely disappear at any time, potentially resulting in loss of computed data. If the asynchronous process is used to solve the main linear system, these events would either severely hamper convergence or destroy convergence completely. Either way, by using the asynchronous iteration as a preconditioner — assuming that the outer iteration is performed on reliable hardware — the whole iteration process may temporarily slow down in the worst case, but is otherwise unaffected.

In addition, the proposed algorithm has several highly favorable properties.

- *No expensive asynchronous convergence detection.* By spending a fixed amount of time on preconditioning in each outer iteration step, there is no need for a — possibly complicated and expensive — convergence detection algorithm in the asynchronous preconditioning iteration.
- *Highly flexible and extendible iteration scheme.* The algorithm allows for many different implementation choices. For example, highly recursive iteration schemes may be used. That is, it could be possible to solve a sub–block from a block Jacobi iteration step in parallel on some distant non–dedicated cluster. Another possibility is that the processors that perform the preconditioning iteration do not need to be equal to the nodes performing the outer iteration.
- *The potential for efficient multi–level preconditioning.* The spectrum of a coefficient matrix is the set of all its eigenvalues. Generally speaking, the speed at which a problem is iteratively solved depends on three key things: the iterative method, the preconditioner, and the spectrum of the coefficient matrix. The second and third component are closely related in the sense that a good preconditioner should transform (or *precondition*) the linear system into a problem that has a more favorable spectrum. Many important large–scale applications involve solving linear systems that have highly unfavorable spectra, which consist of many large and many small eigenvalues. The large eigenvalues can be efficiently handled by the asynchronous iteration. On the other hand, the small and more difficult eigenvalues require advanced preconditioners, which can be neatly incorporated in the outer iteration. In this way, both small and large eigenvalues may be efficiently handled by the combined preconditioner. This is just one example of the possibilities.

Naturally, the algorithm is far from perfect and there are several potential drawbacks. The main bottlenecks are the following.

- *Robustness issues.* There are several parameters which have a significant impact on the performance of the complete iteration process. Determining the optimal parameters for a specific application may be a difficult issue. For example, finding the ideal amount of time to spend on preconditioning is highly problem–dependent. Furthermore, it may be advantageous to vary the amount of preconditioning in each iteration step.
- *Algorithmic and parallel efficiency issues.* The preconditioning operator varies in each outer iteration step. In most cases this implies that a flexible method has to be used, which can introduce additional overhead in the outer iteration. In

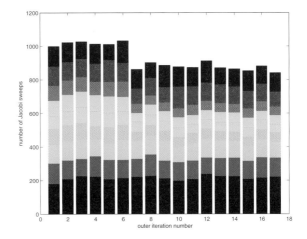

Fig. 3 This experiment is performed using ten servers on a large heterogeneous and non–dedicated local cluster during a typical workday. The figure shows the number of Jacobi iterations — broken down for each server — during each outer iteration step. Here, a fixed amount of time is devoted to each preconditioning step. After the sixth outer iteration several nodes began to experience an increased workload. Its effect on the number of Jacobi iterations is clearly noticeable.

order to avoid potential computational bottlenecks, the outer iteration has to be performed in parallel as well. In addition, it is well–known that block Jacobi–type methods are slowly convergent for a large number of subdomains. In the current context of large–scale scientific computing, this problem needs to be addressed as well.

Despite these crucial issues, the proposed algorithm has the potential to be highly effective in Grid computing environments.

4.2 Some experimental results

In order to give a rough idea on the effect a heterogeneous computing environment may have on the performance of the proposed algorithm, two illustrative experiments will be discussed. Figure 3 shows the effect heterogeneity can have on the number of Jacobi iterations performed by each server. The effect of the variability in computational environment on the amount of work is clearly visible.

The second experiment illustrates the potential gain of desynchronising part of a subspace method, i.e., in this case the preconditioner. In Fig. 4 some problem is solved using both an asynchronous and a synchronous preconditioner. For this particular application, the use of asynchronous preconditioning nearly cuts the total computing time in half.

These experiments conclude the first and general part of the chapter. The second part of the chapter contains more advanced topics and deals with specific implementation issues.

Fig. 4 In this experiment a comparison is made between synchronous and asynchronous preconditioning. The problem to be solved consists of one million equations using four servers within a heterogeneous computing environment. Each point represents a single outer iteration step. By devoting a significant (and fixed) amount of time to asynchronous preconditioning, the number of expensive outer iterations is reduced considerably, resulting in reduced total computing time.

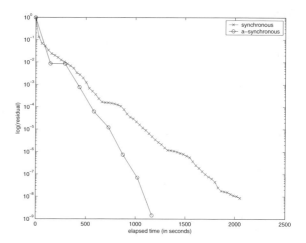

5 Efficient numerical algorithms in Grid computing

The implementation of numerical methods on Grid computers is a complicated process that uniquely combines many concepts from mathematics, computer science, and physics. In the second part of this chapter the various facets of the whole process will be discussed in detail. Most of the concepts given here are taken from [17, 18, 19].

Four key ingredients may be distinguished when implementing numerical algorithms on Grid computers: (i) the numerical algorithm, (ii) the Grid middleware, (iii) the target hardware, and last but not least, (iv) the application. Choosing one particular component can have great consequences on the other components. For example, some middleware may not be suitable for particular type of hardware. Another possibility is that some applications require that specific features are present in the algorithm.

The discussion will take place within the general framework of the aforementioned proposed algorithm, i.e., a flexible method in combination with an asynchronous iterative method as a preconditioner. As previously argued, it possesses many features that make it perfectly suitable for Grid computing. Furthermore, two important classes of Grid middleware will be discussed and correspondingly, two types of target hardware. Although the current approach is applicable to a wide range of scientific applications, the main focus will be on problems originating from large–scale computational fluid dynamics.

The exposition is concluded by briefly mentioning several more advanced techniques.

Table 3 Several characteristics of two types of Grid middleware.

CRAC	GridSolve
dedicated hardware	non–dedicated hardware
direct communication	bridge communication
asynchronous iterative algorithms	general algorithms
miscellaneous applications	embarrassingly parallel problems
data persistence	non–persistent data
no fault tolerance	fault tolerant

5.1 Grid middleware

One of the primary components in Grid computing is the *middleware*. It serves as the key software layer between the user and the computational resources. The middleware is designed to facilitate client access to remote resources and to cope with issues like heterogeneity and volatility. In which manner and to what extent the middleware handles these important issues will be briefly discussed.

Although Grid middleware comes in many different shapes and sizes, the focus will be on two leading examples, i.e., GridSolve [28, 62] and CRAC [21]. Table 3 lists some *prototypical* classifications pertaining to both middleware. Some of these classifications are directly related in the sense that some types of middleware are better suited for particular applications than others. As an example, the bridge communication used in GridSolve would make it more appropriate for embarrassingly parallel problems.

5.1.1 Brief description of GridSolve

GridSolve is a distributed programming system which uses a client–server model for solving complex problems remotely on global networks. It is an instantiation of the GridRPC model, a standard for a Remote Procedure Call (RPC) mechanism on Grid computers [45]. The GridRPC Application Programming Interface (API) is defined within the Global Grid Forum [37]. Other projects that implement the GridRPC API are DIET [15], NetSolve [44], Ninf–G [52], and OmniRPC [43].

Software environments such as GridSolve are often called Network Enabled Servers (NES). These systems typically consist of six components: clients, agents, servers, databases, monitors, and schedulers. In the context of the current version of GS[1] (see Fig. 5) these components will be discussed in detail. The GS servers (component 3) are software components that are started on each computational node which may consist of a single CPU or a cluster. The server monitors the workload of the node and keeps an updated list of the services (or *tasks*) that are installed on the

---

[1] Latest version is v0.17.0 as of May 4th, 2008.

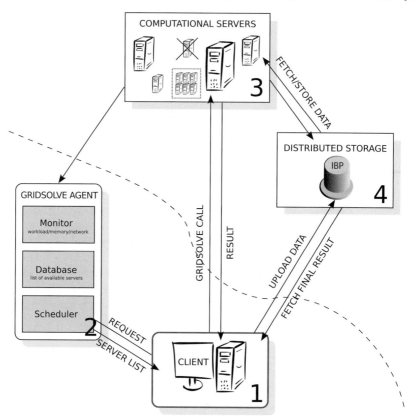

Fig. 5 Schematic overview of GridSolve. The dashed line symbolises (geographical) distance between the client and servers.

server. For example, a task can be a single dgemm or a parallel MPI job. Services can be easily added or modified without restarting the server.

A single GridSolve agent (component 2) actively monitors the server properties such as CPU speed, memory size, computational services, and availability. These properties are stored in a database on the agent node and are periodically updated. When a GridSolve client program (component 1) written in either C, Fortran, or Matlab uses the GridRPC API to initiate a GS call to a remote problem, the GS middleware first contacts the agent. Based on the problem complexity, size of the input parameters, and the available computational resources, the agent then returns a list of servers sorted by minimum completion time. The client resorts the list after performing a quick network performance test. Input parameters are sent to the first server on the list and the task, which can be either blocking or non–blocking, is executed on the server. The result (if any) is then sent back to the client. If a task should fail it is transparently resubmitted to the next server on the list.

The main advantages of GridSolve are that it is easy to use, install, maintain, and that it is a standard for programming on Grid environments. Nevertheless, the current implementation has several limitations. For example, the remote servers cannot communicate directly. In the current GridSolve model, separate tasks communicate data through the client, resulting in bridge communication. As a result, input and output data associated with a task are continuously being sent back and forth between the client and the server using a possibly slow network connection. Also, any data that are read or generated locally during the execution of a task is lost after it is completed. Several strategies such as data persistence and data redistribution have been proposed to tackle these deficiencies for different implementations of the GridRPC API [14, 13, 36, 63, 24]. Furthermore, a proposal for a Data Management API within the GridRPC is currently being developed.

In GridSolve there is a partial solution to the data management problem called the Distributed Storage Infrastructure (DSI). At the Logistical Computing and Internetworking (LoCI) Laboratory of the University of Tennessee the IBP (Internet Backplane Protocol) middleware has been developed based on this approach [6]. To avoid multiple transmissions of the same data between the client and the server, the client can upload data to an IBP data depot which is in close proximity to the computational servers. Subsequently a data handle is sent to the server and the task can fetch and update the data on the IBP depot (see component (4) in Fig. 5). Using the DSI can be considered as programming for a shared memory model.

5.1.2 Brief description of CRAC

The Grid middleware CRAC (*Communication Routines for Asynchronous Computations*) was developed by Stéphane Domas at Laboratoire d'Informatique de Franche–Comté (LIFC) and is specifically designed for efficient implementation of parallel asynchronous iterative algorithms. It allows for direct communication between the processors, both synchronous and asynchronous.

The CRAC library is primarily intended for dedicated parallel systems consisting of geographically separated computational resources. For this reason there are no built–in facilities for detecting properties like varying workload or other types of heterogeneity in computational hardware. However, the object–oriented approach of the software ensures that such functionalities can be easily incorporated.

In the current version of CRAC^2, there are no countermeasures in place for handling resources that have completely failed. It is the responsibility of the algorithm designer to make sure that such an event does not destroy the convergence process. Furthermore, it is not yet possible to add or remove computational resources during an iteration process.

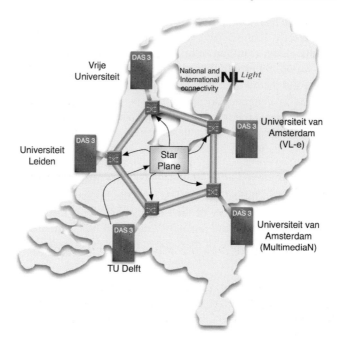

Fig. 6 The DAS–3 supercomputer and StarPlane.

5.2 Target hardware

Numerous computing platforms exist that may be qualified as Grid computing hardware. However, for the purpose of this chapter the focus will be on the following two architectures.

1. *Local networks of non–dedicated computers associated with organisations,* such as universities and companies. These networks typically consist of the computers used daily by employees. Such hardware may considerably differ in speed, memory size, and availability. An example of such a cluster is the network at the Numerical Analysis department at the Delft University of Technology.
2. *Cluster of dedicated clusters linked by a high–speed network.* For example, the Dutch DAS–3 national supercomputer is a cluster of five clusters, located at four academic institutions across the Netherlands, connected by specialised fiber optic technology (i.e., *StarPlane* [50]). It is designed for dedicated parallel computing and although each cluster separately is homogeneous, the system as a whole can be considered heterogeneous. For more specific details on the architecture see Tab. 4 and Fig. 6.

---

[2] Latest version is v1.0 as of May 4th, 2008.

Table 4 DAS–3: five clusters, one system.

Cluster	Nodes	Type	Speed	Memory	Storage	Node HDDs	Network
VU	85	dual–core	2.4 GHz	4 GB	10 TB	85 × 250 GB	Myri-10G and GbE
LU	32	single–core	2.6 GHz	4 GB	10 TB	32 × 400 GB	Myri-10G and GbE
UvA	41	dual–core	2.2 GHz	4 GB	5 TB	41 × 250 GB	Myri-10G and GbE
TUD	68	single–core	2.4 GHz	4 GB	5 TB	68 × 250 GB	GbE (no Myri-10G)
UvA-MN	46	single–core	2.4 GHz	4 GB	3 TB	46 × 1.5 TB	Myri-10G and GbE

Not surprisingly, the most likely candidates for these types of Grid hardware are GridSolve and CRAC, respectively.

5.3 Parallel iterative methods: building blocks revisited

The next vital step in implementing numerical algorithms on Grid computers is to revisit the four building blocks of subspace methods as mentioned in Sect. 4. Where appropriate, each item will be discussed in the context of the aforementioned types of target architectures.

Dividing the work is an essential aspect of parallel iterative methods. Traditional load balancing aims to divide the computational work as evenly as possible under the constraint of minimal communication. In most cases, this is achieved by a form of hypergraph partitioning algorithm, such as *Mondriaan* [58]. In addition, the current methodology dictates that the load balancer incorporates properties related to the heterogeneity of the computational hardware into the partitioning process [53]. Also, the computational effort involved with the partitioning process itself is far from negligible and may be performed in parallel as well [25].

It is not unlikely that the preconditioning iteration is performed on completely different hardware as the outer iteration. Taking the DAS–3 architecture as an example, the outer iteration may be performed on a single cluster, while the preconditioning iteration is performed utilising all five clusters. The point is that the data distribution used in the outer iteration may be different from the data distribution used in the preconditioning iteration.

Depending on the type of Grid middleware, it may be advantageous to perform the preconditioning iteration on the same hardware as the outer iteration in order to preserve data locality.

5.3.1 Matrix–vector multiplication

Partitioning the matrix–vector multiplication may be done in numerous ways. In the current type of application, the number of non–zeros on each row of the coefficient matrix is roughly the same. A simple but effective distribution is the one–

Fig. 7 Heterogeneous one–dimensional block–row partitioning for four servers of a two–dimensional *Poisson* problem. The input (shown at the top) and output (shown left) vectors are partitioned identically.

dimensional block–row partitioning, depicted in Fig. 7. When performing the parallel matrix–vector multiplication only nearest–neighbour communication is required. Nevertheless, nothing prevents the algorithm designer from using more advanced partitioning algorithms in the outer iteration, such as the aforementioned hypergraph partitioner.

The bulk of the computational work in the outer iteration is comprised of the matrix–vector multiplication. Taking into account the fact that the general idea is to minimise the total number of outer iterations, it is unlikely that this operation will be the computational bottleneck of the complete algorithm. As a result, efficient load balancing of the matrix–vector multiplication appears less crucial.

5.3.2 Vector operations

In every subspace method, a newly obtained vector from a preconditioning step is orthogonalised against one or more previous vectors. This is done by an orthogonalisation procedure, such as classical Gram–Schmidt. Although this procedure has good parallel properties, it may suffer from numerical instabilities. This may be remedied by using a selective reorthogonalisation procedure [11, 22].

5.3.3 Preconditioning step

An efficient and robust preconditioner is crucial for rapid convergence of iterative methods. Generally speaking, preconditioners fall into three different classes.

1. *Algebraic techniques.* These methods exploit algebraic properties of the coefficient matrix, such as sparsity patterns and size of matrix elements. For example, incomplete factorisations such as Incomplete Cholesky and block ILU [40].
2. *Domain decomposition techniques.* Many applications in scientific computing involve solving some partial differential equation on a computational domain. Often, the domain can be divided quite naturally into subdomains that may be handled more efficiently. Examples include block Jacobi and alternating Schwarz methods [48].
3. *Multilevel techniques.* Solutions often contain both slowly–varying and fast–varying components. By solving the same problem at different scales in a recursive manner, both components can be efficiently captured. Examples of such methods are multigrid, deflation, and domain decomposition with coarse grid correction [32, 60].

Efficient parallelisation of a preconditioner is a difficult problem, especially in extremely heterogeneous computational environments. A possible solution is to use an asynchronous iterative method as a preconditioner. In addition, by using a flexible method as the outer iteration, the preconditioning operator is allowed to vary in each outer iteration step and the preconditioning iteration may be performed on unreliable computational hardware.

In the context of asynchronism, efficient load balancing of the preconditioning iteration appears less important. Nevertheless, significant desynchronisation of the Jacobi processes may result in suboptimal convergence rates and some form of load balancing may be appropriate.

The bulk of the computational work in the preconditioning iteration consists of solving the block diagonal system in each Jacobi iteration step. As opposed to the amount of work performed by the outer iteration, this amount is difficult to predict. The reason is that the local linear systems are solved iteratively and in most cases inexactly. Furthermore, problem characteristics may cause highly erratic convergence rates. These issues make efficient load balancing problematic.

5.3.4 Convergence detection

The final but essential component of iterative methods is knowing when to stop. In the proposed algorithm a distinction has to be made between convergence detection in the preconditioning iteration and convergence detection in the outer iteration. In most cases, the outer iteration is performed on reliable hardware in a local manner and as a result, convergence detection in the outer iteration is relatively straightforward.

Matters are far more complicated for the preconditioning step. If the preconditioning iteration is performed on unreliable computational hardware as may be the case with GridSolve in combination with a local network of non–dedicated hardware, it is difficult to construct a robust and efficient convergence detection algorithm. In this case, some form of time–dependent stopping criteria may be more appropriate. An obvious disadvantage is that determining the ideal amount of said time may be extremely problem–dependent.

On the other hand, if the preconditioning iteration is performed on dedicated but geographically separated hardware such as the DAS–3 architecture, some sophisticated decentralised convergence detection algorithm may have to be employed [3, 12, 59]. In analogy to the aforementioned case, determining how accurate one should solve the preconditioning iteration is far from trivial.

5.4 Applications

Many important large–scale problems from computational fluid dynamics are solved on highly refined meshes in conjunction with large jumps in the coefficients. The arising linear systems are often severely ill–conditioned and finding efficient (parallel) preconditioners for these systems is vital to fast solution methods. Some examples of said applications are water flow around swimming fish (Fig. 8), air flow around wind turbine blades, and bubbly flow.

The presence of many large and many small eigenvalues severely deteriorates convergence rates, which can only be remedied by using sophisticated multilevel preconditioners. As previously mentioned, such preconditioners can be efficiently incorporated in the proposed algorithm.

For the aforementioned types of flow applications, the so–called *Immersed Boundary Method* (IBM) is particularly appropriate. Although IBMs come in many different flavours, they all share one common characteristic. Instead of adapting the computational mesh to the (possibly complex and moving) boundary, an IBM immerses the boundary on simple Cartesian meshes and modifies the governing equations in the vicinity of the boundary. The use of fixed and structured meshes expedites the implementation of numerical algorithms immensely, particularly in a parallel context. For a more thorough discussion on IBMs the reader is kindly referred to the chapter by Hassen and Koren in this book.

5.5 Advanced techniques

Block Jacobi iterations and domain decomposition techniques are closely related. Combined with the large–scale size of the linear systems involved, some type of *coarse grid correction* within the asynchronous preconditioning iteration may become appropriate. However, the inherently global nature of these techniques may not suit the current context of asynchronism. Nevertheless, this approach warrants further investigation.

There exists a large number of multi–level preconditioning methods, some more robust than others. Finding the most efficient technique for the current application is also a vital research question.

6 Concluding remarks and further reading

In the early days of iterative methods, Jacobi and Gauss–Seidel iterations for solving linear systems were quite popular. However, their slow convergence rates and strict convergence conditions severely limited the applicability of such methods to the constantly increasing pool of computational problems. This was followed by the discovery of subspace methods in conjunction with incomplete factorisations as *preconditioners*, which immensely boosted the popularity of iterative methods for solving large sparse linear systems from a wide variety of applications.

Then came the era of parallel and vector processing, which rekindled the interest in classical methods as highly parallel block preconditioners. The need for increasingly realistic simulations motivated using the aggregated power of computational resources, which introduced the problem of dealing with heterogeneity. The lack of any synchronisation and coarse–graininess in parallel *asynchronous* classical iterations motivated the idea of using these methods for solving linear systems on large heterogeneous networks of computers.

Nowadays, history is repeating itself and said asynchronous iterations are being used — again as parallel preconditioners — in *flexible* subspace methods, where the preconditioner is allowed to change in each iteration step. By combining the best of both worlds, extremely large sparse linear systems may be solved on extremely large heterogeneous networks of computers.

Designing efficient numerical algorithms for Grid computing is a complex process that brings together many different scientific disciplines. By using an asynchronous iterative method as a preconditioner in a flexible iterative method, an algorithm is obtained that has the potential to reap the benefits and awards of both cluster and Grid computing. In this chapter a comprehensive study was made of the various advantages and disadvantages of said approach.

Some of the advantages of desynchronising the preconditioning phase in this manner are that: (i) the preconditioner can be easily and effectively parallelised on Grid computers, (ii) no additional synchronisation points are introduced, and (iii) by devoting the bulk of the computational effort to the preconditioner, the computation to communication ratio can be improved significantly, while reducing the number of expensive (outer) synchronisations considerably. However, potential robustness issues require further investigation.

In addition, the efficient implementation of these algorithms on Grid computers depends on many aspects related to the type of target hardware, Grid middleware, and the application. Some of these aspects were also discussed in detail. It is believed that the proposed algorithm has the potential to perform efficient large–scale numerical simulations on loosely coupled networks of computers in various fields of science.

Large sparse linear systems are emerging from a constantly growing number of scientific applications and finding efficient preconditioners for these problems is becoming increasingly important. This observation has partly motivated the decision of using an asynchronous iterative method as a preconditioner. However, there are many other potential applications of this kind of preconditioner. For example, an asynchronous iterative method could be used as a so–called *hybrid smoother* in *multigrid*, which in itself is often used as a preconditioner. Another possibility is using an asynchronous method to approximate the *correction equation* in large–scale eigenvalue problems.

It is evident that there are many interesting applications and that much research is still needed. It is hoped that the reader has gained some understanding of the complexities related to the design of efficient numerical algorithms for Grid computers.

For the interested reader, the book by Dimitri Bertsekas and John Tsitsiklis contains a wealth of information on parallel asynchronous iterative algorithms for various applications [9]. Furthermore, more extensive discussions on various aspects of parallel scientific computing may be found in the excellent book by Rob Bisseling [10].

For a comprehensive discussion on iterative methods for solving linear systems, the classic book by Gene Golub and Charles van Loan is greatly recommended [34], as well as the more recent book by Henk van der Vorst [57]. More on domain decomposition techniques can be found in [48, 54]. For more technical details on Grid hardware and Grid software technologies, the reader is referred to [8, 27, 26, 31]. The recent overview article on iterative methods by Valeria Simoncini and Daniel Szyld is also highly recommended [47]. Another excellent overview article by Michele Benzi discusses various types of preconditioning techniques [7].

Extensive experimental results and specific implementation details pertaining to implementing numerical algorithms on Grid computers may be found in [17, 18, 19].

Acknowledgements The work of the first author was financially supported by the Delft Centre for Computational Science and Engineering. This work is performed as part of the research project *"Development of an Immersed Boundary Method, Implemented on Cluster and Grid Computers, with Application to the Swimming of Fish."* and is joint work with Barry Koren and Yunus Hassen. The Netherlands Organisation for Scientific Research (NWO) is gratefully acknowledged for the use of the DAS–3. The authors would like to thank the GridSolve team for their prompt response pertaining to our questions and also Stéphane Domas for his prompt and extensive responses pertaining to our questions regarding the CRAC programming system. They also thank Hans Blom for information on the performance of the DAS–3 network system and Kees Verstoep for answering questions regarding DAS–3 inner workings. Figure 6 was kindly donated by Xu Lin, whilst Fig. 8 has been provided by Barry Koren. Paulo Anita kindly provided information on the communication patterns induced by the algorithm on the DAS–3 cluster.

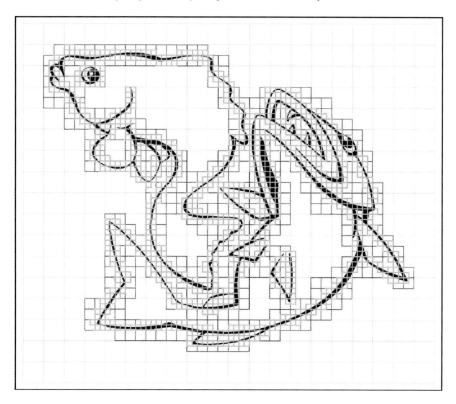

Fig. 8 Artist's impression of fish with Immersed Boundaries (artwork by Tobias Baanders).

References

1. David P. Anderson, Jeff Cobb, Eric Korpela, Matt Lebofsky, and Dan Werthimer. SETI@home: an experiment in public–resource computing. *Commun. ACM*, 45(11):56–61, 2002.
2. Owe Axelsson. *Iterative Solution Methods*. Cambridge University Press, New York, NY, USA, 1994.
3. Jacques M. Bahi, Sylvain Contassot-Vivier, and Raphaël Couturier. Evaluation of the asynchronous iterative algorithms in the context of distant heterogeneous clusters. *Parallel Comput.*, 31(5):439–461, 2005.
4. D. El Baz. A method of terminating asynchronous iterative algorithms on message passing systems. *Parallel Algorithms and Applications*, 9:153–158, 1996.
5. Didier El Baz, Pierre Spiteri, Jean Claude Micllou, and Didier Gazen. Asynchronous iterative algorithms with flexible communication for nonlinear network flow problems. *J. Parallel Distrib. Comput.*, 38(1):1–15, 1996.
6. Micah Beck, Dorian Arnold, Alessandro Bassi, Fran Berman, Henri Casanova, Jack Dongarra, Terry Moore, Graziano Obertelli, James Plank, Martin Swany, Sathish Vadhiyar, and Rich Wolski. Middleware for the use of storage in communication. *Parallel Comput.*, 28(12):1773–1787, 2002.

7. Michele Benzi. Preconditioning techniques for large linear systems: a survey. *J. Comput. Phys.*, 182(2):418–477, 2002.
8. Fran Berman, Geoffrey Fox, and Anthony J. G. Hey. *Grid Computing: Making the Global Infrastructure a Reality*. John Wiley & Sons, Inc., New York, NY, USA, 2003.
9. Dimitri P. Bertsekas and John N. Tsitsiklis. *Parallel and Distributed Computation: Numerical Methods*. Prentice–Hall, Englewood Cliffs, N.J., 1989. republished by Athena Scientific, 1997.
10. Rob H. Bisseling. *Parallel Scientific Computation: A Structured Approach Using BSP and MPI*. Oxford University Press, 2004.
11. Å. Björck. Solving linear least squares problems by Gram–Schmidt orthogonalization. *BIT*, 7:1–21, 1967.
12. Kostas Blathras, Daniel B. Szyld, and Yuan Shi. Timing models and local stopping criteria for asynchronous iterative algorithms. *Journal of Parallel and Distributed Computing*, 58(3):446–465, 1999.
13. T. Brady, E. Konstantinov, and A. Lastovetsky. SmartNetSolve: High level programming system for high performance Grid computing. Rhodes Island, Greece, 25-29 April 2006 2006. IEEE Computer Society. CD-ROM/Abstracts Proceedings.
14. Eddy Caron, Bruno Del-Fabbro, Frédéric Desprez, Emmanuel Jeannot, and Jean-Marc Nicod. Managing data persistence in network enabled servers. *Sci. Program.*, 13(4):333–354, 2005.
15. Eddy Caron and Frédéric Desprez. DIET: A scalable toolbox to build network enabled servers on the Grid. *International Journal of High Performance Computing Applications*, 20(3):335–352, 2006.
16. A. T. Chronopoulos and C. W. Gear. S–step iterative methods for symmetric linear systems. *J. Comput. Appl. Math.*, 25(2):153–168, 1989.
17. Tijmen P. Collignon and Martin B. van Gijzen. Implementing the Conjugate Gradient Method on a grid computer. In *Proceedings of the International Multiconference on Computer Science and Information Technology, Volume 2, October 15–17, 2007, Wisla, Poland*, pages 527–540, 2007.
18. Tijmen P. Collignon and Martin B. van Gijzen. Solving large sparse linear systems efficiently on Grid computers using an asynchronous iterative method as a preconditioner. Technical report, Delft University of Technology, Delft, the Netherlands, 2008. DUT report 08–08.
19. Tijmen P. Collignon and Martin B. van Gijzen. Two implementations of the preconditioned Conjugate Gradient method on a heterogeneous computing grid with applications to 3D bubbly flow problems. Technical report, Delft University of Technology, Delft, the Netherlands, 2008. DUT report 08–xx.
20. Raphaël Couturier, Christophe Denis, and Fabienne Jézéquel. GREMLINS: a large sparse linear solver for grid environment. *Parallel Computing*, December 2008.
21. Raphaël Couturier and Stéphane Domas. CRAC: a Grid Environment to solve Scientific Applications with Asynchronous Iterative Algorithms. In *21th IEEE and ACM Int. Symposium on Parallel and Distributed Processing Symposium, IPDPS'2007*, page 289 (8 pages), Long Beach, USA, March 2007. IEEE computer society press.
22. J. Daniel, W. B. Gragg, L. Kaufman, and G. W. Stewart. Reorthogonalization and stable algorithms for updating the Gram–Schmidt QR factorization. *Mathematics of Computation*, 30:772–795, 1976.
23. Eric de Sturler. Truncation strategies for optimal Krylov subspace methods. *SIAM J. Numer. Anal.*, 36(3):864–889, 1999.
24. Frederic Desprez and Emmanuel Jeannot. Improving the GridRPC model with data persistence and redistribution. In *ISPDC '04: Proceedings of the Third International Symposium on Parallel and Distributed Computing/Third International Workshop on Algorithms, Models and Tools for Parallel Computing on Heterogeneous Networks (ISPDC/HeteroPar'04)*, pages 193–200, Washington, DC, USA, 2004. IEEE Computer Society.
25. K.D. Devine, E.G. Boman, R.T. Heaphy, R.H. Bisseling, and U.V. Catalyurek. Parallel hypergraph partitioning for scientific computing. In *Proc. of 20th International Parallel and Distributed Processing Symposium (IPDPS'06)*. IEEE, 2006.

26. J. Dongarra and A. Lastovetsky. An overview of heterogeneous high performance and Grid computing. *Engineering the Grid: Status and Perspective*, February 2006 2006.
27. Jack Dongarra, Ian Foster, Geoffrey Fox, William Gropp, Ken Kennedy, Linda Torczon, and Andy White, editors. *Sourcebook of Parallel Computing*. Morgan Kaufmann, 2003.
28. Jack Dongarra, Yinan Li, Zhiao Shi, Don Fike, Keith Seymour, and Asim YarKhan. Homepage of NetSolve/GridSolve, 2007. http://icl.cs.utk.edu/netsolve/.
29. Stanley C. Eisenstat, Howard C. Elman, and Martin H. Schultz. Variational iterative methods for nonsymmetric systems of linear equations. *SIAM J. Numer. Anal.*, 20:345–357, 1983.
30. Folding. Folding@home distributed computing. http://folding.stanford.edu/.
31. I. Foster and C. Kesselman. *The Grid: Blueprint for a new Computing Infrastructure*. Morgan Kaufman Publishers, second edition, 2004.
32. J. Frank and C. Vuik. On the construction of deflation–based preconditioners. *SIAM J. Sci. Comput.*, 23(2):442–462, 2001.
33. Andreas Frommer and Daniel B. Szyld. Asynchronous iterations with flexible communication for linear systems. *Calculateurs Parallèles Réseaux et Systèmes Répartis*, 10:421–429, 1998.
34. Gene H. Golub and Charles F. Van Loan. *Matrix Computations (Johns Hopkins Studies in Mathematical Sciences)*. The Johns Hopkins University Press, October 1996.
35. Magnus R. Hestenes and Eduard Stiefel. Methods of Conjugate Gradients for solving linear systems. *Journal of Research of National Bureau Standards*, 49:409–436, 1952.
36. A. Lastovetsky, X. Zuo, and P. Zhao. A non–intrusive and incremental approach to enabling direct communications in RPC–based grid programming systems. Technical report, 2006.
37. Craig Lee, Hidemoto Nakada, and Yusuke Tanimura. GridRPC Working Group, 2007. http://forge.ogf.org/sf/projects/gridrpc-wg/.
38. J. C. Miellou, D. El Baz, and P. Spiteri. A new class of asynchronous iterative algorithms with order intervals. *Math. Comput.*, 67(221):237–255, 1998.
39. Y. Notay. Flexible Conjugate Gradients. *SIAM Journal on Scientific Computing*, 22:1444–1460, 2000.
40. Y. Saad. *Iterative Methods for Sparse Linear Systems*. Society for Industrial and Applied Mathematics, Philadelphia, PA, USA, 2003.
41. Youcef Saad. A flexible inner–outer preconditioned GMRES algorithm. *SIAM J. Sci. Comput.*, 14(2):461–469, 1993.
42. Youcef Saad and Martin H. Schultz. GMRES: a generalized minimal residual algorithm for solving nonsymmetric linear systems. *SIAM J. Sci. Stat. Comput.*, 7(3):856–869, 1986.
43. Mitsuhisa Sato, Taisuke Boku, and Daisuke Takahashi. OmniRPC: a Grid RPC system for parallel programming in cluster and Grid environment. In *CCGRID '03: Proceedings of the 3rd International Symposium on Cluster Computing and the Grid*, pages 206–213, Washington, DC, USA, 2003. IEEE Computer Society.
44. K Seymour, A YarKhan, S Agrawal, and J Dongarra. NetSolve: Grid enabling scientific computing environments. In L. Grandinetti, editor, *Grid Computing and New Frontiers of High Performance Processing*. Elsevier, 2005.
45. Keith Seymour, Hidemoto Nakada, Satoshi Matsuoka, Jack Dongarra, Craig Lee, and Henri Casanova. Overview of GridRPC: A Remote Procedure Call API for Grid Computing. In *GRID '02: Proceedings of the Third International Workshop on Grid Computing*, pages 274–278, London, UK, 2002. Springer–Verlag.
46. Valeria Simoncini and Daniel B. Szyld. Flexible inner–outer Krylov subspace methods. *SIAM J. Numer. Anal.*, 40(6):2219–2239, 2002.
47. Valeria Simoncini and Daniel B. Szyld. Recent computational developments in Krylov subspace methods for linear systems. *Numerical Linear Algebra with Applications*, 14:1–59, 2007.
48. Barry F. Smith, Petter E. Bjørstad, and William Gropp. *Domain Decomposition: Parallel Multilevel Methods for Elliptic Partial Differential Equations*. Cambridge University Press, Cambridge, 1996.
49. Peter Sonneveld and Martin B. van Gijzen. IDR(s): a family of simple and fast algorithms for solving large nonsymmetric linear systems. Technical report, Delft University of Technology, Delft, the Netherlands, 2007. DUT report 07–07.

50. StarPlane. Application-specific management of photonic networks, 2007. http://www.starplane.org/.
51. Thomas Sterling, Ewing Lusk, and William Gropp, editors. *Beowulf Cluster Computing with Linux*. MIT Press, Cambridge, MA, USA, 2003.
52. Y. Tanaka, H. Nakada, S. Sekiguchi, T. Suzumura, and S. Matsuoka. Ninf–G: A reference implementation of RPC–based programming middleware for Grid computing. *Journal of Grid Computing*, 1(1):41–51, 2003.
53. James D. Teresco, Karen D. Devine, and Joseph E. Flaherty. *Numerical Solution of Partial Differential Equations on Parallel Computers*, chapter Partitioning and Dynamic Load Balancing for the Numerical Solution of Partial Differential Equations. Springer–Verlag, 2005.
54. A. Toselli and O. B. Widlund. *Domain Decomposition: Algorithms and Theory*, volume 34. Springer Series in Computational Mathematics, Springer, Berlin, Heidelberg, 2005.
55. H. A. van der Vorst. Bi–CGSTAB: A fast and smoothly converging variant of Bi–CG for the solution of nonsymmetric linear systems. *SIAM Journal on Scientific and Statistical Computing*, 13(2):631–644, 1992.
56. H.A. van der Vorst and C. Vuik. GMRESR: a family of nested GMRES methods. *Num. Lin. Alg. Appl.*, 1(4):369–386, 1994.
57. Henk A. van der Vorst. *Iterative Krylov Methods for Large Linear systems*. Cambridge University Press, Cambridge, 2003.
58. Brendan Vastenhouw and Rob H. Bisseling. A two–dimensional data distribution method for parallel sparse matrix-vector multiplication. *SIAM Rev.*, 47(1):67–95, 2005.
59. Flavien Vernier, Jacques M. Bahi, Sylvain Contassot-Vivier, and Raphael Couturier. A decentralized convergence detection algorithm for asynchronous parallel iterative algorithms. *IEEE Trans. Parallel Distrib. Syst.*, 16(1):4–13, 2005.
60. P. Wesseling. *An Introduction to Multigrid Methods*. John Wiley & Sons, Chichester, 1992.
61. Samuel Williams, John Shalf, Leonid Oliker, Shoaib Kamil, Parry Husbands, and Katherine Yelick. The potential of the cell processor for scientific computing. In *CF '06: Proceedings of the 3rd conference on Computing frontiers*, pages 9–20, New York, NY, USA, 2006. ACM.
62. Asim YarKhan, Keith Seymour, Kiran Sagi, Zhiao Shi, and Jack Dongarra. Recent developments in GridSolve. *International Journal of High Performance Computing Applications (IJHPCA)*, 20(1):131–141, 2006.
63. X. Zuo and A. Lastovetsky. Experiments with a software component enabling NetSolve with direct communications in a non–intrusive and incremental way. In *Proceedings of the 21st International Parallel and Distributed Processing Symposium (IPDPS 2007)*, Long Beach, California, USA, 26-30 March 2007 2007. IEEE Computer Society.

Data Assimilation Algorithms for Numerical Models

A.W. Heemink, R.G. Hanea, J. Sumihar, M. Roest, N. Velzen and M. Verlaan

Abstract To understand and predict the behavior of a system one can use measurements or one can develop physically based numerical models. In many applications however neither of these approaches is able to provide an accurate description of the dynamic behavior of the system. A model is always a simplification of the real world while measurements seldom produce a complete picture of the system behaviour. Using data assimilation techniques measurements and model results are both used to obtain an optimal estimate of the state of the system. In this article we present an overview of methods available to assimilate data into a numerical model. Attention is concentrated on variational methods and on Kalman filtering. The main problem of using these advanced data assimilation schemes is the huge computational burden that is required for solving real life problems. For variational methods the adjoint model implementation is essential to obtain an efficient data assimilation algorithm. For Kalman filtering problems a number of approximate algorithms have been introduced recently: Ensemble Kalman filters and Reduced Rank filters. These algorithms make the application of Kalman filtering to large-scale data assimilation problems feasible. After a brief introduction to the most important data assimilation approaches we will discuss the advantages and disadvantages of the various methods and present a number of real life applications.

1 Introduction

Measurements can be used to develop statistical models for predicting the behavior of environmental processes. However these types of models are derived from the data and do not include physical knowledge of the process. Furthermore, measurements alone do generally not provide a complete picture of the process. Especially in case of processes that vary in space and time it is very hard to reconstruct the

A.W. Heemink
Delft University of Technology, e-mail: A.W.Heemink@tudelft.nl

spatial and temporal patterns only from data. Physically based numerical models produce results that are spatially and temporally consistent. However these models are usually not able to accurately reproduce the measurements that are available. The information provided by the models and by the measurement information is often complementary. Therefore it is important to study a methodology for integrating measurements and physically based mathematical models. This methodology is called data assimilation. By using numerical models that are based on physical laws and that are continuously adapted by the measurements available the two sources of information of the process, model information and measurement information, can be integrated.

Data assimilation can be defined as a procedure to incorporate data into a model simulation so as to improve the predictions. However, assimilating data into a numerical model is far from trivial. The simplest data assimilation procedure is to overwrite the model values at the measurement locations with the observed data. Inserting the data in this way into a numerical model is in general not a satisfactory method. It leaves the model dynamically unbalanced and introduces spurious waves into the model. These short waves may even cause instabilities of the underlying numerical model.

A common data assimilation technique used in numerical weather prediction is optimal interpolation. Here some estimates of the error statistics of the numerical model are used to correct the results of the model using the measurements. However, since these error statistics have to be determined by adopting some ad-hoc statistical assumptions, the correction produced by optimal interpolation is again not consistent with the underlying numerical model. As a consequence the use of optimal interpolation often still yields unrealistic correction or instabilities.

More accurate data assimilation methods are variational data assimilation and Kalman filtering. The basic idea of these data assimilation methods is to use the data to only correct the weak points in the model. Weak parts of the model may be due to uncertainty in initial and boundary conditions or imperfectly known model parameters. The data is not allowed to modify the accurate parts of the model. As a result these types of data assimilation problems are in fact, inverse problems. The specified inputs (model uncertainties) have to be reconstructed from the output (measurements). A variational approach or Kalman filtering solves these inverse problems accurately. For linear problems it can be shown that both approaches produce exactly the same results for the same problem formulation. Optimal interpolation does not solve an inverse problem. It produces corrections for the model output without reconstructing model uncertainties. As a result the variational method and Kalman filtering are superior to optimal interpolation.

In the last decennium the variational approach and Kalman filtering have gained acceptance as powerful frameworks for data assimilation. However, both methods require a very large computational burden, at least an order of magnitude larger than the computational effort required for the underlying numerical model. This is the main disadvantage of these methods compared to optimal interpolation that requires only a small increase in computer time.

Starting point for the data assimilation methodology is a state space representation of the numerical model and the measurements. Let us assume that modeling techniques have provided us with a deterministic state space representation of the form:

$$X_{k+1} = f(X_k, k) + B(k)u_k, \quad X_0 = x_0. \tag{1}$$

Here the X_k is the system state, u_k is the input of the system, f is a nonlinear function, and $B(k)$ is an input matrix. For a numerical model that describes the behavior of a physical process in space and time, the state consists of all the variables in all the grid points of the model at a certain time, while the function f in this case represents one time step of the numerical scheme of the model.

The measurements taken from the actual system are assumed to be available according to the relation:

$$Z_k = m(X_k, k), \tag{2}$$

where Z_k is a vector containing the measurements and m is a nonlinear function that specifies the relation between the model results and the measurements.

In this chapter we describe in Section 2 the basic idea of the variational approach and discuss a number of extensions. In Section 3 we introduce the Kalman filter as data assimilation framework. Here we present a number of filter algorithms for solving large-scale data assimilation problems. In Section 4 a software environment for data assimilation is presented and the advantages of having a general framework for applying different algorithms for different applications. Some of the large scale real life applications are presented in the following chapters. In Section 5 applications in coastal sea modeling are presented, in the specific cases of storm surge prediction and assimilation of high frequency (HF) radar data into a coastal ocean model. A large scale atmospheric chemistry application is presented in Section 6 together with a comparison of the performance of different sequential algorithms.

2 Variational data assimilation

2.1 Data assimilation formulated as a minimization problem

If it can be assumed that the only uncertainties of the model (1)-(2) are introduced by a number of poorly known parameters, the data assimilation problem can be formulated as a deterministic parameter estimation problem. Rewrite the model according to:

$$\begin{aligned} X_{k+1} &= f(X_k, p, k) + B(k, p)u_k, \quad X_0 = x_0, \\ Z_k &= m(X_k, k), \end{aligned} \tag{3}$$

where p is a vector containing the uncertain parameters. Uncertain parameters may be model parameters, initial conditions or inputs. The state X_k is now dependent on the parameters $X_k(p)$. In order to estimate the parameters we first define a criterion

$J(p)$ as a measure for the distance between the measurements and the model results:

$$J(p) = \sum_{k=1}^{K} \Big(Z_k - m(X_k(p),k)\Big)^T R(k)^{-1} \Big(Z_k - m(X_k(p),k)\Big). \tag{4}$$

Here the generalized least squares criterion has been chosen to define J. The covariance matrix $R(k)$ is a weighting matrix that takes into account the errors associated with the measurements. This formulation is used very often in practice. The optimal parameter p is found by minimizing the criterion $J(p)$.

Prior information about the parameter values p_0 can be included by adding a regularization term to the criterion:

$$\begin{aligned} J(p) = &\sum_{k=1}^{K} \Big(\big(Z_k - m(X_k(p),k)\big)^T R(k)^{-1} \big(Z_k - m(X_k(p),k)\big) \Big) \\ &+ (p-p_0)^T P_0^{-1}(p-p_0). \end{aligned} \tag{5}$$

Here P_0 is the covariance matrix of the prior information p_0, modeling the uncertainty associated with this prior information. Usually p_0 is the first guess of the uncertain parameters and the starting value for the optimization procedure. The regularization term in the criterion prevents that parameter estimates become unrealistic if the measurement information is limited. In this case the estimates will simply remain close to the first guess (as they should be).

2.2 The adjoint model

In most practical data assimilation problems the number of uncertain parameters is very large. For example, if p represents the initial condition of a numerical model, the number of parameters is equal to the dimension of the system state. Most optimization techniques are not able to solve these very large-scale problems. In this case the only feasible approach is to use a gradient-based optimization method and to use a variational approach to efficiently compute the gradient of the criterion.

For every parameter p the system state X_k has to satisfy the model constraint (3). Therefore we can rewrite the criterion (4) as:

$$\begin{aligned} J(p) = &\sum_{k=1}^{K} \big(Z_k - m(X_k(p),k)\big)^T R(k)^{-1} \big(Z_k - m(X_k(p),k)\big) + \\ &\sum_{k=0}^{K-1} L_{k+1}^T \big(X_{k+1} - f(X_k,p,k) - B(k,p)u_k\big), \end{aligned} \tag{6}$$

where L_k represents a Lagrange multiplier or adjoint state. Note that expression (6) holds for any choice of L_k. It is easy to show that if L_k satisfies the system of adjoint equations:

$$L_k^T = L_{k+1}^T \frac{\partial f(X_k,p,k)}{\partial X_k} + 2(Z_k - m(X_k,k))^T R(k)^{-1} \frac{\partial m(X_k,k)}{\partial X_k}, \qquad (7)$$

for $k = K-1, K-2, \ldots, 1$, with end condition $L_K = 0$, the gradient of the criterion can be computed by using the very simple expression:

$$\frac{\partial J}{\partial p} = -\sum_{k=0}^{K-i} L_{k+1}^T \left(\frac{\partial f(X_k,p,k)}{\partial p} + \frac{\partial B(k,p)}{\partial p} u_k \right). \qquad (8)$$

Using this expression the gradient of the criterion can be determined from the results of one forward simulation to evaluate the terms:

$$\frac{\partial f(X_k,p,k)}{\partial p} + \frac{\partial B(k,p)}{\partial p} u_k, \qquad (9)$$

for $k = 1, 2, \ldots, K$, and the results L_k, for $k = K, K-1, \ldots, 1$ of one adjoint model simulation. The computational effort required to determine the gradient is almost independent of the number of parameters. Therefore, by combining this idea with a gradient based optimization scheme, a very efficient implementation can be obtained, especially for data assimilation problems with many uncertain parameters. This approach is often called variational data assimilation or the adjoint method.

2.3 Discussion

Variational data assimilation schemes are iterative schemes to minimize a criterion $J(p)$. The number of iterations needed depends on the optimization scheme used. In practice quasi-Newton schemes are often used for data assimilation problems. For these types of schemes it is known that if the criterion is exactly quadratic, the iteration process converges in $d_p + 1$ steps, with d_p as the dimension of p. In most practical data assimilation problems the actual number of iterations can be chosen significantly less since the largest improvement is often obtained in the first few iterations.

One problem with the application of variational data assimilation schemes is that the results of one complete forward simulation have to be stored. This is for most real life data assimilation applications a serious problem. Therefore one generally only stores the system state at a limited number of time steps. If results at intermediate time steps are necessary the states are recomputed using the forward model again. This reduces the storage problem at the expense of additional computations.

Another problem with variational data assimilation is the implementation of the adjoint model. Although the mathematical derivation of this model from the original forward model is straightforward, the coding of the adjoint model can be very difficult. Especially in case when the forward model has been developed step by step over a long period by different programmers and many parts of the code are not well understood (although they produce reliable results). Moreover, forward models are often still under development and are being improved continuously. As a result the adjoint should also be updated continuously. Recently adjoint compilers have become available that produce adjoint code from the code of the forward model (Kaminski et al., 2003). However, for complicated forward models these compilers still produce unsatisfactory results. Probably the best way to develop an adjoint implementation is to redesign and rewrite the forward code such that the adjoint compiler is able to generate the adjoint. An advantage of this approach is that modifications in the forward model can easily be included in the adjoint implementation.

Another approach to variational data assimilation with a comparable computational efficiency is based on model reduction (Ravindran, 2002, Vermeulen et al., 2004, 2006). This approach does not require the implementation of the adjoint of the tangent linear approximation of the original model. Using an ensemble of forward model simulations an approximation of the covariance matrix of the model variability is determined. A limited number of leading eigenvectors (EOF's) of this matrix are selected to define a model subspace. By projecting the original model onto this subspace an approximate linear model is obtained. Once this reduced model is available, its adjoint can be implemented very easily and the minimization process can be solved completely in reduced space with negligible computational costs. If necessary, the procedure can be repeated a few times by generating new ensembles close to the most recent estimate of the parameters.

The approach described in this section is a strong constraint variational method. This means that the only uncertainties in the model are the specified parameters. The model is considered to be perfect. A weak constraint variational approach allows for model uncertainties too (Bennett, 2002, Heemink and Metzelaar, 1995). Like with Kalman filtering model uncertainties are modeled by including stochastic forcing term in the model equations. Algorithms based on a weak constraint formulation are (even) more time consuming then the strong constraint approach and, therefore, are not used very often yet.

3 Kalman filtering

3.1 The linear Kalman filter

In Section 2 model uncertainties were only due to poorly known parameters. Uncertainties in a mathematical model can also be modeled by embedding the model in a stochastic environment. For a linear system a stochastic state space representation can be formulated as follows:

$$X_{k+1} = F(k)X_k + B(k)u_k + G(k)W_k, \tag{10}$$
$$Z_k = M(k)X_k + V_k. \tag{11}$$

A Gaussian system noise process W_k with zero mean and covariance matrix $Q(k)$ is introduced to take into account model uncertainties. $G(k)$ is the noise input matrix.

The Gaussian measurement noise V_k with zero mean and covariance matrix $R(k)$ represents uncertainties in the measurements. The initial condition X_0 is assumed to be Gaussian with mean x_0 and covariance matrix P_0. W_k, V_k and X_0 are assumed to be mutually independent.

Having defined the general stochastic state space representation of the model (10) and the measurement relation (11) it is desired to combine the measurements with the information provided by the model to obtain an optimal estimate of the system state. To solve this filtering problem the probability density of the state X_k conditioned on the history of the measurements $Z_0, Z_1, \ldots, Z_l$ taken, has to be determined. Under the assumption described above it can be shown that this conditional density function is Gaussian. Therefore it is completely characterized by the mean $X(k|k)$ and covariance matrix $P(k|k)$. Moreover, the mean is in this case equivalent to the least squares estimate or any other meaningful estimate. Recursive equations to obtain these quantities can be summarized as follows:

Time update:

$$X(k \mid k-1) = F(k-1)X(k-1 \mid k-1) + B(k-1)u_{k-1}, \tag{12}$$

$$P(k \mid k-1) = F(k-1)P(k-1 \mid k-1)F(k-1)^T + G(k-1)Q(k-1)G(k-1)^T. \tag{13}$$

Measurement update:

$$K(k) = P(k \mid k-1)M(k)^T \left(M(k)P(k \mid k-1)M(k)^T + R(k)\right)^{-1}, \tag{14}$$

$$X(k \mid k) = X(k \mid k-1) + K(k)\left(Z_k - M(k)X(k \mid k-1)\right), \tag{15}$$

$$P(k \mid k) = \left(I - K(k)M(k)\right)P(k \mid k-1), \tag{16}$$

where $K(k)$ is called the Kalman gain.

The initial conditions for the recursive equations are $X(0 \mid 0) = x_0, P(0 \mid 0) = P_0$.

The Kalman filter has a predictor-corrector structure. Based on all previous information, a prediction of the state vector at time t is made by means of the equations (12) and (13). Once this prediction is known it is possible to predict the next measurement Z_k by:

$$M(k)X(k \mid k-1).$$

When the measurement Z_k has become available the difference between this measurement and its predicted value is used to update the estimate of the state vector by means of the equations (14)-(16). Note that the Kalman filter gain matrix $K(k)$ does not depend on the measurements and may be precomputed.

The Kalman filter does not only produce the optimal estimate of the system state, but also computes the covariance matrix of the estimation error (13,16). This gives insight into the accuracy of the estimate. If the covariance of the state estimate is computed with and without the use of measurements, the difference of the two results indicates the improvement of the state estimate due to the measurement information.

The variational approach described in Section 2 is based on minimizing a least squares criterion. Since the Kalman filter also produces least squares estimates, the filter results will be identical to the results of the variational approach for the same linear problem formulation. This is, e.g., the case if the initial condition is chosen as uncertain parameter in the variational approach and in the Kalman filter the system noise is set to zero. The Kalman filter algorithm does not, however, explicitly computes the estimate of the initial condition, but only the optimal estimate of the final state. If necessary the filter algorithm can be modified to also produce estimates of previous states (Kalman smoother). This would produce the same estimate of the initial condition as obtained with the variational approach. A discussion on optimal smoothing is, however, out of the scope of this chapter.

For linear systems the Kalman filter is optimal for Gaussian noise processes. Linearity of the system results in a Gaussian conditional probability density function for X_k given the measurements available up to and including time k. As a result the mean $X(k|k)$ and covariance matrix $P(k|k)$ completely determine this probability density function. Higher order moments do not have to be taken into account to derive the optimal filter. This is not the case for non-linear systems. Due to the non-linearities the density function for X_k is non-Gaussian and has to be described in general by an infinite number of moments. This makes optimal filters in the non-linear case computationally infeasible. However, it is possible to derive approximate filters for many non-linear systems using linearization techniques.

3.2 Kalman filtering for large-scale systems

3.2.1 Square root filtering

Using the standard Kalman filter algorithm for systems with a very large dimension n of the state vector would impose an unacceptable great computational burden. The most time consuming part of the filter algorithm for large-scale systems is the first term in the time update of the covariance matrix(13):

$$F(k-1)P(k-1 \mid k-1)F(k-1)^T.$$

For large-scale systems it is not practical to store the matrix $F(k-1)$ and to determine its transpose. $F(k-1)$ merely represents a sequence of linear operations. Therefore the time update equation of the covariance matrix is rewritten as:

$$F(k-1)\bigl(F(k-1)P(k-1 \mid k-1)\bigr)^T.$$

From this equation we can see that for every time step k the operator $F(k-1)$ has to be applied $2n$ times: first for updating the n columns of the covariance matrix, then for updating the n rows of the resulting matrix. As a result the computational effort required for the algorithm is approximately equivalent with $2n$ model simulations.

In order to obtain a computationally feasible filter, simplifications have to be introduced. However, there are serious problems with the more obvious simplifications that one may consider. E.g., modifying the filter algorithm by forcing correlations at large distances to be zero usually results in a covariance matrix that is not semi positive definite. The last condition is physically not possible and often results in a total failure of the filter algorithm. To avoid this problem a square root algorithm can be used. This algorithm is based on the fact that a covariance matrix P can be factorized as:

$$P(k-1 \mid k-1) = L(k-1 \mid k-1)L(k-1 \mid k-1)^T,$$

where L is a square root of P. The factorization is not unique and there exist many square roots of P. By rewriting the filter algorithm in terms of L instead of P a more robust algorithm is obtained because approximations of L will never result in a covariance matrix that is not semi positive definite. Most filter algorithms that have been developed for large-scale systems are based on a square root representation of the covariance matrix P where the square root of P is approximated by a matrix with a reduced number of columns r. Hence, we have:

$$F(k-1)P(k-1 \mid k-1)F(k-1)^T = $$
$$F(k-1)L(k-1 \mid k-1)\bigl(F(k-1)L(k-1 \mid k-1)\bigr)^T.$$

As a result only the r columns of $L(k-1 \mid k-1)$ have to be updated yielding a computational effort that is approximately equivalent with only r model simulations.

3.2.2 Classical Ensemble Kalman filter

The Ensemble Kalman filter (EnKF) introduced by Evensen (1994) is based on a representation of the probability density of the state estimate by a finite number N of randomly generated system states

$$\xi_i(k-1 \mid k-1), i = 1, ..., N.$$

The optimal estimate and the square root of the covariance matrix of the estimation error are now given by:

$$X(k-1 \mid k-1) = \frac{1}{N} \sum_{i=1}^{N} \xi_i(k-1 \mid k-1), \qquad (17)$$

$$L(k-1 \mid k-1) = $$
$$[\xi_1(k-1 \mid k-1) - X(k-1 \mid k-1) \quad \cdots \quad \xi_N(k-1 \mid k-1) - X(k-1 \mid k-1)]. \qquad (18)$$

The square root $L(k-1 \mid k-1)$ defines an approximation of the covariance matrix $P(k-1 \mid k-1)$ with rank N:

$$P^N(k-1 \mid k-1) = \frac{1}{N-1} L(k-1 \mid k-1) L(k-1 \mid k-1)^T. \tag{19}$$

$P^N(k-1 \mid k-1)$ is however never actually computed. Using the algorithm, first the initial ensemble of state vectors is generated with mean x_0 and covariance matrix P_0. Then for updating the ensemble, realizations of the system noise and measurement noise processes are generated too. The Ensemble Kalman filter algorithm can be summarized as follows:

Time update:

$$\xi_i(k \mid k-1) = f\big(\xi_i(k-1 \mid k-1), k-1\big) + G(k) w_k^i, \tag{20}$$

$$X(k \mid k-1) = \frac{1}{N} \sum_{i=1}^{N} \xi_i(k \mid k-1), \tag{21}$$

$$L(k \mid k-1) = [\xi_1(k \mid k-1) - X(k \mid k-1) \quad \cdots \quad \xi_N(k \mid k-1) - X(k \mid k-1)]^T. \tag{22}$$

Measurement update:

$$K(k) = \tfrac{1}{N-1} L(k \mid k-1) L(k \mid k-1)^T M(k)^T \\ \big(\tfrac{1}{N-1} M(k) L(k \mid k-1) L(k \mid k-1)^T M(k)^T + R(k)\big)^{-1}, \tag{23}$$

$$\xi_i(k \mid k) = \xi_i(k \mid k-1) + K(k)\big(Z_k - M(k)\xi_i(k \mid k-1) + v_k^i\big). \tag{24}$$

The time update equation is computationally dominant. As a result the computational effort required for the Ensemble Kalman filter is approximately N model simulations. The standard deviation of the statistical errors in the state estimate converges very slowly with the sample size ($\approx 1/N$). This is one of the very few drawbacks of this Monte Carlo type approach. Note that for the time update only simulations with the original non-linear model are used. The tangent linear model is not required.

3.2.3 Reduced Rank Square Root Kalman filter

The reduced rank square root filter algorithm (Verlaan and Heemink 1997) is based on a factorization of the covariance matrix of the state estimate according to:

$$P^Q(k-1 \mid k-1) = L(k-1 \mid k-1) L(k-1 \mid k-1)^T,$$

where $L(k-1 \mid k-1)$ is a matrix with the q leading eigenvectors $l_i(k-1 \mid k-1)$

Data Assimilation Algorithms for Numerical Models

of $P(k-1 \mid k-1)$ as columns. These eigenvectors are scaled with the square root of the corresponding eigenvalue. The computationally dominating part of the filter algorithm is to determine:

$$\bar{F}(k-1)L(k-1 \mid k-1).$$

Here the tangent linear model has to be applied to every column $l_i(k-1 \mid k-1)$ of $L(k-1 \mid k-1)$:

$$\bar{F}(k-1)l_i(k-1 \mid k-1), \quad i = 1, \cdots, q.$$

To avoid the use of the tangent linear model we use the approximation:

$$\bar{F}(k-1)l_i(k-1 \mid k-1) \approx$$
$$\frac{1}{\varepsilon}\Big(f(\hat{X}(k-1 \mid k-1) + \varepsilon l_i(k-1 \mid k-1), (k-1) - f(\hat{X}(k-1 \mid k-1), k-1)\Big),$$

where ε represents a small perturbation. This approximation results in a time update where only simulations with the original nonlinear model are required. The algorithm can now be summarized as follows:

Time update:

$$\hat{X}(K \mid k-1) = f\big(\hat{X}(k-1 \mid k-1), k-1\big) + B(k-1)u_{k-1}, \tag{25}$$

$$\tilde{l}_i(k \mid k-1) = \frac{1}{\varepsilon}\Big(f(\hat{X}(k-1 \mid k-1) + \varepsilon l_i(k-1 \mid k-1), (k-1) - f(\hat{X}(k-1 \mid k-1), k-1)\Big), \tag{26}$$

$$\tilde{L}(k \mid k-1) = \begin{bmatrix} \tilde{l}_1(k \mid k-1) & \tilde{l}_2(k \mid k-1) & \cdots & \tilde{l}_N(k \mid k-1) & G(k-1)Q^{\frac{1}{2}}(k-1) \end{bmatrix}^T. \tag{27}$$

By computing:

$$\begin{aligned}
\tilde{L}(k \mid k-1)\tilde{L}(k \mid k-1)^T &= \bar{F}(k-1)L(k-1 \mid k-1)\big(\bar{F}(k-1)L(k-1 \mid k-1)\big)^T + \\
&\quad G(k)Q(k)G(k)^T \\
&= \bar{F}(k-1)L(k-1 \mid k-1)L(k-1 \mid k-1)^T\bar{F}(k-1)^T + \\
&\quad G(k)Q(k)G(k)^T,
\end{aligned} \tag{28}$$

it is easy to show that equation (28) is equivalent to the time update (16) of the original Kalman filter algorithm and that $\tilde{L}(k \mid k-1)$ is a square root approximation of the covariance matrix $P(k \mid k-1)$. However, a serious problem with this square root is that its dimension has increased with respect to the dimension of the square root $L(k-1 \mid k-1)$ of $P(k-1 \mid k-1)$. This would increase the computational effort of the algorithm in every time step. This can be avoided by computing a new approximate square root of $P(k \mid k-1)$ with reduced dimension:

$$L(k \mid k-1) = \Pi(k \mid k-1)\tilde{L}(k \mid k-1), \tag{29}$$

where $\Pi(k \mid k-1)$ is a projection onto the q leading eigenvectors of the matrix

$$\tilde{L}(k \mid k-1)\tilde{L}(k \mid k-1)^T.$$

Using this reduction step we obtain again an approximate square root $L(k \mid k-1)$ of $P(k \mid k-1)$ with only q columns.

Measurement update (for scalar measurements only):

$$\beta(k) = \left(\frac{1}{N-1}M(k)L(k \mid k-1)L(k \mid k-1)^T M(k)^T + R(k)\right)^{-1}, \tag{30}$$

$$K(k) = L(k \mid k-1)L(k \mid k-1)^T M(k)^T \beta(k), \tag{31}$$

$$\hat{X}(k \mid k) = \hat{X}(k-1 \mid k-1) + K(k)\left(Z_k - \hat{X}(k-1 \mid k-1)\right), \tag{32}$$

$$L(k \mid k) = L(k \mid k-1) - \frac{k(k)M(k)L(k \mid k-1)}{1 + \beta(k)R(k)^{\frac{1}{2}}}. \tag{33}$$

It is easy to show that for scalar measurements this measurement update is equivalent to the measurement update (14)-(16) of the original Kalman filter algorithm. Independent measurements ($R(k)$ is diagonal) can be processed one by one using the update equations (30)-(33). In case measurements are not independent of each other they have to be transformed.

3.2.4 Discussion

One of the main practical advantages of the Kalman filter based algorithms is that they are completely separate from the model implementation. The underlying model $f(X_k, k)$ is used a number of times for different system states in order to compute approximations of the mean and covariance of the state estimate. Neither a tangent linear model nor an adjoint model implementation is required. As a result the coding of the algorithm is very simple. Only an interface between the model code and the filter code has to be developed.

For both the Ensemble Kalman filter as the Reduced Rank approach the choice of the ensemble size is very essential for the performance of the algorithms. The computational effort is approximately proportional with the ensemble size, so the ensemble size has to be chosen as small as possible. However, if the ensemble size is chosen too small the algorithm can become very inaccurate or unstable. Usually a simple "trial and error" procedure is the only way to get insight into an appropriate value for the ensemble size.

Recently, many new variants of the ensemble type filter algorithms have appeared in literature. Heemink et al. (2001) have been examining different approaches which combine ideas from RRSQRT filtering and the EnKF to derive computationally more efficient methods. The new algorithm, COFFEE (Complementary Orthogonal subspace Filter For Efficient Ensembles) is a RRSQRT algorithm where the truncated modes are not neglected, but taken into account by a number of random ensemble members (see also 6.8). Bishop et al. (2001) used an implementation of the EnKF in an observation system simulation experiment. Ensemble-predicted error statistics were used to determine the optimal configuration of future targeted observations. The methodology was named Ensemble Transform Kalman Filter. The EnKF can also be related to some other sequential filters such as the Singular Evolutive Extended Kalman (SEEK) filter by Pham et al. (1998) and the Error Subspace Statistical Estimation (ESSE) by Lermusiaux and Robinsin (1999a,b) and Lermusiaux (2001), which can be interpreted as an EnKF where the analysis is computed in the space spanned by the EOFs of the ensemble. Anderson (2001) proposed a method denoted the Ensemble Adjustment Kalman Filter, where the analysis is computed without adding perturbations to the observations. Whitaker and Hamill (2002) proposed another version of the EnKF where the perturbation of observations are avoided. The scheme provides a better estimate of the analysis variance by avoiding the sampling errors of the observation perturbations. This is essentially a Monte Carlo implementation of the square root filter and was named EnSRF. A summary of the square root filters by Bishop et al. (2001), Anderson (2001), and Whitaker and Hamill (2002) has been given by Tippet et al. (2003).

4 A software environment for data assimilation: COSTA

Data assimilation and calibration techniques are widely used in various modeling areas such us meteorology, oceanography and chemistry. Unfortunately it is very hard to reuse the existing software implementing these techniques because they are in general very model specific. Because existing software cannot be reused it is necessary to program them from scratch in order to extend models with data assimilation and model calibration techniques. COSTA [1](Velzen, 2006 and Velzen and Verlaan, 2007) is a modular framework for data assimilation and model calibration. Within the COSTA framework it is possible to combine models with the available data assimilation and model calibration methods without the need of additional software development. The usage of COSTA significantly reduces the development costs for implementing data assimilation for simulation models, especially when a model has to be combined with various methods.

When new data assimilation methods are developed in COSTA they can be tested on various models. Therefore their performance can be better determined and they can be compared to other methods. Finally, data assimilation methods can be reused and do not need to be programmed. COSTA simplifies the application of data assimilation methods such that it becomes available to a wider group of users and application areas. Data assimilation and model calibration software that is specially

---

[1] http://costapse.org

developed for a single model on the other hand are in general computationally efficient. COSTA is set up in order to be as computationally efficient as possible, without losing its generic properties. The aim is that applications developed with COSTA have a computational performance comparable to tailor made implementations. COSTA is set up such that it can be used for a range of different models varying from small to large scale models and in sequential as well as parallel environments.

4.1 COSTA components

COSTA defines a number of building blocks called components. A data assimilation application is created by combining these components. The components can be split into two groups; the high level components; the model, the observations and the data assimilation method, and the low level components including among others the vector, matrix, time and tree-vector.

A data assimilation application is created by combining the high level components as illustrated in Figure 1. In most cases one or more of these components are created using existing software. For that purpose some interfacing between the existing code and COSTA needs to be provided. This is illustrated by the red parts of the components in Figure 1. The low level components provide a generic way for data transfer between the high level components.

The principle of COSTA is that the programmer can choose whether to use a generic COSTA implementation of a component or program a specialized implementation of a component. The COSTA development strategy is to quickly try out alternative methods or models using the generic implementations of the components and in a later stage improve performance or functionality by replacing some of the components by specialized implementations, when necessary.

The components in COSTA are similar to classes in object-oriented languages. This means that there can be multiple instances of a component and the data inside a component is not directly accessible from outside. The data of a component can only be manipulated by a set of functions, together called the interface of the component. Different components with the same interface can therefore be interchanged and reused for other applications. An additional advantage of shielding the data inside a component is that the data does not need to be directly accessible. The data can therefore e.g. be stored on file or in memory on another computer. It is therefore possible to use the same interface for a black box model, where all interaction takes place by reading box model output and writing COSTA input files as models that use parallel computing.

Data Assimilation Algorithms for Numerical Models   121

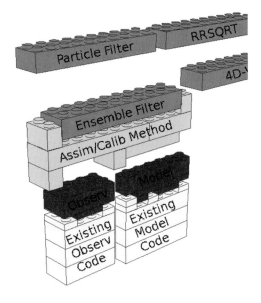

Fig. 1 *Creating COSTA components and combine them into an application. In this example existing model code and observation handling code is transformed into COSTA components. For this purpose an interface layer (red) needs to be created. Thereafter it becomes possible to combine the model with various data assimilation and calibration methods without any additional adjustments.*

4.2 The COSTA software

COSTA is an open source software package developed at Delft University of Technology. The software can be freely downloaded at sourceforge.net, project costapse and it is released under the LGPL license. Under this license it is possible to freely use COSTA in combination with open source models and also in combination with proprietary models. The COSTA project exists of two parts; the definition of components and their interfaces and a software library, containing various data assimilation methods, calibration methods, default implementation of various COSTA components and tools for creating new COSTA components. The design of COSTA is object oriented. This does not imply that COSTA can only be used in an object oriented language. COSTA is developed in C and Fortran to ensure high performance and currently it can be used from C, Fortran and Java.

The COSTA software environment is built up in different layers. The bottom layer is not specific for data assimilation but provides the tools for programming the COSTA components and contains software written by third parties. This layer contains basic components like vectors, matrices and time but also robust and efficient numerical methods, input/output facilities and data storage. On top of this layer, the data assimilation and calibration specific components are constructed like tree-vector, model, observations, and component builders. The top layer contains complete (utility) programs. This is illustrated in Figure 2.

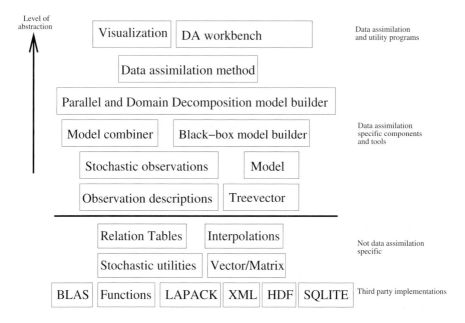

Fig. 2 *An overview of the layers in COSTA. The bottom layer contains non data assimilation and calibration specific software. The middle layer contains data assimilation components and the top layer contains utility programs*

A generic implementation is available for most COSTA components. The programmer can always choose to create his own implementation of a component. This can be necessary for increasing the performance or the addition of new models and methods. The interface of a component can consist of a large number of methods. Fortunately this does not mean that all these methods need to be implemented when programming a new version of a component. Some of the methods are redundant and can be derived from others. The interface of the vector for example contains a method for setting all values at once and a method for setting a single value in the vector. Setting the whole vector can, although less efficiently, be implemented using the method for setting a single value. COSTA will, when possible, automatically implement missing methods in the case an implementation of a component does not provide all the methods, The interfaces of the components contain this redundancy because they simplify the usage of the component saving programming time and yielding better software. The extra methods also provide flexibility for optimizing the implementation of the component.

We will illustrate why redundant methods can be very useful using the following example. At the analysis step of a Kalman filter, the state is adjusted towards the observations. A possible implementation, in pseudo code is:

```
state=model.getstate
state=state+delta_state
model.setstate(state)
```

A disadvantage of this implementation is that it does not provide much flexibility. The preferred implementation in COSTA is:

```
model.axpy(1.0,delta_state) //note axpy: y=alpha*x+y
```

If the model state contains for example concentrations that must be positive at all time, the model programmer can handle this nicely by implementing the axpy method. An additional advantage of the axpy-method is that this eliminates the need for the unnecessary copying of state vectors. When the model interface does not implement the axpy method it will be implemented by COSTA using the getstate and setstate methods of the model component. In some situation it is not not possible for COSTA to fill in the missing methods. However this is not always a problem. For example a model that does not implement an adjoint can still work with the sequential data assimilation methods in COSTA. It is always possible to add the missing methods when needed.

COSTA is set up such that it can be used for a wide range of data assimilation and calibration methods. Currently COSTA contains the following sequential data assimilation methods: Ensemble Kalman Filter (EnKF) (Evensen, 2003), Ensemble Square Root Filter (EnSRF) (Whitaker and Hamill, 2002), Complementary Orthogonal subspace Filter For Efficient Ensembles (COFFEE) (Heemink et al., 2001) and RRSQRT Kalman filter (Verlaan and Heemink, 1997). The model calibration in COSTA where time independent model parameters are calibrated. The available methods are: the Conjugate Gradient (CG) method (Fletcher and Reeves, 1964), a Quasi Newton methods using the LBFGS method (Nocedal, 1980 and Byrd et al.,1994) and the Simplex method (Nelder and Mead, 1965).

5 Applications in coastal sea modeling

5.1 Storm surge prediction using Kalman filtering

Large areas of the Netherlands lie below the mean sea water level, which give rise to a continuous risk of storm surge flooding. A number of severe floods of this kind have occurred in the past. The last disaster in 1953 killed more than 1800 people and flooded over 340.000 acres of land in the southwest of the Netherlands. In order to prevent such disaster from recurring, a number of projects concerning safety have been conducted, which include heightening the dikes and constructing some movable storm surge barriers. In connection to this, accurate storm surge predictions at least six hours ahead, are necessary to decide what precautionary actions have to be taken to protect the dikes and for proper closure of the movable storm surge barriers. Since the mid-1980s, these predictions are based on a numerical hydrodynamic model called the Dutch Continental Shelf Model (DCSM). To increase the predictive capability, in the operational system observed water level is assimilated to the DCSM. In this section we describe the DCSM and data assimilation technique applied to it, which is used operationally in the Netherlands.

The DCSM covers the area of the north-east European continental shelf, i.e. 12^oW to 13^oE and 48^oN to 62^oN as shown in Figure 3. It is based on the two dimensional non-linear shallow water equations, which describe the large-scale motion of water level and depth-integrated horizontal flow. (Gerritsen, et.al. [1995]). At closed boundaries (coastlines) the flow velocity perpendicular to the coastline is set to zero. On the other hand, at open boundaries, the flow velocity parallel to the boundaries is set to zero and tidal harmonic water set-up is specified with ten tidal constituents: M2, S2, N2, K2, O1, K1, Q1, P1, NU2, and L2.

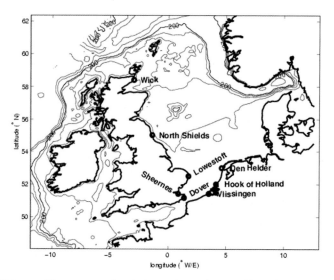

Fig. 3 DCSM area with some water-level observation stations used in assimilation

To solve the system of equations which describes the model, discretization using an Alternating Directions Implicit (ADI) method and a staggered C-grid is used. The numerical scheme is developed by Stelling (1984) based on the work of Leendertse (1967), solved in spherical grid (e.g. Verboom, et.al. [1992]). The resolution of the spherical grid is $1/8^o \times 1/12^o$, which is approximately 8×8 km. This configuration results in 201×173 computational grid cells with 19809 active cells. The time step is 10 minutes. This model uses forecasts of the KNMI (Netherlands Meteorological Institute) meteorological high-resolution limited area model (HiRLAM) as input.

Since 1992, a special feature of DCSM has been added, which is the use of assimilation of real-time measurements from tide gauges using a steady-state Kalman filter (Heemink and Kloosterhuis, 1990). The objective of the implementation of this filter is to improve the initial state from which to start the forecast. This filter assimilates the selected water level observations from eight tide gauges at the British and Dutch coasts (see Figure 3). Observations from British gauges during the period of surge coming from the Atlantic, contain information on the surge in the Southern North Sea several hours ahead. If they can be assimilated into the model, external surges can be included and errors due to imperfect meteorology during this period can be corrected for (Gerritsen et.al., 1995).

The implementation of the Kalman filter requires a conceptual change from the deterministic model into a stochastic model. This is done by adding error terms to the deterministic model. In the operational DCSM, the errors in the model are assumed to be introduced by the uncertainty in the wind input. In the operational system, this uncertainty is accounted for by adding error terms to the u and v depth-average velocities (Heemink and Kloosterhuis, 1990).

The implementation of the Kalman filter on the DCSM is done by exploiting the fact that the observations come from a fixed network and by assuming that the error statistics in the model and the observations vary only slightly in time (Gerritsen et.al., 1995). This means that the solution for the covariance equations in Section 3.2 will become constant after several recursions. Since the propagation of the error covariance is independent from the real observation, it is possible to compute the steady-state Kalman gain off-line. The use of the steady-state Kalman filter has the advantage of being computationally cheap. It leads to only 10% extra computational cost (Gerritsen et.al., 1995). For the operational system, the computation of the steady-state Kalman gain is carried out by using a Chandrasekhar-type algorithm (Heemink and Kloosterhuis, 1990). Another ensemble type algorithm for computing the steady-state Kalman gain implemented on DCSM is proposed by Sumihar, et.al [2008]. Once the Kalman gain is computed, the original non-linear model is used to propagate the mean for prediction.

To illustrate the effect of implementing steady-state Kalman filter on DCSM, in Figure 4 we show the time series of water level on the whole day of 12 January 2005, obtained from observation, model results with and without assimilation at four locations. Two of these stations (Wick and Lowestoft) are located on the British coast, of which the data is used for assimilation. On the other hand, the other two stations, Cadzand and Huibertgat, are located in the South-west and North of the Netherlands respectively. The data from the latter are not used for the assimilation. By visual check on Figure 4, we see that by assimilation, the model results are pulled closer towards the observation. To acquire a better idea about the filter performance in assimilating observed data, the root-mean-square (RMS) of water level innovation is computed over the simulation period of 1 November 2004 - 1 February 2005. The RMS innovation of water level is computed both for assimilation and validation stations. Assimilation station is the station whose data is used for assimilation, while validation station is the one where data is used for validation but not for assimilation. The results are presented in Figures 5-6. It is clear from these pictures that the Kalman filter reduces the RMS innovation of water level in both assimilation and validation stations. This indicates that the correction is propagated spatially consistent as it also applies to locations where data is not used for assimilation. This demonstrates that the filter succeeds in forcing the model closer to the observation. The steady-state Kalman filter is used in operational system for reinitializing the model before making forecast. The output is evaluated every few months, after every severe storm and if the hydrologist or meteorologist indicates any flaws. To illustrate the forecast performance, we show in Figure 7 an example of operational validation, consisting the RMS forecast errors with and without Kalman filter (Verlaan, et.al. 2005).

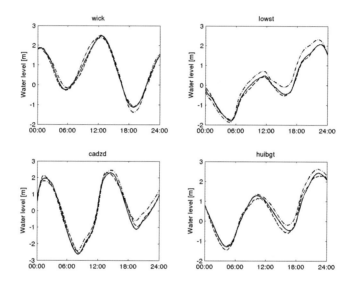

Fig. 4 Water level time series at Wick, Lowestoft, Cadzand, and Huibertgat on 12 January 2005. The dashed lines represent observed water level, the dash-dotted lines model results without assimilation, and the full-lines model results with assimilation.

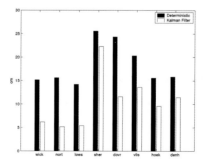

Fig. 5 RMS innovation of water level at assimilation stations

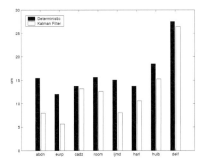

Fig. 6 RMS innovation of water level at validation stations

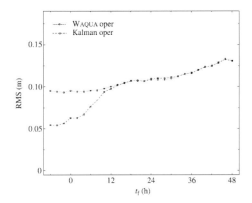

Fig. 7 Operational evaluation for location Scheveningen. RMS errors for different forecast ranges with and without Kalman filter (picture is taken from Verlaan, et.al [2005])

5.2 Using an RRSQRT Kalman filter to assimilate HF radar data into a 3D coastal ocean model

5.2.1 Introduction

This section presents extensive experiments that have been done with a time-dependent Kalman filter for a large-scale, three dimensional numerical model of the coastal region near IJmuiden in the Netherlands (the IJmond model). The filter has been used to assimilate high frequency(HF)-radar measurements and ADCP measurements into the model. The experiments have been done both to show that Kalman filtering for such large-scale models is feasible and to investigate the 3D effects in the data-assimilation. The IJmond region is the region around IJmuiden, on the North-West coast of the Netherlands. It is the entrance to the Noordzee channel, which leads to the port of Amsterdam, the capital of the Netherlands. Apart from this, the IJmond region is economically important in its own right.

In regions like IJmond, 3D models are far superior to 2D models because they are capable of representing stratification and the related flow phenomena. A proper representation of these phenomena is needed to bring in large ships under strong flow conditions. Without such models, large ships have less opportunities to safely enter the harbor or the Noordzee channel, which has an obvious negative economic effect.

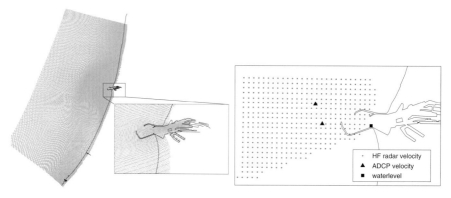

Fig. 8 Left: The computational grid for the IJmond model.
Right: Locations of the observation points.

The region of interest covers some 70 km along the coast and stretches some 40 km into the sea (see Figure 8). It is a rather shallow, sandy region with a maximum depth of about 28 meters. The main feature is a channel in the bottom to allow taller ships to enter the harbor, and a very deep pit just before the entrance to the harbor. The flow is predominantly along-shore. The entrance to the harbor itself is protected by two piers, one to the north and one to the south. The harbor is known for significant short-term variations in water level within the harbor. Sweet water enters the region from the Noordzee channel at moments when the sluices are opened. This means that there is a very relevant interaction between sweet and salt water in the IJmond region.

5.2.2 The TRIWAQ model for the IJmond region

The numerical model used for this study is the 3D flow model TRIWAQ. It is the 3D variant of the WAQUA model, which is an operational software system used for a wide range of applications by the Directorate for Public Works and Water Management. This directorate is responsible for the management of the major water systems in the Netherlands, including the coastal region.

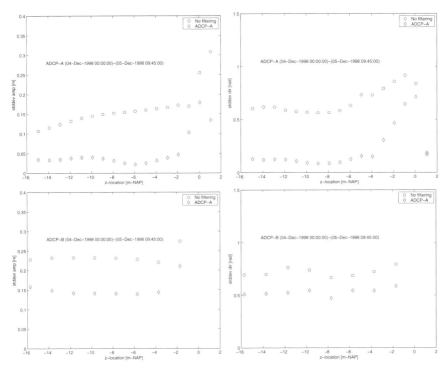

Fig. 9 Standard deviation of the difference between observed and predicted velocity in ADCP-point A (top) and ADCP-point B (bottom) with and without assimilation of actual data from ADCP A. The left-hand plots apply to the magnitude of the velocity, the right-hand plots apply to the direction.

TRIWAQ is based on a finite-difference discretization of the shallow water equations and includes a transport model for salinity and temperature. Forcing is done by boundary conditions which can be composed of different combinations of water level, flow velocity and discharge. The wind can be applied uniformly or through a spatially varying wind velocity and pressure. Bottom-friction is modeled through Chézy coefficients computed according to the Manning or the White-Colebrook equations. TRIWAQ has extensive features for the modeling of drying and flooding and for special hydrodynamic objects such as barriers, weirs and sluices.

The IJmond region is modeled in TRIWAQ by a curvilinear grid with a maximum extension of 188 grid-points along the coast and 230 grid-points in the cross-shore direction (see Figure 8). The grid is highly contracted around the harbor area to give a high resolution there. In vertical direction, the flow field is modeled with four equidistant sigma layers. This is much less than what is common for coastal models (which usually apply more than ten layers). The main reason for using such a small number of layers is to keep the amount of computations within limits. At the same time, it should be able to represent at least some of the vertical dynamics. The time step is half a minute, mainly to accommodate for the fine mesh inside the harbor. Boundary conditions are given along the sea-perimeter of the model. The boundary conditions are determined from a 2D model of the entire Dutch coastline.

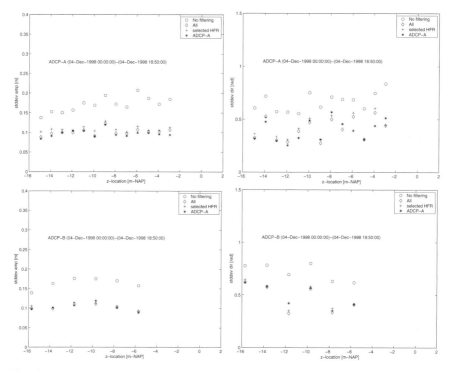

Fig. 10 Standard deviation of the difference between observed and predicted velocity in ADCP-point A (top) and ADCP-point B (bottom) with and without assimilation of different kinds of artificial data. The left-hand plots apply to the magnitude of the velocity, the right-hand plots apply to the direction.

5.2.3 Kalman filtering for large 3D models

The quality of model predictions can be improved by applying data-assimilation: using observations of the actual flow situation to correct the model state so that the ensuing predictions are better. One of the methods to perform data-assimilation is Kalman filtering.

The full Kalman covariance matrix in the case of the IJmond model would be about 700.000×700.000 elements. The RRSQRT algorithm reduces this to a matrix of 700.000×50 elements. Even with the RRSQRT approximation, computing requirements are still excessive. But they can be handled on large scale parallel computing facilities. Computing facilities that employ some form of parallelism are quickly becoming a commodity. Modern PC's use almost exclusively multi-core processors and the use of computing-clusters is far from exceptional. In order to use this type of hardware, a parallel version has been developed of the RRSQRT Kalman filter implementation that is available for TRIWAQ. This parallel version shows excellent speedup figures on up to 32 processors.

To show the feasibility of data-assimilation for a large scale 3D model, experiments have been done using data from an extensive observation campaign that has been done in 1998. This campaign involved the already existing tide gauge at one of the harbor piers, two ADCP moorings just outside the harbor and an HF radar system (see Figure 8). The ADCP's are 1200 kHz workhorse ADCP's, providing velocity data for bins of 1 meter (Point A) and 2 meters (Point B). ADCP A is located just outside the south pier, ADCP B is located somewhat further to the north-west. The HF radar system has been located at the Harbor Operation Center (HOC) next to the harbor and at Bloemendaal aan Zee, some five kilometers to the south. It has provided observations on a grid of 250×250 meters. There are about 450 grid points, covering an area of about 7 ×4 km.

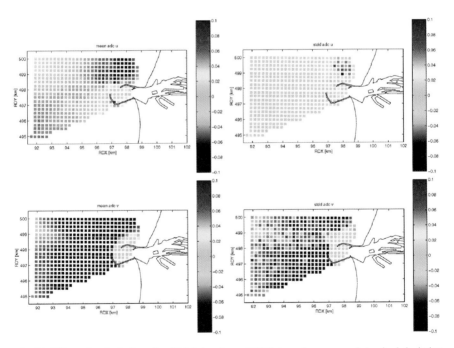

Fig. 11 Effect of assimilation of artificial data from ADCP A on the mean and standard-deviation of the difference between predicted and observed velocities in HF-radar points. Colors mark values in $[\text{ms}^{-1}]$. Red colors indicate a deterioration, blue colors an improvement.

5.2.4 Results

The data-assimilation scheme itself works properly. Assimilating observations from only ADCP A gives improvements also at the location of ADCP B some distance away (see Figure 9). This illustrates that the adaptations made by the filter have physical meaning. But the errors in HF-radar points do not benefit from the adaptations. This is most likely to be attributed to the poor quality of the HF-radar data itself. Analysis of this data has shown that it often displays rather erratic flow patterns. Even during the observation campaign, there were doubts about the quality of the observations. Hence, any definite conclusions with respect to the HF-radar data can not be drawn from these experiments. To determine whether better-quality HF-radar observations could be beneficial, an experiment has been done in which artificial HF-radar observations have been assimilated. The artificial observations have been obtained from a run of the IJmond model with modified forcings. The modifications to the forcings have been made in such a way that they match the assumptions of the filter about the errors in the model. In this way, when the model is run with the original, unmodified forcings, the filter should be able to reconstruct the state from the distorted model by assimilating the artificial observations.

The experiments with artificial data show more consistent results than those using the actual observations. In this case, the errors in the predictions at HF-radar observation locations are reduced by the assimilation of ADCP observations (see Figure 11). Furthermore, Figure 10 shows that assimilation of a selection of HF-radar observations also improves the predictions below the sea-level. The improvements are comparable to those obtained by assimilating the ADCP observations. So, under certain assumptions it is possible to reconstruct a 3D flow pattern based on 2D HF-radar observations. But the experiment is highly idealized. The uncertainty model used to distort the forcings exactly matches the uncertainty model assumed by the Kalman filter, which is usually not the case. Also, the HF-radar observations are assumed to have independent and small observation errors. In particular the assumption of independent errors is probably not correct in real HF-radar systems.

In conclusion, the experiment has primarily shown that data-assimilation based on RRSQRT Kalman filtering for large-scale 3D coastal ocean models is feasible with current technology.

6 Applications in atmospheric chemistry modeling

Elevated concentrations of ozone in the boundary layer (first 3 km of the atmosphere) can cause adverse affects to humans and ecosystems (EC, 2002). Ozone in the boundary layer is formed by chemical reactions of the ozone precursors, nitrogen oxides (NO_x), volatile organic compounds (VOCs), and carbon monoxide and methane under the influence of sunlight. The impact of ozone is not limited to the area close to where the ozone precursors are emitted. Transboundary fluxes transport these precursors over distances of hundreds of kilometers (Lelieveld et al.,

2002). Atmospheric chemistry transport models have been developed to understand the processes controlling the formation of ozone, to study the potential effects of ozone on ecosystems and humans, and to assess the effects of emission reductions on the ozone concentration. Atmospheric models and observations are usually applied separately to obtain information about ozone and air-pollutant concentrations in the boundary layer. Assimilating ozone observations in the model can improve the information on the state of the atmosphere. One method to assimilate measurement data into a model simulation to get a better estimate of the real ozone concentration is data assimilation.

6.1 Eulerian chemistry transport model EUROS

The Eulerian atmospheric chemistry transport model EUROS (EURopean Operational Smog, Rheineck et al., 1990, Pul et al. 1996, Matthijsen et al, 2001) was developed at RIVM (National Institute of Public Health and the Environment, The Netherlands). The model contains parameterizations of the various chemical, dynamical and radiative processes in the atmosphere, as well as information about the emissions of ozone precursors, nitrogen oxides, VOCs, carbon monoxide and methane. The model can be used to examine the time and spatial behavior of SO_x, NO_x, O_3, and VOCs in the lower troposphere over Europe. The model area extends over a large part of Europe and uses a shifted pole coordinate system, i.e., the real North Pole is at 30° northern latitude and the equator has been shifted to 600 northern latitude in the new coordinate system. In these shifted pole coordinates the model domain is $[-8.25^o, 20.35^o] \times [-23.1^o, 7.15^o]$. The grid consists of 52×55 grid cells with a longitude-latitude resolution of $0.55^o \times 0.55^o$ (about 60×60 km). The vertical stratification of EUROS consists of four layers from the surface up to 3000 m.

6.2 State space representation

In order to derive a state space representation of the EUROS model, a state vector, defined as the vector containing the ozone concentrations at each grid point of the model, is necessary. The EUROS model can now be rewritten as a state space model in the form:

$$c_{k+1}^t = M(c_k^t). \tag{34}$$

The stochastical model required for the Kalman filter is based on a specification of the errors of the deterministic model EUROS and of the observations. Knowledge of the uncertainties in the EUROS model and in the observations is crucial for a successful data assimilation. The description of the errors strongly affects the results of the assimilation. One of the goals of this study is to gain insight in the uncertain parameters of the model and the impact of these parameters on the assimilation. One method to accomplish this is to treat these parameters as model input and to define them as stochastic parameters. A disadvantage of modeling uncertainties using a white noise input w is that there is no memory in the system, i.e., the w is uncorrelated in time. At a certain hour t the noise parameter may indicate an emission increase of 30% with respect to the original field, whereas it estimates a decrease of 30% at $t+1$. Such irregular and unrealistic behavior can be prevented by the use of colored noise. The white noise input w forces a colored noise process λ, which operates on the parameters in the model. The exact size of λ depends on which parameter in the model is considered to be stochastic. Each element of λ allows use of different values for the time correlation parameter $a = exp(\frac{-1}{\tau})$ and standard deviation $\sigma_a = \sigma \cdot \sqrt{1-a^2}$. The parameter τ is the time correlation length in hours. Five parameters are considered uncertain. In the implementation of the Kalman filter with the EUROS model, the noise parameters form a part of the model state (Segers, 2002), therefore making the use of the augmented state vector necessary. The general stochastical model (equation (2)) takes the following form:

$$\begin{pmatrix} c(t_{k+1}) \\ \lambda_{k+1}(t_{k+1}) \\ \lambda_k(t_{k+1}) \end{pmatrix} = \begin{pmatrix} M(c(t_{k+1}), \lambda_k(t_k)) \\ a\lambda_k(t_k) \\ \lambda_k(t_k) \end{pmatrix} + \begin{pmatrix} 0 \\ \sigma_a I \\ 0 \end{pmatrix} w(t_k), \qquad (35)$$

or:

$$x^t(t_{k+1}) = M(x^k(t_k)) + Gw(t_k), \qquad (36)$$

where the left-hand side of equation (35) is the state vector, c is the concentration array of all species in the model, and the matrix G distributes the impact of the noise values for all the parameters considered uncertain in the model in specific areas.

6.3 Ground based observations

To improve the estimation of the assimilated concentrations of ozone and to reduce the uncertainties in the emissions the model is combined with available measurements. Only ozone observations at background stations are used in the assimilation, since the output of the EUROS model represents the ozone concentrations on the same scale. For the period studied (summer 1996) data was available for in total 135 of those stations in Europe. (Figure 12). For each ozone measurement it is assumed that the standard deviation of the error is equal to 10 μgm^{-3}. This error accounts for both the uncertainty in the actual concentration measured at the specific station as well as for a representation error, reflecting the size of the grid cell.

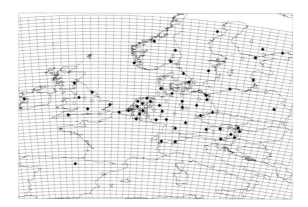

Fig. 12 Stations with ozone observations in Europe in June 1996. Stations used (default) for assimilation (set 1) are marked by filled diamonds, the stations used (default) for validation marked with open diamonds (set 2) and open triangles (set 3). Part of the EUROS grid is shown as well. If there is more than one station per grid cell, only one is used for the assimilation, the other is indicated by an open triangle. For the year 1996 there is only one observation in France in the databases.

6.4 Results

To quantify the effect of the various simulations, we calculated the residual between the simulated and the measured ozone concentrations. We have introduced the term Absolute Average Residual (*AAR*), defined as:

$$AAR = \frac{1}{N}\sum_{i=1}^{N} |c_i^{observed} - c_i^{simulated}|, \tag{37}$$

where N is the number of valid measurements in the time series and the simulated concentration can refer to the EUROS model alone or the KF results.

Figure 13 shows the Ensemble Kalman filter mean (KF mean) ozone concentration without (panel a) and with (panel b) assimilation. Also shown are the ozone concentrations from the EUROS model and from the observations. Without data assimilation (panel a), no observations are used in the simulation and the KF mean is simply the statistical mean of the 20 separate model calculations. The improvement in ozone concentration is clear in the case where observations are assimilated. The assimilated time series for this station are in better agreement with the observations than the ones calculated by the model alone. This improvement cannot only be observed at measurement sites the observations of which are used in the assimilation procedure (i.e., so-called assimilation stations); it also holds for validation stations the observations of which are not used in the assimilation. In panel c of Figure 13 the concentration for the validation station Hellendoorn is shown. A similar

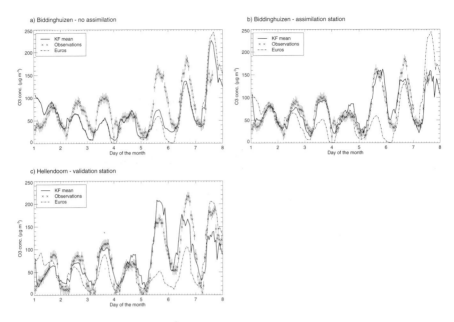

Fig. 13 Ozone concentrations (μgm^{-3}) at station Biddinghuizen (the Netherlands) without (a) and with (b)data assimilation for the Kalman filter mean (solid line) for June 1-7, 1996, calculated with ENKF with 20 modes. Ozone concentrations at validation station Hellendoorn (the Netherlands), of which the observations are not used in the assimilation process, are shown in panel (c) Data from the EUROS model (dash-dot) and the observations (crosses) are also shown. The vertical bar of the crosses represents the uncertainty in the observations given as input in the assimilation.

improvement of the concentration given by the Kalman filter mean in the direction of the measurements is shown. In Figure 14 the absolute average residuals (*AARs*), averaged over a month, are shown for all the validation stations in the Netherlands. The first bar represents a run without assimilation, the second bar the EUROS run, and the last one the KF mean. The difference between the first two bars is due to the stochastic noise that was implemented in the model and the slightly better performance of the first is a consequence of the averaging. In all the stations the improvement given by the KF mean is evident. Since an improvement at the assimilated stations is expected anyway, in most simulations the analysis was based on the data for the validation stations.

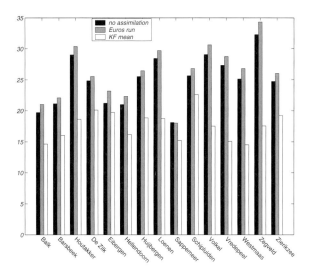

Fig. 14 AARs (μgm^{-3}) of an assimilation (ENKF, 20 modes) for all the validation stations in the Netherlands for the EUROS model calculation, the Kalman filter and the mean without assimilation. Simulation for June 1996.

6.5 Performance of RRSQRT algorithm, ENKF algorithm and COFFEE (Complementary Orthogonal subspace Filter For Efficient Ensembles) algorithm

A parameter that is important for the assimilation in Kalman filter algorithms is the number of modes used in assimilation. The AAR's values for all three algorithms are presented in Figure 15. The behavior of the algorithms with a real life large scale model is consistent with the specific theoretical characteristics of these algorithms discussed in Section 3, like the robustness of RRSQRT filter in comparison with the ENKF for a small number of modes and the convergence of both of them when the ensemble size (number of modes for RRSQRT) are increasing. This instability is shown in Figure 15, where the RRSQRT algorithm has higher AAR values for 10 modes than for 3 and 5 modes.

The following filter set-ups were examined to gain insight into the efficiency during the assimilation process: ENKF and RRSQRT filter with $N=5, 10, 20, 50, 100, 200$ ensemble members/modes and a COFFEE with $q = 5, 10, 20, 40$ modes and $N = 5, 10, 20, 40$ ensemble members. The number of model evaluations required comes to N for the ENKF, $1+q$ for the RRSQRT filter and $1+q+N$ for the COFFEE filter. The assimilation period was the month of July 2000. The results are

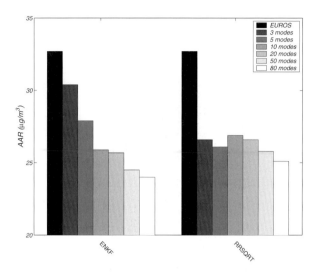

Fig. 15 Comparison of the performance of the ENKF algorithm and the RRSQRT algorithm using different numbers of modes/ensembles. The AARs ($\mu g m^{-3}$) are averaged over all the validation stations in Europe. Noise was applied to all five parameters

shown in Figure 16. The COFFEE results fall somewhere between the ENKF and RRSQRT results for less than about 30 model evaluations. For more than about 30 model evaluations, the COFFEE results are, in all cases, better than, or in a single case, equal to, the ENKF and RRSQRT results. The results of the COFFEE simulations with more than about 60 model evaluations turn out significantly better than all the ENKF and RRSQRT simulations, also those with 100 and 200 modes or ensemble members (not shown in the Figure).

7 Conclusions

In this Chapter we have introduced the main ideas of data assimilation. Using data assimilation techniques, measurements and model results are both used to obtain an optimal estimate of the state of the system. The two main directions for data assimilation have been presented: variational and sequential methods. The basic theoretical background was given and some new algorithms were introduced with their advantages and disadvantages: variational methods (adjoint based approach) and sequential methods (Ensemble Kalman filter(EnKF), Reduced Rank Square Root Kalman

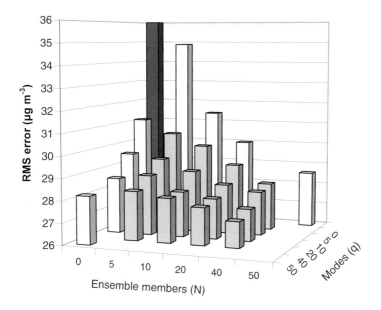

Fig. 16 RMS errors ($\mu g\ m^{-3}$), averaged over all validation stations in Europe, calculated with the ENKF, RRSQRT, and COFFEE filters for various numbers of modes (q) and ensemble members (N). The columns with $q = 0$ contain the ENKF results and the columns with $N = 0$, the RRSQRT results. The RMS error of the EUROS model calculation ($N = 0$ and $q = 0$) is 42.4 $\mu g\ m^{-3}$; it is truncated in the figure.

filter(RRSQRT), and Complementary Orthogonal subspace Filter For Efficient Ensembles(COFFEE).

We also presented a number of real life applications of these methods. Application in coastal sea modeling (storm surge prediction) and coastal ocean modeling (assimilation of high frequency radar data). The last application was in the field of large scale atmospheric chemistry transport models. For each new application one should decide first what kind of assimilation algorithm might work for that specific problem. So, it would be very helpful if one can have the possibility to easily switch between the data assimilation methods and find the one that it suits best for the specific application under study. One solution is presented in Section 4: the COSTA framework. It is a modular framework for data assimilation and model calibration. Within the COSTA framework it is possible to combine models with the available data assimilation and model calibration methods without the need of additional software development. The usage of COSTA significantly reduces the development costs for implementing data assimilation for simulation models.

References

Anderson, J.L., An ensemble adjustment Kalman filter for data assimilation, Mon. Weather. Rev., Vol. 129, 2001, 2884-2903

Bennett, A.F., Inverse modeling of the ocean and atmosphere, Cambridge University Press, UK, 2002.

Bishop, C.H., and Etherton, B.J., and Majumdar, S.J., Adaptive sampling with the ensemble transform Kalman filter Part I: Theoretical aspects, Monthly Weather Review, Vol. 129, 2001, 420-436.

Booij, N., R.C. Ris, L.H. Holthuijsen, A third-generation wave model for coastal regions: 1. model description and validation, J. of Geoph. Research, Vol. 104, 1999, 7649-7666.

Byrd, R.H., J. Nocedal, R.B. Schnabel, Representations of quasi-Newton matrices and their use in limited memory methods, Mathematical Programming, Vol. 63, 1994, 129-156.

EC, Council directive 96/62/ec of 27 September 1996 on ambient air quality assessment and management from EU air quality framework directive, Off. J. Eur. Commun., L, Legis, 1996, 296:55-63.

Evensen G, Sequential data assimilation with a nonlinear quasi-geostrophic model using Monte Carlo methods to forecast error statistics, JGR, 99(c5), 1994, 10143-10162.

Evensen, G., The Ensemble Kalman Filter: theoretical formulation and practical implementation, Ocean Dynamics, Vol. 53, 2003, 343-367.

Gerritsen, H., de Vries, J., Philippart, M., The Dutch Continental Shelf model, in: Quantitative Skill Assessment for Coastal Ocean models, D. Lynch and A. Davies (eds.), Coastal and Estuarine Studies, Vol 47, 1995, pp. 425-267.

Fletcher, R., C.M. Reeves, Function minimization by conjugate gradients, Computer Journal, Vol. 7, 1964, 149-154.

Hanea, R. G., G. J. M. Velders, A. Heemink, Data assimilation of ground-level ozone in Europe with a Kalman filter and chemistry transport model, J. Geophys, Vol. 109, 2004, D10302, doi:10.1029/2003JD004283.

Heemink, A. W., H. Kloosterhuis, Data assimilation for non-linear tidal models. –International Journal for Numerical Methods in Fluids, Vol. 11, 1990, 1097-1112.

Heemink, AW, and Metzelaar IDM, Data assimilation into a numerical shallow water flow model: a stochastic optimal control approach, J. Mar. Sys, Vol. 6, 1995, 145-158.

Heemink, A. W., M. Verlaan, A. J. Segers, Variance reduced Ensemble Kalman filtering, Mon. Weather Rev., Vol. 129, 2001, 1718-1728.

Kaminski T, et al., An example of an automatic differentiation-based modelling system, Lecture Notes in Computer Science 2668, 2003, 95-104.

Lelieveld, J., H. Berresheim, S. Borrmann, P.J. Crutzen, F.J. Dentener, H. Fischer, J. Feichter, P.J. Flatau, J. Heland, R. Holzinger, R. Korrmann, M.G. Lawrence, Z. Levin, K.M. Markowicz, N. Mihalopoulos, A. Minikin, V. Ramanathan, M. de Reus, G.J. Roelofs, H.A. Scheeren, J. Sciare, H. Schlager, M. Schultz, P. Siegmund, B. Steil, E.G. Stephanou, P. Stier, M. Traub, C. Warneke, J. Williams, and H. Ziereis, Global air pollution crossroads over the Mediterranean. Science, Vol. 201, 2002, 794-799.

Lermusiaux, P.F.J., and Robinson, A.R., Data assimilation via error subspace statistical estimation. Part I: Theory and schemes, Monthly Weather Review, Vol. 127, 1998, 1385-1407.

Lindström G., B. Johansson, M. Persson, M. Gardelin, S. Bergström, Development and test of the distributed HBV-96 hydrological model, Journal of Hydrology, Vol. 201, 1997, 272-288.

Matthijsen, J., L. Delobbe, F. Sauter, and L. de Waal, Changes of surface ozone over Europe upon the Gothenburg protocol abatement of 1990 reference emissions, Springer-Verlag, New York, 2001, 1384-1388.

Mouthaan, E. E. A., A. W. Heemink, K. B. Robaczewska, Assimilation of ERS-1 altimeter data in a tidal model of the continental shelf, Deutsche Hydrographische Zeitschrift, Vol. 46, 285-319.

Nelder, J. A., R. Mead, A Simplex Method for Function Minimization, Computer Journal, Vol. 7, 1965, 308-313.

Nocedal, J., Updating quasi-Newton matrices with limited storage, Mathematics of Computation, Vol. 35, 1980, 773-782.

Pham, D. T. and J. Verron, and C.M. Roubaud, A singular evolutive extended Kalman filter for data assimilation in oceanography, Journal of Marine Systems, Vol. 16, 1998, 323-340.

Pul, W.A.J. van, J.A. van Jaarsveld, and C.M.J. Jacobs, Deposition of persistent organic pollutants over Europe. Air Pollution modelling and its application XI 1995, Baltimore, 1996.

Ravindran SS, Control of flow separation over a forward-facing step by model reduction, Comp. Meth. in Applied Mechanics and Eng., Vol. 191, 2002, 4599-4617.

Reggiani, P., J. Schellekens, Invited Commentary: Modelling of hydro-logic responses: The Representative Elementary Watershed (REW) approach as an alternative blueprint for watershed modelling, Hydrological Processes, Vol. 17, 2004, 3785-3789, DOI 10.1002/hyp.5167.

Rheineck Leyssius, H.J. van, F.A.A.M. de Leeuw, and B.H. Kessenboom, A regional scale model for the calculation of episodic concentrations and depositions of acidifying components, Water, Air and Soil Pollution, Vol. 51, 1990, 327–344.

Schaap, M., H. A. C. Van Der Gon, F. J. Dentener, A. J. H. Visschedijk, M. Van Loon, H. M. ten Brink, J.-P. Putaud, B. Guillaume, C. Liousse, P. J. H. Builtjes, Anthropogenic black carbon and fine aerosol distribution over Europe, J. Geophys. Res., Vol. 109, D18207, doi:10.1029/2003JD004330.

Segers, A.J., Data assimilation in atmospheric chemistry models using Kalman filtering, PhD thesis, Delft University of Technology, Delft, Netherlands, 2002.

Stelling, G. S., On the construction of computational methods for shallow water flow problems. – Ph.D. thesis, Delft University of Technology, Rijkswaterstaat communications no. 35., 1984.

Tippett M.K., Anderson J.L., Bishop C.H., et al., Ensemble square root filters Monthly Weather Rev., Volume 131, 2003, 1485-1490

Velzen, N. van, 2006: COSTA a Problem Solving Environment for Data Assimilation, Paper presented at CMWR XVI -Computational Methods in Water Resources, Copenhagen, Denmark, 2006.

Velzen, N, van, M. Verlaan, COSTA a problem solving environment for data assimilation applied for hydrodynamical modelling, Meteorologische Zeitschrift, Vol. 16, No. 6, 2007, 777-793, DOI 10.1127/0941-2948/2007/0241.

Verlaan, M., Efficient Kalman Filtering Algorithms for Hydrodynamic Models, Ph.D. Thesis, Delft University of Technology, Netherlands, 1998.

Verlaan, M., Heemink A.W., Tidal flow forecasting using Reduced-Rank Square Root filters, Stoch. Hydrology and Hydraulics, Vol. 11, 1997, 349-368.

Verlaan M., Heemink A.W., Nonlinearity in data assimilation applications: A practical method for analysis, Mon. Weather Rev., Vol. 129, 2001, 1578-1589.

Verlaan, M., E. E. A. Mouthaan, E. V. L. Kuijper, M.E. Philippart, Parameter estimation tools for shallow water flow models, Hydroinformatics, Vol. 96, 1996, 341-347.

Vermeulen P.T.M., Heemink A.W., Model-reduced variational data assimilation, Mon. Weather Rev., Vol. 134 (10), 2006, 2888-2899.

Vermeulen P.T.M., Heemink A.W., Stroet C.B.M.T. te, Reduced models for linear groundwater flow models using EOFs, Adv. in Water Res. 27 (1), 2004, 57-69.

Whitaker, J. S., T. M. Hamill, 2002: Ensemble data assimilation without perturbed observations, Mon. Wea. Rev., Vol. 130, 2002, 1913-1924.

Radial Basis Functions for Interface Interpolation and Mesh Deformation

A. de Boer, A.H. van Zuijlen, and H. Bijl

1 Introduction

Many engineering applications involve fluid-structure interaction (FSI) phenomena. For instance light-weight airplanes, long span suspension bridges and modern wind turbines are susceptible to dynamic instability due to aeroelastic effects. FSI simulations are crucial for an efficient and safe design. Computers and numerical algorithms have significantly advanced over the last decade, such that the simulation of these problems has become feasible.

In FSI computations it is required that pressure loads are transmitted from the fluid side of the fluid-structure interface to the structural nodes on that interface. Once the motion of the structure has been determined, the motion of the fluid mesh points on the interface has to be imposed. In FSI simulations it is usually not desirable to generate matching meshes at the fluid-structure interface, because different solvers may take care of the different physical domains. In addition, also the flow generally requires a much finer mesh than the structure. This means that the discrete interface between the domains may not only be non-conforming, but there can also be gaps and/or overlaps between the meshes. The exchange of data over the discrete interface becomes then far from trivial. In Section 2 we introduce a method for data transfer at the interface using radial basis functions and consider conservation and consistency properties.

A. de Boer
Deltares, P.O. Box 177, 2600 MH, Delft, The Netherlands, e-mail: aukje.spruyt@deltares.nl

A.H. van Zuijlen
Delft University of Technology, P.O. Box 5058, 2600 GB, Delft, The Netherlands, e-mail: a.h.vanzuijlen@tudelft.nl

H. Bijl
Delft University of Technology, P.O. Box 5058, 2600 GB, Delft, The Netherlands, e-mail: h.bijl@tudelft.nl

Since the fluid domain is deforming in FSI computations, a mesh deformation algorithm is required. To be able to perform the unsteady flow computations accurately and efficiently, a fast and reliable method is needed to adapt the computational grid to the new domain. Regenerating a grid each time step in an unsteady computation is a natural choice. However, the generation of a complex grid is a time-consuming and nontrivial task. Therefore, a fast and accurate algorithm is needed to update the grid automatically. In Section 3 we introduce a mesh deformation method based on radial basis function interpolation, which can automatically deform a mesh with high accuracy even for large deformations of the mesh.

2 Non-matching meshes

In FSI computations data has to be transferred over an interface of generally non-matching meshes. In Figure 1 a 2D example of a non-matching discrete interface between a flow and structure domain is shown. When the meshes are non-matching, an interpolation/projection step has to be carried out to enable transfer of information between the two domains. In literature different methods can be found to transfer data between non-matching meshes, such as nearest neighbour interpolation [34], projection methods [13, 27, 29] and methods based on interpolation by splines [3, 31, 32]. In this chapter we only focus on radial basis function interpolation methods.

The general opinion is that energy should be conserved over the interface. The overall conservation properties depend both on the time and the spatial coupling used, which cannot be investigated separately if the system is solved in a partitioned way. By sub-iterating the partitioned scheme the partitioning error in time can be reduced [18]. High order accurate convergence in time can be obtained without a need for sub-iterating the partitioned algorithm by using mixed implicit-explicit higher order schemes [41, 42, 40], which are an extension of the multi-stage Runge-Kutta schemes already applied in computational fluid dynamics simulations [4, 5]. In combination with multi-level techniques the efficiency can be increased even more [44].

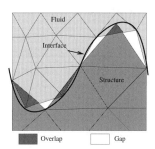

Fig. 1 Non-matching meshes in 2D.

In this section we focus only on the spatial coupling. In [17] a conservative coupling approach in space is introduced. This approach is based on the global conservation of virtual work over the interface, where a transformation matrix performs the transfer of displacements and the transposed of this matrix the transfer of pressure loads between the two discrete interfaces. However, for a general coupling method this can lead to unphysical oscillations in the pressure forces received by the structure as is briefly mentioned by *Ahrem et al* [1]. Especially for highly flexible structures (like airbags) this can have a large negative influence on the accuracy of the solution.

Instead of using the same transformation matrix for both transferring the displacement and pressure loads over the interface, two different transformation matrices can be defined, which we will define as the consistent approach. This leads to a coupling approach without unphysical oscillations in the pressure forces. However, conservation of energy over the interface is not guaranteed. When a partitioned coupling technique is used to advance in time this does not have to be a problem. In unsteady partitioned computations energy is generally already not conserved due to errors caused by the time lag between flow and structure. When the coupling error introduced by the information transfer is smaller than the spatial and temporal discretization error, this coupling error does not have to affect the stability and accuracy of the computation, especially when the spatial and time discretization themselves are very dissipative. However, when stability must be ensured, the variation of energy caused by the non-matching interface should be negative.

In this chapter we investigate the difference in accuracy and efficiency between the conservative and consistent approach for the radial basis function methods described in [10]. First the two approaches are presented followed by a short description of the coupling method. The difference in the interpolation properties between the two approaches is investigated using two analytical test problems. A simple steady quasi-1D FSI problem is used to investigate the performance of the methods in FSI computations with multiple transfers between flow and structure side.

2.1 Conservative and consistent coupling approach

In this section the conservative and consistent coupling approach are presented. With the conservative approach the total energy is conserved when transferring displacement and pressure forces over the interface. The consistent approach ensures that a constant displacement and a constant pressure are exactly interpolated over the interface. The starting point of both approaches are the kinematic and dynamic boundary conditions at the interface, which are commonly used to couple the fluid and structure equation, and are given by

$$\mathbf{u}_f = \mathbf{u}_s \quad \text{on} \quad \Gamma, \tag{1a}$$
$$p_s \mathbf{n}_s = p_f \mathbf{n}_f \quad \text{on} \quad \Gamma, \tag{1b}$$

with $\mathbf{u}_{f,s}$ the displacement, $p_{f,s}$ the pressure or stress tensor and $\mathbf{n}_{f,s}$ the outward normal of the flow and structure interface, respectively. The continuous interface between the flow and structure is represented by Γ. The first of these two boundary equations, (1a), expresses the compatibility between the displacement fields of the structure and the fluid at the fluid-structure interface. The second equation, (1b), states that the tractions of the wet surface of the structure are in equilibrium with those on the fluid side. In the continuous formulation and for steady state problems either displacement, velocity or acceleration can be used for the kinematic boundary condition (1a). However, for dynamic problems this changes the stability properties [19].

Whichever coupling method is chosen to define the discrete form of these conditions, its outcome can be formulated as

$$\mathbf{U}_f = H_{fs}\mathbf{U}_s, \qquad (2a)$$
$$\mathbf{P}_s = H_{sf}\mathbf{P}_f, \qquad (2b)$$

with the $(n_f^u \times n_s^u)$ matrix H_{fs} and $(n_s^p \times n_f^p)$ matrix H_{sf} transformation matrices between the flow and structure interface, where $n^{u,p}$ is the number of unknowns on the interface for the displacement and pressure, respectively and the subscript f or s denotes whether this is on the flow or the structure side. The discrete values in the interface points of the displacement and pressure are contained in $\mathbf{U}$ and $\mathbf{P}$, respectively. They are defined by the approximations

$$\mathbf{u}(\mathbf{x}) = \sum_{j=1}^{n^u} N^j(\mathbf{x})\mathbf{U}_j, \qquad p(\mathbf{x})\mathbf{n}(\mathbf{x}) = \sum_{i=1}^{n^p} D^i(\mathbf{x})\mathbf{P}_i, \qquad (3)$$

where $N(\mathbf{x})$ is a function that depends on the spatial discretization method used for the displacement (for example, a step function in the finite volume formulation or the basis function in the finite element formulation) and $D(\mathbf{x})$ is a function that depends on the discretization method used for the pressure. When the row-sum of H is equal to one, constant values are interpolated exactly. In the following analysis we only consider the usual case that $n_f = n_f^u = n_f^p$ and $n_s = n_s^u = n_s^p$.

2.1.1 Conservative approach

The general opinion is that energy should be conserved over the interface leading to a conservative coupling approach [17]. In this paper we focus only on the spatial coupling and therefore we look at the limit of very small time steps (virtual displacements, or steady state solution). In this case energy is globally conserved over the interface when

$$\int_{\Gamma_f} \mathbf{u}_f \cdot p_f \mathbf{n}_f \, ds = \int_{\Gamma_s} \mathbf{u}_s \cdot p_s \mathbf{n}_s \, ds, \qquad (4)$$

with $\mathbf{u}$ the displacement of the interface.

Writing out the left hand side of (4) using (3) gives

$$\int_{\Gamma_f} \mathbf{u}_f \cdot p_f \mathbf{n}_f \, ds = \sum_{i=1}^{n_f} \left[\sum_{j=1}^{n_f} \mathbf{U}_{fj}^T \left(\int_{\Gamma_f} N_f^j D_f^i \, ds \right) \right] \mathbf{P}_{fi} = \mathbf{U}_f^T M_{ff} \mathbf{P}_f. \qquad (5)$$

In a similar way we find for the right hand side of (4)

$$\int_{\Gamma_s} \mathbf{u}_s \cdot p_s \mathbf{n}_s \, ds = \mathbf{U}_s^T M_{ss} \mathbf{P}_s, \tag{6}$$

where matrices M_{ff} ($n_f \times n_f$) and M_{ss} ($n_s \times n_s$) are defined as follows

$$M_{ff}^{ij} = \int_{\Gamma_f} N_f^i D_f^j \, ds, \quad M_{ss}^{ij} = \int_{\Gamma_s} N_s^i D_s^j \, ds. \tag{7}$$

Substituting (5) and (6) into (4) shows that energy is globally conserved when

$$\mathbf{U}_f^T M_{ff} \mathbf{P}_f = \mathbf{U}_s^T M_{ss} \mathbf{P}_s \quad \Rightarrow \quad \mathbf{U}_s^T H_{fs}^T M_{ff} \mathbf{P}_f = \mathbf{U}_s^T M_{ss} \mathbf{P}_s, \quad \forall \mathbf{U}_s$$
$$\Rightarrow M_{ss} \mathbf{P}_s = H_{fs}^T M_{ff} \mathbf{P}_f. \tag{8}$$

So choosing

$$H_{sf} = M_{ss}^{-1} H_{fs}^T M_{ff} \tag{9}$$

in (2b) for the transformation of pressure over the interface results in global conservation of energy over the interface.

Note that this is, not surprisingly, the same result as obtained in [17] for the forces. If we define the forces $\mathbf{F}_s = M_{ss}^T \mathbf{P}_s$ and $\mathbf{F}_f = M_{ff}^T \mathbf{P}_f$ we can rewrite (8) as $\mathbf{F}_s = H_{fs}^T \mathbf{F}_f$ and the transposed transformation matrix has to be used to exchange forces to obtain global conservation of energy over the interface.

2.1.2 Consistent approach

In order to obtain a consistent interpolation, a constant displacement and constant pressure should be exactly interpolated over the interface (similar to the patch test criterion in domain decomposition methods [20]). This means that in the conservative approach the row-sum of both H_{fs} and $H_{sf} = \left[M_{ff} H_{fs} M_{ss}^{-1} \right]^T$ should be equal to one. For a general transformation matrix H_{fs} this is not the case as we will see in the following section.

To ensure a consistent interpolation of the pressure the matrix H_{sf} can be directly obtained by using one of the coupling methods, as is done to obtain matrix H_{fs}. In this way H_{sf} is independent of H_{fs} and both can be created with a row-sum equal to one. However in this way global conservation of energy over the interface is not guaranteed and in the remainder of the paper we will address this as the consistent approach. The main question is whether global conservation of energy or a consistent interpolation is preferred in fluid-structure interaction computations.

2.2 Radial basis function interpolation (RBFI)

Interpolation with radial basis functions (RBF's) has become a very powerful tool in multivariate approximation theory through scattered data, because of its excellent approximation properties [11]. Radial basis functions have been successfully applied to areas as diverse as computer graphics [12], geophysics [7, 6], mesh deformation [9], error estimation [24] and the numerical solution of partial differential equations [22, 23]. They can also be used to interpolate between non-matching meshes in FSI computations [3, 31, 32]. The quantity to be transferred from mesh A to mesh B is approximated by a global interpolation function which is a sum of basis functions

$$\mathbf{w}(\mathbf{x}) = \sum_{j=1}^{n_A} \gamma_j \phi(||\mathbf{x} - \mathbf{x}_{A_j}||) + q(\mathbf{x}), \qquad \mathbf{w} = \{\mathbf{u}, p\mathbf{n}\}, \qquad (10)$$

where $\mathbf{x}_{A_j}$ are the centres in which the values are known, in this case the nodes at the interface of mesh A, q a polynomial, and ϕ a given radial basis function with respect to the Euclidean distance $||\mathbf{x}||$. The coefficients γ_j and the polynomial q are determined by the interpolation conditions

$$\mathbf{w}(\mathbf{x}_{A_j}) = \mathbf{W}_{A_j}, \qquad (11)$$

with $\mathbf{W}_A$ containing the known discrete values of $\mathbf{w}$ at the interface of mesh A, and the additional requirements

$$\sum_{j=1}^{n_A} \gamma_j s(\mathbf{x}_{A_j}) = 0, \qquad (12)$$

for all polynomials s with a degree less than or equal to that of polynomial q. The minimal degree of polynomial q depends on the choice of the basis function ϕ. A unique interpolant is given when the basis function is a conditionally positive definite function (Definition 3.1 of [3]). If the basis functions are conditionally positive definite of order $m \leq 2$, as is the case for the functions used in this paper, a linear polynomial can be used [3]. A consequence of using a linear polynomial is that constant values are exactly interpolated leading to a consistent interpolation. The interpolation function (10) is defined in the whole domain in contrast to for example [16], where spline-like polynomials are used for the Lagrange multiplier to glue together nonconforming meshes. The spline-function is then only defined on the non-conforming interface.

The interpolation conditions (11) and (12) can be written in matrix form as follows

$$\begin{bmatrix} \mathbf{W}_A \\ 0 \end{bmatrix} = \begin{bmatrix} \Phi_{AA} & Q_A \\ Q_A^T & 0 \end{bmatrix} \begin{bmatrix} \gamma \\ \beta \end{bmatrix}, \qquad (13)$$

with γ containing the coefficients γ_j, β the coefficients of the linear polynomial q, Φ_{AA} an $n_A \times n_A$ matrix containing the evaluation of the basis function $\phi_{A_iA_j} = \phi(||\mathbf{x}_{A_i} - \mathbf{x}_{A_j}||)$. The matrix Q_A is an $n_A \times (d+1)$ matrix with row j given by $[1 \; x_{A_1} \; x_{A_2} \; \cdots \; x_{A_d}]$ and d the dimension of the problem.

In order to obtain the values for the unknown quantity at the interface of mesh B we have to evaluate (10) in the nodes on the interface of mesh B which can be written in matrix form as

$$\mathbf{W}_B = \begin{bmatrix} \Phi_{BA} & Q_B \end{bmatrix} \begin{bmatrix} \gamma \\ \beta \end{bmatrix}. \tag{14}$$

Combining (13) and (14) gives the relation

$$\mathbf{W}_B = \underbrace{\begin{bmatrix} \Phi_{AB} & Q_B \end{bmatrix} \begin{bmatrix} \Phi_{AA} & Q_A \\ Q_A^T & 0 \end{bmatrix}^{-1}}_{\tilde{H}} \begin{bmatrix} \mathbf{W}_A \\ 0 \end{bmatrix}. \tag{15}$$

We now can define the transformation matrix H_{BA} as the first n_B rows and n_A columns of matrix $\tilde{H}$ to obtain $\mathbf{W}_B = H_{BA}\mathbf{W}_A$. Contrary to the weighted residual method and nearest neighbour method, no orthogonal projection and search algorithm is needed to obtain H_{BA}. This is because the radial basis functions are defined in all space and not only on the interface. The computation of H_{BA} only involves the inversion of a relatively small matrix. The number of rows and columns of this matrix is equal to the number of flow or structure points on the fluid-structure interface, which is usually very small compared to the total number of structure and flow points. However, this is a full matrix when the radial basis function does not have compact support. In practice matrix H_{BA} is not computed explicitly, because we are only interested in the value of $\mathbf{W}_B$ which can be obtained by solving system (13) and then evaluating the matrix vector product (14).

2.2.1 Radial basis functions

RBFs can be divided into two groups, functions with compact support and functions with global support. Beckert and Wendland [3] use for their FSI computations compactly supported radial basis functions based on polynomials to interpolate between non-matching meshes. Their C^2 radial basis function, which is defined as

$$\phi(||\bar{\mathbf{x}}||) = (1 - ||\bar{\mathbf{x}}||/r)_+^4 \, (4||\bar{\mathbf{x}}||/r + 1), \tag{16}$$

gives the best result. The subscript $+$ means that only positive values are taken into account (this function is in the remainder of the paper abbreviated by CP). The distance between two points is normalized with the largest distance between two points, so $||\bar{\mathbf{x}}|| = ||\mathbf{x}||/||\mathbf{x}||_{\max}$. The radius r defines the support of the radial basis function. A large support radius yields a good approximation order, but then a full matrix system has to be solved. Moreover, too large radii lead to nearly singular matrices, because then all the entries of Φ_{AA} are approximately equal to one. A small support radius leads to a well conditioned system with a band matrix that can be easily solved, but the interpolation is less accurate than with a large support radius. For an accurate computation the support radius for a fluid-structure interaction problem should be chosen at least as large as the normalized distance between the centre which is most far from its neighbours and its nearest neighbour. This nearest neighbour can be in either of the two meshes.

Several global radial basis functions have been tested and evaluated for analytical interpolation tests as well as real fluid-structure interaction computations by Smith et al. [31, 32]. From this work the following two functions are shown to be the most robust, cost effective and accurate of the methods tested:

- Multi-quadric Biharmonic spline (MQ)

$$\phi(||\bar{\mathbf{x}}||) = \sqrt{||\bar{\mathbf{x}}||^2 + a^2}. \qquad (17)$$

- Thin-plate spline (TPS)

$$\phi(||\bar{\mathbf{x}}||) = ||\bar{\mathbf{x}}||^2 \ln ||\bar{\mathbf{x}}||. \qquad (18)$$

Both functions do not vanish when $||\bar{\mathbf{x}}||$ goes to infinity as is the case for the compact supported basis function. The MQ-method uses a parameter a that controls the shape of the basis functions. A large value of a gives a flat sheetlike function, while a small value of a gives a narrow cone-like function. In literature it is still an open question how to find the optimal value of a. Smith et al. choose a typically in the range $10^{-5} - 10^{-3}$ when a domain of size 1 is used [31, 32]. In this paper we use the value $a = 10^{-3}$. In contrast with the radial basis functions used by Beckert and Wendland, these two functions are defined on the entire domain. As a result, always a full matrix system has to be solved.

2.2.2 Conservative approach

Due to the addition of the linear polynomial, constant values are exactly recovered, and therefore the interpolation is consistent. However, when the conservative transfer approach is used, assuming that RBF interpolation is used for the displacement, the interpolation is not consistent for the transformation of pressure values (being understood that we take the usual assumption that the interpolation is performed for the displacements). The reason for this is shown below for positive definite functions (as for example the MQ). For positive definite functions the addition of the linear polynomial is not necessary to make the system uniquely solvable. In this case we can write, according to equation (15), for the transformation of displacements:

$$\mathbf{U}_f = \Phi_{fs}\Phi_{ss}^{-1}\mathbf{U}_s. \qquad (19)$$

According to (8) the following requirement for the pressure must hold with the conservative approach

$$M_{ss}\mathbf{P}_s = \Phi_{ss}^{-1}\Phi_{fs}^T M_{ff}\mathbf{P}_f. \tag{20}$$

We replace the pressure vectors $\mathbf{P}_s$ and $\mathbf{P}_f$ with the vectors with all constant components β_s and β_f, respectively, and obtain

$$M_{ss}\beta_s = \Phi_{ss}^{-1}\Phi_{sf} M_{ff}\beta_f. \tag{21}$$

If non-equidistant grids or higher order basis functions are involved in the flow domain, the pressure force on the flow side, $M_{ff}\beta_f$, can be highly oscillatory. The multiplication with $\Phi_{ss}^{-1}\Phi_{sf}$ is a smoothing operation, leading to an overall smooth right hand side. However, the pressure force on the structure side, $M_{ss}\beta_s$, can also be highly oscillatory if non-equidistant grids or higher order basis functions are used in the structure domain. In this case equation (21) does not hold.

If, for example, third order basis functions are used on a 1D equidistant grid, as is the case for the test cases in this thesis, the pressure force on the flow side $M_{ff}\beta_f$ alternates with the values $\frac{2}{3}\beta\Delta x_f$ and $\frac{4}{3}\beta\Delta x_f$, where Δx_f is the grid size on the flow side. The multiplication with $\Phi_{ss}^{-1}\Phi_{sf}$ results in a vector with all constant values $\beta\Delta x_s$, with Δx_s the grid size on the structure side. However, for a constant pressure the pressure force on the structure side, $M_{ss}\beta_s$, should be alternating with the values $\frac{2}{3}\beta\Delta x_s$ and $\frac{4}{3}\beta\Delta x_s$. This means that the error in pressure forces $\varepsilon_F = M_{ss}\beta_s - \Phi_{ss}^{-1}\Phi_{sf}M_{ff}\beta_f$ is alternating with values $\pm\frac{1}{3}\beta\Delta x_s$. The amplitude of this error does decrease when the structure grid is refined, but the error in pressure itself $\varepsilon_p = \beta_s - M_{ss}^{-1}\Phi_{ss}^{-1}\Phi_{sf}M_{ff}\beta_f$ is alternating with values $\pm\frac{1}{3}\beta$ and its amplitude does not decrease by refining the flow and structure grid simultaneously, only the frequency increases. The error ε_p only converges if the meshes converge to a mesh that is both matching and conforming at the interface, whereby we mean with conforming that also the underlying discretization of the flow and structure mesh must be the same.

The addition of the linear polynomial for conditionally positive definite radial basis functions does not solve this problem as is shown in sections 2.3 and 2.4. The conclusion on the convergence of ε_p actually holds for any transfer algorithm that uses an interpolation scheme and does not incorporate information of the underlying basis functions.

2.3 Analytical test problems

In this section the different transfer methods are compared for a smooth and non-smooth analytical problem, to be able to investigate their general interpolation properties. For all the methods both the conservative and consistent approach are employed.

The tests consist of a single transfer of a displacement and pressure field over an interface. This interface has the form $q_e = 0.2\sin(2\pi x)$, with $x \in [-0.5, 0.5]$. The profile of the pressure and displacement fields is either smooth or non-smooth. The procedure of the tests is as follows:

1. Start with the continuous form of the displacement and pressure field at the flow or structure side of the interface, respectively.
2. Discretize the continuous fields using a third order order finite element method. Because the number of cells that are used to discretize the interface differ between the flow and structure side, the interface becomes non-matching.
3. Transfer the discretized displacement field from the discrete structure side of the interface to the discrete flow side using one of the transfer methods.
4. Compare the obtained results at the discrete flow interface to the exact values of the displacement field at this interface by looking at the relative transfer error.
5. Transfer the discretized pressure field from the discrete flow side of the interface to the discrete structure side using the conservative or consistent approach.
6. Compare the obtained results at the discrete structure interface to the exact values of the pressure field at this interface by looking at the relative transfer error.

The relative L_2 transfer error is defined as

$$\varepsilon = \sqrt{\frac{\sum_{i=1}^{n_\alpha} ||\mathbf{w}_{ex}^i - \mathbf{w}_\alpha^i||^2}{\sum_{i=1}^{n_\alpha} ||\mathbf{w}_{ex}^i||^2}}, \tag{22}$$

where $\mathbf{w}_\alpha^i$ is the vector containing the values received at the flow ($\alpha = f$) or structure ($\alpha = s$) side using one of the transfer methods and $\mathbf{w}_{ex}^i$ the vector with the exact values on that side of the interface.

By discretizing the displacement and pressure fields, already an error is made with respect to the continuous field. The relative L_2 discretization error of a continuous function $w_{ex}(x)$ on the flow ($\alpha = f$) or the structure ($\alpha = s$) interface is defined as

$$\varepsilon_{disc} = \sqrt{\frac{\int_{\Gamma_\alpha} ||w_{ex}(x) - \sum_{i=1}^{n_\alpha} N_\alpha^i(x) \mathbf{w}_{ex}^i||^2 \, dx}{\int_{\Gamma_\alpha} ||w_{ex}(x)||^2 \, dx}}, \tag{23}$$

where $\sum_{i=1}^{n_\alpha} N_\alpha^i(x) \mathbf{w}_{ex}^i$ is the discretized form of $w_{ex}(x)$ using the basis functions of the flow, N_f, or structure, N_s. The integrals are computed with Gauss integration. As long as the transfer error is smaller than the spatial discretization error, the transfer error does not affect the spatial discretization order of the total system.

The applications we are interested in (wing flutter, deforming wind turbine blades) typically have 5 till 10 times more cells on the flow interface than on the structure interface. Therefore we use $n_f = 26 \cdot 2^k$ flow cells and $n_s = 5 \cdot 2^k$ structure cells, with $k \in \{0, 1, 2, 3, 4, 5\}$, leading to a ratio of about 20%.

The transfer error in the displacement field received by the discrete flow side and in the pressure field received by the discrete structure side for both the conservative and consistent approach is investigated for the projection of a smooth and a non-smooth field in the following two sections.

2.3.1 Transferring a smooth field

In this section a smooth displacement field is transferred from the structure to the flow interface and a smooth pressure field from flow to structure and compared to the exact values of the field. Both fields have the form $w(x) = 0.01\cos(2\pi x)$.

The L_2-error of the displacement (22) in the flow points versus the number of structure points after one interpolation step is depicted in Figure 2 for the RBFI methods (CP ($r = 0.25$ and $r = 2$), TPS and MQ). Note that the CP method is actually no longer compactly supported with $r = 2$, because then all basis functions cover the whole domain which has length one.

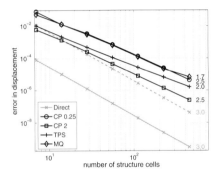
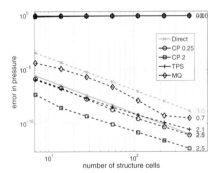

Fig. 2 Error in displacement - RBFI methods.

Fig. 3 Error in pressure - RBFI methods (—: conservative – –: consistent).

The number next to a line represents the order of the method represented by the line. The gray lines (Direct) represent the discretization error (23); the solid gray line is the discretization error on the flow interface and the dashed line on the structure interface. Above these lines the coupling error of a method is higher than the discretization error.

The CP method has an order of about 2.5, but the accuracy depends on the value of the radius: the larger the radius r, the more accurate the method. The MQ and TPS method are approximately second order accurate where the accuracy of the TPS method is higher. With $r = 2$, CP is more accurate than TPS and with $r = 0.25$ it is comparable to MQ.

The interpolation of the displacement is the same for both the conservative and consistent approach. The difference is in the way the pressure is interpolated. The L_2-error of the pressure versus the number of points on the structure interface after one interpolation step is depicted in Figure 3. The solid lines are obtained with the conservative and the dashed lines with the consistent approach. It can be seen that, as we expected, the interpolation methods do not converge when the conservative approach is used. When the consistent approach is used, the interpolation error is smaller than the discretization error of the structure. This is due to the fact that the pressure is transferred from the finer flow grid to the coarser structure grid.

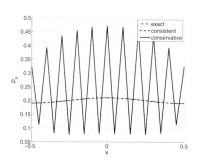

Fig. 4 Pressure received by the structure obtained with the CP method with $r = 2$ for $n_f = 52$ and $n_s = 10$.

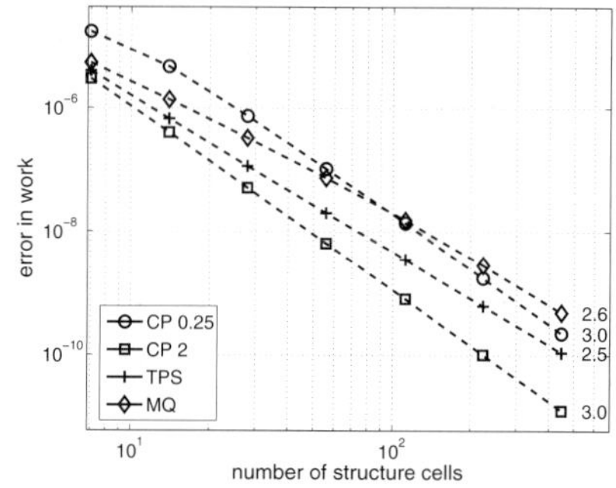

Fig. 5 Difference in work for the consistent approach - RBFI methods.

The reason for the lower order of convergence for the conservative approach can be seen in Figure 4, where the exact pressure together with the pressures obtained with the conservative and consistent approach are shown for CP with $r = 2$. The difference between the exact solution and the one obtained with the consistent approach is barely visible, but large oscillations are visible in the solution obtained with the conservative approach. The amplitude of these wiggles does not decrease when the meshes are refined simultaneously (see section 2.2.2), leading to the zeroth order convergence. The coupling error only decreases when the meshes become conforming, which is not the case when both meshes are refined simultaneously.

In Figure 5 the difference in work exerted on the interface between the flow and structure side is depicted. When energy is conserved over the interface, this difference should be zero, as is by construction the case for the conservative approach (and therefore not shown in the graphs). The error converges with approximately one order higher than that of the error in displacement and pressure. So, even as the consistent approach is not strictly globally conservative for the energy transfer over the interface, the error decreases consistently with the discretization error.

2.3.2 Transferring a non-smooth field

In this section the same calculations are performed as in the previous section, but this time a hat-shape function is transferred instead of a cosine, so

$$w(x) = \begin{cases} 0.01\,(2 - |x|/a) & |x| < a, \\ 0.01 & |x| \geq a. \end{cases} \qquad (24)$$

The value of a is chosen in such a way that the two outer discontinuities are located exactly at a grid point. For the displacement which is defined on the structure interface we use $a = 0.5 - 2/5$ and for the pressure which is located on the flow interface $a = 0.5 - 9/26$. The discontinuity in the centre is always located exactly at a grid point.

The L_2-error of the displacement (22) in the flow points versus the number of structure points after one interpolation step is depicted in Figure 6. The results for the different RBFI methods are all very similar. The discretization error on the

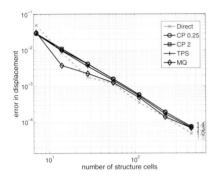

Fig. 6 Error in displacement.

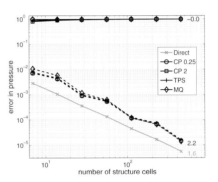

Fig. 7 Error in pressure ($-$: conservative $--$: consistent).

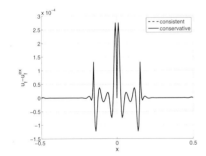

Fig. 8 Error in displacement received by the flow obtained with the CP method with $r = 2$ for $n_f = 104$ and $n_s = 20$.

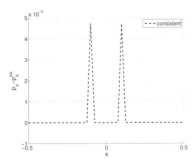

Fig. 9 Error in pressure received by the structure obtained with the CP method with $r = 2$ for $n_f = 104$ and $n_s = 20$.

structure side is by construction equal to zero and is therefore not depicted in the figure. For the discretization error on the flow side the third order convergence is not obtained, because only the discontinuity in the centre coincides with a grid point. The resulting order of convergence for the discretization error is 1.4.

Figure 7 shows the L_2-error in the pressure received by the flow for the conservative and consistent approaches. The solid lines are obtained with the conservative and the dashed lines with the consistent approach. The conservative approach does not converge and for the consistent approach, the RBFI methods have an interpolation error that is almost equal to the discretization error on the flow side, leading to convergence order of 1.5. The small wiggles in the error convergence of the consistent CP are caused by the fact that the discontinuity is not always located in the same position within a structure element when the grids are refined.

For the CP method with $r=2$ the error between the displacement received by the flow and the exact displacement field, $\mathbf{u}_f - \mathbf{u}_f^{ex}$, is plotted in Figure 8. Large oscillations are visible with the CP method with $r=2$. These oscillations are caused by the fact that the RBFI method tries to generate a global smooth function through all grid points and therefore the highest peak in the error is visible close to the largest discontinuity which is located in the centre.

The error between the exact pressure field and the one obtained by the structure, $\mathbf{p}_s - \mathbf{p}_s^{ex}$, is depicted in Figure 9 for the CP method with $r=2$. Only the results for the consistent approach are shown, because with the conservative approach the same large oscillations are obtained as with the transfer of the smooth pressure field and then the error for the consistent approach would not be visible anymore. For the consistent method the transformation matrix is built on the discrete structure interface, which can not exactly represent the pressure at the discontinuities. Hence, we can see that the error is the largest at these locations, but we see less oscillations than for the displacement. This is due to the fact that the structure mesh is much coarser than the flow mesh.

The conclusions for the above described test cases also hold for other parameter settings. Overall it can be concluded that for these simple analytical problems the consistent approach is preferred over the conservative approach. For the RBFI methods CP with $r=2$ is preferred.

2.4 Quasi-1D FSI problem

For the investigation of the behaviour of the methods in FSI simulations, where multiple transfers of pressure and displacement are performed, a quasi-1D problem is used. It is chosen such that it allows for the investigation of the problems arising with non-matching meshes. We consider a quasi-1D channel with a flexible curved wall. The main velocity of the compressible flow is in the x-direction and the structure is modeled as a membrane. The diameter of the channel may vary due to a pressure difference between the pressure in the flow and the pressure behind the wall. Considering only the static case allows us to analyze the coupling in space separately, excluding errors based on coupling in time. To obtain the steady state solution an iterative approach

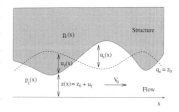

Fig. 10 Configuration of the quasi-1D FSI problem.

2.4.1 Flow equations

A simple flow model is used which is valid for supersonic flow over a panel:

$$p_f = -\rho_0 c_0 V_0 \partial_x z, \tag{25}$$

with ρ_0, c_0 and V_0 the density, speed of sound and velocity, respectively, assumed to be constant, p_f the pressure and $z = z_0 + u_f$ the location of the panel which is equal to the initial location of the panel, z_0, plus the displacement from this initial position, u_f.

For convenience the variables are scaled as follows

$$\bar{x} = \frac{x}{L}, \quad \bar{V}_0 = \frac{V_0}{c_0}, \quad \bar{p}_f = \frac{p_f}{\rho_0 c_0^2}, \quad \bar{z} = \frac{z}{L}, \tag{26}$$

with L the length of the channel. This results in the following non-dimensional equation:

$$\bar{p}_f = -\bar{V}_0 \partial_{\bar{x}} \bar{z}. \tag{27}$$

For ease of notation the bars are dropped in the remainder of the paper. To discretize the equations, a third order finite element discretization is used.

2.4.2 Structure equations

The equation that describes the behaviour of the flexible wall is given by

$$\kappa u_s - T \partial_{xx} u_s = p_s - p_e, \tag{28}$$

where u_s is the displacement from the 'dry' equilibrium position, $q_e(x)$ (when $p_s = p_e$); p_s is the pressure acting on the wall (note that at the continuous level $p_s = p_f$), p_e is the pressure behind the wall, assumed to be constant, κ the elasticity per unit length and T the longitudinal tension per unit length. In this test case the 'dry' equilibrium position is equal to the initial location of the panel, so $q_e = z_0$. During the simulations the structural displacement is zero at the boundaries, i.e. $u_s = 0$ at $x = 0$ and at $x = L$. Again the variables are scaled using the non-dimensional variables of (26) and the additional variables

$$\bar{u}_s = \frac{u_s}{L}, \quad \bar{p}_s = \frac{p_s}{\rho_0 c_0^2}, \quad \bar{p}_e = \frac{p_e}{\rho_0 c_0^2}, \quad \bar{\kappa} = \frac{\kappa L}{\rho_0 c_0^2}, \quad \bar{T} = \frac{T}{L \rho_0 c_0^2}. \tag{29}$$

This results in an equation which has two non-dimensional physical parameters $\bar{\kappa}$ and $\bar{T}$ and has the same form as (28). In the remainder of the paper the bars are dropped. Again a third order finite element discretization is used to discretize the equations.

2.4.3 Coupling procedure

Coupling between the fluid and the structure equations is obtained through the kinematic (1a) and dynamic (1b) boundary conditions at the fluid-structure interface. A simple iterative coupling procedure is implemented to obtain the steady state solution. This iterative approach proceeds as follows

1. Calculate $\mathbf{p}_s = H_{sf}\mathbf{p}_f$.
2. Calculate the new displacement of the structure, $\mathbf{u}_s$ from (28).
3. Obtain $\mathbf{u}_f = H_{fs}\mathbf{u}_s$ and update the location of the wall $\mathbf{z} = \mathbf{z}_0 + \mathbf{u}_f$.
4. Calculate the new pressure in the flow $\mathbf{p}_f$ from (27).

These four steps are repeated until the change in $\mathbf{u}_s$ is smaller than a certain threshold.

To obtain a numerically 'exact' solution the steady state problem is solved at once on very fine matching meshes ($n_f = n_s = 2000$). When the meshes are matching we have $p = p_f = p_s$ and $u = u_f = u_s$ and therefore we can solve the following equation for the displacement

$$\kappa u + V_0 \partial_x u - T \partial_{xx} u = -V_0 \partial_x q_e - p_e, \tag{30}$$

after which the pressure can be evaluated as

$$p = -V_0 \partial_x u. \tag{31}$$

In order to obtain the 'exact' solution a fourth order finite element discretization is used to discretize the equations.

2.4.4 Results

For the test cases the following configuration is used. The boundaries of the domain are $x_{\min} = -0.5$ and $x_{\max} = 0.5$ and the initial shape of the tube wall is given by

$$z_0(x) = a_0 - a_1 e^{-a_2 x^2}, \tag{32}$$

where the parameters have the values $a_0 = 0.5$, $a_1 = 0.25$ and $a_2 = 80$. This means that we have a smooth converging/diverging channel. The 'dry' equilibrium position of the membrane, q_e, is equal to this initial shape. The values used for the non-dimensional structure parameters are: $\kappa = 50$ and $T = 0.04$, which results in a rather flexible membrane. The flow velocity is equal to $V_0 = 3$, corresponding to a supersonic flow of Mach 3. Initially the pressure in the flow, $\mathbf{p}_f$, the pressure behind the wall, $\mathbf{p}_e$, and the displacement $\mathbf{u}_s$ are all equal to zero. We use again $n_f = 26 \cdot 2^k$ flow cells and $n_s = 5 \cdot 2^k$ structure cells, with $k \in \{0,1,2,3,4,5\}$, leading to a ratio of approximately 20%.

The L_2-error (22) of the displacement in the flow points versus the number of structure points is depicted in Figure 11. The solid line is obtained with the conservative and the dashed line with the consistent approach. The gray lines represent the discretization error (23) of the displacement: the solid gray line is the discretization error on the flow interface and the dashed line on the structure interface.

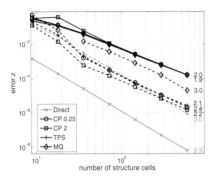
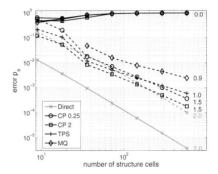

Fig. 11 Error in displacement - RBFI methods (—: conservative - -: consistent).

Fig. 12 Error in pressure - RBFI methods (—: conservative - -: consistent).

The RBFI methods are second order accurate with the consistent approach giving the most accurate results. Only for higher values of n_s the coupling error is larger than the discretization error, when CP or TPS is used. When the discretization of the total system is second order or lower, instead of the third order discretization that is used in the examples in this paper, the coupling error is smaller than the discretization error for all RBFI methods.

The relative L_2-error of the pressure versus the number of points on the structure interface is depicted in Figure 12. Because the value for the pressure is obtained from the spatial derivative of z, the convergence is one order lower than for the displacement. It can be seen that the conservative approach leads again to a zeroth order error for all methods. The consistent RBFI methods are first till 1.5 order accurate where CP with $r = 2$ gives the most accurate results. Only for larger values of n_s the coupling error is larger than the discretization error, when CP or TPS is used.

In Figures 13 and 14, the exact solution and the ones obtained with the conservative and consistent approach using the CP method with $r = 2$ are shown for the pressure received by the structure, p_s, and the displacement of the flow interface u_f, respectively. It can be seen that the large oscillations in the pressure felt by the structure result also in small deviations in the displacement of the flow interface.

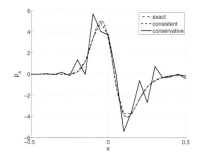 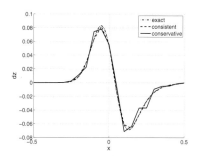

Fig. 13 Pressure obtained with CP ($r = 2$) for $n_f = 52$ and $n_s = 10$.

Fig. 14 Displacement obtained with CP ($r = 2$) for $n_f = 52$ and $n_s = 10$.

The more flexible the structure, the larger these deviations become. The results for the difference in work are similar to the ones obtained by the analytical test case.

The main conclusion is that the RBFI methods show large oscillations in the pressure obtained by the structure when the conservative approach is used. For these coupling methods the consistent approach provides the best accuracy.

3 Mesh movement based on radial basis function interpolation

For structured meshes there are efficient techniques available to deform the mesh, such as Transfinite Interpolation [35]. The displacements of points at the boundaries of the mesh are interpolated along grid lines to points in the interior of the mesh. However, these techniques are unsuitable for unstructured grids. The greater flexibility of unstructured grids is required for the meshing of complex domains and grid adaptation. Therefore, we are in this chapter interested in efficient mesh movement techniques for unstructured grids.

Two different mesh movement strategies are known for unstructured grids: grid-connectivity and point-by-point schemes. The first exploits the connectivity of the internal grid points. The connection between the grid points is represented for example by springs [2, 15, 14] or as solid body elasticity [28]. Special instances of this continuous approach include moving grids based on Laplacian and Biharmonic operators [21]. All the methods based on grid connectivity involve solving a system of equations involving all the flow points and can therefore be quite expensive. In unstructured meshes nodes that define the corner of one cell may be located on the cell face of its neighboring cell. These so-called hanging nodes usually have no resistance against moving out of the cell face plane when a pseudo-structural model is used for the mesh deformation and therefore require special treatment.

The other strategy moves each grid point individually based on its position in space, the so called point-by-point schemes. Hanging nodes are no problem and also the implementation for partitioned meshes, occurring in parallel flow computations, is straightforward. This might be especially useful for Finite Volume flow solvers which do not incorporate efficient algorithms to deform the mesh with a pseudo-structural approach. However, until now point-by-point schemes are only applied to the boundary nodes of multi-grid blocks [30]. The interior mesh of the blocks is adapted with fast techniques available for structured grids.

Radial basis functions (RBF's) have become a well-established tool to interpolate scattered data. They are for example used in fluid-structure interaction computations to transfer information over the discrete fluid-structure interface, which is often non-matching [3, 31] as is shown in Section 2. An interpolation function is used to transfer the displacements known at the boundary of the structural mesh to the boundary of the aerodynamic mesh. But why not interpolate the displacement to all the nodes of the flow mesh, instead of only to the boundary? This idea has already been applied to the block boundaries in multi-block grids [30, 33]. There it was mentioned that applying it to the whole internal grid would be computationally very expensive. This is because for the structured part of multi-block meshes much more efficient techniques are known. We want to investigate if interpolation of the displacement with radial basis functions does result in an efficient point-by-point mesh movement scheme for completely unstructured grids.

The objective of this section is to develop a new mesh movement scheme for unstructured meshes based on interpolation with radial basis functions. We outline the principle of interpolation with RBF's applied to mesh movement in section 3.1. There are various RBF's available in literature that can be used for the new method and we want to determine which one generates the best meshes and which one is the most efficient. To be able to compare the deformed meshes generated with the different RBF's, a mesh quality metric is introduced in section 3.2. This metric is used in section 3.3 to determine the best RBF's for our mesh movement scheme, by applying the method to several severe test cases. For one of the test cases the results are also compared with mesh deformation using semi-torsional springs. Furthermore realistic results on a distorted 2D mesh are presented, where flow computations are performed around a NACA-0012 airfoil. In section 3.3.1 it is investigated if incorporating rotational information is also beneficial for the new mesh movement strategy and the importance of smooth mesh deformation for higher order time-integration schemes is shown in section 3.4. To show the capability of the new method for 3D applications two test cases are considered in section 3.5.

3.1 Radial basis function interpolation

Radial basis function interpolation can be used to derive the displacement of the internal fluid nodes given the displacement of the structural nodes on the interface. The interpolation function, s, describing the displacement in the whole domain, can be approximated by a sum of basis functions

$$s(\mathbf{x}) = \sum_{j=1}^{n_b} \gamma_j \phi(||\mathbf{x} - \mathbf{x}_{b_j}||) + p(\mathbf{x}), \tag{33}$$

where $\mathbf{x}_{b_j} = [x_{b_j}^1, x_{b_j}^2, x_{b_j}^d]$ are the centres in which the values are known, in this case the boundary nodes, with d the dimension, p a polynomial, n_b the number of boundary nodes and ϕ a given basis function with respect to the Euclidean distance $||\mathbf{x}||$. The coefficients γ_j and the polynomial p are determined by the interpolation conditions

$$s(\mathbf{x}_{b_j}) = \mathbf{d}_{b_j}, \tag{34}$$

with $\mathbf{d}_b$ containing the discrete known values of the displacement at the boundary, and the additional requirements

$$\sum_{j=1}^{n_b} \gamma_j q(\mathbf{x}_{b_j}) = 0, \tag{35}$$

for all polynomials q with a degree less or equal than that of polynomial p. The values for the displacement in the interior of the flow mesh $\mathbf{d}_{in}$, can then be derived by evaluating the interpolation function (33) in the internal grid points:

$$\mathbf{d}_{in_j} = s(\mathbf{x}_{in_j}). \tag{36}$$

The displacement can be interpolated separately for each spatial direction. The process of mesh deformation with RBF's for one direction is visualized in Figures 15 and 16. Each individual point is moved individually based on its position in space according to the interpolation function, and this means that no mesh-connectivity information is needed at all. However, to be able to see the effects on the mesh clearly, the points are connected in a very simple mesh as is shown in Figure 15. The block

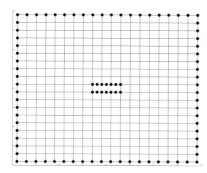

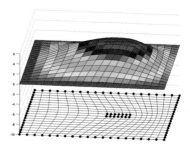

Fig. 15 Initial mesh.

Fig. 16 Interpolation function and resulting deformed mesh.

in the middle is moved to the right and the resulting interpolation function is shown in Figure 16. This function is equal to zero at the outer boundaries and equal to the displacement of the block at the location of the block boundaries. The displacement of the nodes in the interior of the mesh can then be derived from this function and the resulting deformed mesh is shown in Figure 16.

The size of the system that has to be solved for obtaining the coefficients γ and the polynomial p is equal to $(n_b + 4) \times (n_b + 4)$ which is usually very small compared to the systems that have to be solved in mesh-connectivity schemes. The systems encountered there are approximately as large as $n_{in} \times n_{in}$, with n_{in} the total number of mesh points. The total number of mesh points is a dimension higher than the number of points on the boundary of the mesh. The new moving mesh technique is very easy to implement, even for 3D applications, because no mesh-connectivity information is needed. Also the implementation for partitioned meshes, occurring in parallel flow computations, is straightforward.

In this section a vast variety of radial basis functions are evaluated for application to mesh deformation. In Table 1 various radial basis functions with compact support

nr.	name	$f(\xi)$
1	CP C^0	$(1-\xi)^2$
2	CP C^2	$(1-\xi)^4(4\xi+1)$
3	CP C^4	$(1-\xi)^6(\frac{35}{3}\xi^2+6\xi+1)$
4	CP C^6	$(1-\xi)^8(32\xi^3+25\xi^2+8\xi+1)$
5	CTPS C^0	$(1-\xi)^5$
6	CTPS C^1	$1+\frac{80}{3}\xi^2-40\xi^3+15\xi^4-\frac{8}{3}\xi^5+20\xi^2\log(\xi)$
7	CTPS C_a^2	$1-30\xi^2-10\xi^3+45\xi^4-6\xi^5-60\xi^3\log(\xi)$
8	CTPS C_b^2	$1-20\xi^2+80\xi^3-45\xi^4-16\xi^5+60\xi^4\log(\xi)$

Table 1 Radial basis functions with compact support, from [36].

are given. In this section all compact RBF's are scaled with r, so we use $\xi = x/r$. The first four are based on polynomials [36]. These polynomials are chosen in such a way that they have the lowest degree of all polynomials that create a C^n continuous basis function with $n \in \{0,2,4,6\}$. The last four are a series of functions based on the thin plate spline which create C^n continuous basis functions with $n \in \{0,1,2\}$ [36]. There are two possible CTPS C^2 continuous functions which are distinguished by subscript a and b. In Table 2 six globally supported radial basis functions are given which are used in this section.

The new mesh movement scheme based on interpolation with radial basis functions will be tested with all these different RBF's, but first a mesh quality metric is given to be able to compare the quality of the meshes after deformation.

nr.	name	abbrev.	$f(x)$
9	Thin plate spline	TPS	$x^2 \log(x)$
10	Multiquadric Biharmonics	MQB	$\sqrt{a^2+x^2}$
11	Inverse Multiquadric Biharmonics	IMQB	$\sqrt{\frac{1}{a^2+x^2}}$
12	Quadric Biharmonics	QB	$1+x^2$
13	Inverse Quadric Biharmonics	IQB	$\frac{1}{1+x^2}$
14	Gaussian	Gauss	e^{-x^2}

Table 2 Radial basis functions with global support.

3.2 Mesh quality metrics

To be able to compare the quality of different meshes after mesh movement we introduce mesh quality metrics [25]. The mesh quality metrics are based on a set of Jacobian matrices which contain information on basic element qualities such as size, orientation, shape and skewness.

It is assumed that the initial mesh is generated in an optimal way and therefore the element shapes should be changed as little as possible after deformation. This means that both the volume and the angles of the elements should be preserved. These two properties can be measured with the relative size and skew metric.

The relative size metric measures the change in element size. Let τ be the ratio between the current and initial element volume. The *relative size metric* [25] is then given by $f_{size} = \min(\tau, 1/\tau)$. Essential properties of the relative size metric are: $f_{size} = 1$ if and only if the element has the same total area as the initial element and $f_{size} = 0$ if and only if the element has a total area of zero. The relative size metric can detect elements with a negative total area (degenerate) and elements which change in size due to the mesh deformation.

The skew metric measures the skewness and therefore the distortion of an element. When a node of an element possesses a local negative area, this metric value is set to zero. The expressions for the *skew metric* for triangular, quadrilateral, tetrahedral and quadrilateral elements can be found in [25]. Essential properties of the skew metric are: $f_{skew} = 1$ if and only if the element has equal angles and $f_{skew} = 0$ if and only if the element is degenerate.

To measure both the change in element size and the distortion of an element, the *size-skew metric* [25] is introduced which is defined as the weighted product of the relative size and skew metrics: $f_{ss} = \sqrt{f_{size} f_{skew}}$, since changes in element volume have a smaller influence on the mesh quality than element distortion. Essential properties of the quadrilateral size-skew metric are:

- $f_{ss} = 1 \Leftrightarrow$ element has equal angles and same size as the initial element.
- $f_{ss} = 0 \Leftrightarrow$ element is degenerate.

This is the quality metric we will use to measure the quality of a mesh after deformation.

The average value of the metric over all the elements indicates the average quality of the mesh. The higher the average quality of the mesh, the more stable, accurate and efficient the computation will be. The minimum value of the metric over all the elements indicates the quality of the cell with the lowest quality. This value is required to be larger than zero, otherwise the mesh will contain degenerate cells. Degenerate elements have a very negative influence on the stability and accuracy of numerical computations. In the next section we will use both the average and minimal value of the size-skew metric to compare meshes after mesh movement.

3.3 2D mesh movement

The new mesh movement strategy is tested with the 14 radial basis functions introduced in section 3.1. First, three simple 2D test problems are performed to investigate the difference in quality of the mesh obtained with the RBFs after movement of the boundary. The tests include mesh movement due to rigid body rotation and translation of a rectangle block and deformation of an airfoil-flap configuration. After that the efficiency of the most promising RBFs is investigated. Furthermore realistic results on a distorted mesh are presented, where flow computations are performed around a NACA-0012 airfoil.

The displacements from initial to final configuration can be imposed in one step or in a number of incremental displacement steps, using the latest configuration as the new reference configuration. Using multiple steps reduces the displacements imposed in a single step and should improve the accuracy and robustness of the interpolation.

The quality and robustness of the new method depend on the value of the support radius when a radial basis function with compact support is used. When the support radius is chosen large enough, the quality and robustness converge to an optimum. Therefore, a relatively high value, r is 2.5 times the characteristic length of the computational domain, is used in the first three test cases, where we only investigate the accuracy of the different RBFs.

3.3.1 Test case 1: Rotation and translation

The first test case consists of mesh movement due to severe rotation and translation of a block in a small domain. The mesh nodes on the block follow its movement, while the nodes on the outer boundary are fixed. The block has dimension $5D \times 1D$, with D the thickness of the block, and is initially located in the center of a domain which has dimension $25D \times 25D$. The initial mesh is triangular and given in Figure 19. The block is translated $10D$ down and to the left and is rotated 60 degrees around the center of the block. The mesh deformation is performed in a variable number of steps (iterations) between the initial and final location, with a minimum of 1 step and a maximum of 15 steps.

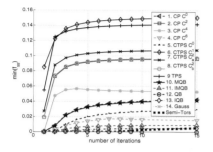

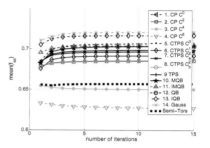

Fig. 17 Quality of the worst cell of the mesh for the different RBF's (test case 1).

Fig. 18 Average quality of the mesh for the different RBF's (test case 1).

The minimum value of f_{ss} after mesh movement with the different RBF's is shown in Figure 17 for an increasing number of intermediate steps. It can be seen that for all RBF's the minimum value of f_{ss} indeed increases when more intermediate steps are taken. Only for the 5 best performing RBF's 2, 6, 7, 8 and 9 the ranking is given in Table 3. Figure 18 shows the average value of the mesh quality metric.

Table 3 Ranking for the test cases.

test-case 1		test case 2		test case 3	
min	mean	min	mean	min	mean
6	9	2, 8	2, 8	2, 7, 8	6, 9
9	7	7	7, 9	6, 9	7
7	6	9	6		2, 8
2, 8	2, 8	6			

The Gaussian basis function (nr. 14), has the best average quality, however, the minimum of f_{ss} for this function is equal to zero. This results in highly distorted meshes as can be seen in Figure 20 where the final mesh generated with the Gaussian basis function with 15 intermediate steps is shown. Only cells at a certain distance from the block are heavily deformed, resulting in a high average mesh quality, but the flow solver will probably crash due to the degenerate cells. The mesh with the highest minimum value for the mesh quality is generated with CTPS C^1 (nr. 6) and is shown in Figure 21. The average value of the mesh quality metric is lower than for the Gaussian function, because all cells are deformed, but this results in a much smoother mesh. This will have a positive effect on the accuracy, stability and efficiency of an unsteady flow computation. In the next two test cases we will only consider the RBF's 6, 9, 7, 2 and 8, because they are the most promising.

The mesh deformation is also performed with a method based on semi-torsional springs [38], which is an improvement of the popular spring analogy [2]. This

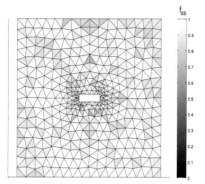

Fig. 19 Initial mesh.

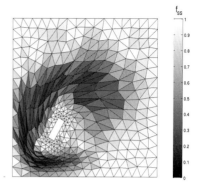

Fig. 20 Final mesh using Gaussian basis function with 15 intermediate steps.

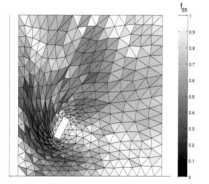

Fig. 21 Final mesh using CTPS C^1 with 15 intermediate steps.

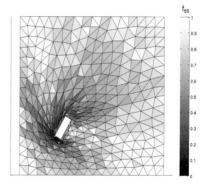

Fig. 22 Final mesh using semi-torsional springs with 15 intermediate steps.

method is based on elastic deformation of element edges where element edges are modeled as springs producing forces to propagate boundary displacements into the interior of the mesh. The formation of mis-shaped elements is penalized by incorporating angle information into the spring stiffness. A boundary improvement technique [8] is implemented which increases the stiffness of springs close to the moving boundary so that surface displacement spreads further into the mesh. In accordance with [38] we imposed one layer of boundary stiffness by increasing the spring stiffness with a factor 3.5. The results for the minimal and average value for the mesh quality metric are added to Figures 17 and 18, respectively (bold dashed line). More than 8 intermediate steps are needed to avoid degenerate cells and the minimum value of the mesh quality metric is very low. The final mesh using 15 intermediate steps is shown in Figure 22. It can be seen that the displacement does not spread very far into the domain and the cells close to the moving block are heavily deformed. The mesh quality obtained with the 5 best RBF's is much higher, because the deformation is more evenly spread through the domain.

3.3.2 Test case 2: Rigid body Rotation

For rigid body translations the interpolation is exact and therefore the initial mesh is recovered when the domain returns to its initial form. However, it is not guaranteed that this is the case when rotations are present. Therefore we investigate the mesh quality when the block is severely rotated and brought back to its initial position. The nodes on the outer boundary can freely move along this boundary. The initial mesh is the same as for test case 1 (Figure 19). First the block is rotated 180° counterclockwise, then 360° clockwise and back to the starting position by rotating it again 180° counterclockwise. The rotation is performed with a variable number of intermediate steps. In Figure 23 again the minimal value and in Figure 24 the aver-

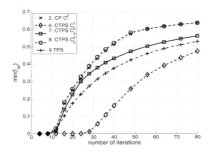 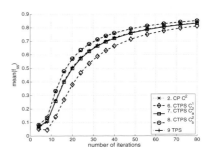

Fig. 23 Quality of the worst cell of the mesh for the different RBF's.

Fig. 24 Average quality of the mesh for the different RBF's.

age value of the mesh quality metric is shown against the number of intermediate steps. The resulting ranking for the RBF's is shown in Table 3.

As can be seen from Figure 23, more than 10 intermediate steps are needed with all functions to obtain a positive value of f_{ss} for all cells. Figures 25 and 26 show the

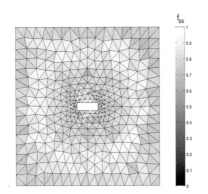 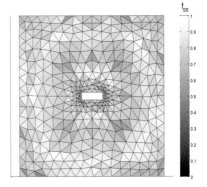

Fig. 25 Final mesh using CTPS C_b^2 after 40 intermediate steps.

Fig. 26 Final mesh using CTPS C^1 after 40 intermediate steps.

final meshes with 40 intermediate steps using the best, CTPS C_b^2 (nr. 8) and the worst RBF, CTPS C^1 (nr. 6), respectively. Figure 26 shows that especially the cells close to the block and boundary are deformed. With the CTPS C_b^2 function the mesh is also distorted compared to the initial mesh. However, this distortion is rather small considering the very large rotation of the block. The final meshes obtained with TPS, CP C^2 and CTPS C_a^2 are very similar to that of CTPS C_b^2.

3.3.3 Test case 3: Airfoil flap

Until now we only studied rigid body rotation and translation of a block. In the third test case we investigate a more realistic situation of an airfoil-flap configuration. The test case starts with the initial mesh given in Figure 27. A close up of the mesh around the gap between the airfoil and flap is shown in Figure 28. The flap is rotated 30 degrees in clockwise direction around an axis a tenth of a chord below the trailing edge of the airfoil. In Figure 29 the minimal value and in Figure 30 the average

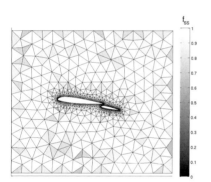

Fig. 27 Initial mesh.

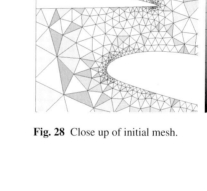

Fig. 28 Close up of initial mesh.

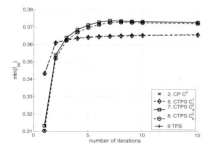

Fig. 29 Quality of the worst cell of the mesh for the different RBF's.

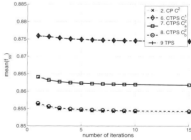

Fig. 30 Average quality of the mesh for the different RBF's.

value of f_{ss} is shown for the remaining RBF's. It can be seen that all the RBF's are able to deform the mesh well, and all functions give meshes of similar quality. The ranking is given in Table 3.

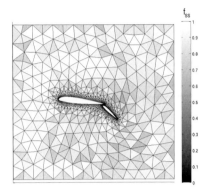
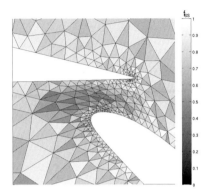

Fig. 31 Final mesh using CTPS C_a^2 with 10 intermediate steps.

Fig. 32 Zoom of final mesh using CTPS C_a^2 with 10 intermediate steps.

The final configuration of the airfoil-flap and the resulting mesh is shown in Figure 31 using 10 intermediate steps with the CTPS C_a^2 function. Figure 32 displays a close up of this mesh and shows that the mesh quality is still very good in the region where the largest deformation takes place and suitable for flow computations. The final meshes obtained with the other functions look very similar.

3.3.4 Test case 4: Flow around airfoil

Until now only the mesh quality of the deformed meshes is investigated without solving a single physical problem. In this section more realistic flow results on a distorted mesh are presented. We perform two calculations of viscid flow around a NACA-0012 airfoil. The airfoil is rotated 8° and moved 5 chords downstream and 2 chords upwards and a steady state solution of Mach 0.3 with Re = 1000 is computed around the wing. In the first computation a new unstructured hexahedral mesh is generated around the moved airfoil and a close-up of the mesh together with the pressure field is shown in Figure 33. In the second computation the mesh is deformed in one step with the thin plate spline from its original position. The result is shown in Figure 34. The resulting pressure distributions over the wing are identical as can be seen in Figure 35. The difference in lift computed on the two different meshes is only 0.8%.

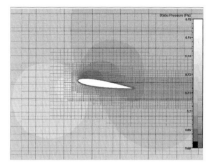

Fig. 33 Pressure field around wing on a new generated mesh.

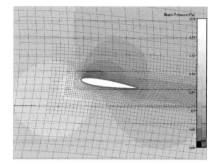

Fig. 34 Pressure field around wing on a deformed mesh.

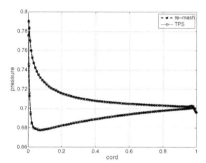

Fig. 35 Pressure distribution over the wing.

3.4 Importance of smooth mesh deformation for higher order time-integration

In order to investigate the effect of the mesh deformation algorithm on the temporal accuracy of higher order time integration schemes we consider the one-dimensional piston problem [39]. The flow equations are solved in a two-dimensional domain on an "imperfect" mesh which has cells that are not perfectly orthogonal. The mesh deformation technique based on RBF interpolation is compared to a technique which solves the Laplace equation to create a displacement field [26]. After displacing the flow vertices with this second method, the mesh is optimized which is necessary to avoid degenerate cells. In Fig. 36 the L_2-norms of the fluid density ρ, pressure p, velocity in x-direction u and y-direction v for the different meshes and mesh deformation schemes are shown.

We use the fourth order IMEX scheme [39] for the partitioned time integration. The results obtained with the Laplace smoothing are not satisfactory. The fourth order of the scheme is not observed and although the test problem is essentially one-dimensional, the v-velocity is not zero due to the imperfect mesh. For the large time steps this perturbation is only small compared to the other errors. However, since the convergence for the v-velocity is clearly not fourth order, its influence becomes more apparent for the smaller time steps. RBF interpolation for the imperfect

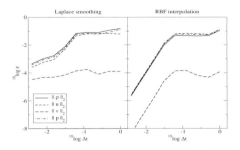

Fig. 36 Convergence for the L_2-norm of the density, pressure and u- and v-velocity components for the piston problem with Laplace smoothing and radial basis function interpolation.

mesh has the same nonzero v-velocity for the large time steps. This time, however, the perturbation does converge with fourth order accuracy and therefore its influence on the solution remains negligible. Therefore we can conclude that the RBF interpolation does not aggravate the imperfections in the flow mesh.

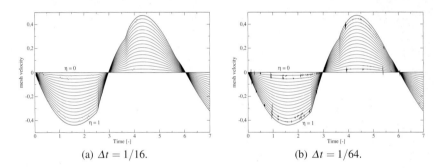

(a) $\Delta t = 1/16$.   (b) $\Delta t = 1/64$.

Fig. 37 Mesh face velocities with Laplace smoothing.

In order to explain the bad convergence with the Laplace smoothing we study the mesh face velocities for the cell faces which are displayed in Fig. 37. It shows that the Laplace smoothing with optimization introduces irregularities (wiggles) in the mesh face velocities which are worse for small Δt. The mesh face velocity does not converge to a consistent solution so the design order of the IMEX scheme can not be expected. Due to the high accuracy and regularity of the displacement field obtained with RBF interpolation, the RBF mesh deformation algorithm does not exhibit these convergence problems.

3.5 3D mesh deformation

To show the capability of the new method for 3D applications we consider in this section a test case similar to test case 1, only this time in a 3D domain and a real FSI computation considering the AGARD 445.6 test case.

3.6 Rotation and translation of 3D block

We consider a test case in a 3D domain similar to test case 1 (section 3.3.1). A cross-section showing the initial mesh and location of the block is given in Figure 38. The block is translated 2.5 times the thickness of the block and rotated 15 degrees in all three directions. A cross-section of the final mesh and location of the block is shown in Figure 39. Visually it is quite hard to judge the quality of the mesh and

Fig. 38 Cross-section of initial mesh.

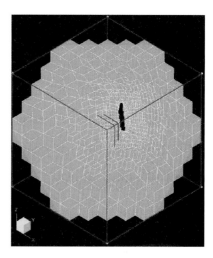

Fig. 39 Cross-section of final mesh.

therefore the values of the mesh quality metric in the cross-sections is displayed in Figures 40 and 41 for the CP C^2 and TPS function, respectively. It can be seen that the mesh quality is everywhere larger than 0.5 and therefore the meshes are suitable for computational analysis. The main difference between the two figures is that with CP C^2 the mesh quality close to the moving structure remains a little higher than with TPS.

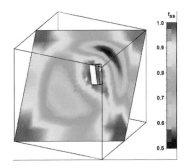
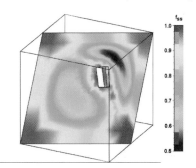

Fig. 40 Mesh quality in cross-section using CP C^2.

Fig. 41 Mesh quality in cross-section using TPS.

3.6.1 Flutter of the AGARD 445.6 wing

To demonstrate the practical applicability for real-world three-dimensional fluid-structure interaction the AGARD 445.6 test case [37, 43] is used. The consistent approach is used for the interpolation of data at the fluid-structure interface. For the investigation into the performance of the third order IMEX scheme [39] with the RBF mesh deformation a time step convergence study is performed. The solution obtained with $\Delta t = 0.001$ is taken as the temporally exact solution. In Fig. 42 the results for the pressure field, velocity in vertical direction w, structural displacement and lift at the end of the simulation are shown. The figures show a clear third order

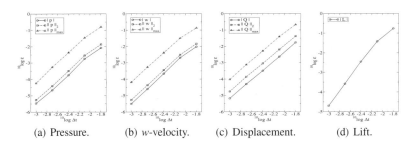

(a) Pressure.   (b) w-velocity.   (c) Displacement.   (d) Lift.

Fig. 42 Time step convergence for the third order IMEX scheme.

convergence for all the properties in all the norms, which shows that the combination of the third order partitioned IMEX scheme with RBF mesh deformation retains the order of the time integration scheme without the necessity to sub-iterate.

The displacements generated in the AGARD 445.6 test case are rather small and do not place a large demand on the mesh deformation method. To investigate the performance of the mesh deformation method for large displacements we also enforced a large deformation on a 3D beam. The resulting mesh on the beam and the mirror plane is shown in Figure 43. It can be seen that the mesh is still regular

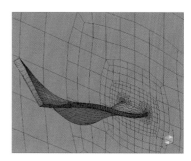

Fig. 43 Mesh obtained with RBF interpolation around a heavily deformed 3D beam.

and also the mesh quality metrics indicated a high mesh quality, such that accurate flow calculations are possible.

4 Conclusions

In this chapter we investigated the use of radial basis functions for application to interpolation of data at non-matching interfaces and for application to mesh deformation.

RBFI for non-matching meshes

In Section 2 the difference in accuracy between a conservative and a consistent approach for different RBFI coupling methods is investigated. The performance is analysed for two analytical test problems as well as a simple quasi-1D FSI problem. For RBFI coupling methods the conservative approach results in large unphysical oscillations in the pressure received by the structure. When the structure is flexible enough these oscillations can result in deviations in the displacement of the flow interface. Therefore the consistent approach provides the best accuracy.

RBFI for mesh deformation

In Section 3 a point-by-point mesh movement algorithm is presented for the deformation of unstructured grids. Radial basis functions (RBF's) are used to interpolate the displacements of the boundary nodes of the mesh to the inner domain. The method requires solving a small system of equations, only involving the nodes on the boundary of the flow domain. The implementation of the method is relatively simple, even for 3D applications, because no grid-connectivity information is needed. Also the implementation for partitioned meshes, occurring in parallel flow computations, is straightforward.

The new algorithm is tested with fourteen RBF's for a variety of problems. The method can handle large deformations of a mesh caused by translation, rotation and deformation of a structure both on 2D and 3D meshes. The performance of the method is not the same for all RBF's. Five RBF's generate meshes of high quality after deformation. However, when efficiency is more important, the CP C^2 RBF with compact support is the best choice, closely followed by the thin plate spline.

In a first comparison the RBF-method produces meshes of higher quality than the popular semi-torsional spring analogy for very large deformations. The quality of the meshes after deformation is high enough to perform accurate flow calculations.

It is shown that working with higher quality meshes can increase the efficiency of the computation. This is due to the fact that the new method preserves the design order of higher order time integration methods, due to its smooth deformation in time. The capability of the method to deform 3D meshes is shown by rotating and translating a 3D block and performing a case study on flutter of the AGARD 445.6 wing.

References

1. Ahrem, R., Beckert, A., Wendland, H.: A new multivariate interpolation method for large-scale coupling problems in aeroelasticity. Conference proceedings IFADS, Munich (2005)
2. Batina, J.T.: Unsteady Euler algorithm with unstructured dynamic mesh for complex-aircraft aeroelastic analysis. Tech. Rep. AIAA-89-1189 (1989)
3. Beckert, A., Wendland, H.: Multivariate interpolation for fluid-structure-interaction problems using radial basis functions. Aerospace Science and Technology **5**(2), 125–134 (2001)
4. Bijl, H., Carpenter, M.H.: Iterative solution techniques for unsteady flow computations using higher order time integration schemes. International Journal for Numerical Methods in Fluids **47**(8–9), 857–862 (2005)
5. Bijl, H., Carpenter, M.H., Vatsa, V.N., Kennedy, C.A.: Implicit time integration schemes for the unsteady compressible Navier-Stokes equations: Laminar flow. Journal of Computational Physics **179**, 313–329 (2002)
6. Billings, S.D., Beatson, R.K., Newsam, G.N.: Interpolation of geophysical data with continuous global surfaces. Geophysics **67**, 1810–1822 (2002)
7. Billings, S.D., Newsam, G.N., Beatson, R.K.: Smooth fitting of geophysical data with continuous global surfaces. Geophysics **67**, 1823–1834 (2002)
8. Blom, F.J.: Considerations on the spring analogy. International Journal for Numerical Methods in Fluids **32**, 647–668 (2000)
9. de Boer, A., van der Schoot, M.S., Bijl, H.: Mesh deformation based on radial basis function interpolation. Computers and Structures **85**(11—14), 784–795 (2007)
10. de Boer, A., van Zuijlen, A.H., Bijl, H.: Review of coupling methods for non-matching meshes. Computer Methods in Applied Mechanics and Engineering **196**(8), 1515–1525 (2006)
11. Buhmann, M.D.: Radial basis functions. Acta Numerica **9**, 1–38 (2000)
12. Carr, J.C., Beatson, R.K., McCallum, B.C., Fright, W.R., McLennan, T.J., Mitchell, T.J.: Smooth surface reconstruction from noisy range data. First International Conference on Computer Graphics and Interactive Techniques (2003)
13. Cebral, J.R., Löhner, R.: Conservative load projection and tracking for fluid-structure problems. AIAA Journal **35**(4), 687–692 (1997)
14. Degand, C., Farhat, C.: A three-dimensional torsional spring analogy method for unstructured dynamic meshes. Computers and Structures **80**, 305–316 (2002)

15. Farhat, C., Degrand, C., Koobus, B., Lesoinne, M.: Torsional springs for two-dimensional dynamic unstructured fluid meshes. Computer Methods in Applied Mechanics and Engineering **163**, 231–245 (1998)
16. Farhat, C., Géradin, M.: On a component mode synthesis method and its application to incompatible substructures. Computers and Structures **51**, 459–473 (1994)
17. Farhat, C., Lesoinne, M., Tallec, P.: Load and motion transfer algorithms for fluid/structure interaction problems with non-matching discrete interfaces: Momentum and energy conservation, optimal discretization and application to aeroelasticity. Computer Methods in Applied Mechanics and Engineering **157**, 95–114 (1998)
18. Felippa, C.A., Park, K.C., Farhat, C.: Partitioned analysis of coupled mechanical systems. Computer Methods in Applied Mechanics and Engineering **190**, 3247–3270 (2001)
19. Gravouil, A., Combescure, A.: Multi-time-step explicit-implicit method for non-linear structural dynamics. International Journal for Numerical Methods in Engineering **50**(1), 199–225 (2001)
20. Heinstein, M.W., Laursen, T.A.: A three dimensional surface-to-surface projection algorithm for non-coincident domains. Communications in Numerical Methods in Engineering **19**, 421–432 (2003)
21. Helenbrook, B.T.: Mesh deformation using the biharmonic operator. International Journal for Numerical Methods in Engineering **56**, 1007–1021 (2003)
22. Kansa, E.J.: Multiquadrics – a scattered data approximation scheme with applications to computational fluid-dynamics – I: Surface approximations and partial derivative estimates. Computers & Mathematics with Applications **19**, 127–145 (1990)
23. Kansa, E.J.: Multiquadrics – a scattered data approximation scheme with applications to computational fluid-dynamics – II: Solutions to parabolic, hyperbolic and elliptic partial differential equations. Computers & Mathematics with Applications **19**, 147–161 (1990)
24. Kee, B.B.T., Liua, G.R., Zhanga, G.Y., Luc, C.: A residual based error estimator using radial basis functions. Finite Elements in Analysis and Design **44**(9–10), 139–181 (2008)
25. Knupp, P.M.: Algebraic mesh quality metrics for unstructured initial meshes. Finite Elements in Analysis and Design **39**, 217–241 (2003)
26. Kovalev, K.: Unstructured hexahedral non-conformal mesh generation. Ph.D. thesis, Vrije Universiteit Brussel (2005)
27. Löhner, R., Yang, C., Cebral, J., Baum, J.D., Luo, H., Pelessone, D., Charman, C.: Fluid-structure interaction using a loose coupling algorithm and adaptive unstructured grids. In: M. Hafez, K. Oshima (eds.) Computational Fluid Dynamics Review. John Wiley (1995)
28. Lynch, D., ONeill, K.: Elastic grid deformation for moving boundary problems in two space dimensions. In: S. Wang (ed.) Finite Elements in Water Resources (1980)
29. Maman, N., Farhat, C.: Matching fluid and structure meshes for aeroelastic computations: A parallel approach. Computers and Structures **54**(4), 779–785 (1995)
30. Potsdam, M.A., Guruswamy, G.P.: A parallel multiblock mesh movement scheme for complex aeroelastic applications. Tech. Rep. AIAA-2001-0716 (2001)
31. Smith, M.J., Cesnik, C.E.S., Hodges, D.H.: Evaluation of some data transfer algorithms for noncontiguous meshes. Journal of Aerospace Engineering **13**(2), 52–58 (2000)
32. Smith, M.J., Hodges, D.H., Cesnik, C.E.S.: Evaluation of computational algorithms suitable for fluid-structure interactions. Journal of Aircraft **37**(2), 282–294 (2000)
33. Spekreijse, S., Prananta, B., Kok, J.: A simple, robust and fast algorithm to compute deformations of multi-block structured grids. Tech. rep. (2002)
34. Thévenza, P., Blu, T., Unser, M.: Interpolation revisited. IEEE Transactions on Medical Imaging **19**(7), 739–758 (2000)
35. Wang, Z.J., Przekwas, A.J.: Unsteady flow computation using moving grid with mesh enrichment. Tech. Rep. AIAA-94-0285 (1994)
36. Wendland, H.: Konstruktion und untersuchung radialer basisfunktionen mit kompaktem träger. Tech. rep. (1996)
37. Yates jr., E.C.: AGARD standard aeroelastic configurations for dynamic response. candidate configuration I.-Wing 445.6. Tech. Rep. Technical Memorandum 100492 (1987)

38. Zeng, D., Ethier, C.R.: A semi-torsional spring analogy model for updating unstructured meshes in 3D moving domains. Finite Elements in Analysis and Design **41**, 1118–11,139 (2005)
39. van Zuijlen, A.H.: Fluid-structure interaction simulations - efficient higher order time integration of partitioned systems. Ph.D. thesis, Delft University of Technology (2006)
40. van Zuijlen, A.H., Bijl, H.: A higher-order time integration algorithm for the simulation of non-linear fluid-structure interaction on moving meshes. Nonlinear Analysis-Theory Methods & Applications **63**, 1597–1605 (2005)
41. van Zuijlen, A.H., Bijl, H.: Implicit and explicit higher order time integration schemes for structural dynamics and fluid-structure interaction computations. Computers and Structures **83**, 93–105 (2005)
42. van Zuijlen, A.H., Bijl, H.: Implicit and explicit higher-order time integration schemes for fluid-structure interaction computations. International Journal of Multiscale Computational Engineering **4**(2), 255–263 (2006)
43. van Zuijlen, A.H., de Boer, A., Bijl, H.: Higher-order time integration through smooth mesh deformation for 3D fluid-structure interaction simulations. Journal of Computational Physics **224**, 414–430 (2007)
44. van Zuijlen, A.H., Bosscher, S., Bijl, H.: Two level algorithms for partitioned fluid-structure interaction computations. Computer Methods in Applied Mechanics and Engineering **196**(8), 1458–1470 (2006)

Least-Squares Spectral Element Methods in Computational Fluid Dynamics

Marc Gerritsma and Bart De Maerschalck

Abstract The least-squares spectral element method (LSQSEM) is a relatively novel technique for the numerical approximation of the solution of partial differential equations. The method combines the weak formulation based on the minimization of a residual norm, the least-squares formulation, with the higher-order spectral element discretization. A well-posed least-squares formulation leads to a symmetric, positive-definite system of algebraic equations which are highly amenable to well-established solvers such as the preconditioned conjugate gradient method. Furthermore, the formulation is very robust in the sense that no stabilization operators are required to acquire convergent solutions. The spectral element discretization renders high order accuracy to the scheme. This new numerical scheme is applied to incompressible, compressible and non-Newtonian flow problems.

1 Introduction

The least-squares spectral element method (LSQSEM) combines the weak formulation obtained from a least-squares minimization problem and a spectral element discretization. In the early 70's the least-squares technique was introduced for the solution of partial differential equations by Bramble, [10, 11]. The method was already described in Russian literature by Džiškariani, [30] and Lučka, [55]. In its initial form it was less competitive than the Galerkin formulation due to the high condition numbers and the more stringent regularity requirements. It was only when it was recognized that partial differential equations need to be rewritten in terms of

Marc Gerritsma
Dept. of Aerospace Engineering, TU Delft, Kluyverweg 1, 2629 HS Delft, The Netherlands e-mail: M.I.Gerritsma@TUDelft.nl

Bart De Maerschalck
Flemish Institute for Technological Research, Boeretang 200, BE-2400 Mol, Belgium e-mail: bart.demaerschalck@vito.be

equivalent first order partial differential equations that the method gained renewed interest.

In 2002/2003 two papers appeared shortly after another in the Journal of Computational Physics: [70] by Proot and Gerritsma, and [61] by Pontaza and Reddy. In both papers the least-squares formulation was combined with higher order spectral element methods. Since then the number of practitioners and publications in this field has grown.

The outline of this chapter is as follows: In Section 2 the Rayleigh-Ritz principle, the Galerkin formulation and the least-squares formulation are presented. In Section 3 a brief overview of the spectral element methods is given. In Section 4 convergence rates for well-posed problems are discussed. Applications of the numerical scheme are presented in the Sections 5 and 6, in which incompressible and compressible flows are discussed, respectively. In Section 7 new developments are briefly described. The final section, Section 8, gives an overview of the relevant literature.

2 The least-squares formulation

Weak formulations of differential equations consist of functionals over function spaces whose stationary points solve the original differential equation. These stationary points can be distinguished in (global) minima or saddle points. If a partial differential equation can be rephrased in a minimization problem, very robust numerical methods can be developed. A well-known form is the Rayleigh-Ritz formulation.

If, on the other hand, no equivalent minimization formulation can be established a more general formulation in terms of stationary points can be devised. The disadvantage of this approach is that the analysis of well-posedness of the weak formulation is much harder to perform and in many instances the well-posedness properties on the continuous level are not inherited by a discrete approximation.

In order to position the least-squares formulation within the framework of weighted residual methods and to introduce some notation and concepts we will briefly describe the Rayleigh-Ritz method, the Galerkin formulation and eventually the least-squares formulation.

2.1 Rayleigh-Ritz method

Consider the Poisson problem

$$\Delta \phi(x) = f(x), \quad x \in \Omega \subset \mathbb{R}^d, \tag{1}$$

with

$$\phi(x) = 0 \quad \text{for } x \in \Gamma_1 \subset \partial\Omega \quad \text{and} \quad \frac{\partial \phi}{\partial n} = 0 \quad \text{for } x \in \Gamma_2 \subset \partial\Omega, \quad \Gamma_1 \cap \Gamma_2 = \emptyset. \quad (2)$$

This partial differential equation with homogeneous boundary conditions can be converted in the following minimization problem:

$$\text{Find } \phi \in H \text{ which minimizes } \mathscr{I}(\psi) = \int_\Omega \left[\frac{1}{2} (\nabla \psi)^2 + f\psi \right] d\Omega. \quad (3)$$

The set H contains all admissible functions for the minimization problem to make sense. Firstly, all functions in H must satisfy the essential boundary condition $\psi(x) = 0$ for $x \in \Gamma_1$. Furthermore, the integral in (3) should exist. The set of functions for which the gradient of a scalar is square integrable, i.e. $\nabla \psi \in \left[L^2(\Omega)\right]^d$ and $\psi \in L^2(\Omega)$, ensures that

$$\mathscr{I}(\psi) \leq C \|\psi\|_H^2, \quad (4)$$

where H is defined as

$$H = \left\{ \psi \,|\, \psi \in L^2(\Omega) \text{ and } \nabla\psi \in \left[L^2(\Omega)\right]^d \text{ and } \psi(x) = 0 \text{ for } x \in \Gamma_1 \right\}, \quad (5)$$

with norm

$$\|\psi\|_H^2 = \|\psi\|_{L^2(\Omega)}^2 + \|\psi\|_{[L^2(\Omega)]^d}^2, \quad \forall \psi \in H. \quad (6)$$

If ϕ minimizes the functional $\mathscr{I}(\psi)$ we need to have that

$$D\mathscr{I}(\phi)\tilde{\psi} = 0, \quad \forall \tilde{\psi} \in H, \quad (7)$$

where the linear operator $D\mathscr{I}(\psi)$ is defined by, [79]

$$\lim_{\varepsilon \to 0} \frac{\mathscr{I}(\psi + \varepsilon\tilde{\psi}) - \mathscr{I}(\psi) - \varepsilon D\mathscr{I}(\psi)\tilde{\psi}}{\varepsilon} = 0. \quad (8)$$

For the functional given by (3) this gives

$$D\mathscr{I}(\psi)\tilde{\psi} = \lim_{\varepsilon \to 0} \frac{\int_\Omega \left[(\nabla\psi + \varepsilon\nabla\tilde{\psi})^2 - 2f(\psi + \varepsilon\tilde{\psi}) - (\nabla\psi)^2 + 2f\psi \right] d\Omega}{2\varepsilon} \quad (9)$$

$$= \lim_{\varepsilon \to 0} \frac{\int_\Omega \left[2\varepsilon \nabla\psi \nabla\tilde{\psi} + \varepsilon^2 (\nabla\tilde{\psi})^2 - 2\varepsilon f\tilde{\psi} \right] d\Omega}{2\varepsilon} \quad (10)$$

$$= \int_\Omega [\nabla\psi \nabla\tilde{\psi} - f\tilde{\psi}] d\Omega \quad (11)$$

$$= 0, \quad \forall \tilde{\psi} \in H. \quad (12)$$

This condition for a minimizer can be abstractly written as

$$a(\psi, \tilde{\psi}) = (f, \tilde{\psi}), \quad \forall \tilde{\psi} \in H. \quad (13)$$

The bi-linear form $a(\cdot,\cdot)$ is in this case symmetric, i.e. $a(u,v) = a(v,u)$.

Definition 1. A bi-linear form $a(\cdot,\cdot)$ on a normed linear space H is said to be **bounded** (or **continuous**) if $\exists C < \infty$ such that

$$|a(u,v)| \leq C\|u\|_H \|v\|_H, \quad \forall u,v \in H, \tag{14}$$

and **coercive** on H if $\exists \alpha > 0$ such that

$$a(u,u) \geq \alpha \|u\|_H^2, \quad \forall u \in H. \tag{15}$$

□

If a partial differential equation can be converted into a minimization problem over a function space H, where H is a Hilbert space and the bilinear symmetric form $a(\cdot,\cdot)$ is bounded and coercive, then the minimization possesses a unique solution. Furthermore, if the bilinear form is only coercive on a proper (closed) subspace $V \subset H$, then the minimization problem possesses a unique solution in V.

This last property, the fact that one may look for a minimizer on a closed subspace of H, allows one to construct finite-dimensional subspaces (finite elements) and look for an approximate solution in such a subspace V.

Furthermore, existence and uniqueness of a solution for the whole function space H are inherited by proper closed subsets $V \subset H$.

An alternative way to derive the conditions for a minimizer (12) is to take the differential equation and multiply it with an arbitrary function $\tilde{\psi} \in H$. After integration over the domain Ω this gives

$$\int_\Omega (-\Delta\phi + f)\tilde{\psi}\, d\Omega = 0. \tag{16}$$

Applying integration by parts and using the boundary conditions gives

$$0 = \int_\Omega (-\Delta\phi + f)\tilde{\psi}\, d\Omega \tag{17}$$

$$= \int_\Omega [-\nabla \cdot (\tilde{\psi}\nabla\phi) + \nabla\phi\nabla\tilde{\psi} + f\tilde{\psi}]\, d\Omega \tag{18}$$

$$= \int_\Omega [\nabla\phi\nabla\tilde{\psi} + f\tilde{\psi}]\, d\Omega - \int_{\partial\Omega} \tilde{\psi}(\nabla\phi, \mathbf{n})\, d\Gamma \tag{19}$$

$$= \int_\Omega [\nabla\phi\nabla\tilde{\psi} + f\tilde{\psi}]\, d\Omega, \tag{20}$$

where $\mathbf{n}$ denotes the outward unit normal at the boundary $\partial\Omega$. The boundary can be decomposed as $\partial\Omega = \Gamma_1 + \Gamma_2$. On Γ_1 $\tilde{\psi} = 0$ and on Γ_2 $\partial\phi/\partial n = 0$, so the boundary integral vanishes. The final result of the exercise shows that if ϕ solves the PDE, the weak formulation (20) is equal to the condition for a minimizer of the functional $\mathcal{J}$, (12).

The variational approach described above was already used by Rayleigh in 1870, [74], Ritz in 1908, [75] and Galerkin in 1915, [32]. But it was only until the 1960's

2.2 Galerkin formulations

It is not always possible to convert a differential equation to an associated minimization problem. A well-known example with applications in fluid dynamics is the steady linear advection equation given by

$$a\frac{du}{dx} = 0, \quad x \in (0,1), \quad a > 0, \tag{21}$$

with boundary condition $u(0) = u_0$. The method described by (16) is still applicable. This gives

$$a(u,v) = \int_\Omega a\frac{du}{dx} v \, d\Omega = 0. \tag{22}$$

This formulation where the partial differential equation is weighted by a suitably chosen set of test functions is called the **Galerkin formulation**. This weak formulation does not correspond to a minimization problem and generally the space of trial solutions, u, and test functions, v, are not the same. For the integral (22) to make sense we need to have that $u \in H^1(0,1)$ and $v \in L^2(0,1)$, so a setting in terms of Hilbert spaces is not possible. Consider the slightly more general case given by

$$\text{Seek } u \in W \text{ such that: } a(u,v) = f(v), \quad \forall v \in V, \tag{23}$$

where W and V are function spaces equipped with the norms $\|\cdot\|_W$ and $\|\cdot\|_V$, respectively. W is called the solution space and V is called the test space. We assume that the bi-linear form is continuous as defined in Definition 1 and $f \in V'$. Then the following theorem establishes the conditions for well-posedness, [31]

Theorem 1. *Let W be a Banach space and let V be a reflexive Banach space. Let $a \in \mathscr{L}(W \times V; \mathbb{R})$ and $f \in V'$, then problem (23) is well-posed if and only if*

$$\exists \alpha, \quad \inf_{w \in W} \sup_{v \in V} \frac{a(w,v)}{\|w\|_W \|v\|_V} \geq \alpha, \tag{24}$$

$$\forall v \in V, \quad (\forall w \in W, \ a(w,v) = 0) \implies (v = 0). \tag{25}$$

□

This condition is referred to as the inf-sup condition, the Ladyzhenskaya-Babuška-Brezzi (LBB) condition or the Banach-Nečas-Babuška condition. It is generally quite hard for a given weak formulation to prove the existence of the positive constant α, the coercivity constant.

In contrast to the minimization problem, well-posedness on the continuous level does not imply well-posedness for discrete conforming subspaces $W_h \subset W$ and $V_h \subset$

V. This means that for every problem discrete function spaces need to be identified such that well-posedness also holds for (W_h, V_h). Different weak formulations then require different approximating subspaces. Such may be the case with multi-physics simulations such as fluid-structure interaction problems.

In order to establish well-posedness and to develop a consistent and convergent scheme a variety of stabilization operators are generally required.

2.3 Least-squares formulation

An alternative approach of establishing equivalence between differential equations and weak formulations is offered by the so-called least-squares formulation. The least-squares formulation transforms partial differential equations into a minimization problem and in this sense extends the Rayleigh-Ritz formulation. The Rayleigh-Ritz formulation is limited to elliptic, self-adjoint problems. The previous subsection, 2.1 and 2.2, demonstrated that it is favorable to have a weak formulation which corresponds to a minimization problem, but that it is not always possible to convert a partial differential equation in terms of a minimization problem. The least-squares formulation provides a framework in which partial differential equations can be cast into a minimization setting, thus retrieving the appealing properties of the Rayleigh-Ritz principle.

This is established by introducing a norm equivalence between the residual and the error. Two norms, $\|\cdot\|_X$ and $\|\cdot\|_Y$, are called equivalent when two positive constants, C_1 and C_2, exist such that

$$C_1 \|u\|_X \leq \|u\|_Y \leq C_2 \|u\|_X, \quad \forall u \in X. \tag{26}$$

If we define the linear spaces to be

$$X = \{u \mid \|u\|_X \leq \infty\} \quad \text{and} \quad Y = \{u \mid \|u\|_Y \leq \infty\}, \tag{27}$$

norm-equivalence states that both sets contain the same elements, $X = Y$. Equivalent norms on a function space X define the same topology and more specifically, Cauchy sequences in both norms are the same. For finite-dimensional spaces all norms are equivalent.

Let the abstract first order partial differential equation be given by

$$\mathscr{L}u(x) = f(x), \quad \forall x \in \Omega, \tag{28}$$

with

$$\mathscr{R}u(x) = g(x), \quad \forall x \in \partial\Omega, \tag{29}$$

where $\mathscr{L}$ is a first order linear partial differential operator and $\mathscr{R}$ is a linear trace operator which prescribes boundary values. In the least-squares formulation, we now look for function spaces, X and Y, with associated norms such that we have the norm-equivalence

$$\alpha \|u\|_{X(\Omega)} \leq \|\mathcal{L}u\|_{Y(\Omega)} + \|\mathcal{R}u\|_{Z(\partial\Omega)} \leq C\|u\|_{X(\Omega)}, \quad \forall u \in X. \tag{30}$$

The space X is called the solution space and Y is called the residual space. Due to the linearity of the differential and trace operator we have that, if $u_{ex} \in X$

$$\alpha \|u - u_{ex}\|_{X(\Omega)} \leq \|\mathcal{L}u - f\|_{Y(\Omega)} + \|\mathcal{R}u - g\|_{Z(\partial\Omega)} \leq C\|u - u_{ex}\|_{X(\Omega)}, \quad \forall u \in X, \tag{31}$$

in which u_{ex} denotes the exact solution of (28-29). This equivalence tells us that small residual norms correspond to small error norms and conversely, small error norms correspond to small residual norms.

Based on this equivalence we can now define the so-called least-squares functional

$$\mathcal{J}(u) = \frac{1}{2}\left\{\|\mathcal{L}u - f\|^2_{Y(\Omega)} + \|\mathcal{R}u - g\|^2_{Z(\partial\Omega)}\right\}. \tag{32}$$

Solving the abstract PDE (28-29) is now equivalent to minimizing the least-squares functional $\mathcal{J}(u)$. So by applying the least-squares formulation, the problem has been recast in terms of a minimization problem thus avoiding stability prerequisites such as inf-sup conditions as discussed in the previous section.

Minimization of the least-squares functional requires setting $D\mathcal{J}(u)v = 0$ for all $v \in X$ analogous to the minimization procedure described in Section 2.1. Note also that the boundary conditions, $\mathcal{R}u = g$, are also incorporated in the minimization process.

Without loss of generality, we can set $g = 0$ in the equation $\mathcal{R}u = g$. This allows us to incorporate the boundary conditions in the space X. When we do so, the boundary conditions are strongly enforced and can be dropped from the variational statement.

If Y is a Hilbert space this gives the weak formulation

$$(\mathcal{L}u, \mathcal{L}v)_{Y(\Omega)} = (f, \mathcal{L}v)_{Y(\Omega)}, \quad \forall v \in X. \tag{33}$$

Note that this formulation is symmetric; interchanging u and v on the left hand side of this equation leaves the expression unchanged. Furthermore, the basic existence and uniqueness criterion, the norm-equivalence, (30), is inherited by conforming subspaces.

Although in the previous discussion general function spaces X and Y were treated, from a practical point of view it is convenient to choose function spaces which allow for easy calculation of the residual norm in the minimization process. So in many applications, the boundary conditions are already incorporated in the function space X and one takes $Y(\Omega) = L^2(\Omega)$. This leads to the least-squares functional

$$\mathcal{J}(u) = \frac{1}{2}\|\mathcal{L}u - f\|^2_{L^2(\Omega)}. \tag{34}$$

Minimization is in the L^2-sense, so point-wise convergence, in the L^∞-sense, is only attained if the exact solution is sufficiently regular. The next section discusses the construction of conforming finite-dimensional subspaces of the solution space X.

3 Spectral element methods

Instead of seeking the minimizer over the infinite-dimensional space X we restrict our search to a conforming subspace $X^h \subset X$ by performing a domain decomposition where the solution within each sub-domain is expanded with respect to a polynomial basis. The domain Ω is sub-divided into K non-overlapping quadrilateral open sub-domains Ω^k:

$$\Omega = \bigcup_{k=1}^{K} \bar{\Omega}^k, \quad \Omega^k \cap \Omega^l = \emptyset, \, k \neq l. \tag{35}$$

Each sub-domain is mapped onto the unit cube $(-1,1)^d$, where $d = \dim(\Omega)$. Within this unit cube the unknown function is approximated by polynomials. Generally a spectral element method based on *Legendre* polynomials, $L_k(x)$ over the interval $[-1,1]$, is employed, [15, 22, 52]. We define the Gauss-Lobatto-Legendre (GLL) nodes by the zeroes of the polynomial

$$\left(1 - x^2\right) L'_N(x), \tag{36}$$

and the Lagrange polynomials, $h_i(x)$, through these GLL-points, x_i, by

$$h_i(x) = \frac{1}{N(N+1)} \frac{(x^2 - 1) L'_N(x)}{L_N(x_i)(x - x_i)} \quad \text{for} \quad i = 0, \ldots, N, \tag{37}$$

where $L'_N(x)$ denotes the derivative of the Nth Legendre polynomial. For multi-dimensional problems tensor products of the one-dimensional basis functions are employed in the expansion of the approximate solution. We can therefore expand the approximate solution in each sub-domain in terms of a truncated series of these Lagrangian basis functions, which for $d = 2$ yields

$$u^N(x,y) = \sum_{i=0}^{N} \sum_{j=0}^{N} \hat{u}_{ij} h_i(x) h_j(y), \tag{38}$$

where the $\hat{u}_{ij}$'s are to be determined by the least-squares method. If we have converted a general higher order PDE to an equivalent first order system, C^0-continuity suffices to patch the solutions on the individual sub-domains together.

The integrals appearing in the least squares formulation, (34), are approximated by Gauss-Lobatto quadrature

$$\int_{-1}^{1} f(x)\,dx \approx \sum_{i=0}^{P} f(x_i) w_i, \tag{39}$$

where w_i are the GL weights given by

$$w_i = \frac{2}{P(P+1)} \frac{1}{L_P^2(x_i)}, \quad i = 0,\ldots,P \geq N. \tag{40}$$

It has been shown in [19] that it is beneficial for non-linear equations possessing large gradients to choose the integration order P higher than the approximation of the solution, N.

The method is not restricted to higher order methods based on Legendre polynomials. In [18] and [71] Lagrangian basis functions based on the Chebyshev polynomials were used for non-linear, time-dependent hyperbolic equations and the incompressible Navier-Stokes equations, respectively. Other systems of orthogonal polynomials can be easily introduced into the least-squares spectral element framework.

From a practical point of view, only least-squares formulations which allow for the use of C^0-finite or spectral elements are usable. Since C^0-finite or spectral element methods are based on piecewise continuously differentiable polynomials, standard finite and spectral elements can be used, which results in a very practical method from an implementational point of view. This can be accomplished by *first* transforming the system into a first order system and subsequently requiring that only (scaled) L^2-norms are used in the quadratic least-squares functional, see (34). The transformation into a first order system has two important consequences. First of all, the continuity requirements between neighboring spectral elements are mitigated such that C^0-finite or spectral elements can be used (in case the residuals are measured by L^2-norms). Secondly, the transformation will keep the condition number of the resulting discrete system under control [3, 14, 16].

4 Convergence and a priori error estimates

The H^1- and L^2-spaces are particularly suitable as the function spaces X and Y in least-squares finite or spectral element methods resulting from the minimization of first order partial differential equations with strongly imposed boundary conditions. To appreciate this, assume that we have the following norm-equivalence

$$\alpha \|u\|_{H^1(\Omega)} \leq \|\mathscr{L}u\|_{L^2(\Omega)} \leq C\|u\|_{H^1(\Omega)}, \quad \forall u \subset X = \{u \in H^1(\Omega) \mid \mathscr{R}u = 0 \text{ on } \partial\Omega\}, \tag{41}$$

where the space X represents the space of functions which already satisfy the homogeneous boundary condition and for which the function itself and its first derivatives are square integrable over the domain Ω. Based on the norm-equivalence (41), the quadratic least-squares functional can be obtained which upon minimization yields the weak formulation of the least-squares problem. If one uses a conforming finite-dimensional subspace $X^h \subset X = H^1(\Omega)$, then one can approximate $u^h \in X^h$ by piecewise continuously differentiable polynomials. In [68], it has been shown that least-squares methods based on the norm-equivalence (41) yield the following error estimate:

$$\left\| u - u^h \right\|_1 \leq \tilde{C} \inf_{v^h \in X^h} \left\| u - v^h \right\|_1, \tag{42}$$

where the constant $\tilde{C}$ is given by $\tilde{C} = 1 + 2C^2/\alpha^2$, with the constants α and C from (41). Here the subscript 1 in the norm $\|\cdot\|_1$ refers to the Sobolev norm $\|\cdot\|_{H^1(\Omega)}$, while $\|\cdot\|_0$ will denote $\|\cdot\|_{H^0(\Omega)} = \|\cdot\|_{L^2(\Omega)}$.

Since the interpolation of the solution u in the space X^h obviously belongs to the finite-dimensional subspace X^h, the right-hand side of equation (42) can be bounded in the following way:

$$\inf_{v^h \in X^h} \left\| u - v^h \right\|_1 \leq \left\| u - \pi_N^h(u) \right\|_1, \tag{43}$$

where $\pi_N^h(u) \in X^h$ represents the interpolation of the solution $u \in X$ in the space X^h consisting of polynomials of degree N. The right-hand side of expression (43) can be further bounded from above if an $h-$ or $p-$refinement strategy is used to obtain a better approximation of the solution $u \in X$. Since both refinement strategies result in different estimates for the interpolation error (43) and hence in a different error estimate (42), they are discussed separately hereafter.

Combining expression (43) with the interpolation error corresponding to $h-$ refinement, results in the following estimate if the exact solution $u \in H^s(\Omega)$ for some $s \geq 2$

$$\left\| u - u^h \right\|_1 \leq Ch^l |u|_{l+1}, \tag{44}$$

where $l = \min(N, s-1)$ and where N represents the approximating order of the C^0- spectral elements and where h represents a grid parameter (for example, the maximum of the square root of the area of the spectral elements), which decreases with increasing number of spectral elements. Here $|\cdot|_s$ denotes the semi-norm defined by

$$|v|_s^2 = \sum_{|\alpha|=s} \|D^\alpha v\|_{L^2(\Omega)}^2, \tag{45}$$

where α is a multi-index given by the vector

$$\alpha = (\alpha_1, \ldots, \alpha_d), \tag{46}$$

where the α_i are non-negative integers and

$$|\alpha| = \alpha_1 + \cdots + \alpha_d \tag{47}$$

We used the following notation for partial derivatives

$$D^\alpha = \frac{\partial^{|\alpha|}}{\partial x_1^{\alpha_1} \partial x_2^{\alpha_2} \ldots \partial x_d^{\alpha_d}}. \tag{48}$$

For example, if $d = 2$ and $\alpha = (1,2)$, then

$$D^\alpha u = \frac{\partial^3 u}{\partial x_1 \partial x_2^2}. \tag{49}$$

The rate of convergence (44) is *optimal* in the H^1–norm since it provides the highest possible rate of convergence allowed by polynomials of degree N, [73]. Note that the optimal rate of convergence depends on the polynomial degree (N) and the regularity of the exact solution, denoted by the Sobolev exponent s. A similar rate of convergence can be obtained for the L^2–norm. In [73], it is shown that the following L^2–error estimate holds if the exact solution $u \in H^s(\Omega)$ for some $s \geq 2$:

$$\left\| u - u^h \right\|_0 \leq C h^{l+1} |u|_{l+1}. \tag{50}$$

In case of p–refinement, the order N of the spectral elements is increased while keeping the number of the spectral elements constant. Consequently, the grid parameter h remains constant. As a consequence, the convergence rates (44) and (50) are not suitable in this case. In order to obtain useful convergence rates in case of p–refinement, one can use the Legendre interpolation operator to bound the interpolation error $u - \pi_N^h(u)$ from above. Assuming that $u \in H^s(\Omega)$ for some $s \geq 2$, it can be proven [73, page 126] that

$$\left\| u - \pi_N^h(u) \right\|_k \leq C N^{k-s} |u|_s, \quad \text{with } k = 0,1, \tag{51}$$

where $\pi_N^h(u)$ represents the Legendre interpolation operator applied to the exact solution u. Combining the latter expression with (42) and (43) results into the following error estimate

$$\left\| u - u^h \right\|_k \leq C N^{k-s} |u|_s, \quad \text{with } k = 0,1. \tag{52}$$

Note that the rate of convergence is only bounded by the smoothness degree of the solution s and not by any other grid parameter h as it occurs in the finite element case. As a consequence, exponential convergence rates can only be obtained for smooth problems if a p-refinement strategy is used. Since this rate of convergence results from p-refinement, it is called p-convergence hereafter.

Least-squares formulations based on the norm-equivalence (41) are called *fully H^1-coercive* formulations and they have optimal h–convergence rates in the H^1- and L^2-norms. Moreover, if the *fully H^1-coercive* least-squares method is supplemented with strongly imposed boundary conditions, one can use standard finite and spectral elements. Consequently, it is not so surprising that *fully H^1-coercive* formulations with strongly imposed boundary conditions are the preferred setting for least-squares finite and spectral element methods since these methods are both practical and optimally accurate.

Note that the above a priori estimates only depend on continuity and coercivity of the differential operator $\mathscr{L}$ and interpolation estimates. In [69, 70] these error estimates have been confirmed numerically with the use of Legendre polynomials. But since the estimates are independent of the type of orthogonal basis functions used, similar results will hold for polynomials such as Chebyshev polynomials. These error estimates will play a role in hp-adaptive schemes to be discussed in Section 7.1.

Unfortunately, not all differential equations can be converted into H^1-coercive least-squares formulations. Especially compressible, inviscid flows which exhibit shocks or contact discontinuities and models like Burgers' equation do not fit in the H^1-coercive framework and therefore require a different treatment to be discussed in Section 6.

5 Incompressible flows

For incompressible flows we can distinguish three regimes, the case where the Reynolds number is so low that the convective terms can be neglected (Stokes flow), the case where the Reynolds number is sufficiently small to yield steady solutions and for slightly higher Reynolds numbers, we have time-dependent flow.

Stokes flow has been thoroughly discussed by Proot, [69, 70]. All function spaces for which norm-equivalence can be established have been identified in these papers. In this chapter we will show some applications to steady and unsteady incompressible, viscous flow problems.

5.1 Governing equations

In terms of the primitive variables $(\mathbf{u}, p)$ the governing equations read

$$\frac{\partial \mathbf{u}}{\partial t} + (\mathbf{u} \cdot \nabla)\mathbf{u} = -\nabla p + \frac{1}{Re}\Delta \mathbf{u} + \mathbf{f} \quad \text{in } \Omega, \tag{53}$$

$$\nabla \cdot \mathbf{u} = 0 \quad \text{in } \Omega, \tag{54}$$

where $\mathbf{u}$ represents the velocity vector, p the kinematic pressure, $\mathbf{f}$ the forcing term per unit mass (if applicable) and Re the Reynolds number.

5.2 The first order formulation of the Navier-Stokes equations

In order to obtain an equivalent first order formulation of the unsteady Navier-Stokes equations, the vorticity ω has been introduced as an auxiliary variable. By using the identity $\nabla \times \nabla \times \mathbf{u} = -\Delta \mathbf{u} + \nabla(\nabla \cdot \mathbf{u})$ and by using the incompressibility constraint $\nabla \cdot \mathbf{u} = 0$, the governing equations subsequently read

$$\frac{\partial \mathbf{u}}{\partial t} + \mathbf{u} \cdot \nabla \mathbf{u} + \nabla p + \frac{1}{Re} \nabla \times \omega = \mathbf{f} \quad \text{in } \Omega, \tag{55}$$

$$\omega - \nabla \times \mathbf{u} = \mathbf{0} \quad \text{in } \Omega, \tag{56}$$

$$\nabla \cdot \mathbf{u} = 0 \quad \text{in } \Omega, \tag{57}$$

where, in the particular case of the two-dimensional problem, $\mathbf{u}^T = [u_1, u_2]$ represents the velocity vector, p the kinematic pressure, $\mathbf{f}^T = [f_1, f_2]$ the forcing term per unit mass (if applicable) and Re is the Reynolds number.

5.3 Linearization of the non-linear terms

Before the least-squares principles can be applied and the corresponding weak form discretized with spectral elements, the convective term $\mathbf{u} \cdot \nabla \mathbf{u}$ must be linearized. To this end, one can use a Picard (e.g. successive substitution)

$$\mathbf{u} \cdot \nabla \mathbf{u} \approx \mathbf{u}_0 \cdot \nabla \mathbf{u}, \tag{58}$$

or a Newton linearization

$$\mathbf{u} \cdot \nabla \mathbf{u} \approx \mathbf{u}_0 \cdot \nabla \mathbf{u} + \mathbf{u} \cdot \nabla \mathbf{u}_0 - \mathbf{u}_0 \cdot \nabla \mathbf{u}_0. \tag{59}$$

In the latter two equations, the subscript "0" indicates that the value of the corresponding variable is known from the previous iteration step.

If Picard linearization is used the following linearized momentum equation is obtained:

$$\frac{\partial \mathbf{u}}{\partial t} + (\mathbf{u}_0 \cdot \nabla \mathbf{u} + \nabla p + \nu \nabla \times \omega - \mathbf{f}) = 0. \tag{60}$$

For steady problems the time derivative can be dropped. For time-dependent calculations either a time-stepping scheme needs to be selected or space-time elements can be used, [19, 20, 21, 26, 63].

The least-squares formulation now becomes: Find $\mathbf{u}, p, \omega \in H^1(\Omega)$ which minimizes the functional in the absence of body forces

$$\mathscr{I}(\mathbf{u}, p, \omega) = \|\nabla \cdot \mathbf{u}\|_{L^2}^2 + \|\omega - \nabla \times \mathbf{u}\|_{L^2}^2 +$$

$$\left\| \frac{\partial \mathbf{u}}{\partial t} + (\mathbf{u}_0 \cdot \nabla \mathbf{u} + \nabla p + \nu \nabla \times \omega) \right\|_{L^2}^2 . \tag{61}$$

Variational analysis with respect to the four unknowns $\mathbf{u}$, p and ω leads to the symmetric positive definite system.

5.4 Steady flow around a cylinder at low Reynolds

One of the test cases considered for LSQSEM is the steady flow around a circular cylinder. This research was performed by De Groot, [41], with the conventional least-squares method discussed above and Direct Minimization, see Section 7.2.

Here we are simulating the flow around a circular cylinder placed perpendicularly in a uniform parallel flow, see Fig. 1.

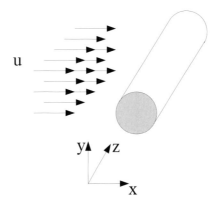

Fig. 1 Description of the flow problem

It is known that for flows with Reynolds numbers $Re < 45$ a steady solution exists and for Reynolds number higher than $Re = 45$ time-dependent behaviour sets in and the well-known Von Karman vortex street develops. The investigation of the flow will be in the Reynolds range: $1 < Re < 45$. In this range two different flow types are observed. For $Re < 6$ the flow is completely attached to the cylinder as can be seen in Fig. 2 (a). For $Re > 6$ the flow starts to separate somewhere on the cylinder forming two attached eddies behind the cylinder, Fig. 2 (b). For a benchmark between LSQSEM on the one hand and experimental data on the other, the estimation of the Re number for which the twin vortices appear, is an important property of the flow. This Re number is from here on referred to as Re_{onset}. The twin vortices grow in length and change shape as Re increases. The accurate prediction of this relation is a common benchmark. Fig. 3 gives a schematic overview of the flow where two steady vortices exist behind the cylinder.

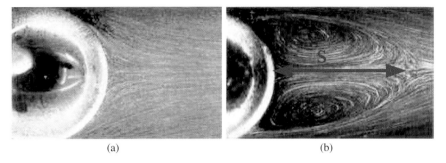

Fig. 2 Types of flow for low Re numbers, $Re < 6$ (a) and $Re > 6$ (b)

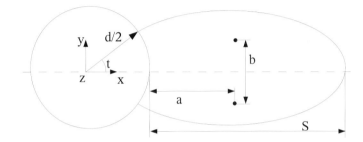

Fig. 3 Flow characteristics for a low Re flow around a circular cylinder

5.5 Comparison of numerical data with experimental data

Results of Taneda [81], Coutanceau [17] and Tritton [82] are used to benchmark the results of LSQSEM.

5.5.1 S/d: LSQSEM vs. Experiment

The LSQSEM values are compared to experimental values from Coutanceau and Taneda in Table 1. The same data are depicted in Fig. 4. For the LSQSEM simulations a mesh consisting of 62 elements was used with varying polynomial degree. The LSQSEM results given in Fig. 4 were obtained with polynomial degree $P = 10$ in both x- and y-direction. This corresponds to an 11th-order scheme.

The Onset Reynolds number The onset Reynolds number is the lowest Reynolds number for which the twin vortices behind the cylinder can be observed. By using polynomials of degree $N = 10$, Re_{onset} is sought on the mesh where symmetry is explicitly imposed. Fig. 5 displays S/d for various Reynolds numbers.

Vortex center The vortex center, of which the location is indicated by a and b in Fig 3, is also a property commonly investigated for this type of flow. The result for a and b can be found in Table 3 for the simulations as well as the experimental values of Coutanceau [17].

Table 1 Recirculation length S/d for LSQSEM and data from Coutanceau

Re	S/d Groot	Coutanceau	Taneda
10	0.27	0.34	0.3
20	0.94	0.93	0.9
30	1.57	1.53	1.5
40	2.17	2.13	2.1

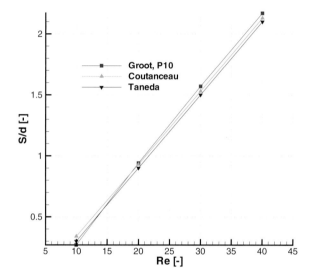

Fig. 4 Recirculation length S/d form LSQSEM and data form Coutanceau

Table 2 Recirculation length S/d, obtained by the least squares method and other numerical methods

Re	Groot	Groot 2003	Den & Shi 1965	Kaw & Jai 1966	Tak & Kel 1969	Nie & Kel 1973
10	0.27	0.25	0.56	0.3	0.25	0.217
20	0.94	0.88	1.06	1	0.935	0.803
30	1.57	1.44	1.16	1.75	1.611	1.543
40	2.17	1.94	0.94	2.515	2.325	2.179

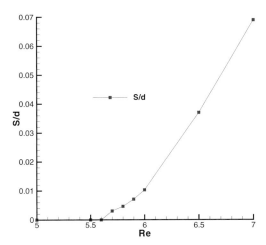

Fig. 5 Recirculation length in the vicinity of Re_{onset}

Table 3 Position of vortex core as a function of Re

Re	Groot, $P=10$		Coutanceau	
	a	b	a	b
10	0.12	0.24	0.12	0.32
20	0.35	0.44	0.33	0.47
30	0.53	0.52	0.55	0.54
40	0.68	0.58	0.76	0.59

Separation angle The separation angle is also a characteristic of the flow around a circular cylinder. For the definition of the separation angle see Fig. 3. Table 4 shows the results for the separation angle in the range, $10 \leq Re \leq 40$, for two different settings. Table 4 shows in the third column the experimentally obtained values of Coutanceau [17].

Table 4 Separation angle as function of Re

Re	Groot		Coutanceau
	$P=7$	$P=10$	
10	29.75	30.14	32.5
20	43.50	43.93	44.8
30	49.57	49.90	50.1
40	53.38	53.77	53.5

Drag coefficient Besides the flow field features as separation angle and recirculation length, also an integral quantity of the flow is investigated, the drag coefficient (C_D).

Almost every textbook on the fundamentals of Aerodynamics gives "accurate" formulae to predict C_D for a cylinder in uniform parallel flow. Let us investigate how well the least squares solution matches these curve fitted formulae. The first formula is derived by Tritton [82] and valid around $Re \approx 1.0$,

$$C_D \approx 1 + \frac{10.0}{Re_D^{2/3}} . \qquad (62)$$

The second formula, from Sucker and Brauer [80], is valid over a much wider range, $10^{-4} < Re < 2.0 \cdot 10^5$,

$$C_D \approx 1.18 + \frac{6.8}{Re_D^{0.89}} + \frac{1.96}{Re_d^1/2} - \frac{0.0004 \cdot Re_D}{1 + 3.64 \cdot e - 7 \cdot Re_D^2} . \qquad (63)$$

White [83] emphasizes the good accuracy of (63). Fig. 6 shows the converged results from the simulation together with the curve fits. Despite the claimed accuracy of the curve fits a significant gap with the numerical data emerges. To further investigate the accuracy of the numerically obtained C_D values, experimental data of Tritton [82] was used. Fig. 7 shows the numerical results obtained with the least squares method and the experimental results from Tritton [82] over the range: $1 \leq Re \leq 45$.

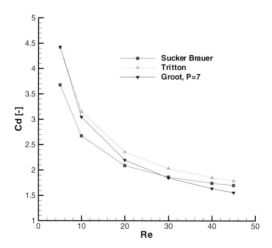

Fig. 6 Numerical results for C_D and curve fits form experimental data

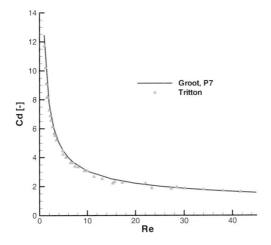

Fig. 7 Drag coefficient (C_D) from simulation and experiments

5.6 Unsteady flow around a cylinder at low Reynolds

For time dependent flows one can either employ time stepping methods or discretize the system of governing equations in space-time. The latter approach was investigated by De Maerschalck, [18, 19, 20, 21]. Kwakkel, [49, 53] performed a series of numerical simulations for the flow around a cylinder in a channel. A BDF3 scheme was used for the time integration.

The final test case is the periodic flow around a moving circular cylinder in a narrow channel. To be able to simulate the flow around the cylinder, the cylinder is fixed and the channel is moved (change of reference frame). The boundary conditions are $u = 1$ and $v = 0$ at the inflow and channel walls and a no-slip condition ($u = v = 0$) at the cylinder surface. The outflow boundary condition is the same as in the work of Pontaza, [58] and is described in Section 5.6.1.

5.6.1 Outflow boundary condition

The outflow boundary conditions are defined as

$$\left(-p + \frac{1}{\text{Re}}\frac{\partial u}{\partial x}\right)n_x + \frac{1}{\text{Re}}\frac{\partial u}{\partial y}n_y = 0, \tag{64}$$

$$\frac{1}{\text{Re}}\frac{\partial v}{\partial x}n_x + \left(-p + \frac{1}{\text{Re}}\frac{\partial v}{\partial y}\right)n_y = 0, \tag{65}$$

where n_x and n_y are the x and y-components of the outward unit normal along the boundary. This boundary condition is enforced in a weak sense through the least-squares functional, see (32).

Vortex shedding is described by the dimensionless Strouhal number

$$\text{St} = \frac{fl}{V}, \tag{66}$$

where f is the frequency of the vortex shedding, l the characteristic length (diameter of the cylinder) and V the free stream velocity of the fluid. Pontaza, [58], reports a Strouhal number of St=1/1.88=0.5319, see Fig. 8. The Reynolds number was set to $Re = 100$.

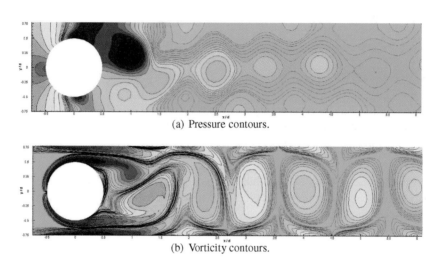

(a) Pressure contours.

(b) Vorticity contours.

Fig. 8 Instantaneous pressure and vorticity contours for $P_x = 8$.

5.6.2 Base pressure coefficient

The base pressure coefficient C_{pb} is defined by

$$C_{pb} = \frac{p_b - p_\infty}{q_\infty}, \tag{67}$$

where p_b is the base pressure, p_∞ the free stream pressure and where

$$q_\infty = \frac{1}{2}\rho_\infty V_\infty^2$$

is the dynamic pressure.

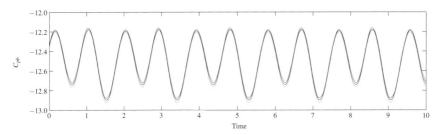

Fig. 9 Time history of C_{pb} $P = 6$ (red) $P = 8$ (blue) and the reference value from [58] with $P = 6$ (green)

Fig. 9 shows the time history of the base pressure coefficient C_{pb} for two polynomial degrees and the reference solution from [58].

5.6.3 Drag and lift coefficients

The lift C_L and drag C_D coefficients are often used to compare the results of the flow around a circular cylinder. These coefficients are defined by

$$C_D = \frac{F_x}{dq_\infty}, \quad C_L = \frac{F_y}{dq_\infty}, \tag{68}$$

where F is the force in the direction indicated, q_∞ the dynamic pressure and d the diameter of the cylinder, which is equal to unity. The forces are calculated by

$$\bar{F} = -\int_\Gamma p\,\mathrm{d}\bar{\Gamma} + \int_\Gamma \tau \cdot \mathrm{d}\bar{\Gamma}, \tag{69}$$

where Γ is the boundary of the cylinder and τ the extra stress tensor. In components this can be written as

$$F_x = -\int_\Gamma p n_x \mathrm{d}\Gamma + \int_\Gamma (\tau_{xx} n_x + \tau_{xy} n_y)\,\mathrm{d}\Gamma, \tag{70}$$

$$F_y = -\int_\Gamma p n_y \mathrm{d}\Gamma + \int_\Gamma (\tau_{yx} n_x + \tau_{yy} n_y)\,\mathrm{d}\Gamma. \tag{71}$$

The extra stress tensor for a Newtonian incompressible fluid is

$$\tau = \mu \begin{bmatrix} 2\frac{\partial u}{\partial x} & \frac{\partial u}{\partial y} + \frac{\partial v}{\partial x} \\ \frac{\partial v}{\partial x} + \frac{\partial u}{\partial y} & 2\frac{\partial v}{\partial y} \end{bmatrix}, \tag{72}$$

where μ is the dynamic viscosity. For Re < 188 the flow will be periodic in time, so both C_D and C_L must be periodic. The frequencies of the drag and lift curves both depend on the frequency of the vortex shedding. The frequency of the drag curve has twice the frequency of the lift curve however. This is because for the lift it is important *which* vortex is shed off (top or bottom) and for the drag it is only important *if* a vortex is shed. Comparison of Fig. 10 and 11 shows that this is the case.

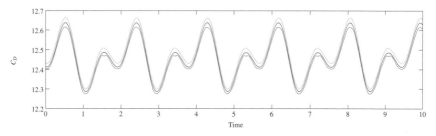

Fig. 10 Time history of C_D; $P = 6$ (red) $P = 8$ (blue) and the reference value from [58] with $P = 6$ (green)

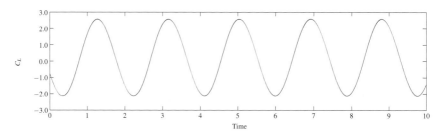

Fig. 11 Time history of C_L; $P = 6$ (red) $P = 8$ (blue) and the reference value from [58] with $P = 6$ (green)

Fig. 11 shows that the results for C_L agree with the reference values. The frequency that follows from (66) is St $= 1/1.88 = 0.5319$, which agrees with the value of [58]. The value of 1.88 is obtained from the time history plot of the lift coefficient and is the time between two peaks, see Fig. 11.

6 Compressible flows

Inviscid, compressible flows may exhibit discontinuous solutions – shocks or contact discontinuities. These solutions are not in $H^1(\Omega)$ and therefore an H^1-coercive formulation is not possible. The proper functional setting for these problems is in the weaker space $H(\text{div};\Omega)$. Since in the finite-dimensional case all norms are equivalent, one can still use the conventional least-squares approach but in this case optimal convergence is not guaranteed.

Compressible flows in the absence of dissipative terms are governed by the Euler equations. There are several ways in which the Euler equations in differential form can be written, but only the conservative form in terms of conserved quantities will be presented.

The two-dimensional Euler equations in conservation form are given by

$$\frac{\partial}{\partial t}\begin{bmatrix}\rho \\ \rho u \\ \rho v \\ \rho E\end{bmatrix} + \frac{\partial}{\partial x}\begin{bmatrix}\rho u \\ \rho u^2 + p \\ \rho uv \\ \rho uH\end{bmatrix} + \frac{\partial}{\partial y}\begin{bmatrix}\rho v \\ \rho uv \\ \rho v^2 + p \\ \rho vH\end{bmatrix} = \begin{bmatrix}0 \\ 0 \\ 0 \\ 0\end{bmatrix}. \tag{73}$$

These equations express conservation of mass, conservation of momentum in the x- and y-direction and conservation of energy, respectively. Here ρ is the local density, p is the pressure and (u,v) denotes the fluid velocity. The total energy per unit mass is denoted by E. The total energy can be decomposed into internal energy e and the kinetic energy per unit mass

$$\rho E = \rho e + \frac{\rho}{2}\left(u^2 + v^2\right) = \frac{p}{\gamma - 1} + \frac{\rho}{2}\left(u^2 + v^2\right), \tag{74}$$

where in the last equality we assume a calorically ideal, perfect gas with γ the specific ratio of heats of the gas.

6.1 Compressible flow over a circular bump

In this section results are given for the flow over a circular bump in a 2D channel. Results will be given for subsonic flow, $M_\infty = 0.5$, transonic flow, $M_\infty = 0.85$ and supersonic flow, $M_\infty = 1.4$. This is a difficult test problem over the entire Mach range for spectral methods due to the presence of stagnation points at the leading and trailing edge of the bump. See [38] for further details of this approach.

6.1.1 General geometry and boundary conditions

The general geometry for the channel flow with a circular bump is shown in Fig. 12. The bump is modeled by curved elements using the transfinite mapping by Gordon and Hall, [40]. All length and height parameters of the channel will be scaled with the chord length c of the bump.

The influence of the mesh is assessed by refining the mesh around the stagnation point. The refined mesh consists of 72 elements, Fig. 13.

The entropy variation s in the domain is calculated with the freestream entropy as a reference:

$$s = \frac{\hat{s} - \hat{s}_\infty}{\hat{s}_\infty}, \quad \text{where } \hat{s} = p\rho^{-\gamma}. \tag{75}$$

Fig. 12 The general geometry of the 2D channel with a circular bump

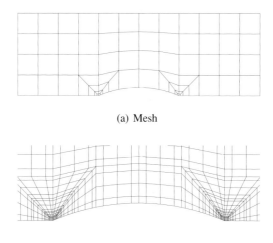

Fig. 13 Refined mesh near stagnation points consisting of 72 spectral elements of polynomial degree $N = 4$. (a) Spectral element mesh, (b) close-up near the bump: spectral elements with Gauss-Lobatto-grid

Pressure contours are given in Fig. 14.

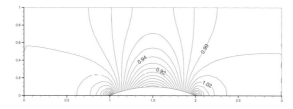

Fig. 14 Pressure contours for the subsonic flow

The results along the lower wall for a polynomial degree $N = 4$, integration order $P = 5$, are displayed in Fig. 15. This figure shows that the resolution of the stagnation points is very pronounced and the flow almost retains its inflow Mach number after the bump. The entropy change remains very small over the bump and the artificial entropy increase is restricted to the location of the stagnation points.

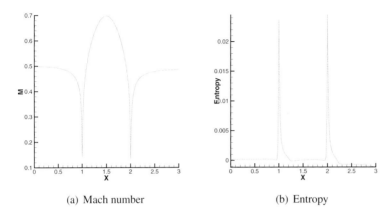

(a) Mach number   (b) Entropy

Fig. 15 The Mach number and entropy distribution for the subsonic flow on a refined mesh

6.1.2 Results for transonic flow

To investigate the transonic flow over a bump the geometry is the same as that for the transonic flow problems described by Spekreijse, [78], and Rizzi and Viviand, [76].

As in the subsonic case, the chord length of the bump is $c = 1$. The length of the channel however, is 5 times the chord length and the height is set at 2.073 times the chord length. The height of the bump is 4.2% of the chord length. The mesh used for this test case is shown in Fig. 16. The polynomial degree is $N = 5$ whereas the integration order is $P = 6$.

In Fig. 17 the Mach contours at an inflow Mach number of $M = 0.85$ are compared to the finite volume results produced by Spekreijse, [78].

The shock is positioned at approximately 86% of the bump with the Mach number just upstream of the shock being $M \approx 1.32$. These results are quantitatively in agreement with the finite volume results obtained by Spekreijse. In Fig. 18 the Mach number distribution along the lower wall of the channel is shown.

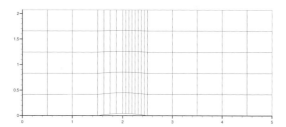

Fig. 16 The mesh used for the transonic test case. The height of the bump is 4.2% of the chord length and 100 elements are used.

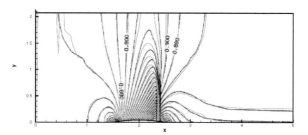

Fig. 17 Iso-Mach lines for transonic flow, $M = 0.85$, obtained by LSQSEM (green) and Finite Volume Method by Spekreijse [78], (black)

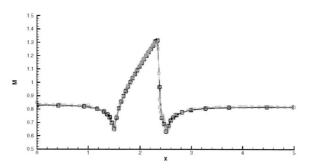

Fig. 18 Mach number distribution along the lower wall of the channel for a $M = 0.85$ flow: LSQSEM solution (green line) and Finite Volume results obtained by Spekreijse, [78], (black line)

6.1.3 Results for supersonic flow

The geometry used for the supersonic test case is similar to that considered for the subsonic flow test case. The only difference is the height of the bump which is 4% of the chord length for this test case. The mesh has a total of 120 elements as can be seen in Fig. 19.

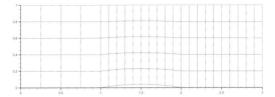

Fig. 19 The mesh used for the supersonic test case. The height of the bump is 4% of the chord length and 120 elements are used.

At inflow the density is set to $\rho = 1.4$ and the pressure to $p = 1$. The Mach number at the inlet boundary is $M = 1.4$.

At the leading edge of the bump a shock develops and runs into the domain until it is reflected by the upper wall. From the trailing edge a shock originates at a slightly smaller angle than the shock at the leading edge. In the region behind the bump the two shocks collide and then merge into a single shock. The iso-Mach contours for this test case are shown in Fig. 20 together with the results obtained by Spekreijse, [78]. This figure reveals that the shock structures agree.

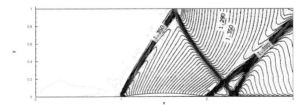

Fig. 20 Iso-Mach lines and shock structure obtained by LSQSEM (green) and Finite Volume Method by Spekreijse [78], (black).

7 Miscellaneous topics

In this section several topics will be discussed which improve the performance of LSQSEM.

7.1 *hp-adaptive LSQSEM*

Although the combination of the least-squares formulation and the spectral element technique leads to a very robust scheme which does not require any form of stabilization and renders highly accurate solutions, using high order polynomials throughout the entire computational domain is very expensive. The following procedure is described in [33].

7.1.1 The mortar element method

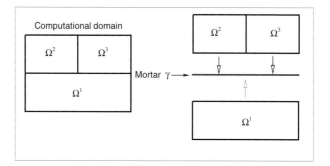

Fig. 21 The mortar element approach - patching the element edges with one approximation space

In the mortar element method (MEM), neighbouring elements in $\mathbb{R}^d$ are patched together by mortar-like elements in $\mathbb{R}^{d-1}$. In $\mathbb{R}^2$ the mortar elements consist of line segments as sketched in Fig. 21. The ith boundary of element k, denoted by Γ_i^k, is associated with a number of mortars, γ_j. The solution on the mortars, ϕ, is connected to the solution at the border of the two neighboring elements. This establishes a connection between the solution at the edge of an element, denoted by u_b, and the mortar solution ϕ. If we choose a polynomial approximation for ϕ on the mortar, we can express the expansion coefficients at the boundary of the element, $\mathbf{u}_b$, in terms of the expansion coefficients of the solution on the mortar, ϕ, as

$$\mathbf{u}_b = \tilde{Z}\phi . \tag{76}$$

The precise relation is inconsequential as long as the matrix $\tilde{Z}$ is of full rank for at least one of the elements associated with the mortar. This condition prevents the appearance of spurious mortar solutions.

Having established the relation between the elemental boundary unknowns and the mortar unknowns, we can express the global system in terms of the inner element unknowns, $\mathbf{u}_i$ and the mortar unknowns ϕ only:

$$\mathbf{u}^k = \begin{bmatrix} \mathbf{u}_b \\ \mathbf{u}_i \end{bmatrix} = \begin{pmatrix} \tilde{Z} & 0 \\ 0 & I \end{pmatrix} \begin{bmatrix} \phi \\ \mathbf{u}_i \end{bmatrix} = \begin{bmatrix} Z^k \end{bmatrix} \tilde{u}^k, \tag{77}$$

where $\tilde{u}^k$ represents the *true unknowns*, i.e. the projected mortar values and the internal unknowns.

This transformation converts the least-squares formulation into

$$L^T W L \mathbf{u}^k = L^T W \mathbf{f}^k \iff \left[Z^k\right]^T L^T W L \left[Z^k\right] \tilde{u}^k = \left[Z^k\right]^T L^T W \mathbf{f}^k. \tag{78}$$

Assembling all element contributions by summing over the projected element matrices gives the global system to be solved.

In this paper the solution on the mortar is defined by an L^2-projection of the element boundary solution

$$\int_{\bar{\Gamma}_l^k} \left(u|_{\Omega^k} - \phi \right) \psi \, ds = 0, \quad \forall \text{ sides } l \text{ and } k = 1, \ldots, K, \tag{79}$$

where $\psi \in P_M\left(\bar{\Gamma}_l^k\right)$ and M is the polynomial degree of the mortar solution. M should be greater than or equal to the degree of the solution in the adjoining elements to prevent spurious mortar solutions. When the Lagrangian basis functions, defined in Section 3, are employed, the vertex condition (Maday et al. [56]) is automatically satisfied. For a more extensive treatment of the mortar element method the reader is referred to [48, 56].

7.1.2 The error estimator

Having discussed how to match spectral elements with different size and polynomial representation, we now have to find a way to detect those elements that need refinement.

In the least-squares formulation we select the solution which minimizes the residual globally over the whole domain Ω. Having obtained such a minimizing solution we can evaluate the least-squares functional locally over every sub-domain $A \subset \Omega$. This gives

$$\eta_A^2 = \|\mathscr{L}u^h - f\|_{L^2(A)}^2 = \|\mathscr{L}\left(u^h - u_{ex}\right)\|_{L^2(A)}^2, \tag{80}$$

due to the linearity of $\mathscr{L}$. Well-posedness of the problem (30) then implies that

$$\alpha^2 \|e\|_{X(A)}^2 \leq \eta_A^2 \leq C^2 \|e\|_{X(A)}^2, \tag{81}$$

where $e = u^h - u_{ex}$. This means that the effectivity index $\theta_{A,X}$, defined by

$$\theta_{A,X} = \frac{\eta_A}{\|e\|_{X(A)}}, \tag{82}$$

which compares the estimated error η_A to the exact error in the X-norm is bounded by

$$\alpha \leq \theta_A \leq C. \tag{83}$$

Alternatively, we may compare the estimated norm η_A with the residual norm $\|R\|_{L^2(A)}$ using the fact that the residual norm is norm equivalent to $\|e\|_X$. Denoting this effectivity index by $\theta_{A,R}$ gives the rather trivial result

$$\theta_{A,R} := \frac{\eta_A}{\|\mathscr{L}u^h - f\|_{Y(A)}} \equiv 1. \tag{84}$$

Based on this observation η_A will be used to identify those regions (elements in case $A = \Omega^k$) which are selected for refinement. This estimator has also been used by Liu, [54] and Berndt et al., [4].

7.1.3 Estimation of the Sobolev regularity

Having found a way to match functionally and geometrically non-conforming elements and an indicator η_{Ω^k} which flags elements for refinement, we now have to determine how to refine. This choice is based on the smoothness of the underlying exact solution. If the exact solution is locally sufficiently smooth, polynomial enrichment is employed. However, if on the other hand, the underlying exact solution has limited smoothness h-refinement is used.

Let κ be a spectral element with size parameter h_κ and polynomial degree p_κ. Let u_{ex}^κ be the exact solution in that element, where $u_{ex}^\kappa \in H^{k_\kappa}$, where $k_\kappa \geq 0$ denotes the Sobolev regularity of the exact solution. Let $u_{p_\kappa}^{h_\kappa}$ denote the LSQSEM solution with $u_{p_\kappa}^{h_\kappa} \in H^q$, $0 \leq q \leq k_\kappa$ then

$$\|u_{ex}^\kappa - u_{p_\kappa}^{h_\kappa}\|_{H^q} \leq C \frac{(h_\kappa)^{s_k - q}}{(p_\kappa)^{k_\kappa - q}} \|u_{ex}^\kappa\|_{H^{k_\kappa}}, \tag{85}$$

where $s_\kappa = \min(p_\kappa + 1, k_\kappa)$ and C is a generic constant. This error estimate tells us that if the solution is very smooth (k_κ very large) then the error decreases more rapidly by increasing p_κ in the denominator. For practical purposes the function is considered smooth if $k_\kappa > p_\kappa + 1$ and non-smooth when $k_\kappa \leq p_\kappa + 1$, in which h-refinement is more effective.

So the choice between h-refinement and p-enrichment is dictated by the Sobolev index of the exact solution. Although the exact solution is in general not available, we can still estimate this index from its numerical approximation. Houston et al., [50] have developed a method to estimate the Sobolev index from a truncated Legendre series. They assume that the one-dimensional solution is in $L^2(-1,1)$ which allows for a Legendre expansion

$$u(x) = \sum_{i=0}^{\infty} a_i L_i(x), \quad \text{with} \quad a_i = \frac{2i+1}{2} \int_{-1}^{1} u(x) L_i(x) \, dx. \quad (86)$$

By Parseval's identity the fact that $u \in L^2(-1,1)$ is equivalent to convergence of the series

$$\sum_{i=0}^{\infty} |a_i|^2 \frac{2}{2i+1}. \quad (87)$$

In [50] it is shown that if

$$\sum_{i=[k+1]}^{\infty} |a_i|^2 \frac{2}{2i+1} \frac{\Gamma(i+k+1)}{\Gamma(i-k+1)}, \quad (88)$$

converges, then $u \in H_w^k(-1,1)$, where

$$H_w^k = \left\{ u \in L^2(-1,1) \mid \sum_{j=0}^{k} \int_{-1}^{1} \left| D^{(j)} u(x) \right|^2 (1-x^2)^j \, dx < \infty \right\}, \quad (89)$$

for integer values of k. By using the Γ-function in (88) this identity can be extended to fractional Sobolev spaces, see [50] for details.

Given the Legendre coefficients a_i, convergence of the series in (88) can be established by well-known techniques such as the ratio test, or the root test.

In this work the root test is employed. This leads to the calculation of

$$l_k = \frac{\log\left(\frac{2k+1}{2|a_k|^2}\right)}{2 \log k}. \quad (90)$$

If $l = \lim_{k \to \infty} l_k > 1/2$ then $u \in H_w^{l-1/2-\varepsilon}(-1,1)$, $\forall \varepsilon > 0$. Else $u \in L^2(-1,1)$. Since in a numerical solution only a finite number of Legendre coefficients a_i are available, this test is applied to the highest Legendre coefficient available in the numerical approximation. Based on the estimated Sobolev index $l - 1/2$, the decision is made whether to refine the mesh, or to increase the polynomial degree locally.

This one-dimensional estimate of the Sobolev index can be extended to multi-dimensional problems by treating each co-ordinate direction separately, see [50] for details. Several tests have been performed to establish the validity of this estimator.

Proposition 1. *If an element κ at refinement level r with characteristic mesh size h_κ^r and polynomial degree p_κ^r is flagged for refinement by the error indicator, we calculate l_{p_κ} by (90). Then for the Sobolev index $k_{p_\kappa} = l_{p_\kappa} - 1/2$ we have*

$$\begin{cases} \text{If } k_{p_\kappa} > p_\kappa^r + 1 \text{ then } p_\kappa^{r+1} \leftarrow p_\kappa^r + 1. \\ \text{If } k_{p_\kappa} \leq p_\kappa^r + 1 \text{ then } h_\kappa^{r+1} \leftarrow h_\kappa^r / 2. \end{cases} \quad (91)$$

7.1.4 Application to the space-time linear advection equation

This section uses the one-dimensional, linear advection problem to validate the presented hp-adaptive theory. The application of LSQSEM to hyperbolic equations has been studied by De Maerschalck, [18, 19, 20, 21].

The model problem is defined as

$$\frac{\partial u}{\partial t} + a \frac{\partial u}{\partial x} = 0 \quad \text{with } 0 \leq x \leq L, t \geq 0, a \in \mathbb{R} \tag{92}$$

$$u(0,t) = 0, \tag{93}$$

$$u(x,0) = \begin{cases} \frac{1}{2} - \frac{1}{2}\cos\left(2\pi \frac{x-x_0}{L_0}\right) & \text{if} \quad x_0 \leq x \leq x_0 + L_0 \\ 0 & \text{elsewhere} \end{cases} \tag{94}$$

where a is the advection speed and L is the length of the domain in spatial direction. On the left boundary of the domain a Dirichlet boundary condition, $u(0,t) = 0$, is imposed. The initial disturbance, $u(x,0)$, is a cosine-hill with offset x_0 and width L_0. We use a space-time formulation that treats the one-dimensional advection problem

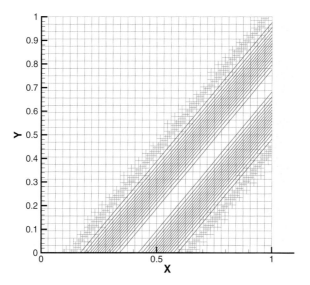

Fig. 22 Illustration of the unstructured mesh and continuous propagation of the cosine-hill with hp-refinement

with a two-dimensional least-squares formulation and considers the time variable t

as second spatial variable. Instead of calculating the solution over the whole domain at once, we use a semi-implicit approach. The domain is subdivided into several space-time strips for which the solution is approximated using the LSQSEM, see [20] for details. In the following, 32 time strips are used within the domain $\Omega = [0,1]$, where each time strip has initially 32 cells. Each time strip uses the solution at the previous strip as initial condition, except for the first strip that uses the prescribed initial condition (94). All time strips use the Dirichlet condition (93) on the left boundary. The advection speed $a = 0.85$, $x_0 = 0.13$ and $L_0 = x_0 + 0.5$. The exact solution of this problem $u_{ex} \in H^{5/2-\varepsilon}(\Omega)$, for all $\varepsilon > 0$. The regularity of the exact solution is limited in space-time over the lines $x - at = x_0$ and $x - at = x_0 + L_0$. For all other points in the space-time domain the solution is infinitely smooth.

7.1.5 Illustration of an *hp*-adaptive strategy

Note that even though only the second derivative is discontinuous, the regularity estimator accurately identifies the region with limited regularity as depicted in Figure 22. No *h*-refinement is used in the smooth parts of the domain.

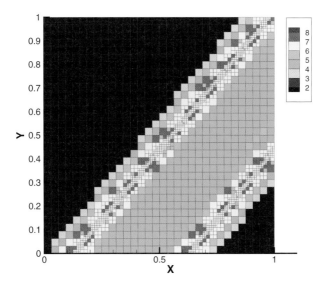

Fig. 23 Polynomial degree distribution for *hp*-adaptivity within each element

Figure 23 shows the final polynomial degree distribution. We imposed the so-called 1-level adaptivity, where the difference in refinement level between neighbouring elements may not be more than one. In the regions where the exact solution is zero neither *p*-enrichment nor *h*-refinement is used. Along the cosine hill polynomials of degree $N = 4$ are used, the green strip. The only place where the algorithm uses higher order elements is along the lines with limited regularity. Along these lines both *p*-enrichment and *h*-refinement are employed.

The reason high order elements are used along these lines is that in order to predict the Sobolev regularity accurately enough, one needs sufficiently high Legendre coefficients.

7.2 Direct Minimization

In this section an alternative method will be described which minimizes the residual directly in contrast to the conventional least-squares formulation where one employs variational analysis to set up the weak formulation. The resulting condition number is only the square root of the condition number that would be obtained if the conventional least squares method had been used. In addition, the new method circumvents a costly matrix-matrix multiplication thus avoiding loss of precision and fill-in in the stiffness matrix. This approach is also described in [49].

7.2.1 Conventional least-squares finite element method

In Section 2 and 3 the approach for the conventional or variational least-squares formulation is described. This approach can be summarized as

$$(\mathscr{L}(u),\mathscr{L}(v)) = (f,\mathscr{L}(v)) \iff \int_\Omega \mathscr{L}(u)\mathscr{L}(v)\,d\Omega = \int_\Omega f\mathscr{L}(v)\,d\Omega \quad \forall v \in X(\Omega). \tag{95}$$

The next step consists of domain decomposition where the integration over Ω is written as the sum of the integrals over the sub-domains Ω^k, $k = 1,\ldots,K$,

$$\sum_k \int_{\Omega^k} \mathscr{L}(u)\mathscr{L}(v)\,d\Omega^k = \sum_k \int_{\Omega^k} f\mathscr{L}(v)\,d\Omega^k \quad \forall v \in X(\Omega^k). \tag{96}$$

Then we insert the finite-dimensional approximation for each element $u^{N,k} = \sum_i u_i^{N,k}\phi_i(x)$, where the ϕ_i are basis functions, which span the finite-dimensional subspace over Ω^k

$$\sum_k \left[\sum_i u_i^{N,k} \int_{\Omega^k} \mathscr{L}(\phi_i)\mathscr{L}(\phi_j)\,d\Omega^k \right] = \sum_k \int_{\Omega^k} f\mathscr{L}(\phi_j)\,d\Omega^k \quad \forall \phi_j, j = 1,\ldots,N. \tag{97}$$

It suffices in (97) to take $v = \phi_j$, because $\mathscr{L}$ is assumed to be a linear operator and since any arbitrary v in the finite-dimensional subspace is a linear combination of these basis functions.

Inserting the Gauss-Lobatto integration then gives

$$\sum_{k}\left[\sum_{i}u_{i}^{N,k}\sum_{p}\mathscr{L}(\phi_{i})(x_{p})\mathscr{L}(\phi_{j})(x_{p})w_{p}\right] = \sum_{k}\sum_{p}f(x_{p})\mathscr{L}(\phi_{j})(x_{p})w_{p},$$
$$\forall \phi_{j}, j = 1,\ldots,N. \tag{98}$$

Here x_p denote the Gauss-Lobatto points and w_p the Gauss-Lobatto weights defined by (36) and (40), respectively. Note that in the multi-dimensional case x_p is a vector, ϕ_i is a tensor product and w_p is the product of the Gauss-Lobatto weights in each direction separately. We now define in each element the matrix A^k by

$$\left(A^{k}\right)_{pi} = \mathscr{L}(\phi_{i})(x_{p}), \tag{99}$$

i.e. the matrix coefficient denotes the application of the differential operator to the ith basis function evaluated at the pth Gauss-Lobatto point. Furthermore we introduce the diagonal weight matrix W^k by

$$\left(W^{k}\right)_{pp} = w_{p}. \tag{100}$$

The discretized least-squares problem (98) can then be written as

$$\sum_{k}\left[\left(A^{k}\right)^{T}WA^{k}\right]u^{N,k} = \sum_{k}\left[\left(A^{k}\right)^{T}WF\right], \tag{101}$$

where the vector F contains the elements $(F)_p = f(x_p)$. The system of algebraic equations obtained in this way, i.e. using variational analysis, is called the *normal equations*. The normal equations reflect on a discrete level the symmetry that was already mentioned at the continuous level. Note that Gauss-Lobatto integration may be performed on a finer grid than the grid on which the unknowns are defined. In this case the matrix A^k is non-square, i.e. there are more rows than columns in the matrix. The resulting normal equations, however, deliver a square, positive definite matrix which possesses a unique solution.

7.2.2 Direct Minimization - LSQSEM-DM

In order to avoid variational analysis we start with the original minimization problem.

$$\text{Find } u \in X(\Omega) \text{ which minimizes } \mathscr{I}(u) = \frac{1}{2}\|\mathscr{L}u - f\|_{Y(\Omega)}^{2}. \tag{102}$$

Since we decompose the computational domain Ω into a union of non-overlapping sub-domains Ω^k, $k = 1,\ldots,K$, we can also write this as

Find all $u^k \in X(\Omega^k)$ which minimize the functional

$$\mathscr{I}(u^1,\ldots,u^K) = \sum_{k=1}^{K} \left\| \mathscr{L}u^k - f \right\|_{Y(\Omega^k)}^2. \tag{103}$$

Now in each domain Ω^k we are going to restrict our search to a finite-dimensional subspace $X^N(\Omega^k) \subset X(\Omega^k)$ using the spectral approximation given by (38)

Find all $u^{N,k} \in X^N(\Omega^k)$ which minimize the functional

$$\mathscr{I}(u^{N,1},\ldots,u^{N,K}) = \sum_{k=1}^{K} \left\| \mathscr{L}u^{N,k} - f \right\|_{Y(\Omega^k)}^2. \tag{104}$$

Next we introduce numerical quadrature to evaluate the integrals which constitute the L^2-norm. This gives

Find all $u^{N,k} \in X^N(\Omega^k)$ which minimize the functional

$$\mathscr{I}(u^{N,1},\ldots,u^{N,K}) \approx \sum_{p=0}^{N_{int}^k} \sum_{k=1}^{K} \left(\mathscr{L}u^{N,k} - f \right)^2 \bigg|_{x_p} w_p, \tag{105}$$

where N_{int}^k denotes the number of integration points in element k. Introducing our matrix notation (99) and (100) this can be written as

Find all $u^{N,k} \in X^h(\Omega^k)$ which minimize the functional

$$\sum_{k=1}^{K} \left(A^k u^{N,k} - F^k \right)^T W^k \left(A^k u^{N,k} - F^k \right) = \sum_{k=1}^{K} \| \sqrt{W^k} \left(A^k u^{N,k} - F^k \right) \|^2. \tag{106}$$

So the procedure of domain decomposition, insertion of an approximate solution and the use of numerical integration has converted the minimization in the function space $L^2(\Omega)$ to a minimization problem in Euclidean space: Find the finite-dimensional vector fields $u = (u^1,\ldots,u^K)^T \in \mathbb{R}^n$ such that the norm in $\mathbb{R}^m$ given by (106) is minimized. If $m = n$, i.e. the number of unknowns in the global system equals the number of equations, the use of the weight matrix W^k is then inconsequential and the problem reduces to a collocation method evaluated in the Gauss-Lobatto-Legendre points, [42, 43, 44], given by

$$\sum_{k=1}^{K} \left(A^k u^{N,k} - F^k \right) = 0. \tag{107}$$

In case $m > n$, we have more equations than unknowns and the solution which minimizes the residual norm of the overdetermined system is then given by

$$\sum_{k=1}^{K} \sqrt{W^k} A^k u^k = \sum_{k=1}^{K} \sqrt{W^k} F^k. \tag{108}$$

Let us for convenience introduce the following notation $B = \sum_{k=1}^{K} \sqrt{W^k} A^k$ and $G = \sum_{k=1}^{K} \sqrt{W^k} F^k$. Then we have the following Theorem, [5, 6]:

Theorem 2. *Let $B \in \mathbb{R}^{m,n}$ and $G \in \mathbb{R}^m$, then the following 2 statements are equivalent:*

- *Determine the vector $u \in \mathbb{R}^n$ which minimizes the Euclidean norm $\|Bu - G\|^2$;*
- *Determine the vector $u \in \mathbb{R}^n$ such that the residual $R = G - Bu \in \mathcal{N}(B^T)$.*

$\square$

See [49] for the proof.

The above Theorem shows that finding the minimizer of the overdetermined system (108) is equal to imposing

$$\left(\sum_{k=1}^{K} \sqrt{W^k} A^k \right)^T \left(\sqrt{W^k} \left(A^k u^k - F^k \right) \right) = 0$$

$$\Longleftrightarrow \tag{109}$$

$$\sum_{k=1}^{K} \left(A^k \right)^T W^k \left(A^k \right) u = \sum_{k=1}^{K} \left(A^k \right)^T W^k F^k ,$$

which is the same equation that we obtained using variational analysis. Therefore, direct minimization given by (108) is equivalent to (101) as a result of the Theorem.

However note that (108) is more appealing to use than (101). Since no pre-multiplication is employed we do not lose the sparsity pattern of the matrix A^k and we prevent fill-in in the global matrix. Bear in mind that W^k is a diagonal matrix and so is its square root. Pre-multiplication with a diagonal matrix amounts to row-scaling which does not affect the sparsity.

7.2.3 Global QR

Let us return to our global system of algebraic equation given by

$$Bu = G \quad \Longleftrightarrow \quad \text{Find } u \text{ which minimizes } \|Bu - G\|^2 , \tag{110}$$

where $B \in \mathbb{R}^{m,n}$, $u \in \mathbb{R}^n$ and $G \in \mathbb{R}^m$. Now for any orthogonal matrix $Q \in \mathbb{R}^{m,m}$ we have

$$\|Q(Bu - G)\|^2 = \|Bu - G\|^2 , \tag{111}$$

since the Euclidean norm is invariant under orthogonal transformations.

We now decompose the $m \times n$ matrix B in a QR-decomposition, $B = QR$, where Q is an orthogonal $m \times m$-matrix and R is an $m \times n$ upper-triangular matrix. The R matrix can be written as

$$R = \begin{pmatrix} \tilde{R} \\ 0 \end{pmatrix} , \tag{112}$$

where $\tilde{R}$ is an upper-triangular $n \times n$ matrix with non-zero diagonal entries, and 0 is an $(m-n) \times n$ matrix with zero entries.

With this decomposition we have

$$\begin{aligned} \|Bu - G\|^2 &= \|Q^T (Bu - G)\|^2 \\ &= \|Ru - Q^T G\|^2 \\ &= \left\| \begin{pmatrix} \tilde{R} \\ 0 \end{pmatrix} u - \begin{pmatrix} c_1 \\ c_2 \end{pmatrix} \right\|^2 \\ &= \|\tilde{R}u - c_1\|^2 + \|c_2\|^2 , \end{aligned} \qquad (113)$$

where c_1 is an n-vector and c_2 is an $m-n$-vector. With this decomposition, minimizing the Euclidean norm is straightforward. The second term, $\|c_2\|^2$, in (113) cannot be minimized. The only terms that can be made small – zero in fact – is the first term on the right hand side of (113). So we have for the least-squares solution

$$u_{LS} = \tilde{R}^{-1} c_1 , \qquad (114)$$

which is just a back-substitution for the upper-triangular matrix $\tilde{R}$. An approximation to the L^2-norm of the residual is given by the second term, $\|c_2\|^2$, and this value is available without solving for u_{LS}. This may be advantageous in hp-adaptive schemes.

Note again, that when exact arithmetic is used the minimizer u_{LS} is equal to the least-squares solution obtained by the conventional least-squares formulation which is obtained by applying variational analysis and solving the normal equations. This algorithm can be improved when one observes that it is not necessary to compute the matrix Q to solve the over-determined system directly.

With a suitable global node numbering this algorithm can be converted into a block-QR algorithm. See [49] for further particulars.

7.2.4 The Poisson equation

In this section a sample problem is presented which consists of a modified Poisson equation given by

$$\kappa \Delta \phi = f(x,y) , \quad (x,y) \in [-1,1]^2 , \qquad (115)$$

where

$$f(x,y) = -2\kappa \sin x \sin y . \qquad (116)$$

The solution in this case is obviously independent of the parameter κ, but the condition number of the resulting system will strongly depend on κ.

In order to apply the least-squares formulation which allows for a C^0 finite element approximation, the governing equation needs to be rewritten as an equivalent first order system

$$u - \nabla \phi = 0 , \qquad (117)$$

$$\kappa \nabla \cdot u = f. \tag{118}$$

Note that there are other equivalent first order systems possible with improved stability estimates, but this model problem is only introduced to compare formulations.

Following Jiang, [51], it is easy to show that this problem is well-posed

$$\kappa^2 C \left(\|\phi\|^2_{H^1(\Omega)} + \|u\|^2_{H(\text{div};\Omega)} \right) \le \|u - \nabla\phi\|^2_{L^2(\Omega)} + \|\kappa\nabla \cdot u\|_{L^2(\Omega)}$$
$$\le \|\phi\|^2_{H^1(\Omega)} + \|u\|^2_{H(\text{div};\Omega)}, \tag{119}$$

for $\kappa \le 1$. So the coercivity constant scales with κ^2 and therefore the condition number is proportional to κ^{-2}. We therefore expect to see differences between the conventional least-squares formulation and direct minimization as proposed in this section for $\kappa \ll 1$. For $\kappa = O(1)$, however, both formulations are expected to give similar results. In order to assess the improved stability of direct minimization the artificially ill-conditioned system is solved on a 5×5 grid with polynomial degree $N = 5$.

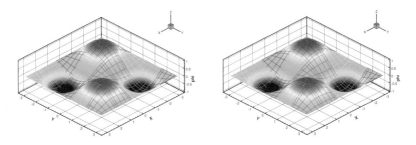

Fig. 24 Solution for $\kappa = 1$ obtained by the conventional least-squares formulation (left) and the result obtained by direct minimization (right)

Figure 24 (left) shows a plot of the solution of the Poisson equation obtained by the conventional least-squares formulation with $\kappa = 1$. Figure 24 (right) gives the solution obtained by Direct Minimization for $\kappa = 1$. The results are indistinguishable. This follows from the observation that both methods are equivalent if exact arithmetic is used.

In Figure 25 results for the case $\kappa = 10^{-5}$ are presented, where the differences between the conventional least-squares formulation and direct minimization become apparent. The conventional least-squares formulation is unable to approximate the exact solution sufficiently accurate due to the loss of precision, whereas Direct Minimization still approximates the solution sufficiently accurate.

The L^2-error for both the conventional least-squares formulation and Direct Minimization versus the parameter κ is depicted in Figure 26. The conventional least-squares formulation (red line) approximates the solution rather well for a κ up to 10^{-4} after which the error grows dramatically. Direct Minimization is capable of

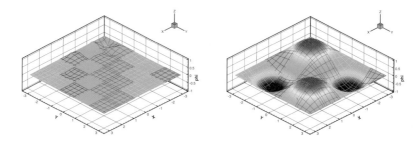

Fig. 25 Solution for $\kappa = 10^{-5}$ obtained by the conventional least-squares formulation (left) and the result obtained by direct minimization (right)

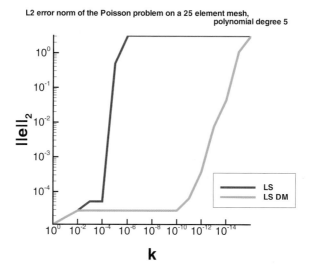

Fig. 26 The L^2-error of the conventional least-squares solution (red) and the solution obtained by direct minimization (green)

approximating the solution up to a κ of $O(10^{-11})$. Note that the solution is independent of κ.

Table 5 shows the condition number of the conventional least-squares and Direct Minimization approach versus polynomial degree in the spectral element method. One observes that the condition number of Direct Minimization is approximately the square root of the condition number associated with the conventional least-squares formulation.

	Condition number	
N	DM	LS
2	5.934	35.215
3	11.914	142.056
4	20.228	409.369
5	30.646	936.737

Table 5 Comparison of the condition numbers obtained from Direct Minimization (DM) and the conventional least-squares method (LS) as a function of the polynomial degree

	Condition number			
	$P=2$		$P=4$	
K	DM	LS	DM	LS
4	4.430	19.635	13.394	197.635
9	5.935	35.222	20.230	409.635
16	7.688	59.111	27.060	732.635
25	9.534	90.891	33.879	1147.786

Table 6 Condition number as a function of the number of elements for Direct Minimization (DM) and the conventional least-squares method (LS)

Table 6 shows the growth of the condition number as a function of the number of elements for two polynomial degrees. One observes that, especially for high order methods which employ much higher polynomial degrees than $N=4$, the difference in condition number between the conventional least-squares method and Direct Minimization grows very fast.

7.3 Application of LSQSEM to viscoelastic fluids

Inspired by the success of the simulation of a Newtonian flow around cylinders attempts were undertaken to solve the flow of a viscoelastic fluid around a cylinder in a channel. The model that was used was the so called Upper-Convected Maxwell (UCM) model. The UCM model is not the most realistic model for viscoelastic flows, but it is the simplest one in terms of number of physical parameters. However, this model is very hard to solve numerically due to its conditional well-posedness, see for instance [35, 39], and therefore is a very good test problem for numerical schemes.

This problem was solved using a discontinuous least-squares formulation, [36]. Furthermore, Direct Minimization was used, see Subsection 7.2.

The Upper Convected Maxwell model is given by conservation of mass for an incompressible flow

$$\nabla \cdot u = 0 \text{ in } \Omega , \qquad (120)$$

where u denotes the velocity vector field.

Conservation of momentum in the Stokes limit yields

$$\nabla \cdot (pI - \tau) = 0 \text{ in } \Omega, \tag{121}$$

where p denotes the pressure field, I the unit tensor in $\mathbb{R}^d$, $d = \dim(\Omega)$ and τ is the extra-stress tensor.

The constitutive equation which relates the extra-stress tensor to the velocity field is given by

$$\lambda \overset{\nabla}{\tau} + \tau = 2\mu d, \tag{122}$$

where λ is the relaxation time and μ the polymeric viscosity of the fluid, $\overset{\nabla}{\cdot}$ denotes the upper convected derivative defined as

$$\overset{\nabla}{A} = \frac{\partial A}{\partial t} + (u, \nabla)A - L\tau - \tau L^T, \tag{123}$$

where $(L)_{ij} = \partial u_i / \partial x_j$ and $2d = L + L^T$. The UCM model (122) describes the fact that the extra-stress does not instantaneously equal the rate of deformation of the flow, but is also convected and deformed along the particle paths as expressed by (123). When the relaxation time $\lambda = 0$, the stress components are no longer convected along the particle paths and Newtonian Stokes flow is retrieved.

Consider the flow past a cylinder placed at the centerline of a channel of width $4R$, where R denotes the radius of the cylinder. The computational domain equals the domain used by Alves, Pinho and Oliveira, [2]. At inflow, $19R$ upstream of the cylinder, a fully developed Poiseuille flow is prescribed for velocity and extra-stress components. The downstream length is taken to be $59R$. The number of spectral elements equals $K = 16$ and the polynomial degree in each element has been set to $N = 16$ for all variables. The topology of the grid and the Gauss-Lobatto grid near the cylinder are shown in Fig. 27. Note the small spectral elements around the cylinder and in the wake of the cylinder. Especially near the rear stagnation point a very small spectral element is placed to capture the high stress-gradients. In order to compare the results obtained with D-LSQSEM-DM the non-dimensional drag coefficient on the cylinder is compared with results reported in [2]. The drag coefficient is defined as

$$C_d := \frac{1}{\mu U} \int_{\text{surf. cyl}} (\tau - pI) \cdot n_x \, dS, \tag{124}$$

where n_x is the x-component of the outward unit normal at the cylinder and U is the bulk velocity. The influence of elasticity in the flow is denoted by the Deborah number, De

$$De = \frac{\lambda U}{R}. \tag{125}$$

Fig. 28 graphically displays the drag coefficients as a function of De. From the results presented above it can be concluded that D-LSQSEM-DM is capable of producing drag coefficients in agreement with those reported in [2]. However, agree-

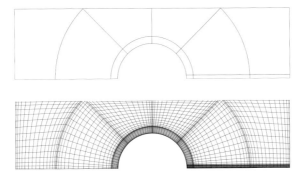

Fig. 27 Topology of the spectral elements in the vicinity of the cylinder (top figure) and the Gauss-Lobatto grid for a polynomial degree $N = 16$

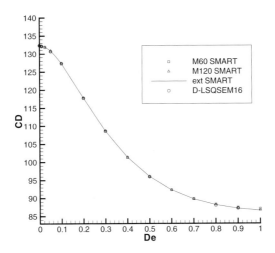

Fig. 28 Non-dimensional drag obtained with D-LSQSEM-DM with a polynomial degree $N = 16$ compared to the finite volume results using SMART discretization reported in [2]

ment of integral quantities does not necessarily imply pointwise agreement. Therefore the contour plot of the extra-stress component in the xx-direction is compared with Fig. 16 taken from [2] in Fig. 29.

This comparison demonstrates that no stabilization terms are required in the least squares method to produce smooth and converged solutions; i.e. no oscillations are present. Both contour plots display a similar pattern. The highest value of the τ_{xx}-component obtained in the D-LSQSEM-DM calculation equals 128.96 at the cylinder. This value is in agreement with the stress levels reported in [2] and later con-

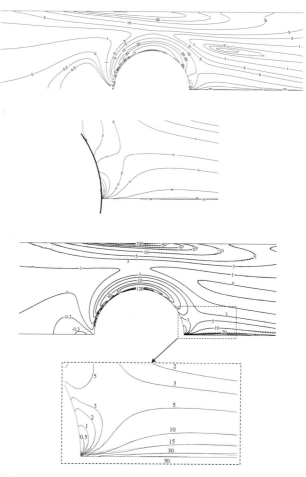

Fig. 29 Contour lines of the τ_{xx}-component of the extra-stress tensor at $De = 0.9$ and detail of the contours just behind the cylinder (top Figures) and comparison of contours of the normal stresses (τ_{xx}) near the cylinder predicted with M60 (dashed line) and M120 (solid line), at $De = 0.9$. Taken from [2] (lower Figure)

firmed in [1]. For further particulars on the use of LSQSEM for viscoelastic flow problems see [34].

8 Further reading

There is a vast amount of literature on weak formulations and finite element methods ranging from very applied to the mathematical theory of variational formulations. We refer to the books by Brenner and Scott, [13] and Ern and Guermond, [31] for the mathematical theory of finite element methods. The two main books on the least-squares finite element method are the books by Jiang, [51] and by Bochev and Gunzburger, [9].

For the use of higher order/spectral method in fluid dynamics the books by Canuto et al., [15], Schwab, [77] and Karniadakis and Sherwin, [52] are excellent introductions.

A simple introduction in the least-squares spectral element method can be found in the VKI lecture series, [37].

Heinrichs, [42, 43, 44, 45, 46, 47], developed least-squares spectral collocation schemes for fluid flow applications. These methods are not directly based on norm-equivalence. Direct Minimization has shown that using the least-squares weak formulation together with Gauss-Lobatto integration can be interpreted as a weighted collocation scheme, thus providing a potential theoretical framework for the least-squares spectral collocation schemes.

Dorao and Jakobsen, [23, 24, 25, 26, 27, 28, 29], applied LSQSEM successfully to population balance equations to model multi-phase flows in chemical engineering.

Pontaza and Reddy, [57, 58, 59, 60, 61, 62, 63, 64, 65, 66, 67], applied LSQSEM to solid mechanics and fluid flow problems.

When only low order finite elements are combined with the least-squares formulation – the so-called least-squares finite element method (LSFEM) – the number of scientific papers and applications is growing at an ever increasing rate signifying the renewed interest in the use of the least-squares formulation in engineering.

References

1. Afonso, A., Alves, M.A., Pinho, F.T., Oliveira, P.J.: Uniform flow of viscoelastic fluids past a confined falling cylinder. Rheol. Acta, **47**, pp. 325-348, (2008)
2. Alves, M.A., Pinho, F.T., Oliveira, P.J.: The flow of viscoelastic fluids past a cylinder: finite-volume high-resolution methods. J. Non-Newtonian Fluid Mech., **97**, pp. 207-232, (2001)
3. Aziz, A.K., Kellogg, R.B., Stephens, A.B.: Least-squares methods for elliptic problems. Math. Comp., **44**, pp. 53-77, (1985)
4. Berndt, M., Manteuffel, T.A., McCormick, S.F.: Local error estimation and adaptive refinement for first-order system least squares (FOSLS). Electron. Trans. Numer. Anal., **6**, pp. 35-43, (1997)
5. Björck, Å.: Least-squares methods. In: P.G. Ciarlet, J.L. Lions (Eds.), Handbook of Numerical Analysis, Solution of Equations in $\mathbb{R}^n$, Part I, Vol. I, Elsevier, North Holland, Amsterdam, pp. 466-647, 1990.
6. Björck, Å.: The calculation of linear least squares problems. Acta Numer., **13**, pp.1-53, (2004)
7. Bochev, P.B., Gunzburger, M.D.: Accuracy of least-squares methods for the Navier-Stokes equations. Comput. Fluids, **22**, pp. 549-563, (1993)

8. Bochev, P.B., Gunzburger, M.D.: Analysis of least-squares finite element methods for the Stokes equations. Math. Comp., **63**, pp. 479-506, (1994).
9. Bochev, P.B., Gunzburger, M.D.: Least-Squares Finite Element Methods. Springer Verlag. Applied Mathematical Series, Vol. 166, 2008
10. Bramble, J.H., Schatz, A.: Rayleigh-Ritz-Galerkin methods for Dirichlet's problem using subspaces without boundary conditions. Comm. Pure Appl. Math., **23**, pp. 653-675, (1970)
11. Bramble, J.H., Schatz, A.: Least-squares for 2mth-order elliptic boundary-value problems. Math. Comp., **25**, pp. 1-32, (1971)
12. Bramble, J.H., Lazarov, R.D., Pasciak, J.E.: Least-squares for second order elliptic problems. Computer Methods in Appl. Mech. Engrg. **152**, pp. 195-210, (1998)
13. Brenner, S.C., Scott, L.R.: The Mathematical Theory of Finite Element Methods. Springer Verlag, New York, 1994.
14. Cai, Z., Lazarov, R., Manteuffel, T.A., McCormick, S.F.: First-order system least-squares for second-order partial differential equations: Part I. SIAM J. Numer. Anal., **31**(6), pp. 1785-1799, (1994)
15. Canuto, C., Hussaini, M.Y., Quarteroni, A., Zang, T.A.: Spectral Methods in Fluids Dynamics. Springer Verlag. 1988
16. Carey, G.F., Jiang, B.-N.: Least-squares finite element method and preconditioned conjugate gradient solution. Int. J. Numer. Methods Eng., **24**, pp. 1283-1296, (1987)
17. Coutanceau, M., Bouard, R.: Experimental determination of the main features of the viscous flow in the wake of a circular cylinder in uniform translation, Part I. Journal of Fluid Mechanics, **79**, (1977)
18. De Maerschalck, B., Gerritsma, M.I.: The use of Chebyshev polynomials in the space-time least-squares spectral element method. Numerical Algorithms, **38**(1-3), pp. 155-172, (2005)
19. De Maerschalck, B., Gerritsma, M.I.: Higher-order Gauss-Lobatto integration for non-linear hyperbolic equations. Journal of Scientific Computing. **27**(1-3), pp. 201-214, (2006)
20. De Maerschalck, B., Gerritsma, M.I., Proot, M.M.J.: Space-time least-squares spectral element methods for convection-dominated unsteady flows. AIAA Journal. **44**(3), pp. 558-565, (2006).
21. De Maerschalck, B., Gerritsma, M.I.: Least-squares spectral element method for non-linear hyperbolic differential equations. Journal of Computational and Applied Mathematics. **215**, pp. 357-367, (2008).
22. Deville, M.O., Fischer, P.F., Mund, E.H.: High Order Methods for Incompressible Fluid Flow. Cambridge University Press, 2002
23. Dorao, C.A., Jakobsen, H.A.: A least-squares method for the solution of population balance problems. Computers and Chemical Engineering, **30**,(3), pp. 535-547, (2005)
24. Dorao, C.A., Jakobsen, H.A.: Application of the least-squares method for solving population balance problems in $\mathbb{R}^{d+1}$. Chemical Engineering Science, **61**(15), pp. 5070-5081, (2006)
25. Dorao, C.A., Jakobsen, H.A.: Numerical calculation of moments of the population balance equation. Journal of Computational and Applied Mathematics, **196**, pp. 619-633, (2006)
26. Dorao, C.A., Jakobsen, H.A.: A parallel time-space least-squares spectral element solver for incompressible flow problems. Applied Mathematics and Computation, **185**(23), pp. 45-58, (2007)
27. Dorao, C.A., Jakobsen, H.A.: Least-squares spectral method for solving advective population balance problems. Journal of Computational and Applied Mathematics, **201**(1), pp. 247-257, (2007)
28. Dorao, C.A., Jakobsen, H.A.: hp-Adaptive least-squares spectral element method for population balance equations. Applied Numerical Mathematics, **58**(5), pp. 563-576, (2008)
29. Dorao, C.A., Fernandino, M.: Simulation of transients in natural gas pipelines. Mathematics and Computers in Simulation. To appear.
30. Džiškariani, A.V.: The least-squares and Bubnov-Galerkin methods. Ž. Vyčisl. Mat. i. Mat. Fiz., **8**, pp. 1110-1116, (1968)
31. Ern, A., Guermond, J.-L.: Theory and Practice of Finite Elements. Springer Verlag, New York, 2004.

32. Galerkin, B.G.: Series solution of some problems in elastic equilibrium of rods and plates. Vestn. Inzh. Tech., **19**, pp. 897-908, (1915)
33. Galvão, Á., Gerritsma, M.I., De Maerschalck, B.: *hp*-Adaptive least squares spectral element method for hyperbolic partial differential equations. Journal of Computational and Applied Mathematics. **215**(2). pp. 409-418, (2008).
34. Gerritsma, M.I.: Direct minimization of the discontinuous least-squares spectral element method for viscoelastic fluids. Journal of Scientific Computing. **27**(1-3), pp. 245-256, (2006).
35. Gerritsma, M.I., Phillips, T.N.: On the use of characteristics in viscoealastic flow problems. IMA J. Appl. Math., **66**, pp. 127-147, (2001)
36. Gerritsma, M.I., Proot, M.M.J.: Analysis of a discontinuous least-squares spectral element method. Journal of Scientific Computing. **17**(1-4), pp. 297-306, (2002).
37. Gerritsma, M.I., De Maerschalck, B.: The least-squares spectral element method. In: CFD – Higher Order Discretization Methods, VKI Lecture Series 2006-01, Von Karman Institute for Fluid Dynamics. Ed. by H. Deconinck and M. Ricchiuto, 2006.
38. Gerritsma, M.I., Van der Bas, R., De Maerschalck, B., Koren, B., Deconinck, H.: Least-squares spectral element method applied to the Euler equations, International Journal for Numerical Methods in Fluids., **57**, pp. 1371-1395, (2008)
39. Gerritsma, M.I., Phillips, T.N.: On the characteristics and compatibility equations for the UCM model fluid. Z. Angew. Math. Mech., **88**(7), 523-539, (2008)
40. Gordon, W.J., Hall, C.A.: Transfinite element methods - blending-function interpolation over arbitrary curved element domains. Numerische Mathematik, **21**(2), pp. 109-129, (1973)
41. De Groot, R.: Direct Minimization of Equation Residuals in Least-Squares *hp*- Finite Element Method – Numerical study of low Reynolds number flow around a circular cylinder. MSc-thesis, TU Delft, 2004.
42. Heinrichs, W.: Least-squares spectral collocation for discontinuous and singular perturbation problems. Journal of Computational and Applied Mathematics, **157**(2), pp. 329-345, (2003)
43. Heinrichs, W.: Least-squares spectral collocation for the Navier-Stokes equations. Journal of Scientific Computing, **21**(1), pp. 81-90, (2004)
44. Heinrichs, W.: Least-squares spectral collocation with the overlapping Schwartz method for the incompressible Navier-Stokes equations. Numerical Algorithms, **43**(1), pp. 61-73, (2006)
45. Heinrichs, W.: An adaptive least-squares collocation method with triangular elements for the incompressible Navier-Stokes equations. Journal of Engineering Mathematics, **56**(3), pp. 337-350, (2006)
46. Heinrichs, W.: An adaptive least-squares scheme for the Burgers' equation. Numerical Algorithms, **47**(1), pp. 63-80, (2007)
47. Heinrichs, W., Kattelans, T.: A direct solver for the least-squares spectral collocation system on rectangular elements for the incompressible Navier-Stokes equations. Journal of Computational Physics, **227**(9), pp. 4776-4796, (2008)
48. Henderson, R.D., Karniadakis, G.E.: Unstructured spectral methods for simulation of turbulent flows. J. Comput. Phys., **122**, pp. 191-217, (1995)
49. Hoitinga, W., De Groot, R., Kwakkel, M., Gerritsma, M.I.: Direct minimization of the least-squares spectral element functional – Part I: Direct Solver. Journal of Computational Physics, **227**, pp. 2411-2429, (2008).
50. Houston, P., Senior, B., Süli, E.: Sobolev regularity estimation for *hp*-adaptive finite element methods. Report NA-02-02, University of Oxford, UK, 2002.
51. Jiang, B.-N.: The Least-Squares Finite Element Method – Theory and Applications in Computational Fluid Dynamics and Electromagnetics. Springer Verlag, Scientific Computation, 1998.
52. Karniadakis, G.E, Sherwin, S.J.: Spectral/*hp* Element Methods for CFD. Oxford University Press, 2002.
53. Kwakkel, M.: Time-dependent flow simulations using the least-squares spectral element – Application to unsteady incompressible Navier-Stokes flows. MSc-thesis, TU Delft, 2007.
54. Liu, J.-L.: Exact a-posteriori error analysis of the least-squares finite element method. Appl. Math. Comput., **116**, pp. 297-305, (2000)

55. Lučka, A.: The rate of convergence to zero of the residual and the error in the Bubnov-Galerkin method and the method of least-squares. In Proc. Sem. Differential and Integral Equations, No. 1, pp. 113-122 (Russian) Akad, Nauk Ukrain. SSR Inst. Mat., Kiev, Ukraine, 1969
56. Maday, Y., Mavriplis, C., Patera, A.: Non-conforming mortar element methods: application to spectral discretizations. Domain Decomposition Methods, Proc. 2nd Int. Symp., Los Angeles, Califormia, pp. 392-418, (1988)
57. Pontaza, J.P.: Least-squares variational principles and the finite element method: Theory, formulations, and models for solid and fluid mechanics. Finite Elements in Analysis and Design, **41**(7-8), pp. 703-728, (2005)
58. Pontaza, J.P.: A least-squares finite element formulation for unsteady incompressible flows with improved velocity-pressure coupling. J. Comput. Phys. **217**(2), pp. 563-588, (2006)
59. Pontaza, J.P.: A spectral element least-squares formulation for incompressible Navier-Stokes flows using triangular elements. Journal of Computational Physics, **221**(2), pp. 649-665, (2007)
60. Pontaza, J.P.: A new consistent splitting scheme for incompressible Navier-Stokes flows: A least-squares spectral element implementation. Journal of Computational Physics, **225**(2), pp. 1590-1602, (2007)
61. Pontaza, J.P., Reddy J.N.: Spectral/*hp* least-squares finite element formulation for the Navier-Stokes equations. J. Comput. Phys., **190**(2), pp. 523-549, (2003)
62. Pontaza, J.P., Reddy J.N.: Mixed plate bending elements based on least-squares formulation. International Journal for Numerical Methods in Engineering, **60**(5), pp. 891-922, (2004)
63. Pontaza, J.P., Reddy J.N.: Space-time coupled spectral/*hp* least-squares finite element formulation for the incompressible Navier-Stokes equations. J. Comput. Phys., **197**(20), pp. 418-459, (2004)
64. Pontaza, J.P., Reddy J.N.: Least-squares finite elements for shear-deformable shells. Computer Methods in Applied Mechanics and Engineering, **194**(21-24), pp. 2464-2493, (2005)
65. Pontaza, J.P., Reddy J.N.: Least-squares finite element formulations for one-dimensional radiative transfer. Journal of Quantitative Spectroscopy and Radiative Transfer, **95**(3), pp. 387-406, (2005)
66. Pontaza, J.P., Reddy J.N.: Least-squares finite element formulations for viscous incompressible and compressible fluid flows. Computer Methods in Applied Mechanics and Engineering, **195**(19-22), pp. 2454-2494, (2006)
67. Prabhhakar, V., Pontaza, J.P., Reddy, J.N.: A collocation penalty least-squares finite element formulation for incompressible flow. Computer Methods in Applied Mechanics and Engineering, **197**(6-8), pp. 449-463, (2008)
68. Proot, M.M.J.: The least-squares spectral element method – Theory, Implementation and Application to Incompressible Flows. PhD thesis, TU Delft, 2003.
69. Proot, M.M.J., Gerritsma, M.I.: A least-squares spectral element formulation for the Stokes problem. Journal of Scientific Computing. **17**(1-4), pp. 285-296, (2002).
70. Proot, M.M.J., Gerritsma, M.I.: Least-squares spectral elements applied to the Stokes problem. Journal of Computational Physics, **181**(2), pp. 454-477, (2002)
71. Proot, M.M.J., Gerritsma, M.I.: Application of the least-squares spectral element method using Chebyshev polynomials to solve the incompressible Navier-Stokes equations. Numerical Algorithms. **38**(1-3), pp. 155-172, (2005).
72. Proot, M.M.J., Gerritsma, M.I.: Mass- and momentum conservation of the least-squares spectral element method for the Stokes problem. Journal of Scientific Computing. **27**(1-3), pp. 389-401, (2006).
73. Quarteroni, A., Valli, A.: Numerical Approximation of Partial Differential Equations. Springer-Verlag, 1997.
74. Lord Rayleigh (J.W. Strut): On the theory of resonance. Trans. Roy. Soc., London, A161, pp. 77-118, 1870
75. Ritz, W.: Uber eine neue Methode zur Lösung gewisses Variationsprobleme der mathematischen Physik. J. Reine Angew. Math., **135**, pp. 1-61, (1908)

76. Rizzi, A., Viviand, H.: Numerical Methods for the Computation of Inviscid Flows with Shock Waves. Vieweg Verlag, 1981.
77. Schwab, Ch.: p- and h-Finite Element Methods. Oxford Scientific Publications. 1998.
78. Spekreijse, S.P.: Multigrid Solution of the Steady Euler Equations. Stichting Mathematisch Centrum, Amsterdam, CWI Tracts, **46**, 1988.
79. Spivak, M.: Calculus on Manifolds – A modern approach to classical theorems of advanced calculus. The Benjamin Cummings Publishing Company, 1965.
80. Sucker, D., Brauer, H.: Fluiddynamik bei quer angeströmten Zylindern, Wärme und Stoffübertragung, **8**, pp. 149-158, (1975)
81. Taneda, S.: Experimental investigation of the wake behind cylinders and plates at low Reynolds numbers. Journal of Physical Society of Japan, **11**, pp. 302-307, (1956)
82. Tritton, D.J.: Experiments on the flow past a circular cylinder at low Reynolds numbers. Journal of Fluid Mechanics, **6**, pp. 547-567, (1959)
83. White, F.M.: Viscous Fluid Flows. McGraw-Hill, Inc., 1991.

Finite-Volume Discretizations and Immersed Boundaries

Yunus Hassen and Barry Koren

Abstract In this chapter, an accurate method, using a novel immersed-boundary approach, is presented for numerically solving linear, scalar convection problems. As is standard in immersed-boundary methods, moving bodies are embedded in a fixed 'Cartesian' grid. The essence of the present method is that specific fluxes in the vicinity of a moving body are computed in such a way that they accurately accommodate the boundary conditions valid on the moving body. To suppress wiggles, tailor-made limiters are introduced for these special fluxes. The first results obtained are very accurate, without requiring much computational overhead. It is anticipated that the method can readily be extended to real fluid-flow equations.

1 Introduction

The immersed-boundary method – or, synonymously, embedded-boundary method – is a method in which boundary conditions are indirectly incorporated into the governing equations. It has undergone numerous modifications, ever since its introduction by Peskin in 1972 [13], and currently many varieties of it exist (see [10] for a review and the references therein for details).

Immersed-boundary methods are very suitable for simulating flows around flexible, moving and/or complex bodies. Basically, the bodies of interest are just em-

---

Yunus Hassen
Centrum Wiskunde & Informatica, Kruislaan 413, 1098 SJ Amsterdam, the Netherlands,
Faculty of Aerospace Engineering, TU Delft, Kluyverweg 1, 2629 HS Delft, the Netherlands.
e-mail: yunus.hassen@cwi.nl

Barry Koren
Centrum Wiskunde & Informatica, Kruislaan 413, 1098 SJ Amsterdam, the Netherlands,
Faculty of Aerospace Engineering, TU Delft, Kluyverweg 1, 2629 HS Delft, the Netherlands,
Mathematical Institute, Leiden University, Niels Bohrweg 1, 2333 CA Leiden, the Netherlands.
e-mail: barry.koren@cwi.nl

bedded in non-deforming Cartesian grids that do not conform to the shape of the body. The governing equations are modified to include the effect(s) of the embedded boundaries. Doing so, mesh (re)generation difficulties associated with body-fitted grids are obviated, and the underlying regular fixed grid allows us to use a simple data structure as well as simpler numerical schemes over a majority of the domain.

Peskin, in his original paper [13], introduced the idea of replacing an object in a flow by a field of forces. This gave rise to the notion of an 'immersed body.' Peskin described the fluid variables in an Eulerian manner and the object in the Lagrangian manner and computed, from the boundary configuration, the elastic forces generated within the object. Since the object is in direct contact with the fluid, these elastic forces affect the fluid motion. Peskin then transmitted the forces to the fluid in the immediate vicinity of the boundaries of the object, i.e., to the governing equations he added a forcing function, which is zero everywhere except near the immersed boundaries. The forcing term enforces the no-slip condition on the boundaries of the (immersed) object and thus the flow field indirectly feels the presence of an object immersed in it. Finally, he discretized the extended equation in the entire computational domain, including inside the immersed body. He successfully implemented this immersed-boundary method to simulate blood flow in and around heart valves [14, 15].

In a similar way, but independently, Goldstein et al. [4] employed a forcing term continuously computed from a feedback loop. They borrowed concepts from linear control theory and formulated the forcing term depending solely on the velocity of the boundary-surface points (immersed boundaries). This forcing term is added to the momentum equation, and recomputed/corrected (using the computed velocity) at each time step until the relative velocity on the desired boundary-surface points has been set to zero. This recursive procedure eventually evolves the (desired) virtual surface. The subjective part of this forcing term is that it requires the choice of two negative constants, α and β, and a problem-dependent parameter k, which are not defined properly; they are interrelated by a stability criterion and estimated heuristically. This technique introduces a severe restriction on the time step and it is unstable for (complex) flow computations that require large time steps. Goldstein et al. used a spectral method solver to simulate two-dimensional flow around stationary cylinders and three-dimensional turbulent channel flow. Saiki and Biringen [16] adopted the same (feedback-forcing-function) method using higher-order finite-difference methods, and achieved a relatively stable solution with no time-step restriction. They computed the feedback-forcing function by integrating the relative flow velocity, with the associated negative constants, on the boundary-surface points, and showed that the feedback-forcing method of Goldstein et al. is also capable of handling moving boundary problems. They successfully implemented it for low-Reynolds number (Re $\leq$ 400) flows around fixed, rotating and oscillating cylinders.

Mohd-Yusof [12] proposed a modified method which is called the direct-forcing method. This method uses a set of points adjacent to the surface and interior to the body and directly imposes the no-slip boundary conditions on the immersed boundary enabling a direct momentum forcing. It is relatively stable, compared to the method of Goldstein et al., does not impose time-step restrictions and does not need the choice of (empirical) negative constants for defining the forcing function. Mohd-Yusof [11] also used the spectral context and simulated three-dimensional flows for complex geometries. Kim et al. [7] used the direct-forcing method and simulated flows over a cylinder and a sphere using the finite-volume approach on a staggered mesh. They introduced external forcing functions to the momentum and continuity equations, to achieve momentum and mass conservation, respectively. In this approach, an unbalanced mass flux results across the body boundary and a source/sink term is added to the continuity equation.

Fadlun et al. [2] extended the direct-forcing method to finite-difference formulation on a staggered grid, and compared the accuracy and efficiency of their method with those of the feedback forcing method, by simulating three-dimensional flows in complex geometries. They found their method to be of the same order of accuracy, but more efficient. They concluded that the direct-forcing approach is more efficient and suitable to simulate unsteady three-dimensional incompressible flows in complex geometries. Most of the feedback and direct-forcing methods are basically similar except for the way of interpolating the fluid velocity to the immersed boundary points and extrapolating the body force back into the computational grid points [2, 17, 19, 20].

More recently, a different class of immersed-boundary methods has started to emerge. Here, no forcing function and spreading is required. Instead, the velocity of grid points around the immersed boundary is interpolated taking the boundary condition (no-slip, for instance) into account. The resulting interpolation equation is then solved along with the (unmodified) Navier-Stokes equations. The main advantage, in this case, is that no extra terms are included in the governing equations and they are solved only in the fluid domain. Ghost-cell [19] and cut-cell [1] methods are typical examples of this class of methods. Some of the major contemporary methods have been described and reviewed in [10].

In this chapter, we follow the forcing-function-free approach and start to build up a new immersed boundary method from scratch, considering for convenience, a simple model equation. Our approach uses a cell-centered finite-volume discretization. The governing partial differential equations are discretized using a standard finite-volume method away from an embedded boundary (EB). Near the EB, a special finite-volume method is derived which takes the prescribed interior boundary conditions into account.

The outline of the chapter is as follows. In § 2, the problem is described, a standard finite-volume method is described and some of the associated results are presented. The special fluxes which take the effects of the embedded boundaries into account are derived and limiters are introduced in § 3. In § 4, the issues associated with temporal discretization, which gives rise to fully discrete equations, are explained. Total-variation diminishing (TVD) regions are defined and tailor-made limiters, for the special fluxes, are also educed from the fully discrete equations. In § 5, some numerical results, based on the present work, are given and a comparison is made with the standard finite-volume results. In § 6, we give a brief account of the possibilities to extend the presented method to more general cases, and finally, concluding remarks are presented in § 7.

2 Model equation and target problems

Many of the partial differential equations that are derived to model physical situations cannot be solved analytically, and are too complex to study their numerics rigorously. It is logical to first develop numerical schemes for appropriate model equations and then to carry these over to the original partial differential equations for which precise analysis is not possible. It is common practice to take model equations that are sufficiently simplified versions of the corresponding physical equations, but still resemble these equations as much as possible.

Here, we consider the one-dimensional, linear advection equation as the model equation for the Euler equations:

$$\frac{\partial c}{\partial t} + \frac{\partial f(c)}{\partial x} = 0, \quad f(c) = uc. \tag{1}$$

Equation (1) is a model of scalar quantity $c(x,t)$ that is advected by the velocity u, which is constant, and which we assume to be positive. $f(c)$ is the flux function, which is linear. The independent variables x and t represent space and time, respectively. The generic domain of the solution is a one-dimensional rod, of finite length L, and the time interval, in principle, is infinitely long.

The advection equation (1) is a very simple partial differential equation, but it is an important one. It models fluid-flow equations and it proves challenging to solve it numerically. It is hyperbolic with a single set of characteristic lines. These are straight lines in the (x,t)-plane, which are determined from the solution of the ordinary differential equation:

$$\frac{\mathrm{d}x}{\mathrm{d}t} = u, \tag{2a}$$

whose integration yields the equation of the characteristic lines $x - ut = \text{constant}$. Notice that, along a characteristic line, the dependent variable $c(x,t)$ satisfies:

$$\frac{\mathrm{d}c}{\mathrm{d}t} = \frac{\partial c}{\partial t} + \frac{\partial c}{\partial x}\frac{\mathrm{d}x}{\mathrm{d}t} \equiv \frac{\partial c}{\partial t} + u\frac{\partial c}{\partial x} = 0, \tag{2b}$$

and thus, it remains constant along these lines.

Therefore, for a given initial solution $c(x,0) = c_0(x)$, the exact solution of (1), at any location x and time t, can be computed by the method of characteristics, as $c(x,t) = c_0(x - ut)$. That is, as time evolves, the initial data simply propagates unchanged with a velocity u: it propagates to the right if $u > 0$ and to the left if $u < 0$.

Hence, by using the exact solution as a benchmark, numerous numerical schemes can be developed and tested for the one-dimensional, linear advection equation.

2.1 Standard finite-volume discretization

In the finite-volume method, the spatial domain of the physical problem is subdivided into non-overlapping cells or control volumes. The cells are considered to be of uniform size. The domain is taken to be of unit length, $L = 1$, on the interval $x \in [0, 1]$ and is divided into N cells, with the grid size being $h = \frac{1}{N}$.

A single node is located at the geometric centroid of the control volume and the cells are represented with nodal indices: $i-1$, i, $i+1$, etc. The coordinates of the nodes are determined as $x_i = (i - \frac{1}{2})h$, $i = 1, 2, ..., N$. Analogously, the coordinates of the cell faces are labeled by indices-with-fractions and are computed as $x_{i+\frac{1}{2}} = ih$, $i = 1, 2, ..., N$. Figure 1 shows the spatial domain with cells and cell faces.

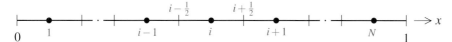

Fig. 1 One-dimensional finite-volume domain

To obtain the finite-volume model, ensuring conservation of c, the model equation is integrated over the control volumes shown in Figure 1. Integrating (1) over the volume Ω of cell i yields:

$$\int_{\Omega_i} \frac{\partial c}{\partial t} d\Omega_i + \int_{\Omega_i} \frac{\partial f(c)}{\partial x} d\Omega_i = 0. \quad (3)$$

We denote the discrete solution in cell i and the flux at cell face $i+\frac{1}{2}$, both at time level n, as $c_i^n := c(x_i, t^n)$ and $f_{i+\frac{1}{2}}^n := f(c(x_{i+\frac{1}{2}}, t^n))$, respectively. We assume c_i^n to be constant in space, in that cell. Applying the Gauss integration theorem, (3) can be rewritten as:

$$h \frac{dc_i}{dt} + (f_{i+\frac{1}{2}} - f_{i-\frac{1}{2}}) = 0. \quad (4)$$

Semi-discrete equation (4) is exact so far in cell Ω_i. It is going to be solved using the method of lines. That is, the fluxes at the cell faces are first approximated and then the temporal part is time-stepped with a suitable time-integration method.

2.2 Initial and boundary conditions

Two initial solutions are considered, each with two interior, moving boundaries. The solution at the left and right of each interior boundary is prescribed. The two interior boundaries represent two infinitely thin bodies that go with the flow. The two moving boundaries have different initial locations (x_1 and x_2, $x_1 \neq x_2$). The solution is discontinuous across both interior boundaries. The two initial solutions are shown in Figure 2, and, in formulae, read:

$$c_0(x) = \begin{cases} 0, & \text{if } x_1 \leq x \leq x_2, \\ 1, & \text{elsewhere;} \end{cases} \quad (5a)$$

$$c_0(x) = \begin{cases} 0, & \text{if } x_1 \leq x \leq x_2, \\ \frac{1}{2}(1 - \cos(2\pi x)), & \text{elsewhere.} \end{cases} \quad (5b)$$

The cosine function in (5b) exploits the advantage that higher-order accurate numerical schemes have in non-constant, smooth solution regions.

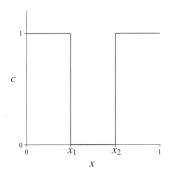
(a) Constant function at peripheries

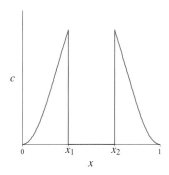
(b) Smooth (cosine) function at peripheries

Fig. 2 Initial solutions with two discontinuous interior boundaries.

The model equation is approximated in a periodic domain. That is, the first and last cell faces are 'glued together' and thus the fluxes in the corresponding faces are readily made equal: $f_{\frac{1}{2}} = f_{N+\frac{1}{2}}$. Apparently, periodicity allows us to time-step for as long as we want for a finite spatial domain.

Fixed-grid finite-volume methods for advection problems with interior moving boundaries are underdeveloped. No rigorous studies exist about numerical properties as accuracy and monotonicity. Here, several finite-volume methods for discontinuous moving interior-boundary problems will be derived, analyzed and tested. The moving interior-boundary conditions will be embedded in the fluxes in the direct neighborhood. The precise way in which this embedding is done is the main theme of this chapter.

2.3 Standard finite-volume schemes

Finite-volume methods distinguish themselves in the way the fluxes are computed. To start, three standard finite-volume methods are considered: first-order accurate

Finite-Volume Discretizations and Immersed Boundaries

upwind, second-order accurate central, and second-order accurate fully one-sided upwind. The latter two can be cast into one general form, the κ-scheme [9].

For positive and constant u, and an equidistant grid, the classical fluxes, at time level n, are computed as follows. The fluxes are given for cell face $i + \frac{1}{2}$ (Figure 1); for the other faces, they are computed analogously.

The general flux at cell face $i + \frac{1}{2}$, dropping the time index n, for convenience, reads:

$$f_{i+\frac{1}{2}} = u c_{i+\frac{1}{2}}, \tag{6a}$$

where $c_{i+\frac{1}{2}}$ is the cell-face state at $i + \frac{1}{2}$, which can be approximated in a variety of ways. For example, for $u > 0$, the first-order upwind flux involves only one cell and takes the form:

$$c_{i+\frac{1}{2}} = c_i. \tag{6b}$$

Equation (6b) shows that the first-order upwind flux is solely based on the information from the upstream side of the cell face.

The second-order central and fully one-sided upwind fluxes involve two cells and they take the form:

$$c_{i+\frac{1}{2}} = c_i + \frac{1}{2}(c_{i+1} - c_i), \tag{7a}$$

$$c_{i+\frac{1}{2}} = c_i + \frac{1}{2}(c_i - c_{i-1}), \tag{7b}$$

respectively. Both are written as the first-order upwind cell-face state (6b) plus a correction term. Equation (7a) is obtained by interpolation, assuming a linear variation of c between points x_i and x_{i+1}. And (7b) is obtained by extrapolation, assuming a linear variation of c between points x_{i-1} and x_i.

By blending these basic second-order accurate schemes, we can reconstruct a general higher-order accurate scheme, as:

$$c_{i+\frac{1}{2}} = \theta \left(c_i + \frac{1}{2}(c_{i+1} - c_i) \right) + (1 - \theta) \left(c_i + \frac{1}{2}(c_i - c_{i-1}) \right), \quad \theta \in [0, 1], \tag{8a}$$

with θ the blending parameter. Formula (8a) can be rewritten as:

$$c_{i+\frac{1}{2}} = c_i + \frac{\theta}{2}(c_{i+1} - c_i) + \frac{1 - \theta}{2}(c_i - c_{i-1}). \tag{8b}$$

Introducing, instead of θ, the parameter κ:

$$\kappa = 2\theta - 1, \quad \kappa \in [-1, 1], \tag{9}$$

equation (8b) turns out to be the well-known Van Leer κ-scheme [9]:

$$c_{i+\frac{1}{2}} = c_i + \frac{1 + \kappa}{4}(c_{i+1} - c_i) + \frac{1 - \kappa}{4}(c_i - c_{i-1}). \tag{10}$$

For $\kappa = 1$, we have the second-order accurate central scheme; and for $\kappa = -1$, we have the second-order accurate fully one-sided upwind scheme. A motivation for the blending is that for the unique value $\kappa = \frac{1}{3}$, we have $\mathcal{O}(h^3)$ net flux accuracy in each cell.

The simplicity and monotonicity of the first-order upwind scheme are appealing. However, it has strong numerical diffusion. On the other hand, the solutions of all κ-schemes, hence also those of the $\kappa = \frac{1}{3}$ scheme, may exhibit wiggles. This recalls Godunov's (1959) theorem [3] which states that there is no linear scheme higher than first-order accurate which is monotone.

We verify this here for the κ-scheme (10), considering the monotonicity requirement:
$$\frac{c_{i+\frac{1}{2}} - c_{i-\frac{1}{2}}}{c_i - c_{i-1}} \geq 0. \tag{11}$$

With the local successive solution-gradient ratios:
$$r_{i+\frac{1}{2}} = \frac{c_{i+1} - c_i}{c_i - c_{i-1}}, \tag{12a}$$
$$r_{i-\frac{1}{2}} = \frac{c_i - c_{i-1}}{c_{i-1} - c_{i-2}}, \tag{12b}$$

and κ-scheme (10), requirement (11) yields:
$$(1+\kappa) r_{i+\frac{1}{2}} - \frac{1-\kappa}{r_{i-\frac{1}{2}}} \geq 2(\kappa - 2). \tag{13}$$

No $\kappa \in [-1, 1]$ exists for which (13) is satisfied for all possible combinations of $r_{i-\frac{1}{2}}$ and $r_{i+\frac{1}{2}}$. It can be directly verified that (13) is not satisfied for $r_{i+\frac{1}{2}} < -1$ in case of $\kappa = 1$, and for $\frac{1}{r_{i-\frac{1}{2}}} > 3$ in case of $\kappa = -1$. For $\kappa = \frac{1}{3}$, requirement (13) is not satisfied for $\frac{1}{r_{i-\frac{1}{2}}} - 2r_{i+\frac{1}{2}} > 5$. The corresponding regions of non-monotonicity in the $(r_{i-\frac{1}{2}}, r_{i+\frac{1}{2}})$-plane are depicted in Figure 3.

Notice that monotonicity requirement (11) is always satisfied for the first-order upwind scheme (6b).

Several algorithms have been proposed in the literature that yield higher-order accurate solutions which are free from wiggles. Most of these algorithms exploit the inherent monotonicity of the first-order upwind scheme. The best known representatives of these algorithms are the limited schemes following Sweby's work [18]. Let us consider limiters that resemble κ-schemes to the largest possible extent within Sweby's TVD domain.

With (12a), the limited form of the cell-face state according to (10) can be written as:

Finite-Volume Discretizations and Immersed Boundaries

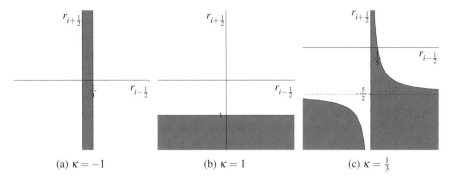

Fig. 3 Non-monotonicity regions for some κ-schemes.

$$c_{i+\frac{1}{2}} = c_i + \frac{1}{2}\phi(r_{i+\frac{1}{2}})(c_i - c_{i-1}), \tag{14a}$$

where $\phi(r)$ is the limiter function, defined as:

$$\phi(r) = \begin{cases} 0, & \text{if } r < 0 \\ 2r, & \text{if } 0 \leq r < \frac{1-\kappa}{3-\kappa}, \\ \frac{1-\kappa}{2} + \frac{1+\kappa}{2}r, & \text{if } \frac{1-\kappa}{3-\kappa} \leq r < \frac{3+\kappa}{1+\kappa}, \\ 2, & \text{if } \frac{3+\kappa}{1+\kappa} \leq r. \end{cases} \tag{14b}$$

Here we specifically adopt the limiter proposed in [8] as the standard limiter, which gives a monotone third-order accurate net flux in a cell, by resembling the $\kappa = \frac{1}{3}$ scheme. This limiter, which is within Sweby's TVD domain, is depicted in Figure 4.

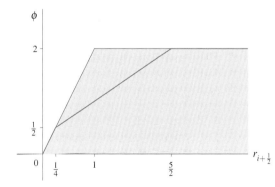

Fig. 4 Standard limiter, which is obtained from (14b) for $\kappa = \frac{1}{3}$.

In the remainder of this chapter, we will derive non-standard finite-volume methods, methods in which the interior boundary conditions are incorporated in the fixed-grid flux formulae. Before doing so, for later comparison purposes, we will show what the solutions are for the standard finite-volume discretizations described above, methods in which no embedded-boundary conditions are imposed, pure capturing methods, in fact. For the time integration, the three-stage Runge-Kutta scheme RK3b from [6] is employed. For both initial solutions given in (5a) and (5b), we consider the initial locations of the EBs to be at $x_1 = \frac{1}{3}$ and $x_2 = \frac{2}{3}$. Furthermore, we take $u = 1$, and we compute the solution at $t_{\max} = 1$, the time at which the solution has made a single full-period. For both the first-order upwind and the $\kappa = \frac{1}{3}$ (unlimited and limited) schemes, the computations are performed on a grid with 20 and 40 cells. The solutions are depicted in Figure 5. The time steps have been taken sufficiently small to ensure that in all cases the time-discretization errors are negligible with respect to the spatial discretization errors.

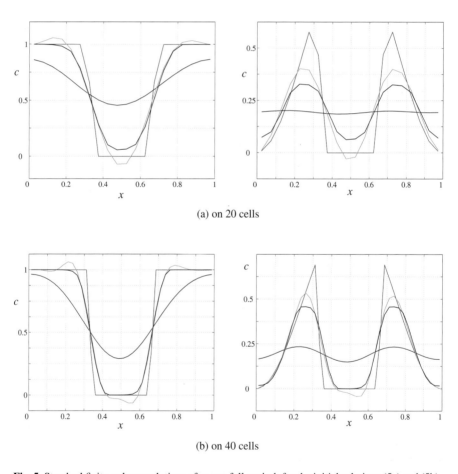

(a) on 20 cells

(b) on 40 cells

Fig. 5 Standard finite-volume solutions after one full-period, for the initial solutions (5a) and (5b). Red: exact discrete, blue: first-order upwind, green: unlimited higher-order upwind-biased, and black: limited higher-order upwind-biased.

3 Fluxes with embedded moving-boundary conditions

As mentioned, the sharp discontinuities of the initial solutions (5a) and (5b), shown in Figure 2, may be considered as infinitely thin bodies going with the flow and the boundary conditions associated with these may be embedded in some fixed-grid fluxes. Here, the EB (embedded boundary) conditions are user-specified and enforced to remain intact to the EB and unchanged at all times. The solution values on the left and right sides of the EB are designated as c_{EB}^l and c_{EB}^r, respectively (see Figure 6).

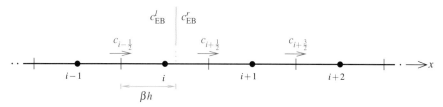

Fig. 6 EB situated in cell i at time t, its associated solution values, and the affected cell-face states.

As shown in Figure 6, for an EB situated in cell i, with its coordinate $x_{EB} = x_{EB}(t)$ given, its relative position with respect to $x_{i-\frac{1}{2}}$ (the left face of cell i) is βh, where $\beta \in [0,1]$ is a (non-dimensional) parameter which is defined as:

$$\beta = \frac{x_{EB} - x_{i-\frac{1}{2}}}{h}. \tag{15}$$

So, $\beta = 0$ when the EB is situated at cell face $i - \frac{1}{2}$, $\beta = \frac{1}{2}$ when the EB is exactly at the centroid i, and $\beta = 1$ when the EB is at cell face $i + \frac{1}{2}$.

There is no information flow across the EB. Fluxes on one side of the EB, at a specific time t, are all computed based on the information on the same side of the EB, at that time, plus the additional interior-boundary condition on the respective side. In general, when considering three-point upwind-biased interpolation for the fluxes, three cell-face states ($c_{i-\frac{1}{2}}$, $c_{i+\frac{1}{2}}$ and $c_{i+\frac{3}{2}}$) are affected by the presence of a single EB (in cell i) and these are the cell-face states of interest for which tailor-made formulae will be derived.

In general, for an EB in a cell, the three affected cell-face states are computed such that the net fluxes in some neighboring cells are as accurate as possible. This shall be discussed in the next section. So far, it is assumed that two successive EBs are sufficiently far apart that no cell-face state exists that is affected by both EBs. Recall that all but the affected cell-face states are computed based on the standard finite-volume schemes discussed in § 2.3.

3.1 Higher-order accurate embedded-boundary fluxes

If a three-point upwind-biased interpolation is considered for computing fluxes, the cell faces $i-\frac{1}{2}$, $i+\frac{1}{2}$ and $i+\frac{3}{2}$ 'feel' the EB situated in cell i (see Figure 6). The higher-order accurate fluxes at these faces are computed from higher-order accurate cell-face states. In principle, all the special cell-face states are written in terms of the blending parameter κ and computed from optimally blended, three-point upwind-biased interpolation formulae. However, for cell-face state $c_{i+\frac{1}{2}}$, no upwind-biased interpolation formula can be derived as we do not draw information across the EB. Hence, there is no blending parameter in the formula for $c_{i+\frac{1}{2}}$, only non-equidistant central interpolation is applied to compute $c_{i+\frac{1}{2}}$. On the other hand, in the formulae for $c_{i-\frac{1}{2}}$ and $c_{i+\frac{3}{2}}$, there will be blending parameters, and $c_{i-\frac{1}{2}}$ and $c_{i+\frac{3}{2}}$ can be taken as optimally weighted averages of two-point central interpolation and two-point fully upwind extrapolation.

Just like away from the EB, also net cell fluxes are optimized for accuracy near the EB. The net fluxes of cells $i-1, i, i+1$ and $i+2$ are affected by the EB. Recalling that only $c_{i-\frac{1}{2}}$ and $c_{i+\frac{3}{2}}$ allow for optimization, only two of the four aforementioned net cell fluxes can be optimized for accuracy: either the net flux in cell $i-1$ or cell i, for $c_{i-\frac{1}{2}}$; and either the net flux in cell $i+1$ or cell $i+2$, for $c_{i+\frac{3}{2}}$.

For the accuracy optimizations, Taylor series expansions are used. Doing so, the net flux in cell i cannot be optimized due to the presence of the EB with its discontinuous solution behavior. Hence, the net flux in cell $i-1$ will be optimized for $c_{i-\frac{1}{2}}$. Secondly, for $c_{i+\frac{3}{2}}$, the net flux in cell $i+2$ will be optimized. The reason why the net flux in cell $i+2$ is optimized, instead of that of cell $i+1$, becomes clear at the end of the derivations in § 3.1.2. We start by first deriving the unlimited EB-affected cell-face states, and after that, EB-sensitive limiters will be derived.

3.1.1 Cell-face states

Here, we derive the unlimited forms of the cell-face states in cells $i-1, i+1$ and $i+2$. These are the EB-affected cell-face states ($c_{i-\frac{1}{2}}$, $c_{i+\frac{1}{2}}$, $c_{i+\frac{3}{2}}$) and the corresponding regular cell-face states ($c_{i-\frac{3}{2}}$, $c_{i+\frac{5}{2}}$).

a. Cell-face states affected by EB

Cell-face state $c_{i-\frac{1}{2}}$: The second-order accurate, non-equidistant, central interpolation, and the second-order accurate, equidistant, fully upwind extrapolation schemes for $c_{i-\frac{1}{2}}$ can be written as:

$$c_{i-\frac{1}{2}} = c_{i-1} + \frac{1}{1+2\beta}(c_{EB}^l - c_{i-1}), \tag{16a}$$

and

$$c_{i-\frac{1}{2}} = c_{i-1} + \frac{1}{2}(c_{i-1} - c_{i-2}), \tag{16b}$$

respectively. The blend of the above two schemes, is:

$$c_{i-\frac{1}{2}} = c_{i-1} + \frac{1}{1+2\beta}\frac{1+\kappa_{i-\frac{1}{2}}}{2}(c_{EB}^l - c_{i-1}) + \frac{1-\kappa_{i-\frac{1}{2}}}{4}(c_{i-1} - c_{i-2}), \tag{16c}$$

with $\kappa_{i-\frac{1}{2}}$ the blending parameter. Note that we get the exact result $c_{i-\frac{1}{2}} = c_{EB}^l$, $\beta = 0$, only for $\kappa_{i-\frac{1}{2}} = 1$. (The accuracy of cell-face states is not our prime interest, the accuracy of net fluxes *is*.)

Cell-face state $c_{i+\frac{1}{2}}$: As mentioned earlier, there are no sufficient number of solution points, on the upstream side of cell face $i + \frac{1}{2}$, up to and including the right face of the EB, to construct a higher-order accurate upwind-biased interpolation scheme. Hence, no κ-scheme is formulated here. Instead, this particular flux is reconstructed with only a, second-order accurate, non-equidistant central interpolation scheme, as:

$$c_{i+\frac{1}{2}} = c_{EB}^r + \frac{2-2\beta}{3-2\beta}(c_{i+1} - c_{EB}^r). \tag{17}$$

Note that we get the expected standard second-order accurate central result for $\beta = \frac{1}{2}$, and the exact result for $\beta = 1$.

Cell-face state $c_{i+\frac{3}{2}}$: The second-order accurate central interpolation and the non-equidistant, second-order accurate, fully upwind extrapolation schemes for $c_{i+\frac{3}{2}}$ can be written as:

$$c_{i+\frac{3}{2}} = c_{i+1} + \frac{1}{2}(c_{i+2} - c_{i+1}), \tag{18a}$$

and

$$c_{i+\frac{3}{2}} = c_{i+1} + \frac{1}{3-2\beta}(c_{i+1} - c_{EB}^r), \tag{18b}$$

respectively. Blending the above two schemes, we get:

$$c_{i+\frac{3}{2}} = c_{i+1} + \frac{1+\kappa_{i+\frac{3}{2}}}{4}(c_{i+2} - c_{i+1}) + \frac{1}{3-2\beta}\frac{1-\kappa_{i+\frac{3}{2}}}{2}(c_{i+1} - c_{EB}^r), \tag{18c}$$

with $\kappa_{i+\frac{3}{2}}$ being the blending parameter.

b. Corresponding regular cell-face states

For cell faces $i - \frac{3}{2}$ and $i + \frac{5}{2}$, the standard $\kappa = \frac{1}{3}$ scheme is applied:

$$c_{i-\frac{3}{2}} = c_{i-2} + \frac{1}{3}(c_{i-1} - c_{i-2}) + \frac{1}{6}(c_{i-2} - c_{i-3}), \qquad (19a)$$

$$c_{i+\frac{5}{2}} = c_{i+2} + \frac{1}{3}(c_{i+3} - c_{i+2}) + \frac{1}{6}(c_{i+2} - c_{i+1}). \qquad (19b)$$

3.1.2 Net cell fluxes

Here, we compute the net cell fluxes and derive the modified equations for cells $i-1$, $i+1$ and $i+2$, from which the blending parameters $\kappa_{i-\frac{1}{2}}$ and $\kappa_{i+\frac{3}{2}}$ will be optimized. Recall that the net flux in cell i cannot be optimized as the solution is discontinuous in there.

Optimal accuracy in cell $i-1$: With (16c) and (19a), we get as semi-discrete equation for cell $i-1$:

$$\frac{dc_{i-1}}{dt} + \frac{u}{h}\left(\frac{1}{1+2\beta}\frac{1+\kappa_{i-\frac{1}{2}}}{2}(c_{EB}^l - c_{i-1}) + \frac{11 - 3\kappa_{i-\frac{1}{2}}}{12}(c_{i-1} - c_{i-2}) - \frac{1}{6}(c_{i-2} - c_{i-3})\right) = 0. \qquad (20a)$$

Substituting Taylor-series expansions of c_{EB}^l, c_{i-2} and c_{i-3} around the point $i-1$ into (20a), we get as modified equation for cell $i-1$, ignoring the index $i-1$:

$$\frac{\partial c}{\partial t} + u\frac{\partial c}{\partial x} + \frac{6\beta - 7 + (9+6\beta)\kappa_{i-\frac{1}{2}}}{48}uh\frac{\partial^2 c}{\partial x^2} + \frac{(3+2\beta)(2\beta-1)(1+\kappa_{i-\frac{1}{2}})}{96}uh^2\frac{\partial^3 c}{\partial x^3} = \mathcal{O}(h^3). \qquad (20b)$$

Equating the leading term of the truncation error to zero, we get:

$$\kappa_{i-\frac{1}{2}} = \frac{7-6\beta}{9+6\beta}, \qquad \kappa_{i-\frac{1}{2}} \in \left[\frac{1}{15}, \frac{7}{9}\right]. \qquad (21)$$

This is the $\kappa_{i-\frac{1}{2}}$ that yields the most accurate net flux in cell $i-1$. It is well within the standard κ-range $[-1,1]$. Its variation for any position of the EB within cell i is depicted in Figure 7(a).

Substituting the optimal value of $\kappa_{i-\frac{1}{2}}$ according to (21) into the modified equation (20b), we get:

$$\frac{\partial c}{\partial t} + u\frac{\partial c}{\partial x} + \frac{2\beta-1}{18}uh^2\frac{\partial^3 c}{\partial x^3} = \mathcal{O}(h^3). \qquad (22)$$

Therefore, in general, we get a second-order (spatial) accuracy in cell $i-1$, with a maximum leading-term truncation-error coefficient of $\pm\frac{1}{18}uh^2$. Evidently, this

Finite-Volume Discretizations and Immersed Boundaries

dispersive term diminishes as the EB is in the immediate vicinity of the center of cell i, $\beta \approx \frac{1}{2}$. For $\beta = \frac{1}{2}$, $\kappa_{i-\frac{1}{2}}$ is restored as $\kappa_{i-\frac{1}{2}} = \frac{1}{3}$ (Figure 7(a)), and then we get third-order spatial accuracy.

Optimal accuracy in cell $i+1$: With (17) and (18c), we get as the semi-discrete equation for cell $i+1$:

$$\frac{dc_{i+1}}{dt} + \frac{u}{h}\left(\frac{3-\kappa_{i+\frac{3}{2}}}{6-4\beta}(c_{i+1}-c^r_{EB}) + \frac{1+\kappa_{i+\frac{3}{2}}}{4}(c_{i+2}-c_{i+1})\right) = 0. \quad (23a)$$

Introducing Taylor-series expansions of c^r_{EB} and c_{i+2} around the point $i+1$, into (23a), we get as modified equation for cell $i+1$, ignoring the index $i+1$:

$$\frac{\partial c}{\partial t} + u\frac{\partial c}{\partial x} + \frac{(6\beta-7)+(5-2\beta)\kappa_{i+\frac{3}{2}}}{16}uh\frac{\partial^2 c}{\partial x^2} + $$
$$\frac{(12\beta^2-36\beta+31)-(4\beta^2-12\beta+5)\kappa_{i+\frac{3}{2}}}{96}uh^2\frac{\partial^3 c}{\partial x^3} = \mathcal{O}(h^3). \quad (23b)$$

Then equating the leading term of the truncation error to zero, we get:

$$\kappa_{i+\frac{3}{2}} = \frac{7-6\beta}{5-2\beta}, \quad \kappa_{i+\frac{3}{2}} \in \left[\frac{1}{3}, \frac{7}{5}\right]. \quad (24)$$

This is the $\kappa_{i+\frac{3}{2}}$ that yields the most accurate net flux in cell $i+1$. This $\kappa_{i+\frac{3}{2}}$ is not within the standard κ-range $[-1,1]$.

Substituting (24) into (23b), we get as modified equation for cell $i+1$, ignoring the index $i+1$:

$$\frac{\partial c}{\partial t} + u\frac{\partial c}{\partial x} + \frac{3-2\beta}{12}uh^2\frac{\partial^3 c}{\partial x^3} = \mathcal{O}(h^3). \quad (25)$$

Note that the leading order error-term in cell $i+1$ is second-order for all β; it does not vanish for $\beta = \frac{1}{2}$.

Moreover, with (18c), (19b) and (24), we get as semi-discrete equation for cell $i+2$:

$$\frac{dc_{i+2}}{dt} + \frac{u}{h}\left(\frac{1-2\beta}{(3-2\beta)(5-2\beta)}(c_{i+1}-c^r_{EB}) + \right.$$
$$\left.\frac{17-2\beta}{30-12\beta}(c_{i+2}-c_{i+1}) + \frac{1}{3}(c_{i+3}-c_{i+2})\right) = 0. \quad (26a)$$

Introducing Taylor-series expansions for c^r_{EB}, c_{i+1} and c_{i+3} around the point $i+2$ into (26a), we get as modified equation for cell $i+2$, ignoring the index $i+2$:

$$\frac{\partial c}{\partial t} + u\frac{\partial c}{\partial x} + \frac{6\beta-7}{24}uh\frac{\partial^2 c}{\partial x^2} = \mathcal{O}(h^2). \quad (26b)$$

Note that the leading order error-term in cell $i+2$ is first-order for all β.

Optimal accuracy in cell $i+2$: With (18c) and (19b), we get as semi-discrete equation for cell $i+2$:

$$\frac{dc_{i+2}}{dt} + \frac{u}{h}\left(\frac{\kappa_{i+\frac{3}{2}}-1}{6-4\beta}(c_{i+1} - c_{EB}^r) + \frac{11-3\kappa_{i+\frac{3}{2}}}{12}(c_{i+2} - c_{i+1}) + \frac{1}{3}(c_{i+3} - c_{i+2})\right) = 0. \quad (27a)$$

Introducing Taylor-series expansions of c_{EB}^r, c_{i+1} and c_{i+3} around the point $i+2$ into (27a), we get as modified equation for cell $i+2$, ignoring the index $i+2$:

$$\frac{\partial c}{\partial t} + u\frac{\partial c}{\partial x} + \frac{(6\beta-15)\kappa_{i+\frac{3}{2}}+(7-6\beta)}{48}uh\frac{\partial^2 c}{\partial x^2} +$$
$$\frac{(4\beta^2-24\beta+35)\kappa_{i+\frac{3}{2}} - (4\beta^2-24\beta+19)}{96}uh^2\frac{\partial^3 c}{\partial x^3} = \mathcal{O}(h^3). \quad (27b)$$

Equating the leading term of the truncation error to zero, now we get:

$$\kappa_{i+\frac{3}{2}} = \frac{7-6\beta}{15-6\beta}, \quad \kappa_{i+\frac{3}{2}} \in \left[\frac{1}{9}, \frac{7}{15}\right]. \quad (28)$$

This is the value of $\kappa_{i+\frac{3}{2}}$ that yields the most accurate net flux in cell $i+2$. As opposed to $\kappa_{i+\frac{3}{2}}$ according to (24), this $\kappa_{i+\frac{3}{2}}$ is well within the standard κ-range $[-1,1]$. Its variation for any position of the EB within cell i is depicted in Figure 7(b).

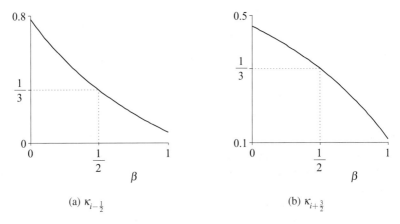

Fig. 7 Variation of the optimal κ values for any position of the EB within cell i.

Substituting the optimal $\kappa_{i+\frac{3}{2}}$ according to (28) into (27b), we get as modified equation for cell $i+2$:

$$\frac{\partial c}{\partial t} + u\frac{\partial c}{\partial x} + \frac{2\beta-1}{36}uh^2\frac{\partial^3 c}{\partial x^3} = \mathcal{O}(h^3). \tag{29}$$

In contrast to the dispersive error in (25), the dispersive term in (29) does vanish as the EB gets in the vicinity of the center of cell i, $\beta \approx \frac{1}{2}$. We, thereby, get third-order spatial accuracy, and $\kappa_{i+\frac{3}{2}}$ according to (28) becomes $\kappa = \frac{1}{3}$ (see Figure 7(b), and the Appendix for a more detailed comparison).

With (17), (18c) and (28), we get as semi-discrete equation for cell $i+1$:

$$\frac{dc_{i+1}}{dt} + \frac{u}{h}\left(\frac{19-6\beta}{(9-6\beta)(5-2\beta)}(c_{i+1}-c^r_{EB}) + \frac{11-6\beta}{30-12\beta}(c_{i+2}-c_{i+1})\right) = 0. \tag{30a}$$

And, substituting Taylor-series expansions for c^r_{EB}, and c_{i+2} around the point $i+1$, into (30a), we get as modified equation for cell $i+1$, ignoring the index $i+1$:

$$\frac{\partial c}{\partial t} + u\frac{\partial c}{\partial x} + \frac{6\beta-7}{24}uh\frac{\partial^2 c}{\partial x^2} = \mathcal{O}(h^2). \tag{30b}$$

Equation (30b) shows that we get a first-order spatial accuracy in cell $i+1$ with a maximum leading-term truncation-error coefficient of $-\frac{7}{24}uh$. Coincidentally, (30b) is the same as (26b); the leading-order error terms in both equations are identical. The accuracy loss in the net flux of a neighboring cell is unavoidable. If the cell-face states were to be first-order accurate, i.e. $c_{i+\frac{1}{2}} = c^r_{EB}$ and $c_{i+\frac{3}{2}} = c_{i+1}$, the modified equation for cell $i+1$, ignoring the index $i+1$, would become:

$$\frac{\partial c}{\partial t} + \frac{3-2\beta}{2}u\frac{\partial c}{\partial x} - \frac{(3-2\beta)^2}{8}uh\frac{\partial^2 c}{\partial x^2} = \mathcal{O}(h^2), \tag{31}$$

which, for all β, except $\beta = \frac{1}{2}$, is even zeroth-order accurate.

As the optimal $\kappa_{i+\frac{3}{2}}$ we choose (28), the one that gives the highest accuracy in cell $i+2$. In summary, the reasons why we choose this $\kappa_{i+\frac{3}{2}}$, instead of the one yielding the highest accuracy in cell $i+1$ ($\kappa_{i+\frac{3}{2}}$ according to (24)), are the following:

- For $\beta = \frac{1}{2}$, we get a third-order (spatial) accuracy in cell $i+2$ with (28) (see (29)). But with (24) we do not get this in cell $i+1$ for any β (see (25)). The truncation error with (28) is much less than that with (24), for any β (see Appendix).
- Noting that the solution is discontinuous across an EB, with (28) we have a dissipative leading-error term in cell $i+1$, which is the cell adjacent to cell i (where the EB is situated), and this makes the solution near the EB less prone to numerical oscillations.

With (24) however, we get the leading-error term in the same cell to be dispersive and this makes the solution near the EB to be more susceptible to numerical oscillations, numerical oscillations which may be hard to suppress because construction of a limiter for cell-face state $c_{i+\frac{1}{2}}$ is hard.

- With (28), the accuracy deterioration due to the presence of an EB in cell i is more confined to the vicinity of the EB. (We get first-order (spatial) accuracy in cell $i+1$, and second-order accuracy in cell $i+2$; whereas with (24), we get second-order accuracy in cell $i+1$, and first-order accuracy in cell $i+2$.)
- $\kappa_{i+\frac{3}{2}}$ according to (28) is well within the standard κ-range $[-1,1]$, but with (24) we get $\kappa_{i+\frac{3}{2}} \in [\frac{1}{3}, \frac{7}{5}]$.

3.1.3 Formulae for cell-face states affected by EB

Here, the formulae for all the special cell-face states that are affected by the EB, in cell i, viz. $c_{i-\frac{1}{2}}$, $c_{i+\frac{1}{2}}$ and $c_{i+\frac{3}{2}}$, are summarized.

With (16c) and (21), $c_{i-\frac{1}{2}}$ can be rewritten as:

$$c_{i-\frac{1}{2}} = c_{i-1} + \frac{8}{(3+6\beta)(3+2\beta)}(c_{EB}^l - c_{i-1}) + \frac{1+6\beta}{18+12\beta}(c_{i-1} - c_{i-2}). \quad (32a)$$

Further, we have:

$$c_{i+\frac{1}{2}} = c_{EB}^r + \frac{2-2\beta}{3-2\beta}(c_{i+1} - c_{EB}^r). \quad (32b)$$

And, with (18c) and (28), $c_{i+\frac{3}{2}}$ can be rewritten as:

$$c_{i+\frac{3}{2}} = c_{i+1} + \frac{11-6\beta}{30-12\beta}(c_{i+2} - c_{i+1}) + \frac{4}{(9-6\beta)(5-2\beta)}(c_{i+1} - c_{EB}^r). \quad (32c)$$

Verify that, in (32a) and (32c), for $\beta = \frac{1}{2}$, we get exactly the same coefficients as in the the standard $\kappa = \frac{1}{3}$ scheme (see (19a) and (19b)).

3.2 Spatial monotonicity domains and limiters

Recalling Godunov's (1959) theorem, all the linear higher-order accurate fluxes, constructed earlier, may yield wiggles. One negative aspect of wiggles is that they may cause the solution c to be negative. If c is a physical quantity that should not become negative (say, density or temperature), this may be highly undesirable. Wiggles can be avoided by carefully constraining or 'limiting' the advective fluxes calculated by the scheme. By limiting the fluxes, they may become first-order accurate in some solution regions.

A limiter is a nonlinear function that acts like a continuous control between the higher-order and first-order schemes. Obviously, limiters may reduce the overall accuracy of the scheme to some extent, albeit in non-smooth flow regions only. Limited schemes are called 'monotonicity-preserving.'

For the cell-face states that are computed by the standard $\kappa = \frac{1}{3}$ scheme (§ 2.3), the standard $\kappa = \frac{1}{3}$ limiter (14b) will be used. In this section, special limiters will be introduced for the EB-affected cell-face states $c_{i-\frac{1}{2}}$ and $c_{i+\frac{3}{2}}$ according to formulae (32a) and (32c). The cell-face state $c_{i+\frac{1}{2}}$, however, will not be limited. This shall be explained later on.

3.2.1 Spatial monotonicity domain and limiter for cell-face state $c_{i-\frac{1}{2}}$

Referring to Figure 6, for cell face $i - \frac{1}{2}$, we define the non-equidistant local successive solution-gradient ratio $\tilde{r}_{i-\frac{1}{2}}$, as:

$$\tilde{r}_{i-\frac{1}{2}} = \frac{c_{EB}^l - c_{i-1}}{\frac{1+2\beta}{2}h} \bigg/ \frac{c_{i-1} - c_{i-2}}{h} \equiv \frac{2}{1+2\beta} \frac{c_{EB}^l - c_{i-1}}{c_{i-1} - c_{i-2}}. \tag{33}$$

Notice that for $\beta = \frac{1}{2}$, EB in the center of cell i, $\tilde{r}_{i-\frac{1}{2}}$ reduces to the standard equidistant solution-gradient ratio known from the theory of standard limiters.

We proceed by rewriting (16c) as:

$$c_{i-\frac{1}{2}} = c_{i-1} + \frac{1}{2}\tilde{\phi}(\tilde{r}_{i-\frac{1}{2}})(c_{i-1} - c_{i-2}), \tag{34a}$$

with

$$\tilde{\phi}(\tilde{r}_{i-\frac{1}{2}}) = \frac{1 - \kappa_{i-\frac{1}{2}}}{2} + \frac{1 + \kappa_{i-\frac{1}{2}}}{2}\tilde{r}_{i-\frac{1}{2}}. \tag{34b}$$

Substituting the optimal $\kappa_{i-\frac{1}{2}}$ according to (21) into (34b), we get:

$$\tilde{\phi}(\tilde{r}_{i-\frac{1}{2}}) = \frac{1+6\beta}{9+6\beta} + \frac{8}{9+6\beta}\tilde{r}_{i-\frac{1}{2}}. \tag{34c}$$

The family of possible $\tilde{\phi}(\tilde{r}_{i-\frac{1}{2}})$ schemes, depending on the position of the EB within cell i ($0 \leq \beta \leq 1$), is represented in Figure 8. The function $\tilde{\phi}(\tilde{r}_{i-\frac{1}{2}})$ will be constrained to yield a monotonicity-preserving scheme and to define the appropriate limiter for the special cell-face state $c_{i-\frac{1}{2}}$. The argument $\tilde{r}_{i-\frac{1}{2}}$ measures the local monotonicity of the solution.

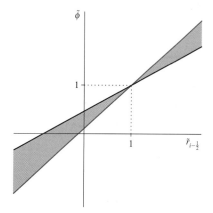

Fig. 8 Family of possible β-schemes according to (34c): the blue line is for $\beta = 1$, the red line is for $\beta = 0$, and the enclosed (colored) region is for all other $\beta \in (0, 1)$.

The local solution-gradient ratio for cell face $i - \frac{3}{2}$ is defined as:

$$r_{i-\frac{3}{2}} = \frac{c_{i-1} - c_{i-2}}{c_{i-2} - c_{i-3}}. \tag{35a}$$

And, the limited form of cell-face state $c_{i-\frac{3}{2}}$ is:

$$c_{i-\frac{3}{2}} = c_{i-2} + \frac{1}{2}\phi(r_{i-\frac{3}{2}})(c_{i-2} - c_{i-3}), \tag{35b}$$

where $\phi(r)$ is standard limiter (14b) with $\kappa = \frac{1}{3}$.

The following monotonicity requirement is enforced, to constrain the function $\tilde{\phi}(\tilde{r}_{i-\frac{1}{2}})$:

$$\frac{c_{i-\frac{1}{2}} - c_{i-\frac{3}{2}}}{c_{i-1} - c_{i-2}} \geq 0. \tag{36a}$$

Substituting (34a) and (35b) into (36a), using (35a), we get as constraint relation:

$$1 + \frac{1}{2}\tilde{\phi}(\tilde{r}_{i-\frac{1}{2}}) - \frac{1}{2}\frac{\phi(r_{i-\frac{3}{2}})}{r_{i-\frac{3}{2}}} \geq 0. \tag{36b}$$

The standard limiter already satisfies $1 - \frac{1}{2}\frac{\phi(r_{i-\frac{3}{2}})}{r_{i-\frac{3}{2}}} \geq 0, \forall r_{i-\frac{3}{2}}$; therefore, the (in)equality (36b) holds good if:

$$\tilde{\phi}(\tilde{r}_{i-\frac{1}{2}}) \geq 0, \quad \forall \tilde{r}_{i-\frac{1}{2}}. \tag{36c}$$

Moreover, enforcing the additional monotonicity requirement:

$$\frac{c_{EB}^l - c_{i-\frac{1}{2}}}{c_{EB}^l - c_{i-1}} \geq 0, \tag{37a}$$

and substituting (34a) into (37a), using (34c) and (33), we get:

$$\frac{\tilde{\phi}(\tilde{r}_{i-\frac{1}{2}})}{\tilde{r}_{i-\frac{1}{2}}} \leq 1 + 2\beta, \quad \forall \tilde{r}_{i-\frac{1}{2}}. \tag{37b}$$

The (in)equalities (36c) and (37b) define the spatial monotonicity domain for the special limiter function $\tilde{\phi}(\tilde{r}_{i-\frac{1}{2}})$. They delineate the left and lower bounds of the domain. The upper bound is still to be derived in § 4 from the fully discrete equation. Once we also have defined this upper bound, the respective limiter will be introduced, in § 4.1.

3.2.2 Limiter for cell-face state $c_{i+\frac{1}{2}}$

Regarding cell-face state $c_{i+\frac{1}{2}}$, a regular monotonicity argument $r_{i+\frac{1}{2}}$ can not be defined here. A regular monitor uses two solution values upstream of cell faces. In this case, since we do not want to use solution values from the other side of the EB, and therefore not c_i, we have only one upstream solution, c_{EB}^r, too little to introduce the regular smoothness monitor. Therefore, $c_{i+\frac{1}{2}}$ is not limited.

3.2.3 Spatial monotonicity domain and limiter for cell-face state $c_{i+\frac{3}{2}}$

Referring to Figure 6, the monotonicity argument $\tilde{r}_{i+\frac{3}{2}}$ is defined, as:

$$\tilde{r}_{i+\frac{3}{2}} = \frac{c_{i+2} - c_{i+1}}{h} \bigg/ \frac{c_{i+1} - c_{EB}^r}{\frac{3-2\beta}{2}h} = \frac{3-2\beta}{2} \frac{c_{i+2} - c_{i+1}}{c_{i+1} - c_{EB}^r}. \tag{38}$$

As expected, for $\beta = \frac{1}{2}$, $\tilde{r}_{i+\frac{3}{2}}$ according to (38) reduces to the known equidistant-grid ratio. Similar to the rewriting of expression (16c) for $c_{i-\frac{1}{2}}$, here we rewrite (18c) for $c_{i+\frac{3}{2}}$, as:

$$c_{i+\frac{3}{2}} = c_{i+1} + \frac{1}{3-2\beta} \tilde{\phi}(\tilde{r}_{i+\frac{3}{2}})(c_{i+1} - c_{EB}^r), \tag{39a}$$

with

$$\tilde{\phi}(\tilde{r}_{i+\frac{3}{2}}) = \frac{1 - \kappa_{i+\frac{3}{2}}}{2} + \frac{1 + \kappa_{i+\frac{3}{2}}}{2}\tilde{r}_{i+\frac{3}{2}}. \tag{39b}$$

Substituting the optimal $\kappa_{i+\frac{3}{2}}$ according to (28) into (39b), we get:

$$\tilde{\phi}(\tilde{r}_{i+\frac{3}{2}}) = \frac{4}{15 - 6\beta} + \frac{11 - 6\beta}{15 - 6\beta}\tilde{r}_{i+\frac{3}{2}}. \tag{39c}$$

The family of possible $\tilde{\phi}(\tilde{r}_{i+\frac{3}{2}})$ schemes is given in Figure 9. Just as $\tilde{\phi}(\tilde{r}_{i-\frac{1}{2}})$, in § 3.2.1, the function $\tilde{\phi}(\tilde{r}_{i+\frac{3}{2}})$ will be constrained to yield a monotonicity-preserving scheme and to define the appropriate limiter for the special cell-face state $c_{i+\frac{3}{2}}$.

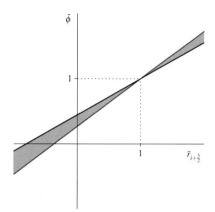

Fig. 9 Family of possible β-schemes according to (39c): the blue line is for $\beta = 1$, the red line is for $\beta = 0$, and the enclosed (colored) region is for all other $\beta \in (0, 1)$.

The monotonicity argument $r_{i+\frac{5}{2}}$ is defined as:

$$r_{i+\frac{5}{2}} = \frac{c_{i+3} - c_{i+2}}{c_{i+2} - c_{i+1}}. \tag{40a}$$

And, the limited form of $c_{i+\frac{5}{2}}$ is:

$$c_{i+\frac{5}{2}} = c_{i+2} + \frac{1}{2}\phi(r_{i+\frac{5}{2}})(c_{i+2} - c_{i+1}), \tag{40b}$$

where $\phi(r_{i+\frac{5}{2}})$ is limiter (14b) with $\kappa = \frac{1}{3}$.

To constrain $\tilde{\phi}(r_{i+\frac{3}{2}})$, the following monotonicity requirements are enforced:

$$\frac{c_{i+\frac{3}{2}} - c_{i+\frac{1}{2}}}{c_{i+1} - c_{EB}^r} \geq 0, \tag{41a}$$

$$\frac{c_{i+\frac{5}{2}} - c_{i+\frac{3}{2}}}{c_{i+2} - c_{i+1}} \geq 0. \tag{41b}$$

Substituting (17) and (39a) into (41a), we get as restriction for $\tilde{\phi}(\tilde{r}_{i+\frac{3}{2}})$:

$$\tilde{\phi}(\tilde{r}_{i+\frac{3}{2}}) \geq -1, \quad \forall \tilde{r}_{i+\frac{3}{2}}. \tag{42a}$$

And, substituting (39a) and (40b) into (41b), we get:

$$\frac{\tilde{\phi}(\tilde{r}_{i+\frac{3}{2}})}{\tilde{r}_{i+\frac{3}{2}}} - \phi(r_{i+\frac{5}{2}}) \leq 2. \tag{42b}$$

Since the standard limiter satisfies $\phi(r_{i+\frac{5}{2}}) \geq 0, \forall r_{i+\frac{5}{2}}$, the (in)equality (42b) holds good if:

$$\frac{\tilde{\phi}(\tilde{r}_{i+\frac{3}{2}})}{\tilde{r}_{i+\frac{3}{2}}} \leq 2, \quad \forall \tilde{r}_{i+\frac{3}{2}}. \tag{42c}$$

The (in)equalities (42a) and (42c) define the spatial monotonicity domain for the special limiter function $\tilde{\phi}(\tilde{r}_{i+\frac{3}{2}})$. They delineate the left and lower bounds of the domain. Once again, after defining the upper bound of the domain in § 4.1, the respective limiter for the special cell-face state $c_{i+\frac{3}{2}}$ will be introduced. Note that as opposed to the spatial monotonicity domain for $c_{i-\frac{1}{2}}$ (see (36c) and (37b)), the monotonicity domain for $c_{i+\frac{3}{2}}$ is independent of β.

4 Temporal discretization

The semi-discrete equation (4), after substituting the appropriate discretizations for the spatial operator, is discrete in space but still continuous in time. It can be compactly written as:

$$\frac{dc_i}{dt} = -\frac{u}{h}(c_{i+\frac{1}{2}} - c_{i-\frac{1}{2}}) \equiv F(c), \tag{43}$$

which is an ordinary differential equation that can be discretized in time using a variety of explicit and implicit time integration methods, to get a fully discrete system of equations. Here, only two explicit schemes are considered: the Forward Euler method and the three-stage Runge-Kutta, RK3b, scheme [6]. The latter gives third-order accuracy in time.

For the Forward Euler method, (43) becomes:

$$c_i^{n+1} = c_i^n + \tau F(c^n) \equiv c_i^n - \nu(c_{i+\frac{1}{2}}^n - c_{i-\frac{1}{2}}^n), \tag{44}$$

where $\nu = u\tau/h$ is the CFL number, and τ the time step. Similarly, for the RK3b scheme, we have:

$$c_i^{n+1} = c_i^n + \frac{1}{6}(R_1 + R_2 + 4R_3), \tag{45a}$$

where the R_j's ($j = 1, 2, 3$) are internal vectors that are computed as:

$$R_1 = \tau F(c^n),$$
$$R_2 = \tau F(c^n + R_1), \tag{45b}$$
$$R_3 = \tau F(c^n + \frac{1}{4}R_1 + \frac{1}{4}R_2).$$

4.1 TVD conditions and time step

The limited numerical flux conditions, as derived in § 3.2, are still insufficient to guarantee monotonicity during time integration. Harten's theorem [5] provides additional conditions that are necessary for the convergence of the fully discrete solutions to the exact, monotone solutions. These conditions define the upper bounds for the limiter functions $\tilde{\phi}(\tilde{r}_{i-\frac{1}{2}})$ and $\tilde{\phi}(\tilde{r}_{i+\frac{3}{2}})$, and consequently result into more stringent restrictions on the CFL number, than the condition for stability.

The theorem in [5] states that any consistent scheme for a conservation law, written in the conservative form:

$$c_i^{n+1} = c_i^n - D_{i-\frac{1}{2}}^-(c_i^n - c_{i-1}^n) + D_{i+\frac{1}{2}}^+(c_{i+1}^n - c_i^n), \tag{46a}$$

where the D's are solution-dependent coefficients, is total-variation diminishing (TVD) if, for all i:

$$D_{i+\frac{1}{2}}^\pm \geq 0, \tag{46b}$$

$$D_{i+\frac{1}{2}}^- + D_{i+\frac{1}{2}}^+ \leq 1. \tag{46c}$$

The total variation (TV) at time level n is defined, in discrete form, by:

$$\mathrm{TV}(c^n) = \sum_i |c_{i+1}^n - c_i^n|, \tag{47}$$

and, any scheme is said to be TVD if $\mathrm{TV}(c^{n+1}) \leq \mathrm{TV}(c^n)$.

Both conditions, (46b) and (46c), can be interpreted as positive coefficient requirements. To do so, we rewrite (46a) as

$$c_i^{n+1} = D_{i-\frac{1}{2}}^- c_{i-1}^n + (1 - D_{i-\frac{1}{2}}^- - D_{i+\frac{1}{2}}^+)c_i^n + D_{i+\frac{1}{2}}^+ c_{i+1}^n. \tag{48}$$

The positive coefficient requirements for (48) are:

$$D^-_{i-\frac{1}{2}} \geq 0, \tag{49a}$$

$$1 - D^-_{i-\frac{1}{2}} - D^+_{i+\frac{1}{2}} \geq 0, \tag{49b}$$

$$D^+_{i+\frac{1}{2}} \geq 0. \tag{49c}$$

Equation (48) holds for any i, so also for $i+1$:

$$c^{n+1}_{i+1} = D^-_{i+\frac{1}{2}} c^n_i + (1 - D^-_{i+\frac{1}{2}} - D^+_{i+\frac{3}{2}}) c^n_{i+1} + D^+_{i+\frac{3}{2}} c^n_{i+2}. \tag{50}$$

The positive coefficient requirements applied to (50) yield, among others,

$$D^-_{i+\frac{1}{2}} \geq 0. \tag{51}$$

So, with (49c) and (51) we have already interpreted (46b) as a positive coefficient requirement. From (49b) it follows

$$D^-_{i-\frac{1}{2}} + D^+_{i+\frac{1}{2}} \leq 1. \tag{52}$$

Combining (52), (49a) and (49c), it may be written:

$$0 \leq D^-_{i-\frac{1}{2}} \leq 1 - \gamma, \tag{53a}$$

$$0 \leq D^+_{i+\frac{1}{2}} \leq \gamma, \tag{53b}$$

with γ some constant in the range $[0,1]$. We assume that the upper bound $1 - \gamma$ holds for all i, hence also for $D^-_{i+\frac{1}{2}}$:

$$0 \leq D^-_{i+\frac{1}{2}} \leq 1 - \gamma. \tag{54}$$

Summation of (53b) and (54) gives

$$0 \leq D^-_{i+\frac{1}{2}} + D^+_{i+\frac{1}{2}} \leq 1. \tag{55}$$

Combined with (49c) and (51) this may be reduced to

$$D^-_{i+\frac{1}{2}} + D^+_{i+\frac{1}{2}} \leq 1, \tag{56}$$

which is TVD requirement (46c).

With this we have shown that TVD requirements (46b) and (46c) are positive coefficient requirements.

It can also be verified that condition (46b) is identical to the monotonicity requirements that we have already considered in § 3.2 (the conditions of which (36a), (37a), (41a) and (41b) are examples). Condition (46c) though has not been considered yet. It will yield the sought upper bounds for our specific limiters under construction. These bounds will be v-dependent. Here, we will impose TVD requirement (46c) to $\tilde{\phi}(\tilde{r}_{i-\frac{1}{2}})$ and $\tilde{\phi}(\tilde{r}_{i+\frac{3}{2}})$.

We consider the Forward Euler scheme and write the fully discrete equations, in the form (46a). Referring to Figure 6, cell $i-1$ is considered for $\tilde{\phi}(\tilde{r}_{i-\frac{1}{2}})$, and cells $i+1$ and $i+2$ are analyzed for $\tilde{\phi}(\tilde{r}_{i+\frac{3}{2}})$. Recall that the EB is situated in cell i and, as in all preceding sections, we will not consider this particular cell for any analysis.

4.1.1 TVD conditions for limiter function $\tilde{\phi}(\tilde{r}_{i-\frac{1}{2}})$

Using (34a) and (35b) and writing the fully discrete equation for cell $i-1$ in the conservative form, we get:

$$c_{i-1}^{n+1} = c_{i-1}^n - \frac{v}{2}\left(2 + \tilde{\phi}(\tilde{r}_{i-\frac{1}{2}}) - \frac{\phi(r_{i-\frac{3}{2}})}{r_{i-\frac{3}{2}}}\right)(c_{i-1}^n - c_{i-2}^n). \qquad (57a)$$

Thus from (57a), we have as the corresponding coefficients:

$$D_{i-\frac{3}{2}}^- = \frac{v}{2}\left(2 + \tilde{\phi}(\tilde{r}_{i-\frac{1}{2}}) - \frac{\phi(r_{i-\frac{3}{2}})}{r_{i-\frac{3}{2}}}\right), \qquad (57b)$$

$$D_{i-\frac{1}{2}}^+ = 0. \qquad (57c)$$

Enforcing condition (46c), we get:

$$\tilde{\phi}(\tilde{r}_{i-\frac{1}{2}}) - \frac{\phi(r_{i-\frac{3}{2}})}{r_{i-\frac{3}{2}}} \leq \frac{2}{v} - 2. \qquad (58a)$$

Because the standard limiter satisfies $0 \leq \frac{\phi(r_{i-\frac{3}{2}})}{r_{i-\frac{3}{2}}} \leq 2, \forall r_{i-\frac{3}{2}}$, the above inequality reduces to:

$$\tilde{\phi}(\tilde{r}_{i-\frac{1}{2}}) \leq \frac{2}{v} - 2, \quad \forall \tilde{r}_{i-\frac{1}{2}}. \qquad (58b)$$

Taking the (in)equalities (36c), (37b) and (58b) into account, the TVD domain of the special limiter for cell-face state $c_{i-\frac{1}{2}}$ is graphically illustrated in Figure 10.

(In)equality (58b) confirms that the upper bound of the special limiter function $\tilde{\phi}(\tilde{r}_{i-\frac{1}{2}})$ depends on the CFL number v. The upper bound increases when lowering v. Note that the choice $v = 1$, the stability bound for Forward Euler,

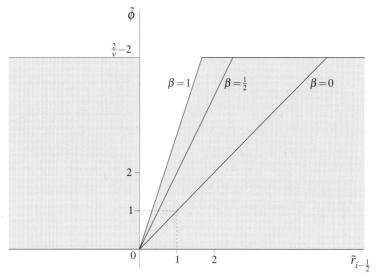

Fig. 10 TVD domain for cell-face state $c_{i-\frac{1}{2}}$, for some characteristic values of β. Note that $\nu \leq \frac{1}{2}$.

yields $\tilde{\phi}(\tilde{r}_{i-\frac{1}{2}}) = 0$. Hence, with $\nu = 1$, the second-order accuracy requirement $\tilde{\phi}(\tilde{r}_{i-\frac{1}{2}} = 1) = 1$ cannot be satisfied for this limiter. $\nu \leq \frac{2}{3}$ allows to meet this accuracy requirement. $\nu < \frac{2}{3}$ even allows for third-order accuracy in space. We take $\nu = \frac{1}{2}$ as the upper bound; it gives sufficient room for good spatial accuracy and does not bound the time step too much. Moreover, it will appear to be the hard upper bound for ν in using limiter $\tilde{\phi}(\tilde{r}_{i+\frac{3}{2}})$.

When we impose this bound for ν on $\tilde{\phi}(\tilde{r}_{i-\frac{1}{2}})$, the ν-dependence of the upper bound in (58b) is avoided; the (in)equality (58b) can then be simplified, $\forall \tilde{r}_{i-\frac{1}{2}}$, to the more practical inequality:

$$\tilde{\phi}(\tilde{r}_{i-\frac{1}{2}}) \leq 2, \qquad \text{for } \nu \leq \frac{1}{2}. \tag{58c}$$

With this, the TVD domain in Figure 10 simplifies to that given in Figure 11.

We now strive for a practical limiter $\tilde{\phi}(\tilde{r}_{i-\frac{1}{2}})$ which coincides with the target scheme (34) to the maximal possible extent. An algorithm for computing this limiter $\tilde{\phi}(\tilde{r}_{i-\frac{1}{2}})$ reads:

1. Compute β according to (15).
2. Compute the actual value of the monotonicity argument $\tilde{r}_{i-\frac{1}{2}}$ according to (33).
3. Compute the values $\tilde{r}^*_{i-\frac{1}{2}}$ and $\tilde{r}^{**}_{i-\frac{1}{2}}$ for which the target function $\tilde{\phi}(\tilde{r}_{i-\frac{1}{2}})$ according to (34c) intersects the simplified TVD domain's bounds, $(1+2\beta)\tilde{r}_{i-\frac{1}{2}}$ and 2, respectively.

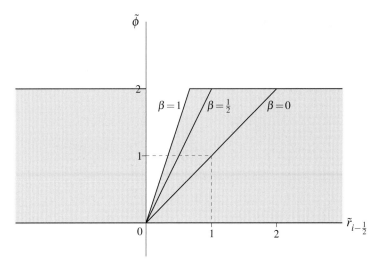

Fig. 11 Simplified TVD domain for cell-face state $c_{i-\frac{1}{2}}$, for some characteristic values of β.

$$\tilde{r}^{*}_{i-\frac{1}{2}} = \frac{1+6\beta}{1+24\beta+12\beta^2}, \tag{59a}$$

$$\tilde{r}^{**}_{i-\frac{1}{2}} = \frac{17+6\beta}{8}. \tag{59b}$$

4. Then, the special limiter for the cell-face state $c_{i-\frac{1}{2}}$ reads:

$$\tilde{\phi}(\tilde{r}_{i-\frac{1}{2}}) = \begin{cases} 0, & \text{if } \tilde{r}_{i-\frac{1}{2}} < 0, \\ (1+2\beta)\tilde{r}_{i-\frac{1}{2}}, & \text{if } 0 \leq \tilde{r}_{i-\frac{1}{2}} < \tilde{r}^{*}_{i-\frac{1}{2}}, \\ \frac{1+6\beta}{9+6\beta} + \frac{8}{9+6\beta}\tilde{r}_{i-\frac{1}{2}}, & \text{if } \tilde{r}^{*}_{i-\frac{1}{2}} \leq \tilde{r}_{i-\frac{1}{2}} < \tilde{r}^{**}_{i-\frac{1}{2}}, \\ 2, & \text{if } \tilde{r}_{i-\frac{1}{2}} \geq \tilde{r}^{**}_{i-\frac{1}{2}}. \end{cases} \tag{60}$$

The EB-sensitive limiter (60) is depicted in Figure 12 for the two extreme values of β, $\beta = 0$ and $\beta = 1$, together with the corresponding TVD domains.

4.1.2 TVD conditions for limiter function $\tilde{\phi}(\tilde{r}_{i+\frac{3}{2}})$

Similarly, to fully constrain the limiter function $\tilde{\phi}(\tilde{r}_{i+\frac{3}{2}})$, the fully discrete equations for cells $i+1$ and $i+2$ are analyzed. Substituting (32b), (39a) and (40b) into (46a), rewritten for c_{i+1}^{n+1} and c_{i+2}^{n+1}, the fully discrete equations can be written as:

Finite-Volume Discretizations and Immersed Boundaries

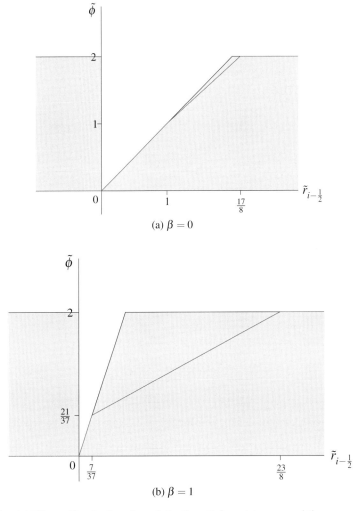

Fig. 12 Special EB-sensitive limiters (in red) for the cell-face state $c_{i-\frac{1}{2}}$ and the corresponding TVD domains for the two extreme values of β.

$$c_{i+1}^{n+1} = c_{i+1}^n - \frac{v}{3 - 2\beta}\left(1 + \tilde{\phi}(\tilde{r}_{i+\frac{3}{2}})\right)(c_{i+1}^n - c_{EB}^r), \tag{61a}$$

$$c_{i+2}^{n+1} = c_{i+2}^n - \frac{v}{2}\left(2 + \phi(r_{i+\frac{5}{2}}) - \frac{\phi(\tilde{r}_{i+\frac{3}{2}})}{\tilde{r}_{i+\frac{3}{2}}}\right)(c_{i+2}^n - c_{i+1}^n). \tag{61b}$$

Thus from (61a) and (61b), we have as the corresponding Harten coefficients, for cells $i+1$ and $i+2$:

$$D^-_{i+\frac{1}{2}} = \frac{\nu}{3-2\beta}\left(1 + \tilde{\phi}(\tilde{r}_{i+\frac{3}{2}})\right), \tag{62a}$$

$$D^+_{i+\frac{3}{2}} = 0, \tag{62b}$$

and

$$D^-_{i+\frac{3}{2}} = \frac{\nu}{2}\left(2 + \phi(r_{i+\frac{5}{2}}) - \frac{\tilde{\phi}(\tilde{r}_{i+\frac{3}{2}})}{\tilde{r}_{i+\frac{3}{2}}}\right), \tag{62c}$$

$$D^+_{i+\frac{5}{2}} = 0, \tag{62d}$$

respectively. Enforcing condition (46c), we get:

$$\tilde{\phi}(\tilde{r}_{i+\frac{3}{2}}) \leq \frac{3-2\beta}{\nu} - 1, \quad \forall \tilde{r}_{i+\frac{3}{2}}, \tag{63a}$$

and

$$\phi(r_{i+\frac{5}{2}}) - \frac{\tilde{\phi}(\tilde{r}_{i+\frac{3}{2}})}{\tilde{r}_{i+\frac{3}{2}}} \leq \frac{2}{\nu} - 2. \tag{63b}$$

The standard limiter satisfies $0 \leq \phi(r_{i+\frac{5}{2}}) \leq 2$, $\forall r_{i+\frac{5}{2}}$, and therefore (in)equality (63b) reduces to:

$$\frac{\tilde{\phi}(\tilde{r}_{i+\frac{3}{2}})}{\tilde{r}_{i+\frac{3}{2}}} \geq 4 - \frac{2}{\nu}, \quad \forall \tilde{r}_{i+\frac{3}{2}}. \tag{63c}$$

Taking the (in)equalities (42a), (42c), (63a) and (63c) into account, the TVD domain of the special limiter for cell-face state $c_{i+\frac{3}{2}}$ is depicted in Figure 13.

Concerning the just derived ν-dependent bounds, $\frac{3-2\beta}{\nu} - 1$ and $4 - \frac{2}{\nu}$, we notice that here $\nu = \frac{1}{2}$ is the maximum value that still allows to meet the second-order accuracy requirement $\tilde{\phi}(\tilde{r}_{i+\frac{3}{2}} = 1) = 1$, for $\beta = 1$. Hence, $\nu \leq \frac{1}{2}$ is the CFL restriction for the fully discrete systems (61a) and (61b) to be TVD and second-order accurate. For $\nu \leq \frac{1}{2}$, the (in)equalities (63a) and (63c) can be simplified to:

$$\tilde{\phi}(\tilde{r}_{i+\frac{3}{2}}) \leq 5 - 4\beta, \quad \forall \tilde{r}_{i+\frac{3}{2}}, \tag{63d}$$

and

$$\frac{\tilde{\phi}(\tilde{r}_{i+\frac{3}{2}})}{\tilde{r}_{i+\frac{3}{2}}} \geq 0, \quad \forall \tilde{r}_{i+\frac{3}{2}}. \tag{63e}$$

Doing so, the TVD domain in Figure 13 simplifies to the one shown in Figure 14.

In analogy to the algorithm for $\tilde{\phi}(\tilde{r}_{i-\frac{1}{2}})$ given in § 4.1.1, here an algorithm is also given for limiter $\tilde{\phi}(\tilde{r}_{i+\frac{3}{2}})$:

1. Compute β according to (15).
2. Compute the actual value of the monotonicity argument $\tilde{r}_{i+\frac{3}{2}}$ according to (38).

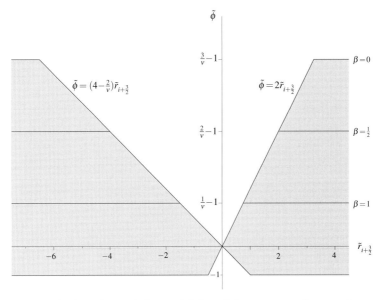

Fig. 13 TVD domain (drawn into scale for $v = 0.4$) for cell-face state $c_{i+\frac{3}{2}}$, for some characteristic values of β.

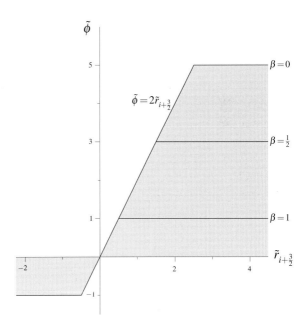

Fig. 14 Simplified TVD domain for cell-face state $c_{i+\frac{3}{2}}$, for some characteristic values of β.

3. Compute the values $\tilde{r}^*_{i+\frac{3}{2}}$, $\tilde{r}^{**}_{i+\frac{3}{2}}$, $\tilde{r}^{***}_{i+\frac{3}{2}}$ and $\tilde{r}^{****}_{i+\frac{3}{2}}$ for which the target function $\tilde{\phi}(\tilde{r}_{i+\frac{3}{2}})$ according to (39c) equals -1 and 0, and intersects the bounds $2\tilde{r}_{i+\frac{3}{2}}$ and $5 - 4\beta$ of the simplified TVD domain (Figure 14), respectively:

$$\tilde{r}^*_{i+\frac{3}{2}} = -\frac{19 - 6\beta}{11 - 6\beta}, \tag{64a}$$

$$\tilde{r}^{**}_{i+\frac{3}{2}} = -\frac{4}{11 - 6\beta}, \tag{64b}$$

$$\tilde{r}^{***}_{i+\frac{3}{2}} = \frac{4}{19 - 6\beta}, \tag{64c}$$

$$\tilde{r}^{****}_{i+\frac{3}{2}} = \frac{71 - 90\beta + 24\beta^2}{11 - 6\beta}. \tag{64d}$$

4. Then, the special limiter for the cell-face state $c_{i+\frac{3}{2}}$ reads:

$$\tilde{\phi}(\tilde{r}_{i+\frac{3}{2}}) = \begin{cases} -1, & \text{if } \tilde{r}_{i+\frac{3}{2}} < \tilde{r}^*_{i+\frac{3}{2}}, \\ \frac{4}{15-6\beta} + \frac{11-6\beta}{15-6\beta}\tilde{r}_{i+\frac{3}{2}}, & \text{if } \tilde{r}^*_{i+\frac{3}{2}} \leq \tilde{r}_{i+\frac{3}{2}} < \tilde{r}^{**}_{i+\frac{3}{2}}, \\ 0, & \text{if } \tilde{r}^{**}_{i+\frac{3}{2}} \leq \tilde{r}_{i+\frac{3}{2}} < 0, \\ 2\tilde{r}_{i+\frac{3}{2}}, & \text{if } 0 \leq \tilde{r}_{i+\frac{3}{2}} < \tilde{r}^{***}_{i+\frac{3}{2}}, \\ \frac{4}{15-6\beta} + \frac{11-6\beta}{15-6\beta}\tilde{r}_{i+\frac{3}{2}}, & \text{if } \tilde{r}^{***}_{i+\frac{3}{2}} \leq \tilde{r}_{i+\frac{3}{2}} < \tilde{r}^{****}_{i+\frac{3}{2}}, \\ 5 - 4\beta, & \text{if } \tilde{r}_{i+\frac{3}{2}} \geq \tilde{r}^{****}_{i+\frac{3}{2}}. \end{cases} \tag{65}$$

In Figure 15, we give the limiter (65) for the two extreme values of β, $\beta = 0$ and $\beta = 1$, with the corresponding TVD domains.

4.2 Local adaptivity in time

Consider the stencil in the (x,t)-plane in Figure 16. The EB is situated in cell i at t^n in such a way that it migrates to the next cell $i+1$ at some time in between t^n and t^{n+1}. Apparently, the solutions c_i^n and c_{i+1}^n are updated, in Forward Euler, using the modified cell-face states $c_{i-\frac{1}{2}}^n$, $c_{i+\frac{1}{2}}^n$ and $c_{i+\frac{3}{2}}^n$. However, as the EB crosses the cell face at $x_{i+\frac{1}{2}}$, there is an abrupt change in the state at this face. Before the crossing, the state at this cell face must be computed based on the data to the right of the EB; whereas, after the crossing, it must be computed based on the (different) data to the left of the EB. The two solutions c_i^{n+1} and c_{i+1}^{n+1}, which are affected by the flux across this particular cell face, need to 'feel' this reversal, i.e., the abrupt change in $c_{i+\frac{1}{2}}$.

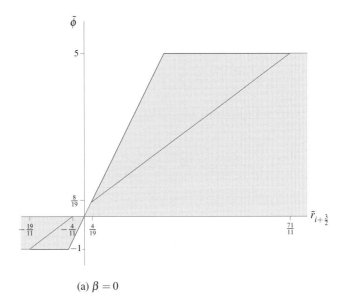

(a) $\beta = 0$

Fig. 15 Special EB-sensitive limiters (in red) for the cell-face state $c_{i+\frac{3}{2}}$ and the corresponding TVD domains for the two extreme values of β.

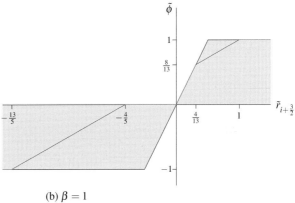

(b) $\beta = 1$

Referring to Figure 16, the time adaptivity or splitting is carried out in the following manner:

1. Compute β at t^n according to (15) and next the cell-face state $c^n_{i+\frac{1}{2}}$ according to (32b).
2. Calculate the time fraction α at which the EB crosses the cell face at $x_{i+\frac{1}{2}}$, i.e:

$$\alpha = \frac{x_{i+\frac{1}{2}} + \varepsilon - x^n_{EB}}{u\tau}, \qquad \alpha \in (0,1), \tag{66}$$

where x^n_{EB} is the location of the EB at time level n. Note that the EB is placed at infinitesimal distance ε off $x_{i+\frac{1}{2}}$, in the direction of the flow.

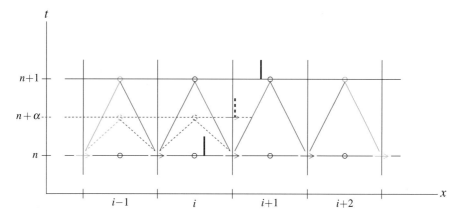

Fig. 16 Stencil for local adaptivity in time. The standard, modified and the intermediate cell-face states are designated in green, blue, and red, respectively.

3. Update the solution values c_{i-1}^n and c_i^n to time level $n+\alpha$, i.e., compute $c_{i-1}^{n+\alpha}$ and $c_i^{n+\alpha}$.
4. Compute the intermediate cell-face state $c_{i+\frac{1}{2}}^{n+\alpha}$ according to (32a), with all indices in (32a) increased with 1, using $\beta = 0$ (formally $\beta = \varepsilon/h$), $c_{i-1}^{n+\alpha}$ and $c_i^{n+\alpha}$.
5. Take the weighted average of $c_{i+\frac{1}{2}}^n$ and $c_{i+\frac{1}{2}}^{n+\alpha}$, and recompute the time-adapted cell-face state at $x_{i+\frac{1}{2}}$, as:

$$c_{i+\frac{1}{2}}^n := \alpha c_{i+\frac{1}{2}}^n + (1-\alpha) c_{i+\frac{1}{2}}^{n+\alpha}. \tag{67}$$

6. Use the time-adapted cell-face state $c_{i+\frac{1}{2}}^n$ and continue updating the solution everywhere with the regular time step τ.

Besides the above approach, in which only the jumping cell-face state $c_{i+\frac{1}{2}}$ is recomputed at $t^{n+\alpha}$, spatially more elaborate ways of doing the time adaptivity might be investigated. For instance, all cell-face states that stop or start to be affected by the EB, viz. $c_{i-\frac{1}{2}}, c_{i+\frac{1}{2}}, c_{i+\frac{3}{2}}$ and $c_{i+\frac{5}{2}}$, might be recomputed at $t^{n+\alpha}$. Or even, the cell-face states of all cells that start or stop to feel the EB might be recomputed, i.e., $c_{i-\frac{3}{2}}, c_{i-\frac{1}{2}}, c_{i+\frac{1}{2}}, c_{i+\frac{3}{2}}, c_{i+\frac{5}{2}}$ and $c_{i+\frac{7}{2}}$. However, the gain we achieve in accuracy, as we consider more intermediate cell-face states than only $c_{i+\frac{1}{2}}$, is marginal for the given cost increase. As expected, recomputation of only the jumping cell-face state $c_{i+\frac{1}{2}}$ is necessary and sufficient for significantly improving the solution accuracy.

For RK3b, we do not yet resort to the temporal local-adaptivity procedure devised, for Forward Euler, above. We instead split the regular time step τ into smaller time steps, depending on the number of EBs crossing cell faces, and update the intermediate solutions everywhere. For instance, for a single EB crossing a cell face, we divide τ into two smaller time steps $\alpha\tau$ and $(1-\alpha)\tau$.

5 Numerical examples

Numerical results are given to validate the immersed-boundary approach presented in this work. We take the same data as in § 2, i.e., the initial solutions (5); initial EB locations $x_1 = \frac{1}{3}$ and $x_2 = \frac{2}{3}$; flow speed $u = 1$; and final time $t_{\max} = 1$. Further, we consider again a grid of 20 and 40 cells.

The results obtained, shown in Figure 17, are remarkably accurate. The results show a significant improvement in resolution, without much computational overhead, over those computed using the standard methods, Figure 5. For the more discriminating initial solution (5b) (the cosine with cavity), the numerical results of the limited higher-order upwind-biased schemes are slightly deficient at the peripheries. This is due to the property of standard limiters that they clip physically correct extrema. Apparently, the deficiency becomes smaller with decreasing mesh width.

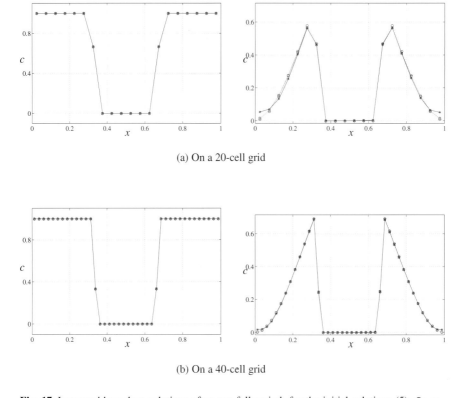

(a) On a 20-cell grid

(b) On a 40-cell grid

Fig. 17 Immersed-boundary solutions after one full-period, for the initial solutions (5). ○: exact discrete, □: unlimited higher-order upwind-biased with Forward Euler, ∗: limited higher-order upwind-biased with Forward Euler, ◇: unlimited higher-order upwind-biased with RK3b, ×: limited higher-order upwind-biased with RK3b.

In Figure 18, we show the error E, which is computed as the difference between the exact and numerical solutions, for the solutions given in Figure 17. As can be seen, there is relatively more discrepancy near the EBs. This near-EB discrepancy is of the same order for both test cases. However, as mentioned earlier, the discrepancy is significantly larger, about five times more, in the cosine-with-cavity case at and near the extrema, due to the limiters. In the former case, the results obtained with the RK3b scheme are superior to those obtained with the Forward Euler scheme, for the obvious reason. In the latter case, both schemes, limited and/or unlimited, yield almost the same accuracy.

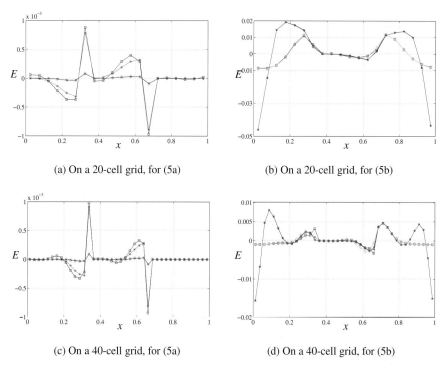

(a) On a 20-cell grid, for (5a)

(b) On a 20-cell grid, for (5b)

(c) On a 40-cell grid, for (5a)

(d) On a 40-cell grid, for (5b)

Fig. 18 Errors after one full-period, for the initial solutions (5). □-blue: unlimited higher-order upwind-biased with Forward Euler, ∗-red: limited higher-order upwind-biased with Forward Euler, ◇-green: unlimited higher-order upwind-biased with RK3b, ×-black: limited higher-order upwind-biased with RK3b.

6 Extension to more general cases

First extensions to the method presented so far, which must and will be made, are: (i) to higher dimensions and (ii) to higher-order accuracy in time. We already performed some work into this direction. Here follows a brief account of the ideas we are currently pursuing.

6.1 Extension to higher dimensions

Extension to 2D and 3D of the current 1D space discretization is done by dimensional splitting. For this purpose, a multi-D embedded boundary is to be projected first on each of its separate coordinate directions (Figure 19). The details of the projection step are crucial. Dimensional splitting has been applied with success to the standard κ-scheme and, as such, is widely spread in CFD. We presume that it will be successful here as well.

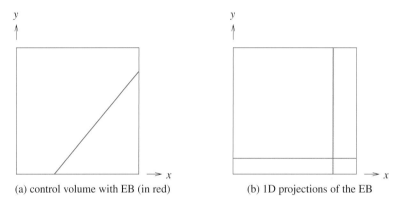

(a) control volume with EB (in red)   (b) 1D projections of the EB

Fig. 19 Example of 1D projections of a 2D embedded boundary.

6.2 Higher-order accuracy in time

Forward Euler can readily be made second-order accurate by following the Modified Euler approach. Given (43), Modified Euler reads:

$$\hat{c}_i^{n+1} = c_i^n + \tau F(c^n), \tag{68a}$$

$$c_i^{n+1} = c_i^n + \tfrac{1}{2}\tau \left(F(\hat{c}^{n+1}) + F(c^n) \right). \tag{68b}$$

Forward Euler step (68a) is the predictor and (68b) the corrector step. Modified Euler is still explicit. Extension of the local adaptivity in time, introduced in § 4.2, from Forward Euler to Modified Euler is rather straightforward, given the close similarity of the two schemes. Details about our local time adaptivity method and Modified Euler will be given in future work.

7 Conclusion

A novel immersed-boundary approach, for solving advection problems, has been introduced. The essence of the approach is that moving bodies are embedded in a regular fixed grid and specific fluxes in the vicinity of the embedded boundary (EB) are computed in such a way that they accurately and monotonously accommodate the boundary conditions valid on the moving body. To suppress the wiggles that exhibit near discontinuities, tailor-made limiters are introduced for the fluxes that are especially modified. Then, over the majority of the domain, where we do not have influence of the embedded boundaries, we can readily use standard methods on the underlying regular fixed grid. Excellent results are achieved, without much computational overhead.

In summary:

- A generalized κ-scheme that uses EB information (EB location and EB solution values), and which is an optimally accurate upwind-biased finite-volume discretization, has been proposed. This near-EB spatial discretization is a generalization of a well-proven finite-volume discretization, which allows us to accommodate EBs.
- Generalized limiters that use EB information have been proposed. These limiters satisfy the spatial monotonicity requirement and Harten's TVD requirement. To be consistent with standard limiters, the generalized limiters are made independent of the CFL number.
- Locally adaptive splitting of the time step, near EBs, has been proposed; a two-stage approach which requires the least computational time and memory for a given gain in accuracy.

We foresee that the numerical methods, developed so far and still to be developed, can readily be extended to realistic flow problems.

Acknowledgements The first author's research is funded by the Delft Centre for Computational Science and Engineering (DCSE).

Appendix

Referring to (25) and (29), the local truncation error terms e in cells $i+1$ and $i+2$, when the net fluxes in cells $i+1$ and $i+2$ are considered to optimize $\kappa_{i+\frac{3}{2}}$, respectively, are:

$$e_{i+1} = \frac{3-2\beta}{12} uh^2 \frac{\partial^3 c}{\partial x^3}, \tag{69a}$$

and

$$e_{i+2} = \frac{2\beta-1}{36} uh^2 \frac{\partial^3 c}{\partial x^3}. \tag{69b}$$

For a given grid size h, (69) can be rewritten, as a function of $\beta \in [0,1]$, as:

$$\frac{e_{i+1}}{b} = 9 - 6\beta, \tag{70a}$$

and

$$\frac{e_{i+2}}{b} = 2\beta - 1, \tag{70b}$$

where

$$b = \frac{uh^2}{36}\frac{\partial^3 c}{\partial x^3}. \tag{70c}$$

The scaled error terms (70a) and (70b) are plotted in the $(e/b, \beta)$-diagram given in Figure 20. We see that $|e_{i+1}| \geq 3|e_{i+2}|, \forall \beta$. Also notice that $e_{i+2} = 0$ for $\beta = \frac{1}{2}$; i.e., we get a third-order accurate net flux in cell $i+2$ when the EB is situated at the center of cell i.

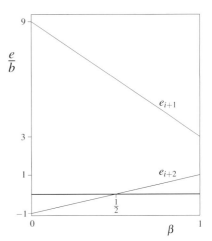

Fig. 20 Variation of the scaled, leading local truncation error terms in cells $i+1$ and $i+2$ with β.

References

1. Calhoun, D.: A Cartesian grid method for solving the two-dimensional streamfunction-vorticity equations in irregular regions. Journal of Computational Physics **176**(2), 231–275 (2002)
2. Fadlun, E.A., Verzicco, R., Orlandi, P., Mohd-Yusof, J.: Combined immersed-boundary methods for three-dimensional complex flow simulations. Journal of Computational Physics **161**, 35–60 (2000)
3. Godunov, S.K.: Finite difference method for numerical computation of discontinuous solutions of the equations of fluid dynamics. Matematicheskii Sbornik **44**, 271–306 (1959). Translated from Russian at the Cornell Aeron. Lab.
4. Goldstein, D., Handler, R., Sirovich, L.: Modeling a no-slip flow boundary with an external force field. Journal of Computational Physics **105**, 354–366 (1993)

5. Harten, A.: On a class of high resolution total-variation-stable finite-difference schemes. SIAM Journal on Numerical Analysis **21**, 1–23 (1984)
6. Hundsdorfer, W., Koren, B., van Loon, M., Verwer, J.G.: A positive finite-difference advection scheme. Journal of Computational Physics **117**, 35–46 (1995)
7. Kim, J., Kim, D., Choi, H.: An immersed-boundary finite-volume method for simulations of flow in complex geometries. Journal of Computational Physics **171**, 132–150 (2001)
8. Koren, B.: A robust upwind finite-volume method for advection, diffusion and source terms. In: Vreugdenhil, C.B., Koren, B. (eds.) Notes on Numerical Fluid Mechanics, **45**, pp. 117-138. Vieweg, Braunschweig (1993)
9. Leer, B. van: Upwind-difference methods for aerodynamic problems governed by the Euler equations. In: Lectures in Applied Mathematics, **22 - 2**, pp. 327-336. American Mathematical Society, Providence, RI (1985)
10. Mittal, R., Iaccarino, G.: Immersed boundary methods. Annual Review of Fluid Mechanics **37**, 239–261 (2005)
11. Mohd-Yusof, J.: Combined immersed-boundary/B-spline methods for simulations of flow in complex geometries. In: CTR Annual Research Briefs, pp. 317-327. Center for Turbulence Research, NASA Ames/Stanford University (1997)
12. Mohd-Yusof, J.: Development of immersed boundary methods for complex geometries. In: CTR Annual Research Briefs, pp. 325-336. Center for Turbulence Research, NASA Ames/Stanford University (1998)
13. Peskin, C.S.: Flow patterns around heart valves: a numerical method. Journal of Computational Physics **10**, 252–271 (1972)
14. Peskin, C.S.: Numerical analysis of blood flow in the heart. Journal of Computational Physics **25**, 220–252 (1977)
15. Peskin, C.S.: The fluid dynamics of heart valves: experimental, theoretical and computational methods. Annual Review of Fluid Mechanics **14**, 235–259 (1982)
16. Saiki, E.M., Biringen, S.: Numerical simulation of a cylinder in uniform flow: application of virtual boundary method. Journal of Computational Physics **123**, 450–465 (1996)
17. Su, S.-W., Lai, M.-C., Lin, C.-A.: An immersed boundary technique for simulating complex flows with rigid boundary. Computers & Fluids **36**, 313–324 (2007)
18. Sweby, P.: High resolution schemes using flux limiters for hyperbolic conservation laws. SIAM Journal on Numerical Analysis **21**, 995–1011 (1984)
19. Tseng, Y.H., Ferziger, J.H.: A ghost-cell immersed boundary method for flow in complex geometry. Journal of Computational Physics **192**, 593–623 (2003)
20. Zhang, N, Zheng, Z.C.: An improved direct-forcing immersed-boundary method for finite difference applications. Journal of Computational Physics **221**, 250–268 (2007)

Large Eddy Simulation of Turbulent Non-Premixed Jet Flames with a High Order Numerical Method

S. van der Hoeven, B.J. Boersma, and D.J.E.M. Roekaerts

Abstract In this chapter we will report on the Large Eddy Simulation (LES) of a turbulent non premixed jet flame. The numerical model used for the LES is based on a discretization with high order compact schemes. These schemes have a negligible amount of numerical dissipation. The subgrid terms in the compressible Navier-Stokes equations are modelled with simple eddy viscosity models. The LES model is used to simulate the Sandia Flame D.

1 Introduction

Large-scale industrial burners typically have a cross section of the order of 0.1 meter or larger. The smallest length scale in the reacting flow generated using such a burner is in general associated with the flame thickness. The laminar flame thickness of a premixed flame is to a good approximation independent of the size of the geometry and depends on material properties such as viscosity, thermal diffusivity, activation energy, etc. The flame thickness is in general less than a millimeter. A full three dimensional numerical simulation (or direct numerical simulation) in which both the smallest and largest scales of the flow are well resolved is not possible with present-day computers. A possible solution to this problem is the use of Large Eddy Simulation (LES), in which the small scales are removed from the spectrum by a filtering procedure. This filtering splits the energy spectrum, see Figure 1, into two parts, a resolved part (large scales) and an unresolved part (small scales). The large scales are resolved on a computational grid. The effect of the small scales which are removed by the filtering is compensated for by a subgrid model. Depending on

---

S. van der Hoeven & D.J.E.M. Roekaerts
Department of Multi-Scale Physics , Delft University of Technology, the Netherlands

B.J. Boersma
Laboratory for Aero and Hydrodynamics, Mekelweg 2, 2628 CA, the Netherlands,
e-mail: b.j.boersma@tudelft.nl

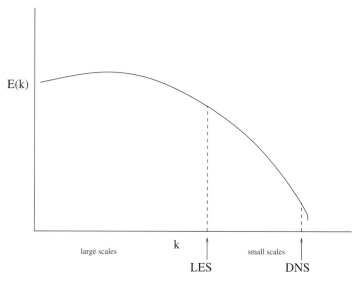

Fig. 1 A sketch of a 1D energy spectrum in a turbulent flow. In Direct Numerical Simulation (DNS) the full energy spectrum is resolved. In Large Eddy Simulation (LES) only the large scale part of the energy spectrum is resolved, and the small scale part is modelled with a subgrid model.

the cut-off length of the filter we can have a well resolved LES or a poorly resolved LES. The fact that flame structure in general belongs to the smallest scales, makes LES modeling of combustion more difficult than modeling of inert flow.

In general dissipation is associated with the small scales in a flow. These small scales have been removed by the LES filter and the subgrid scale model has to provide the required dissipation. Most numerical models also introduce a certain amount of dissipation. The amount of numerical dissipation strongly depends on the type of scheme. Low order central schemes have in general a higher numerical dissipation than high order schemes. Central (symmetric) schemes have in general less dissipation than upwind (asymmetric) schemes. In an actual Large Eddy Simulation the dissipation is always a combination of dissipation given by the sub-grid model and by the numerics of the LES model. This makes the validation of a sub-grid model complicated because dissipative effects can be contributed to the numerics or the sub-grid model. For the proper validation of the sub-grid model numerical dissipation should be minimized.

In the literature many low dissipative methods are reported, using high order finite differences, compact finite differences and orthogonal polynomials, see for instance Lele (1992), Chu & Fan (1998). The disadvantage of these schemes is that they are not very stable in general. Recently, Boersma (2005) proposed a staggered variant of the compact finite difference scheme as developed by Lele (1992). This new scheme is stable because it has certain symmetry properties, see for instance Verstappen & Veldman (2003). In this chapter we will explore the possibility of this scheme for combustion simulations. The model is used to simulate the Sandia Flame

D, which is one of the test cases of the International Workshop on Measurement and Computation of Turbulent Non-Premixed Flames [6].

2 Governing equations

The derivation of the governing equations for combustion is discussed in many textbooks, such as those of Peters [16] and Kuo [17]. Here we will present the Large-Eddy filtered version of these equations. A LES filtered quantity is given by an overbar and is obtained using the following filter function:

$$\overline{f}(x) = \int_V G(\mathbf{x} - \mathbf{x}') f(x') d\mathbf{x}'. \tag{1}$$

In this equation f is a flow variable, V is the volume of the LES filter and G is a filter kernel. The exact form of G is at this stage not important. In a compressible flow it is customary to use Favre-averaged (or density weighted) variables. This is also the approach we will follow in the present paper. The Favre average, denoted by a tilde, is defined as follows

$$\tilde{f} = \frac{\overline{\rho f}}{\overline{\rho}}. \tag{2}$$

Variations with respect to the Favre average are denoted with a double prime, i.e.

$$f'' = f - \tilde{f}.$$

With help of the LES and Favre filters we can write for the equation for conservation of mass:

$$\frac{\partial \overline{\rho}}{\partial t} + \frac{\partial}{\partial x_i} \overline{\rho} \tilde{u}_i = 0, \tag{3}$$

where $\overline{\rho}$ is the LES averaged density, and $\tilde{u}_i$ the Favre-averaged velocity vector. The LES filtered momentum equations read:

$$\frac{\partial \overline{\rho} \tilde{u}_i}{\partial t} + \frac{\partial \overline{\rho} \tilde{u}_i \tilde{u}_j}{\partial x_j} = -\frac{\partial \overline{p}}{\partial x_i} + \frac{\partial}{\partial x_j} \overline{\tau}_{ij} + \frac{\partial}{\partial x_j} \left(\overline{\rho} \tilde{u}_i \tilde{u}_j - \overline{\rho u_i u_j} \right). \tag{4}$$

In this equation $\overline{p}$ is the averaged pressure, $\overline{\rho} \tilde{u}_i \tilde{u}_j - \overline{\rho u_i u_j}$ the so-called subgrid stress, which will be discussed in more detail later, and $\overline{\tau}_{ij}$ the averaged stress tensor with components

$$\overline{\tau}_{ij} = \overline{\mu} \left(\frac{\partial \tilde{u}_i}{\partial x_j} + \frac{\partial \tilde{u}_j}{\partial x_i} \right) - \frac{2}{3} \overline{\mu} \delta_{ij} \frac{\partial \tilde{u}_k}{\partial x_k}, \tag{5}$$

in which $\overline{\mu}$ is the averaged dynamic viscosity of the fluid and δ_{ij} the Kronecker delta function. The dynamic viscosity is a function of the temperature. The temperature dependence of μ will be specified later. The equation for the mean of the total non-chemical energy $E = C_v T + u_i u_i / 2$, with T the temperature and C_v the specific heat at constant volume, reads [18]:

$$\frac{\partial \bar{\rho}\tilde{E}}{\partial t} + \frac{\partial}{\partial x_j}\tilde{u}_j\bar{\rho}\tilde{H} = \frac{\partial}{\partial x_j}\left[\tilde{u}_i\overline{\tau_{ij}} + \lambda(\tilde{T})\frac{\partial \tilde{T}}{\tilde{x}_j} + \bar{\rho}\left(\widetilde{u_jH} - \tilde{u}_j\tilde{H}\right)\right] + \bar{\omega}_f. \quad (6)$$

Here H is the specific enthalpy, λ the thermal conductivity which is a function of the temperature and $\bar{\omega}_f$ is the heat release rate of the combustion process. In this equation again a subgrid term appears, i.e. $\widetilde{u_jH} - \tilde{u}_j\tilde{H}$, which has to be modelled. In combustion modelling of non-premixed flames it is customary to use an equation for the mixture fraction $\tilde{Z}$. This equation reads:

$$\frac{\partial \bar{\rho}\tilde{Z}}{\partial t} + \frac{\partial \bar{\rho}\tilde{u}_i\tilde{Z}}{\partial x_i} = \frac{\partial}{\partial x_i}\bar{\rho}D\frac{\partial \tilde{Z}}{\partial x_i} + \frac{\partial}{\partial x_j}\bar{\rho}\left(\widetilde{u_jZ} - \tilde{u}_j\tilde{Z}\right), \quad (7)$$

where D is the diffusion coefficient of the mixture, and again a subgrid term appears. The mean pressure is obtained from the mean thermal equation of state (8):

$$\bar{p} = \bar{\rho}R\tilde{T}\sum_{i=1}^{N_s}\frac{\tilde{Y}_i}{W_i}, \quad (8)$$

where at the right hand side unclosed terms containing correlations between temperature and composition have been left out. In this equation R is the universal gas constant, N_s is the number of species and $\sum_{i=1}^{N_s}\frac{\tilde{Y}_i}{W_i}$ is the inverse of the mean molar mass of the mixture.

The subgrid fluxes in the LES filtered Navier Stokes equation, equation (4) are closed with a simple model based on an eddy viscosity assumption:

$$\bar{\rho}\left[\widetilde{u_iu_j} - \tilde{u}_i\tilde{u}_j\right] = -2\bar{\rho}v_t\widetilde{S}_{ij}, \quad (9)$$

where $\widetilde{S}_{ij}$ is the strain-rate tensor given by

$$\widetilde{S}_{ij} = \frac{1}{2}\left(\frac{\partial \tilde{u}_i}{\partial x_j} + \frac{\partial \tilde{u}_j}{\partial x_i}\right). \quad (10)$$

The turbulent viscosity v_t is given by a model proposed by Vreman [19]:

$$v_t = C_{vr}\sqrt{\mathbb{Z}/\alpha_{ij}^2}. \quad (11)$$

The model constant is given by $C_{vr} = 2.5C_s$, where C_s is the Smagorinsky constant. The functional dependence of the eddy viscosity on local conditions is given by:

$$\mathbb{Z} = \beta_{11}\beta_{22} - \beta_{12}^2 + \beta_{11}\beta_{33} - \beta_{13}^2 + \beta_{22}\beta_{33} - \beta_{23}^2, \quad (12)$$

where

$$\beta_{ij} = \Delta_m^2\alpha_{mi}\alpha_{mj}, \quad (13)$$

$$\alpha_{ij} \equiv (\nabla \widetilde{\mathbf{u}})_{ij} = \frac{\partial \widetilde{u}_i}{\partial x_j}. \tag{14}$$

In the present work we use an implicit LES filter. The filter width Δ is the same in all three Cartesian directions and proportional to the cubic root of the grid volume, i.e.

$$\Delta = (\Delta x \Delta y \Delta z)^{1/3}. \tag{15}$$

The Vreman model was found to perform better than a standard Smagorinsky model, because its dissipation is relatively small in the transitional region in the lower part of the jet.

The subgrid term in the energy equation $\widetilde{u_j H} - \tilde{u}_j \tilde{H}$ is modelled via the temperature gradient, i.e.

$$\widetilde{u_j H} - \tilde{u}_j \tilde{H} = -\bar{\rho} C_p \mathrm{d}_t \frac{\partial \tilde{T}}{\partial x_j}, \tag{16}$$

$$\mathrm{d}_t = v_t / \mathrm{Pr_t}, \tag{17}$$

where d_t is the turbulent thermal diffusivity, v_t the eddy viscosity given by equation (12), and Pr_t the turbulent Prandtl number. Similarly in the resolved mixture fraction equation

$$\widetilde{u_j Z} - \tilde{u}_j \tilde{Z} = -\bar{\rho} \Gamma_t \frac{\partial \tilde{Z}}{\partial x_j}, \tag{18}$$

$$\Gamma_t = v_t / \mathrm{Sc_t}. \tag{19}$$

In this study the Smagorinsky constant, C_s was set to 0.08. Following [7] the turbulent Prandtl number was set to 0.7 and the turbulent Schmidt number was set to 0.4.

The dynamic viscosity μ and the thermal conductivity λ are a function of the temperature. We use the following simple relations

$$\frac{\mu}{\mu_0} = \left(\frac{T}{T_0}\right)^{0.76}, \frac{\lambda}{\lambda_0} = \left(\frac{T}{T_0}\right)^{0.76}. \tag{20}$$

In regions with large temperatures the increase in dynamic viscosity or thermal diffusivity can be substantial. For the stability of the energy equations it is essential that the increased thermal conductivity is taken into account.

3 Chemistry model

For the chemistry we use a steady flamelet model [16], with a library consisting of only one flamelet with strain rate $a = 100s^{-1}$. The low-Mach number resolved properties are tabulated as a function of resolved mixture fraction and variance of mixture fraction.

The variance of mixture fraction is estimated with a simple gradient model, which can be derived from local equilibrium arguments [8]:

$$\widetilde{Z''^2} = C_Z \Delta^2 \frac{\partial \widetilde{Z}}{\partial x_i} \frac{\partial \widetilde{Z}}{\partial x_i}, \tag{21}$$

with C_Z being 0.2. Resolved temperature, mean molecular weight, specific heat and temperature dependent viscosity are stored in the look up table. The steady flamelet model (without extinction model) is sufficiently accurate to describe the Sandia flame D, discussed below. At this stage our main focus is on testing the compressible code in combination with scalar transport rather than on the chemistry model itself. When using (adiabatic) flamelet methods in low-Mach number codes it is sometimes possible to replace the energy equation with the flamelet temperature solution. However, for compressible formulations this cannot be done in general, due to the presence of finite acoustic time scales (influence of pressure variations on the energy). In order to couple the flamelet solution in the compressible code while retaining acoustics we use a simple relaxation method to bring the system at the flamelet temperature solution. The energy equation is solved with the heat release source term given by a relaxation term:

$$\bar{\dot{\omega}}_f = -\bar{\rho} \bar{c}_V \left(\widetilde{T} - \widetilde{T}_f \right) / \tau_f, \tag{22}$$

where $\widetilde{T}_f$ is the flamelet temperature and τ_f is the relaxation time constant. The effect of using such a flamelet source term was investigated using 1D simulations. Different values for τ_f were tested. It was observed that when a too small time constant is used stability problems arise due to large flow dilatation and on the other hand using time constants larger than the integral time scale of the flow results in inactive behavior (hot gas mixing). It was observed that a time constant of approximately $15u_0/D$, where u_0 is the maximum jet velocity at the inlet and D the nozzle diameter, works most efficiently. Therefore we use this value in the 3D simulations.

4 Numerical method

The LES filter introduces a certain cut-off in the wave number spectrum, see Figure 1. Wave numbers which are smaller than the cut-off wave number are resolved by the numerical scheme. Wave numbers which are larger than the cut-off frequency have to be modelled with a subgrid model. A common approach is to take the cut-off wave number equal to the grid resolution, the so-called implicit LES filter. A better but more elaborate approach is to use an explicit filter which has a cut-off wavelength which is larger than the grid resolution. In this paper we will follow the first approach, i.e. the grid resolution equals the cut-off wavenumber and hence no explicit filtering will be performed.

The unresolved (or subgrid scales) are in general responsible for the dissipation of kinetic energy and also for the diffusion of temperature and species fluctuations. The main task of any subgrid model is to provide the required dissipation or diffusion rate. Unfortunately, any numerical method will also provide a certain amount of dissipation of diffusion. In LES the dissipation or diffusion is in general a mixture of numerical diffusion and model diffusion. Numerical diffusion can be very useful to stabilize computations, but it is in general unwanted, because it does not only depend on gradients in the flow but also on the quality of the grid.

In this study we will use methods that minimize the numerical diffusion. This more or less rules out all (low order) upwind methods. Schemes with a purely central discretization in general do not introduce significant numerical dissipation.

4.1 The derivative

In the equations given previously only first order spatial derivatives appear. Therefore, we will only discuss first order derivatives. The simplest central scheme for the 1st derivative is the classical second order scheme, which reads for a function $f_i = f(i\Delta x)$:

$$f'_i = \frac{f_{i+1} - f_{i-1}}{2\Delta x} + O(\Delta x^2), \tag{23}$$

where Δx is the grid spacing. Here we will assume that the function f_i can be written as a Fourier series

$$f_i = f(i\Delta x) = \sum_k \hat{f}(k) \exp(jki\Delta x), \tag{24}$$

in which $j = \sqrt{-1}$ is the imaginary unit and k a wave number. With help of this Fourier series it is easy to derive the error of the finite difference scheme as a function of the wavenumber k. The result reads:

$$error(k\Delta x) = 1 - \frac{\sin(k\Delta x)}{k\Delta x}. \tag{25}$$

Is is clear that the discretization error made by the scheme is small for small values of the wavenumber k and rather larger for large values of k. If we use this scheme for our Large Eddy Simulation, we will have a large error close to the cut-off wavelength of the LES filter and therefore this is scheme is unwanted. A simple extension of the scheme would be to add two points on the right hand side, i.e:

$$f'_i = \frac{8}{12} \frac{f_{i+1} - f_{i-1}}{\Delta x} - \frac{1}{12} \frac{f_{i+2} - f_{i-2}}{\Delta x} + O(\Delta x^4). \tag{26}$$

This is the classical fourth order scheme. The error as a function of the wavenumber k is given by

$$error(k\Delta x) = 1 - \frac{4}{3}\frac{\sin(k\Delta x)}{k\Delta x} + \frac{1}{6}\frac{\sin(2k\Delta x)}{k\Delta x}. \qquad (27)$$

The errors for the second and fourth order scheme, equations (22) and (24), are plotted in Figure 2. The fourth order scheme is better than the second order scheme but both are fairly inaccurate for wavenumbers larger than 1.5.

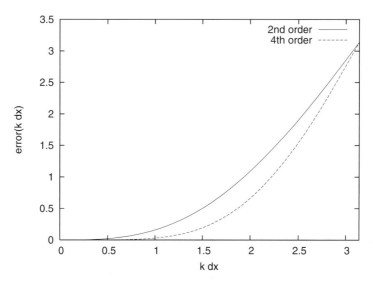

Fig. 2 The error as a function of the wave number $k\Delta x$ for a 2nd and 4th order finite difference scheme

The numerical accuracy for the large wave numbers can be greatly increased by using a so-called compact formulation of the finite difference formulation. This formulation reads in its most simple form:

$$f'_i + \frac{1}{4}[f'_{i-1} + f'_{i+1}] = \frac{3}{4}\frac{f_{i+1} - f_{i-1}}{\Delta x} + O(\Delta x^4). \qquad (28)$$

The actual evaluation of this scheme requires the solution of a tridiagonal system and is therefore computationally more expensive than the standard fourth order schemes. We can write the following relation for the error

$$error(k\Delta x) = \frac{\frac{18}{12}\sin(k\Delta x)}{1 + \frac{1}{2}\cos(k\Delta x)}. \qquad (29)$$

This error is considerably smaller than the error of the central fourth order scheme, see Figure 3.

In the literature mostly the 6th order compact scheme is used. This scheme is constructed by adding two additional points to the righthand side of equation (25) and evaluating the coefficients by a Taylor expansion.

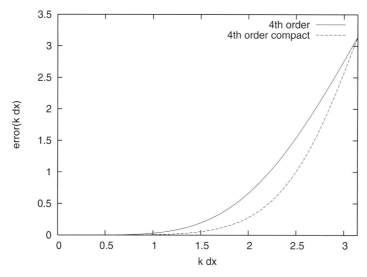

Fig. 3 The error as a function of the wave number $k\Delta x$ for a 4th order central and a fourth order compact scheme.

The drawback of compact schemes for non-linear problems such as the Navier-Stokes equations is the fact that these schemes are fairly unstable. Due to the low dissipation error of the numerical scheme it is not possible to "under resolve" certain wavelengths. A large eddy simulation is by definition under resolved and therefore compact schemes are not very suited for LES. Common practice is to use a high order compact filter to remove the short wave instabilities. This type of approach works but also introduces an artificial viscosity and is therefore unwanted for large eddy simulation.

Recently, staggered variants of the compact schemes have been proposed by [4, 1]. These schemes turn out to be much more stable and can be used when the grid is not sufficiently fine to resolve all scales of motion. In this paper we will adopt the staggered method as proposed in [1]. In this method the vector quantities are stored at the cell faces and all scalar quantities are stored in the cell centers. Compact differentiation and interpolation rules are used to calculate the necessary quantities on the computational grid. This is all done with a 10th order accurate scheme. On non-uniform grids a mapping of the form

$$\frac{df}{dx} = \frac{df}{d\eta}\frac{d\eta}{dx}$$

is used. Here x is the non-uniform coordinate, η the uniform coordinate and the first derivative $d\eta/dx$ is a known analytical function which is non-zero inside the computational domain. In this way we can calculate the derivatives on a non-uniform grid without any additional errors. For the interpolation on the non-uniform grids we use the same rule as for a uniform grid. This is strictly speaking not correct. We have chosen for this approach because only in this way we are sure that the discretization of the first-order derivative is purely symmetric and thus does not introduce any artificial dissipation or diffusion, [5]. The error we make in the interpolation procedure can be controlled by generating grids with very small differences in mesh widths of neighboring cells.

4.2 Time discretization

For the time discretization we use a standard fourth order Runge-Kutta method. For a linear advection diffusion problem the time step limitation for the RK4 scheme is approximately 2.8 based on a CFL type criterion. For the present problem, which is strongly non-linear, we use a fixed time step Δt the value of Δt corresponds roughly to a CFL number of 1.0.

4.3 The discretization of the Navier-Stokes equations for the SANDIA flame D problem

The 3D equations are discretized on a staggered non-uniform Cartesian grid in physical space. A typical grid volume is shown in Figure 4. The non-uniform grid is

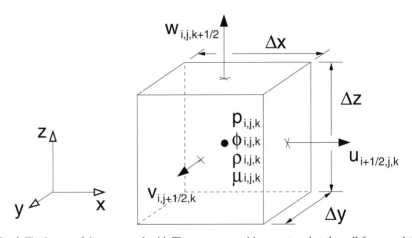

Fig. 4 The layout of the staggered grid. The vector quantities are stored at the cell faces, and the scalar quantities at the cell centers.

constructed with hyperbolic sine and tangent functions to cluster the majority of grid points in the high-strain region of the jet. A typical distribution of the grid is shown in Figure 5. The circular fuel nozzle surrounded by an annulus of pilot

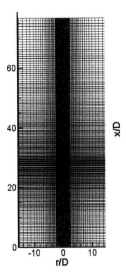

Fig. 5 A typical non-uniform grid used for the combustion simulations.

mixture and an annulus of air are located in the $y-z$ plane at $x=0$. Simulations were performed on a 200×96×96 (X×Y×Z) grid with in x and y direction approximately 40 points in the nozzle region to resolve the strong jet shear-layers. The dimensions of the computational domain are 70×30×30 nozzle diameters. For the calculation of the derivative of a certain variable f on the non-uniform grid we map f on a corresponding uniform grid. The first-order derivative is always calculated on the uniform grid. When a variable is known on the cell faces, $\left[...,i-\frac{3}{2},i-\frac{1}{2},i+\frac{1}{2},i+\frac{3}{2},...\right]$ the following formula is used to find the derivative at the center of the cell, $[...,i-1,i,i+1,...]$:

$$f'_i + a_0\left(f'_{i+1}+f'_{i-1}\right) = \sum_{n=1}^{4} \frac{a_n}{\Delta X}\left(f_{i+(2n-1)/2}-f_{i-(2n-1)/2}\right). \qquad (30)$$

Thus, the variable is implicitly swapped from the cell face to the cell center. In equation (30) f' is the derivative of a certain variable f with respect X, i.e. df/dX in grid point i. ΔX is the uniform grid point distance. The coefficients a_i can be found with Taylor expansions around grid point i. Having the coefficients $a_0,..,a_3$ it is theoretically possible to achieve tenth-order accuracy, but close to the boundary the stencil is made smaller, since one-sided differences are applied there. Interpolation is carried out with the following formula:

$$f_i + b_0\left(f_{i+1}+f_{i-1}\right) = \sum_{n=1}^{4} b_n\left(f_{i+(2n-1)/2}+f_{i-(2n-1)/2}\right). \qquad (31)$$

n	0	1	2	3	4
a_n	49/190	12985/14592	78841/364800	-343/72960	129/851200
b_n	7/18	1225/1536	49/512	-7/1536	1/4608

Table 1 Coefficients for the derivative and interpolation rule with a tenth-order stencil.

5 Boundary conditions

Since the developed code is low-dissipative, (acoustic) waves can reflect at the domain boundaries and disturb the flow. Therefore special attention must be paid to the treatment of the boundaries. Different boundary condition methods were implemented and tested. The class of methods using characteristic boundary conditions is attractive since it allows to control the behavior of characteristic waves in a direct way. We have implemented the Navier-Stokes Characteristic Boundary Conditions (NSCBC) method of Poinsot and Lele [9] in our staggered grid arrangement. The NSCBC method works well for combustion simulations where acoustics are unimportant, but for a staggered grid it appeared to be computationally expensive to define all necessary quantities locally at the cell faces. This holds especially for subsonic outflow conditions. Another drawback of (linearized) characteristic BC's such as the NSCBC method is that it is very difficult to convect a hot jet flow out of the domain without using artificial sponge layers. This is due to the multi-dimensional nature of the structures in jet flows for which a simple pressure relaxation method is not good enough. Yoo *et. al.* [10] proposed to solve the problem by extending the NSCBC method with multi-dimensional corrections.

We follow another procedure which is a variant of the methods proposed by Colonius *et. al.* [11], Ta'asan *et. al.* [12] and Berenger [13]. This method was applied by Freund [14] in aeroacoustics simulations. In this procedure the computational domain contains a non-physical boundary region where extra boundary terms appear in the governing transport equations (Fig. 6). This non-physical region uses nearly one percent of the total amount of grid points. Several approaches can be followed within this boundary domain, and we have combined two of them. The first is to force the solution towards some target solution, by adding a term $\sigma(\phi - \phi_{target})$ to the transport equation for ϕ:

$$\frac{\partial \phi}{\partial t} = RHS + \sigma(\phi - \phi_{target}), \tag{32}$$

where ϕ is a conserved quantity, RHS stands for all terms of the unforced equation and σ is a forcing parameter being nonzero only in the artificial boundary layer. ϕ_{target} can be a time-averaged value or a known solution, e.g. from a RANS calculation [15]. The second approach is to add a term which gives disturbances an increased convection velocity according to a given profile which is supersonic in the unphysical region and decreases to zero at the interface between unphysical and physical region. The latter terms prevent that at the in- and outlet boundaries reflections travel backwards into the domain. This is done in the following way:

$$\frac{\partial \phi}{\partial t} = RHS - U_c \frac{\partial \phi}{\partial x}, \tag{33}$$

where U_c is the given velocity profile, being nonzero only near the inlet and outlet plane. At the lateral boundaries we prescribed a simple reflecting characteristic boundary condition with fixed velocities. Although such a boundary condition allows waves to reflect, any disturbance is damped by forcing the solution to the time-averaged solution as proposed in eq. (32).

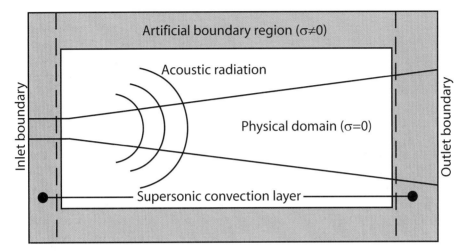

Fig. 6 Schematic view of computational domain with artificial boundary layer.

In order to minimize non-physical behavior at the walls originating from $2\Delta x$ instabilities we applied explicit filtering every 25 time steps in the artificial boundary region. At each boundary of the computational domain, only the component of velocity orthogonal to that boundary is filtered. (We recall that the computational domain is a rectangular box.) We have used a filter of the form:

$$\hat{\phi}_n = \phi_{n-3} - 6\phi_{n-2} + 15\phi_{n-1} + 44\phi_n + 15\phi_{n+1} - 6\phi_{n+2} + \phi_{n+3}, \quad (34)$$

which filters $2\Delta x$ waves from the solution. In the vicinity of the walls the filter stencil is adjusted to one-sided differences.

5.1 Parallel implementation

The model described above has been implemented in Fortran77/Fortran90 and parallelized with the message passing interface MPI. The parallel implementation of the model will be described in this section.

The total calculation consists of $N_x \times N_y \times N_z$ grid points. We use two data distributions, one with $N_x \times N_y \times (N_z)/P$ points, where P is the number of processors and one with $N_x \times (N_y)/P \times N_z$, see Figure 7. On the first distribution we can calculate all the derivatives in the x and y direction. On this distribution only a part of the z direction is available and we can not calculate the z derivatives. For the z derivatives we use the second data distribution. The z derivatives are communicated to the first distribution and the Runge-Kutta (sub)steps on the first distribution. After the Runge-Kutta (sub)step all values are communicated to the second data distribution and a new (sub)step can be performed.

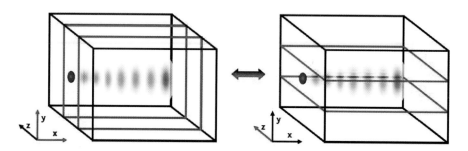

Fig. 7 The decomposition of the computational domain in a distribution with $N_x \times N_y \times (N_z)/P$ and with $N_x \times (N_y)/P \times N_z$ points.

All the communication is performed with the MPI routine MPIALLTOALL. Due to the implicit nature of the compact differences we need all points in a specific direction to be present on a single CPU and we can not use nearest neighbor communications as is used in standard finite difference codes.

Due to global nature of the communication our model only works on parallel computers with very efficient communication hardware. The model will run on a Beowulf cluster but it will be slow. On modern supercomputers, like the IBM-SP5/SP6 the model works well up to a large number of CPUs as can be seen from table 1.

Table 2 Times of the numerical model obtained on an IBM-SP5

# CPU	time/step
2	34 sec
4	21 sec
8	13 sec
16	7 sec

6 Simulation details

Laminar inflow conditions were prescribed at the inlet section. Since the flow is locally supersonic we may prescribe all variables: $\bar{\rho}, \tilde{u}, \tilde{v}, \tilde{w}, \widetilde{E}$ and $\widetilde{Z}$. The axial velocity and mixture fraction profiles were taken from experimental data [6] and reconstructed with polynomial approximations. The inflow profile in the experiment has very large gradients, see Figure 8, and these large gradients could not be reconstructed with the present fairly course grid resolution. The difference between inflow profiles in the simulations and experiments will remain visible throughout the whole simulation.

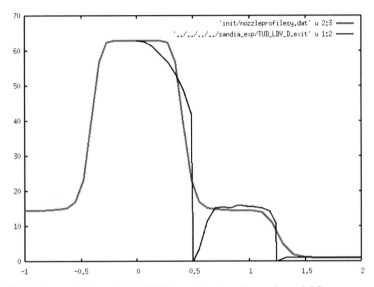

Fig. 8 The inflow profile from the SANDIA experiment and our polynomial fit.

The flame simulations are performed by starting with a cold flow simulation. After the cold jet has developed completely the flamelet relaxation term as specified in Equation (22) is added to the energy equation. With this approach the temperature in the LES model approaches the flamelet temperature as can be seen from Figure 9.

In figure 10 we show contour plots of the Favre-averaged velocity, the eddy viscosity and the Favre-averaged temperature. The region outside the black box is the non-physical region where additional terms to handle boundary conditions have been added. In Figure 11 we compare the predicted radial profiles of time averaged resolved axial velocity and the variance of the axial velocity fluctuations, with the experimental data. Overall the agreement between the simulations and experiments is reasonable. The difference between experiment and simulation can be due to several reasons. The main reason is probably the difference in the inflow profiles as shown in Figure 8, and the fact that laminar inflow profiles were assumed in the simulations. The predicted time averaged resolved temperature profiles and the standard deviation of temperature are compared with experimental results in Figure 12. Again we observe a reasonable agreement. The agreement is not as good as what has been obtained using a RANS model [20, 21] but comparable to earlier low

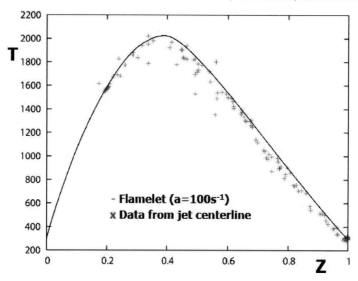

Fig. 9 The flamelet model temperature profile as a function of the mixture fraction Z (solid line), and the local centerline temperature obtained from the LES model (dots).

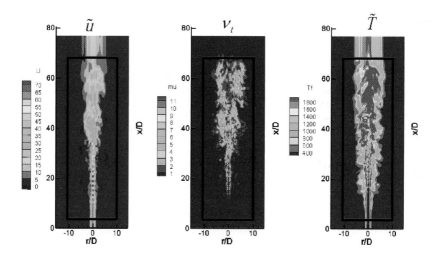

Fig. 10 Instantaneous plots of resolved axial velocity (left), eddy viscosity (middle) and resolved mixture fraction (right).

Mach number LES [7] in combination with steady flamelet modeling. Better results are expected when the present model is combined with a more precise representation of the turbulent inflow.

Large Eddy Simulation of Jet Flames 285

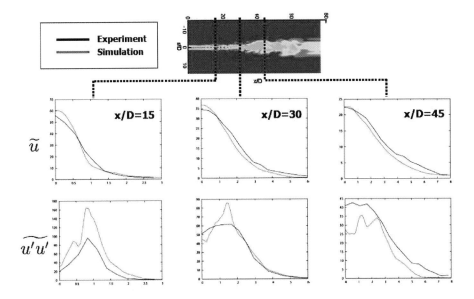

Fig. 11 Radial profiles of the time averaged axial velocity and the variance of axial velocity at three axial positions.

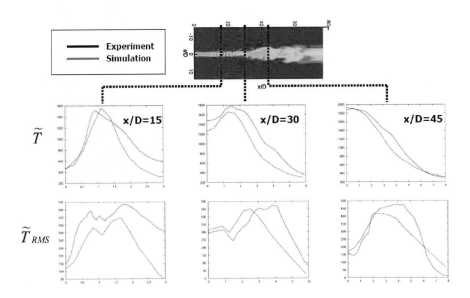

Fig. 12 Radial profiles of the time averaged temperature and the standard deviation of temperature at three axial positions.

7 Summary and conclusions

A new compressible combustion code for DNS/LES of turbulent non-premixed jet flames has been developed, based on a high-order compact finite difference formulation. The Favre-filtered transport equations were solved and the SGS fluxes were closed with a constant Smagorinksy model. The steady flamelet model was used for the chemistry. In order to couple the flamelet solution with the compressible code a relaxation term was added to the total non-chemical energy equation. Special attention was given to the treatment of the computational boundaries. We propose adding a convective term with supersonic velocity in a small region near the in- and outflow boundaries together with explicit filtering of the velocity. Using this approach it was observed that the simulation kept on running stable and reflections where too low to have an effect on the interior domain. Furthermore, the computational costs (measured in time) compared to the NSCBC method were reduced with a factor 1.5-2.0 for this simulation configuration. Nevertheless, it must be mentioned that such a comparison depends strongly on the simulation parameters (number of grid points, number of processors, number of open boundaries) and the implementation (numerical efficiency and used algorithms). To obtain better agreement with experimental results a more accurate representation of the turbulent inflow conditions is needed.

8 Acknowledgements

The Dutch national computing facilities foundation NCF is acknowledged for the allotted computing time on the TERAS/ASTER system, under project nr. SG-07-279.

References

1. B.J. Boersma, A staggered compact finite difference approach for the compressible Navier-Stokes equations, *J. Comp. Phys.*, **208**, 675, (2005)
2. S.K. Lele, Compact finite differences with spectral-like resolution, *J. Comp. Phys.*, **103**, 16, (1992)
3. P. Chu, C. Fan, A three point combined compact difference scheme, *J. Comp. Phys.*, **140**, 370, (1998)
4. S. Nagarajan, S.K. Lele, J.H. Ferziger, A robust high-order compact method for large eddy simulation, *J. Comp. Phys.*, **191**, 392, (2003)
5. R.W.C.P. Verstappen, A.E.P. Veldman, Symmetry-preserving discretization of turbulent flow, *J. Comp. Phys.*, **187**, 343, (2003)
6. *International Workshop on Measurement and Computation of Turbulent Nonpremixed Flames*, http://www.ca.sandia.gov/TNF/DataArch/FlameD.html
7. H. Pitsch, H. Steiner, Large-eddy simulation of a turbulent piloted methane/air diffusion flame (Sandia flame D), *Phys. Fluids.*, **12**, 2541, (2000)
8. N. Branley, W.P. Jones, Large Eddy Simulation of a turbulent non-premixed flame, *Comb. Flame*, **127**, 1914-1934, (2001)

9. T. Poinsot, S.K. Lele, Boundary conditions for direct simulations of compressible viscous flows, *J. Comp. Phys.*, **101**, 1042 (1992)
10. C.S. Yoo, Y. Wang, A. Trouve, H.G. Im, Characteristic boundary conditions for direct simulations of turbulent counterflow flames, *Comb. Th. Mod.*, **9**, 617-646, (2005)
11. T. Colonius, S.K. Lele, P. Moin, Boundary conditions for direct computation of aerodynamic sound computation, *AIAA Journal*, **31**, 1574-1582, (1993)
12. S. Ta'asan, D.M. Nark, An absorbing buffer zone technique for acoustic wave propagation, *AIAA Paper 1995-0146*, (1995)
13. J.P. Berenger, A perfectly matched layer for the absorption of electromagnetic waves, *J. Comp. Phys.*, **114**, 185-200, (1994)
14. J.B. Freund, Proposed inflow/outflow boundary condition for direct computation of aerodynamic sound, *AIAA Journal tech. notes*, **34**, 740-742, (1997)
15. D.J. Bodony, S.K. Lele, Jet noise prediction of cold and hot subsonic jets using Large-Eddy Simulation, *AIAA Paper 2004-3022* (2004)
16. N. Peters, *Turbulent Combustion*, Cambridge University Press (2000)
17. K.K. Kuo, *Principles of Combustion*, John Wiley and sons (2005)
18. T. Poinsot and D. Veynante, *Theoretical and Numerical Combustion*, R.T. Edwards Inc. (2005)
19. R.A. Vreman, An eddy-viscosity subgrid-scale model for turbulent shear flow: Algebraic theory and applications, *Phys. Fluids*, **16**, 3670 (2004)
20. P.J. Coelho, O.J. Teerling and D. Roekaerts, Spectral radiative effects and turbulence/radiation interaction in a non-luminous turbulent jet diffusion flame, *Combust. Flame*, **133**, 75-91, (2003)
21. A. Habibi, B. Merci and D. Roekaerts, Turbulence/radiation interaction in RANS simulations of non-premixed piloted turbulent laboratory scale flames, *Combust. Flame*, **151**, 303-320, (2007)

A Suite of Mathematical Models for Bone Ingrowth, Bone Fracture Healing and Intra-Osseous Wound Healing

F.J. Vermolen, A. Andreykiv, E.M. van Aken, J.C. van der Linden, E. Javierre, and A. van Keulen

Abstract In this paper, some modeling aspects with respect to bone ingrowth, fracture healing and intra-osseous wound healing are described. We consider a finite element method for a model of bone ingrowth into a prosthesis. Such a model can be used as a tool for a surgeon to investigate the bone ingrowth kinetics when positioning a prosthesis. The overall model consists of two coupled models: the biological part that consists of non-linear diffusion-reaction equations for the various cell densities and the mechanical part that contains the equations for poro-elasticity. The two models are coupled and in this paper the model is presented with some preliminary academic results. The model is used to carry out a parameter sensitivity analysis of ingrowth kinetics with respect to the parameters involved. Further, we consider a Finite Element model due to Bailon-Plaza and Van der Meulen for fracture healing in bone. This model is based on a set of coupled convection-diffusion-reaction equations and mechanical issues have not been incorporated. A parameter sensitivity analysis has been carried out. Finally, we consider a simplified model due to Adam to simulate intra-osseous wound healing. This model treats the wound edge as a moving boundary. To solve the moving boundary problem, the level set method is used. For the mesh points in the vicinity of the wound edge, a local adaptive mesh refinement is applied.

F.J. Vermolen, E.M. van Aken, E. Javierre
Delft Institute of Applied Mathematics, Faculty of Electrical Engineering, Mathematics and Computer Science, Delft University of Technology, Mekelweg 4, 2628 CD Delft, The Netherlands
e-mail: F.J.Vermolen@tudelft.nl,

A. Andreykiv, J.C. van der Linden, F. van Keulen
Mechanical, Maritime and Materials Engineering, Delft University of Technology, Mekelweg 2, 2628 CD Delft, The Netherlands e-mail: A.Andreykiv@tudelft.nl

1 Introduction

In osteoporosis, fracture risk is high, after a hip fracture a prosthesis that replaces the joint is often the only remedy. In the case of osteoarthritis and rheumatoid arthritis, the cartilage degrades and moving the joints becomes painful. Ultimately, most patients will receive a prosthesis to restore the function of a diseased joint. Prostheses are usually attached to the host bone by means of surgical screws to obtain sufficient initial stability. A schematic of a prosthesis of the shoulder cavity, embedded within an artificial joint is shown in Figure 1. In the course of time, bone will grow into a porous tantalum layer and hence more stability of the prosthesis is obtained. To investigate the quality and life time of such an artificial joint, one needs to study the effects of the placement of the prosthesis and of the materials that are involved in the joint. At present, these effects are often studied using large amounts of data derived from patients. To predict the life span and performance of artificial joints, numerical simulations are of great value since these simulations give many qualitative insights by means of parameter sensitivity analysis. These insights are hard to obtain by experiments. In Figure 2, an X-ray picture of the prosthesis of the shoulder cavity is shown.

In the case of a shoulder prosthesis, the angle at which the prosthesis is positioned by the surgeon is crucially important. The angle is important for the ability of moving the arm by the patient, but also to have the right strain pattern for (optimal) bone ingrowth. The latter fact is due to the fact that the mammalian bone is only generated if a certain strain is exceeded, but also smaller than a certain upper bound. For the surgery on the shoulder, the incision on the shoulder is made at the front in order to save crucial organs and muscles of the patient. The location of the incision is shown in Figure 3.

As a result of a limited visibility of the orthopaedic surgeon, the angle of placement of the prosthesis is a crucial issue. Currently the stress and strain behavior of the shoulder blade is studied at the Delft University of Technology, using three-dimensional Finite Element simulations as a function of the angle of placement of the prosthesis. An example of a computational domain is shown in Figure 4.

Several studies have been done to simulate bone-ingrowth or fracture healing of bones. To list a few of them, we mention the model due to Adam [1], Ament and Hofer [3], Bailon-Plaza *et al.* [7], Prendergast *et al.* [10] and recently by Andreykiv [4]. The model due to Prendergast *et al.* and LaCroix *et al.* [10, 13] will be treated in more detail, since we expect that this model contains most of the biologically relevant processes, such as cell division and differentiation, tissue regeneration, and cell mobility. Many ideas from modeling fracture healing of bones are used in these models, since bone-ingrowth into a prosthesis resembles the fracture healing process. In the model due to Prendergast, the influence of the mechanical properties on the biological processes are incorporated. Further, we note that Prendergast's model has been compared to animal experiments.

Next to bone-ingrowth into a prosthesis, we present the model due to Bailon-Plaza and Van der Meulen [7] for fracture healing in bone. This model is not coupled with the equations from (poro-)elasticity. Andreykiv [6] applies the model for

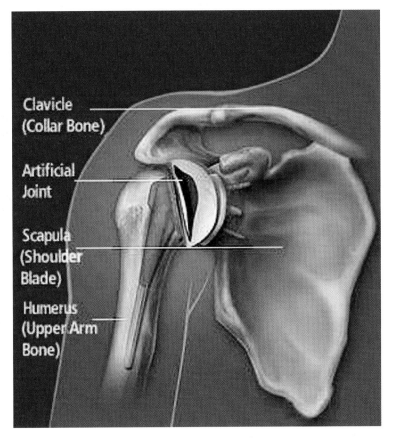

Fig. 1 An example of a schematic of an artificial shoulder joint.

bone-ingrowth to fracture healing in which coupling with mechanics has been accomplished.

Finally, the issue of intra-osseous wound healing is modeled using the simplified formalism due to Adam [1]. An intra-osseous wound may result from surgery (think of a wound on a skull due to brain surgery) or an injury caused by an accident. The equations have a rather simple nature, though obtaining a numerical solution is challenging since the wound edge is treated as a moving boundary.

In this paper, we will see a calibrated existing bone ingrowth model in terms of a system of nonlinearly coupled diffusion-reaction equations, for which the mechanical strain and fluid flow are important input parameters. In order to compute the aforementioned parameters, the poro-elasticity equations are solved. These two classes of models will be treated separately. This paper concerns a compilation of preliminary results, with some data for a shoulder prosthesis. Further, the paper considers a model for fracture healing in bone and finally a model for healing of an intra-osseous wound is described.

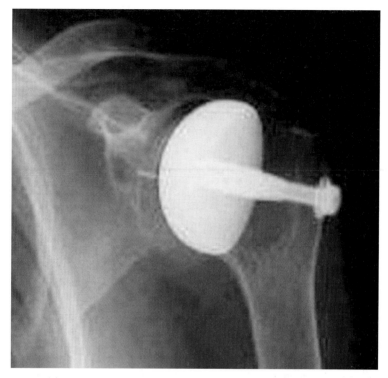

Fig. 2 An example of an X-ray picture of the prosthesis of the shoulder cavity.

2 The bone-ingrowth model

We consider a prosthesis for the shoulder cavity. A sketch of the prosthesis is given in Figure 5. The top part of the prosthesis consists of polyethylene. This part is in actual contact with the upper arm, which exerts a loading on it. The second part consists of a tantalum mesh and polyethylene. In these two parts, we solve an equation for mechanical equilibrium combined with Hooke's Law only. The third part contains the porous tantalum. Into this part, bone ingrowth takes place from the glenoid, which is the part of the scapula, in which the humeral head rotates. We denote the entire computational domain by Ω, which consists of Ω_E (the elastic domain) and Ω_P (the poro-elastic domain), hence $\Omega = \Omega_E \cup \Omega_P \cup (\overline{\Omega}_E \cap \overline{\Omega}_P)$. The overlines indicate the closure of the (sub) domain. Further, we assume Ω_E and Ω_P to be disjoint, that is $\Omega_E \cap \Omega_P = \emptyset$. The domain Ω_E represents the part of the computational domain on which only the elastic equations are solved. On Ω_P, one solves the poro-elastic equations. Further, in Ω_P, we solve the equations for bone-ingrowth, which is referred to as the biological part of the model.

Further, we note that $\Omega_E = \Omega_E^P \cup \Omega_E^B$, where Ω_E^P and Ω_E^B are the elastic part of the prosthesis and bone respectively. Further, $\overline{\Omega}_E^P \cap \overline{\Omega}_E^B$ are separated by $\overline{\Omega}_P$, hence

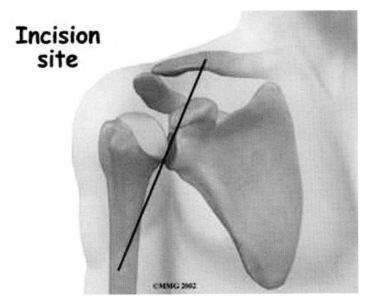

Fig. 3 The location at which the incision to put the shoulder prosthesis is carried out by the surgeon.

$\overline{\Omega}_E^P \cap \overline{\Omega}_E^B = \emptyset$. The boundary of the computational domain, Ω, is denoted by Γ, and of the subdomains Ω_E and Ω_P, their respective boundaries are represented by Γ_E and Γ_P.

The biological part of the model involving cell growth, division, differentiation and formation of bone and cartilage, applies for the porous tantalum. The coefficients in the biological part of the model depend on the local strains and fluid flow velocity, which establishes a nonlinear coupled problem. First, the mechanical model is presented and subsequently we give the biological model.

2.1 The mechanical model

Assuming mechanical equilibrium in both the poro-elastic part and the elastic part of the domain, we solve

$$-\text{div}\,\underline{\underline{\sigma}} = \mathbf{f}, \qquad \mathbf{x} \in \Omega. \tag{1}$$

In the above equation, $\underline{\underline{\sigma}}$ and $\mathbf{f}$ respectively denote the stress tensor $\underline{\underline{\sigma}} = \begin{pmatrix} \sigma_{xx} & \sigma_{xy} \\ \sigma_{yx} & \sigma_{yy} \end{pmatrix}$ in two dimensions and $\mathbf{f}$ represents the internal body force. In our application, we disregard the internal body force, hence we take $\mathbf{f} = \mathbf{0}$. At a part of the top boundary of the prosthesis, which is a part of the computational domain, a quadratic loading is exerted, that is

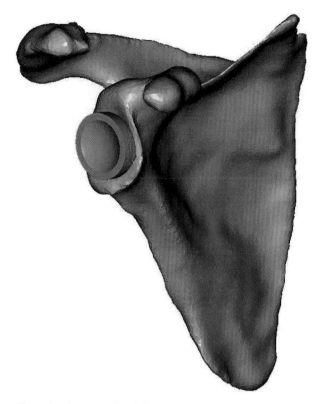

Fig. 4 A three-dimensional computational domain of the shoulder blade.

$$\underline{\underline{\sigma}} \cdot \mathbf{n} = \mathbf{t}, \quad \mathbf{x} \in \Gamma_l. \tag{2}$$

Here, **t** denotes the exerted loading on Γ_l, which is the part of Γ on which the external loading is applied. On the bottom boundary of the computational domain, which is on the glenoid, we assume that the displacement is zero, that is

$$\mathbf{u} = \mathbf{0}, \quad \mathbf{x} \in \Gamma_c. \tag{3}$$

Here Γ_c denotes the part of the outer boundary that is fixed to the host bone. On all the other parts of the boundary, it is assumed that there is no loading, that is

$$\underline{\underline{\sigma}} \cdot \mathbf{n} = \mathbf{0}, \quad \mathbf{x} \in \Gamma \setminus (\Gamma_l \cup \Gamma_c). \tag{4}$$

The material properties vary strongly over the various parts of the computational domain. The displacement and traction are assumed to be continuous.

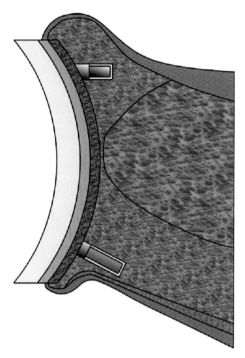

Fig. 5 A schematic of the prosthesis for the glenoid. From left to right: The metal backing, the poro-ethylene part, the porous tantalum, in which bone-ingrowth takes place and the bone.

2.1.1 The elasticity domain

In this part of the domain, only the equations for the mechanical balance are solved. For the link between the stresses and strains, Hooke's Law is used, which reads as follows in the two-dimensional case

$$E\varepsilon_{xx} = \sigma_{xx} - \nu\sigma_{yy},$$
$$E\varepsilon_{yy} = -\nu\sigma_{xx} + \sigma_{yy}, \quad (5)$$
$$E\varepsilon_{xy} = \tfrac{1}{2}(1+\nu)\sigma_{xy}.$$

Here E and ν represent the Young's modulus and Poisson ratio. The strain tensor is denoted by $\underline{\underline{\varepsilon}} = \begin{pmatrix} \varepsilon_{xx} & \varepsilon_{xy} \\ \varepsilon_{yx} & \varepsilon_{yy} \end{pmatrix}$. The relation between the strains and displacements $\mathbf{u} = [u\ v]$ is given by

$$\varepsilon_{xx} = \frac{\partial u}{\partial x}, \quad \varepsilon_{yy} = \frac{\partial v}{\partial y}, \quad \varepsilon_{xy} = \frac{\partial u}{\partial y} + \frac{\partial v}{\partial x}. \quad (6)$$

2.1.2 The porous tantalum

In this part of the domain, the two-phase (poro-elasticity) equations are solved. We write

$$\underline{\underline{\sigma}} = \overline{\underline{\underline{\sigma}}} - p\underline{\underline{I}}, \tag{7}$$

where $\overline{\underline{\underline{\sigma}}}$ is the effective stress that gives the deformations and p is the pressure. Further, $\underline{\underline{I}}$ is the identity tensor. This implies that also in this domain, we solve

$$-\text{div}\,\underline{\underline{\sigma}} = \mathbf{0}, \text{ or } \quad \text{div}\,\overline{\underline{\underline{\sigma}}} = \text{div}\,(p\underline{\underline{I}}), \quad \mathbf{x} \in \Omega_P. \tag{8}$$

Furthermore, for the fluid flow, we get

$$\frac{\partial}{\partial t}(n_f \beta_f p + \text{div}\,\mathbf{u}) - \text{div}\left(\frac{\kappa}{\eta}\,\text{grad}\,p\right) = 0, \quad \mathbf{x} \in \Omega_P. \tag{9}$$

Here, n_f, β_f, κ, η respectively denote the porosity, compressibility of the fluid, permeability of the tantalum and viscosity of the fluid. The Lamé parameters, which are linked to the stiffness and Poisson's ratio of the material are defined by

$$\lambda = \frac{\nu E}{(1+\nu)(1-2\nu)}, \quad \mu = \frac{E}{2(1+\nu)}. \tag{10}$$

Using the abovementioned Lamé parameters, we arrive at the following form of the poro-elastic equations

$$\begin{aligned}
-\text{div}\,(\mu\,\text{grad}\,u) - \frac{\partial}{\partial x}((\lambda+\mu)\,\text{div}\,\mathbf{u}) + \frac{\partial p}{\partial x} &= 0, \mathbf{x} \in \Omega_P, \\
-\text{div}\,(\mu\,\text{grad}\,v) - \frac{\partial}{\partial y}((\lambda+\mu)\,\text{div}\,\mathbf{u}) + \frac{\partial p}{\partial y} &= 0, \mathbf{x} \in \Omega_P, \\
\frac{\partial}{\partial t}(n_f \beta_f p + \text{div}\,\mathbf{u}) - \text{div}\left(\frac{\kappa}{\eta}\,\text{grad}\,p\right) &= 0, \quad \mathbf{x} \in \Omega_P.
\end{aligned} \tag{11}$$

At the bone-implant interface, the displacement and stresses are continuous. The parameters in the equations (E and ν) have to be updated as bone grows into the prosthesis. The Rule of Mixtures is applied to update the mechanical properties (see Lacroix & Prendergast [13]). For more information on the derivation of the above equations, we refer to Bear [8]. As an initial condition p is prescribed and set equal to zero. The boundary conditions for the pressure are

$$\begin{aligned}
p &= 0, \quad \mathbf{x} \in \overline{\Omega}_P \cap \overline{\Omega}_E^B, \\
\frac{\kappa}{\eta}\frac{\partial p}{\partial n} &= 0, \mathbf{x} \in \Gamma_P \setminus (\overline{\Omega}_P \cap \overline{\Omega}_E^B).
\end{aligned} \tag{12}$$

Here Γ_P denotes the boundary of the porous tantalum Ω_P. At the boundary between the porous tantalum and the metal backing, we require continuity of the displacements and traction. In other words, these subdomains are fixed to each other.

Next, we consider a scaled version of equations (11), in which we draw our attention to the third equation. In this scaling argument, we assume that the coefficients in the equations (11) are constant in time and space. Division of this equation by $n_f \beta_f$ (under the assumption that n_f and β_f are constant), and using the dimensionless variables

$$X, Y := \frac{x,y}{L}, \quad \tau := \frac{\kappa}{\eta \beta_f n_f} \frac{t}{L^2}, \quad \text{and } U, V := \frac{u,v}{L}, \tag{13}$$

where L is a characteristic length, such as the length or width of the prosthesis. Then equations (11) change into

$$-\overline{\nabla} \cdot (\mu \overline{\nabla} U) - \frac{\partial}{\partial X}((\lambda + \mu) \overline{\nabla} \cdot \mathbf{U}) + \frac{\partial p}{\partial X} = 0,$$
$$-\overline{\nabla} \cdot (\mu \overline{\nabla} V) - \frac{\partial}{\partial Y}((\lambda + \mu) \overline{\nabla} \cdot \mathbf{U}) + \frac{\partial p}{\partial Y} = 0, \tag{14}$$
$$\frac{\partial}{\partial \tau}(\overline{\nabla} \cdot \mathbf{U}) = n_f \beta_f \left(\overline{\Delta} p - \frac{\partial p}{\partial \tau} \right),$$

where $\overline{\nabla}(.) := \frac{1}{L} \nabla(.)$, $\overline{\Delta}(.) := \frac{1}{L^2} \Delta(.)$ and $\mathbf{U} := \frac{1}{L} \mathbf{u}$. We see that as $n_f \beta_f \to 0$, then, we reach the incompressible limit, which gives a saddle-point problem, similar to the Stokes equations, where one has to consider *LBB condition* satisfying elements or a stabilization. Note also that for the incompressible limit, the boundary conditions for the pressure vanishes. The situation becomes analogous to the Stokes' equations.

2.2 The biological part

Prendergast *et al.* [14] consider the behavior of mesenchymal cells, that originate from the bone marrow and differentiate into fibroblasts, chondrocytes and osteoblasts. These newly created cell types respectively generate fibrous tissue, cartilage and bone. In Prendergast's model, it is assumed that fibroblasts may differentiate into chondrocytes, chondrocytes may differentiate into osteoblasts. The differentiation processes are assumed to be irreversible. The differentiation pattern has been sketched in Figure 6. This biological model applies for the porous tantalum. The accumulation at a certain location of all the cell types is determined by cell mobility, cell division and cell differentiation. Let c_m, c_c, c_f and c_b respectively denote the cell density of the mesenchymal cells, chondrocytes, fibroblasts and osteoblasts, in the poro-elastic tantalum of the prosthesis in which bone ingrowth takes place. Then, the dynamics of the mesenchymal cell density is described by

$$\frac{\partial c_m}{\partial t} = \text{div } D_m \text{ grad } c_m + P_m(1 - c_{\text{tot}})c_m$$
$$- F_f(1 - c_f)c_m - F_c(1 - c_c)c_m - F_b(1 - c_b)c_m, \quad \mathbf{x} \in \Omega_P, \tag{15}$$

where $D_m - D_m^0(1 - m_c - m_b)$.

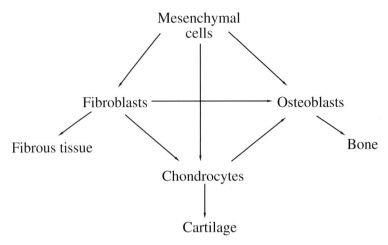

Fig. 6 The scheme of cell differentiation of mesenchymal cells, fibroblasts, chondrocytes and osteoblasts.

The first term in the right-hand side of the above equation represents the transport of mesenchymal cells. The diffusivity of mesenchymal cells, D_m, is determined by the amount of bone and cartilage present. It is assumed that cartilage and bone inhibit diffusion. The second term represents mesenchymal stem-cell production due to cell division, with production rate constant P_m. The other terms incorporate differentiation of mesenchymal stem cells to fibroblasts, chondrocytes and osteoblasts, with their respective differentiation rate constants F_f, F_c and F_b. The dynamics of the fibroblasts, which are the cells that produce fibrous tissue, is represented by

$$\frac{\partial c_f}{\partial t} = \operatorname{div} D_f \operatorname{grad} c_f + P_f(1 - c_{\text{tot}})c_f$$
$$+ F_f(1 - c_f)c_m - F_c(1 - c_c)c_f - F_b(1 - c_b)c_f, \quad \mathbf{x} \in \Omega_P, \quad (16)$$

where $D_f = D_f^0(1 - m_c - m_b)$.

The description of the terms of the right-hand side of the above equation is similar to the previous equation for the mesenchymal stem cells. They express cell division (with production rate constant P_f), transport (with fibroblast diffusivity D_f) and differentiation to other cell types (chondrocytes and osteoblasts with their respective differentiation rate constants F_c and F_b). The mesenchymal stem cells and fibroblasts are the only cell types that are mobile. The chondrocytes and osteoblasts, respectively producing cartilage and bone, are assumed to be immobile. Their reaction processes are modeled by

$$\frac{\partial c_c}{\partial t} = P_c(1-c_{\text{tot}})c_c + F_c(1-c_c)(c_m+c_f) - F_b(1-c_b)c_c,$$
$$\frac{\partial c_b}{\partial t} = P_b(1-c_{\text{tot}})c_b + F_b(1-c_b)(c_m+c_f+c_c). \qquad \mathbf{x} \in \Omega_P. \quad (17)$$

The first terms of the right-hand side in the above equations represent cell division (with production rate constants P_c and P_b for the chondrocytes and osteoblasts respectively), the second term describes the addition due to differentiation from mesenchymal stem cells and fibroblasts. The last term in the top equation represents the differentiation of chondrocytes to osteoblasts. The tissues, fibrous tissue, cartilage and bone are immobile. The volume accumulation of these tissues, denoted by m_f and m_c, respectively, are modeled by

$$\frac{\partial m_f}{\partial t} = Q_f(1-m_{\text{tot}})c_f - (D_b c_b + D_c c_c)m_f m_{\text{tot}},$$
$$\frac{\partial m_c}{\partial t} = Q_c(1-m_b-m_c)c_c - D_b c_b m_c m_{\text{tot}}, \qquad \mathbf{x} \in \Omega_P. \quad (18)$$
$$\frac{\partial m_b}{\partial t} = Q_b(1-m_b)c_b.$$

Here, both production (with rate constants Q_f, Q_c and Q_b for the fibroblasts, chondrocytes and osteoblasts and decay rates (with rate constants D_b and D_c) are incorporated. The quantity m_{tot} denotes the maximum allowable volume fraction of the tissues. In the above equations, a maximum allowable volume fraction of the tissues and decay rates has been incorporated.

The initial concentrations of all tissues and cell types are zero. As boundary conditions, a Dirichlet condition for the mesenchymal cell density at the bone implant and homogeneous Neumann conditions at all other boundaries are applied, that is

$$c_m = 1, \qquad \mathbf{x} \in \overline{\Omega}_P \cap \overline{\Omega}_E^B,$$
$$D_m \frac{\partial c_m}{\partial n} = 0, \; \mathbf{x} \in \Gamma_P \setminus (\overline{\Omega}_P \cap \overline{\Omega}_E^B), \quad (19)$$
$$D_f \frac{\partial c_f}{\partial n} = 0, \; \mathbf{x} \in \Gamma_P.$$

In the present paper, the influence of the micro-motions is neglected. For the fibroblasts homogeneous Neumann boundary conditions are imposed for all boundary segments. The equations for the mesenchymal cells, fibroblasts, chondrocytes and osteoblasts were introduced by Andreykiv [4, 6, 5]. The proliferation, differentiation and diffusion parameters, which are coefficients in the above Prendergast model, depend on the mechanical stimulus. The mechanical stimulus is given by a linear combination of the maximum shear strain and the fluid velocity relative to the rate of displacement of the solid, that is

$$S = \frac{\gamma}{a} + \frac{v}{\beta}, \qquad (20)$$

where γ represents the maximum shear strain and v denotes the relative fluid/solid velocity. Here $\gamma := \frac{1}{2}(\lambda_1 - \lambda_2)$, where $\lambda_{1,2}$ represent the eigenvalues of the strain tensor. The rates of tissue regeneration and differentiation qualitatively depend on the mechanical parameters such that:

- Low strain has a stimulatory effect (in relation to no strain) on the fibroblast proliferation and bone regeneration (if $0 < S < 1$);
- For intermediate values of the strain, cartilage formation is more favorable (if $1 < S < 3$);
- High strains favor the proliferation of fibrous tissue (if $S > 3$).

This gives a coupling of the poro-elasticity model to this biological model.

The above set of partial differential equations poses a nonlinearly coupled set of equations. Standard Galerkin Finite Element methods provide a straightforward method to obtain solutions. For the dependencies of the parameters involved on the mechanical stimulus, that is S, we refer to the thesis due to Andreykiv [4].

2.3 The numerical method for the ingrowth model

To solve the equations, we use a Finite Element method. To derive the weak form for both the metal backing part and the porous tantalum, we express the equations for mechanical equilibrium in terms of the stresses. The equations for poro-elasticity (in the porous tantalum) become the same as for the metal backing. The diagonal entries of the stress tensor change:

$$\sigma_{xx} = \overline{\sigma}_{xx} - p, \qquad \sigma_{yy} = \overline{\sigma}_{yy} - p, \qquad (21)$$

where $\overline{\sigma}_{xx}$ and $\overline{\sigma}_{yy}$ denote the effective stresses as in the metal backing region. A weak form for the equations for mechanical equilibrium is given by

---

Find $u, v \in H^1(\Omega)$, subject to $\mathbf{u} = \mathbf{0}$ on Γ_c, such that

$$-\int_{\Gamma_l} t_1 \phi^u d\Gamma + \int_\Omega \left\{ \sigma_{xx} \frac{\partial \phi^u}{\partial x} + \sigma_{xy} \frac{\partial \phi^u}{\partial y} \right\} d\Omega = 0,$$

$$\int_{\Gamma_l} t_2 \phi^v d\Gamma + \int_\Omega \left\{ \sigma_{xy} \frac{\partial \phi^v}{\partial x} + \sigma_{yy} \frac{\partial \phi^v}{\partial y} \right\} d\Omega = 0,$$

$\forall \phi^u, \phi^v \in H^1(\Omega)$, subject to $\phi^u = \phi^v = 0$ for $\mathbf{x} \in \Gamma_c$.

In the above formulation, we avoided the tensor representation to keep the text readable for researchers that are unfamiliar with mechanical problems. For the pressure, one obtains the following weak form

> Find $p \in H^1(\Omega)$, subject to $p = 0$ on $\overline{\Omega}_P \cap \overline{\Omega}_E^B$, such that
>
> $$\int_\Omega \frac{\partial}{\partial t}(n_f \beta_f p + \mathrm{div}\,\mathbf{u})\psi d\Omega = -\int_\Omega \frac{\kappa}{\eta}\nabla p \cdot \nabla \psi d\Omega,$$
>
> $\forall \psi \in H^1(\Omega)$, subject to $\psi = 0$ whenever p on $\overline{\Omega}_P \cap \overline{\Omega}_E^B$.

For a rather recent comprehensive overview of Finite Element methods applied to solid state mechanics, we refer to the book due to Bræss [9]. The above poro-elasticity equations are often solved using Petrov-Galerkin Finite element methods, such as the Taylor-Hood family: if the pressure is approximated with elements of polynomials of P_n, then, the displacements are approximated using polynomials of P_{n+1}. In the Taylor-Hood elements, one usually uses linear and quadratic basis functions for the pressure and displacements respectively. On the other hand, Crouzeix-Raviart elements, which are often used for Stokes flow problems, are based on a discontinuity of the pressure. Since $p \in H^1(\Omega) \subset C(\Omega)$, the Crouzeix-Raviart elements are not suitable here. As long as the compressibility is sufficiently large, one can also make use of linear-linear elements for the pressure and displacement. This was done successfully in the study due to Andreykiv [4]. If $\beta_f = 0$, which is the incompressible case, then the issue of oscillations and the use of appropriate elements or a stabilization becomes more important. For $\beta_f = 0$, the third equation in equation (11) reduces to the version that is solved by Aguilar et al. [2].

A Galerkin formulation of the above equation with

$$p = \sum_{j=1}^{m} p_j \psi_j(x,y), \text{ and } \mathbf{u} = \sum_{j=1}^{n} \mathbf{u}_j \phi_j(x,y),$$

is applied to equations (11). For consistency, we require $m \leq 2n$ as $n_f \beta_f \to 0$. This case resembles the classical Stokes' equations. For the classical Taylor-Hood elements, we use $\psi_i \in P_1(\Omega)$ and $\phi_i \in P_2(\Omega)$.

By numerical experiments and the argument that the discretization matrix no longer remains an M-matrix if $\Delta t < \frac{h}{6}$, Aguilar et al. [2] demonstrate for the one-dimensional Terzaghi problem that the numerical solution becomes mildly oscillatory. Aguilar et al. [2] use a stabilization term $\gamma \frac{\partial}{\partial t}\Delta p$ (with $\gamma = \frac{\sigma h^2}{4(\lambda + 2\mu)} = O(h^2)$, where $\sigma = 1$) to suppress the spurious oscillations. In our application, the stabilization coefficient is given by $\gamma \approx 1.2 \cdot 10^{-18}$. In this study, we use linear-linear elements to solve poro-elasticity equations. We verified numerically that these elements gave the same results as the Taylor-Hood elements. A possible reason for this is that for our settings the compressibility term is given by $n_f \beta_f \approx 2.5 \cdot 10^{-16}$, which is larger than the stabilization coefficient γ that was introduced by Aguilar et al. [2]. Since this term, and in particular the $\frac{\partial p}{\partial t}$-term (also as $\Delta \tau \to 0$), gives an additional contribution to the diagonal entries of the discretization matrix, the M-matrix property of the discretization matrix is probably preserved. Hence, this term stabilizes the solution. Note that linear-linear elements are always allowable if the stabilization term due to Aguilar is used. Our approach, which is motivated physically, stabilizes in a similar way as Aguilar's term does. We admit that this issue needs more investigation in mathematical rigor. For the concentrations and densities, linear elements are used too.

The equations for poro-elasticity were solved using the Euler backward time integration method in which the data for the material parameters such as the permeability, Young's modulus and Poisson ratio were determined from the bone, cartilage and fibrous tissue densities that were obtained at the previous time step. Using this approach, there is hardly any limitation with respect to the time step. The nonlinear partial differential equations for the differentiation of several cell types were integrated in time using a first order IMEX method. Further, the material properties that depend on local strain and fluid flow were adapted using the data from the previous time step. This approach hardly influences stability as in the previous case of poro-elastic equations. The IMEX method for the reaction-diffusion equations yields good solutions, but here time step with respect to stability becomes more important. The first order IMEX time integration was applied to the reaction terms and to the diffusivity that depends on the cartilage and bone densities. As an example, we present the semi-discretization with respect to the time integration of the equation for the mesenchymal cell density:

$$
c_m^{p+1} = c_m^p + \Delta t \cdot \left\{ \text{div } D_m^p \text{ grad } c_m^{p+1} \right\} +
$$
$$
\Delta t \cdot \left(P_m(1 - c_{tot}^p) - F_f(1 - c_f^p) - F_c(1 - c_c^p) - F_b(1 - c_b^p) \right) c_m^{p+1},
$$
(22)

where p denotes the time index, where $t = p\Delta t$ is the actual time. The maximum allowable time step becomes dependent on the local solution at the time step considered. One can analyze the stability using the eigenvalues of the Jacobi matrix (left multiplied by the mass matrix) from the reaction terms. Using upper bounds and lower bounds of the solution, one can investigate the allowable time steps for the integration. This was not done in this study. We compared the solutions by halving the time step and observed that there was hardly difference when a time step of the order of an hour was taken.

The diffusion part of the equations for the mesenchymal cells and fibroblasts were solved using an IMEX method, where the diffusivities of the mesenchymal cells and fibroblasts were taken from the previous time step. The reaction parts in all the equations were treated using an IMEX time integration method too. The coupling was treated by the use of information from the previous time step. Until now, no iterative treatment of the coupling has been done in the current preliminary simulations. A state-of-the-art book on several numerical time integrators for stiff problems is the work due to Hundsdorfer & Verwer [11].

To determine the stimulus in equation (20), the strain is computed from the spatial derivatives of the displacements. To determine the strains at the mesh points, we proceed as follows: consider the equation for ε_{xx}, then multiplication by a test-function gives

$$\int_\Omega \varepsilon_{xx}\phi\, d\Omega = \int_\Omega \frac{\partial u}{\partial x}\phi\, d\Omega, \text{ for } \phi \in H^1(\Omega), \qquad (23)$$

where $\varepsilon_{xx} \in H^1(\Omega)$. Using the set of basis functions as in our finite element solution, gives

$$\sum_{j=1}^{n} \varepsilon_{xx}^j \int_\Omega \phi_i\phi_j\, d\Omega = \sum_{j=1}^{n} u_j \int_\Omega \frac{\partial \phi}{\partial x}\phi_i\, d\Omega, \qquad \text{for } i \in \{1,\ldots,n\}. \qquad (24)$$

This gives a system of n equations with n unknowns. This is applicable for any type of element. For piecewise linear basis functions, the mass matrix is diagonal (lumped) after applying Newton-Cotes' integration rule. Then, the strains and fluid velocities are used for the mechanical stimulus at the mesh points for the ordinary differential equations, which are solved using an IMEX time integrator only.

2.4 Numerical experiments on the ingrowth model

In Figure 7, the distribution of the stimulus, osteoblast density, mesenchymal stem cell density and the bone fraction in the porous tantalum layer after 100 days have been plotted. The prosthesis is assumed to consist of two parts: the top part being the functional part on which an external force is exerted from the outer motion. The bottom part is the porous tantalum, in which bone is allowed to grow in from the bottom layer. The size of the prosthesis is given by 40 × 10 mm, in which the prosthesis is divided into the top and bottom layer of the same size. The upper force is given by 165.84 N, corresponding to an arm abduction of 30 degrees. In this paper, it is assumed that this force is exerted constantly. In future, we will consider more realistic oscillatory forces. In the top part of the prosthesis, the elasticity equations are solved. The prosthesis has been approximated by a two-dimensional geometry, which can be done with the use of cylindrical co-ordinates. The latter has not been done yet.

It can be seen that the osteoblast density is maximal where the stimulus is maximal. This implies that bone develops at the positions where the osteoblast density and stimulus is maximal. This can be seen clearly from the figures. Furthermore, the mesenchymal cell density shows a decrease where the cells differentiate into osteoblasts. The conditions are such that the model only allows the differentiation into osteoblasts and the development of other cell types and tissues is prohibited. To have bone ingrowth in the other parts of the tantalum, it is necessary that the upper arm moves allowing for the stimulus to increase at various positions within the tantalum. This has been observed to take place in preliminary simulations that are not shown in this paper. For arm abductions of 90 degrees, cartilage is also allowed to develop in the tantalum due to a higher outer force that is exerted on the top of the prosthesis. It can be seen that bone develops in the high stimulus domain. Bone can only remain at locations where it has been generated. Bone resorption has been disregarded in the model since its effect seems to be of second order only.

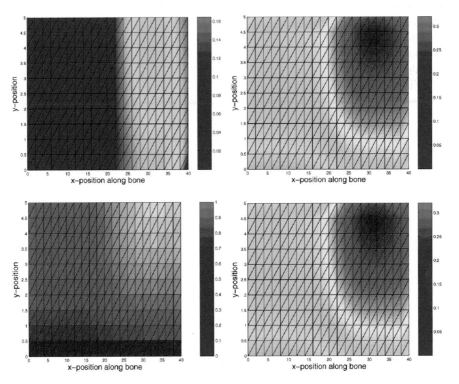

Fig. 7 Some distributions in the porous tantalum after 100 days, from top to bottom: The stimulus, the osteoblasts (bone cells) density, the mesenchymal stem cell density and the bone density.

Some preliminary results reveal that the model is rather insensitive to the diffusion parameters near the current values. There is a high sensitivity with respect to F_b, and Q_b in the present loading regime, where F_b and Q_b can be considered as the differentiation rate of mesenchymal cells to osteoblasts and the bone production rate respectively. Physically, this means that the bone ingrowth pattern is severely influenced by the mesenchymal to osteoblast differentiation rate and osteoblast activity to produce bone. Hence, it is important to have a loading pattern and chemical environment that favor osteoblast and bone production.

3 The fracture healing model due to Bailon-Plaza

An interesting model that is somewhat similar to the previous model due to Prendergast is the model due to Bailon-Plaza and Van der Meulen [7]. Bailon-Plaza and van der Meulen formulated this model and solved the involved partial differential equations. Since this paper concerns an overview of bone ingrowth, fracture healing and intra-osseous healing, the equations due to Bailon-Plaza and van der Meulen are repeated for completeness. This model is used to simulate healing of a bone fracture. The geometry is sketched in Figure 8. This model does not take into account

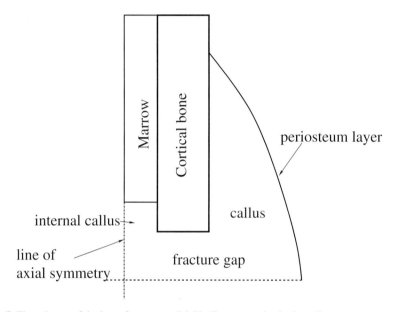

Fig. 8 The scheme of the bone fracture model. Healing proceeds via the callus.

the generation of fibroblasts and fibrous tissue, but it incorporates the production and decay of growth factors that stimulate or inhibit cell production. Growth factors are hormones that influence the rate differentiation of several cell types to other cell types (such as the differentiation of mesenchymal stem cells to chondrocytes). Unlike in the previous model, it is assumed that the formation of bone and cartilage as solid materials introduces an additional nonlinear convective term to the mesenchymal stem cells. The convective velocity is proportional to, and towards the gradient of m, which is defined as $m = m_c + m_b$. This changes the nature of the equation of the mesenchymal cell profile in the following way

$$\frac{\partial c_m}{\partial t} = \text{div}\left\{D_m \text{ grad } c_m - \frac{C_k}{(K_k+m)^2} c_m \text{ grad } m\right\} + \frac{A_{m0}m}{(K_m^2+m^2)^2} c_m (1-\alpha_m c_m)$$

$$-\left(\frac{Y_1 g_b}{H_1+g_b} - \frac{Y_2 g_c}{H_2+g_c}\right) c_m, \qquad \mathbf{x} \in \Omega_P. \tag{25}$$

The first term in the right-hand side models diffusive and convective transport of mesenchymal cells. The quantity D_m is also referred to as the haptotatic cell migration speed. The second term takes care of convective transport resulting from bone and cartilage formation. The quantity C_k is called the haptokinetic migration speed. Further, g_b and g_c are growth factors that enhance the differentiation of mesenchymal cells to chondrocytes and osteoblasts respectively. The quantities C_k, A_{m0}, Y_1 and Y_2 are constants. The diffusion coefficient D_m is given by

$$D_m = \frac{D_h}{K_h^2+m^2} m, \tag{26}$$

where D_h and K_h are constants. The PDE for c_m is a nonlinear convection-diffusion-reaction equation. The PDE for c_m is supplemented with a Dirichlet boundary for c_m at the interface between the internal callus and the marrow region and at the periosteum layer. At the other boundaries of the computational domain, which consists of the internal callus, fracture gap and callus, homogeneous Neumann boundary conditions are applied. Andreykiv et al. [6] use the model as in the previous section to study bone-ingrowth. A comparison between the two models is an interesting topic for future research. The equations for the chondrocytes and osteoblasts look similar to the ones in the model in the previous section, and read as

$$\frac{\partial c_c}{\partial t} = A_c c_c (1-\alpha_c c_c) + \frac{Y_2}{H_2+g_c} g_c c_m - \frac{m_c^6}{B_{ec}^6+m_c^6} \frac{Y_3}{H_3+g_b} g_b c_c,$$

$$\mathbf{x} \in \Omega_P.$$

$$\frac{\partial c_b}{\partial t} = A_b c_b (1-\alpha_b c_b) + \frac{Y_1 g_b}{H_1+g_b} c_m + \frac{m_c^6}{B_{ec}^6+m_c^6} \frac{Y_3}{H_3+g_b} g_b c_c - d_d c_b, \tag{27}$$

The first terms in the above equations represent the logistic growth due to cell mitosis. The second terms take into account the differentiation from mesenchymal cells to chondrocytes and osteoblasts. The quantities A_c, A_b, Y_3, α_c, α_b, H_1, H_2, H_3, d_d and B_{ec} are considered as given constants. These differentiation processes are triggered by the presence of growth factors. In the equations, a maximum for the differentiation rate with respect to the growth factor concentration is incorporated. The other terms take differentiation of chondrocytes to osteoblasts and decay into account. The changes in cartilage and bone density are modeled by

$$\frac{\partial m_c}{\partial t} = P_{cs}(1 - \kappa_c m_c)(c_m + c_c) - Q_{cd} m_c c_b,$$

$$\frac{\partial m_b}{\partial t} = P_{bs}(1 - \kappa_b m_b) c_b, \qquad \mathbf{x} \in \Omega_P. \qquad (28)$$

Here, P_{cs}, P_{bs}, Q_{cd}, κ_c and κ_b are constants. The growth factors for the generation of bone and cartilage are subject to diffusional transport within the callus, formation due to the presence of chondrocytes, osteoblasts and tissues, and decay. The PDE's for the growth factors are the following:

$$\frac{\partial g_c}{\partial t} = \mathrm{div}\,(D_{gc}\,\mathrm{grad}\,g_c) + \frac{G_{gc} g_c}{H_{gc} + g_c} \frac{m}{K_{gc}^3 + m^3} c_c - d_{gc} g_c,$$

$$\frac{\partial g_b}{\partial t} = \mathrm{div}\,(D_{gb}\,\mathrm{grad}\,g_b) + \frac{G_{gb} g_b}{H_{gb} + g_b} c_b - d_{gb} g_b, \qquad \mathbf{x} \in \Omega_P. \qquad (29)$$

Here D_{gc}, D_{gb} are the diffusivities of the cartilage and bone growth factors, respectively. Further, G_{gc}, G_{gb}, H_{gc}, d_{gc}, H_{gb} and d_{gb} are assumed to be known constants. Further, g_c and g_b denote the growth factor concentration for the cartilage and bone regeneration. For the boundary conditions, one uses Dirichlet conditions for g_c at the interface between bone and the fracture gap for $t < \tau$. Here, τ represents the time after which no growth factors appear at the interface between bone and the gap. Typically, it is observed that τ is approximately 24 hours. Further, for g_b a Dirichlet boundary condition is applied along the interface between bone and the (external) callus, at a part away from the fracture gap for $t < \tau$. For $t > \tau$, the Dirichlet conditions are replaced with homogeneous Neumann boundary conditions. At all other boundaries, homogeneous Neumann boundary conditions are applied. The initial conditions for g_c and g_b are $g_b(\mathbf{x},0) = g_c(\mathbf{x},0) = 20$, and $m(\mathbf{x},0) = 0.1 = m_c(\mathbf{x},0)$, which reflect inflammatory conditions.

In Figure 9, the evolution of the integral over the bone- and cartilage density is presented. First, cartilage is developed as an intermediate stage and finally the callus is filled with bone. The disappearance of the callus is not taken into account here. For the values of the parameters involved, we refer to Bailon and Van der Meulen [7]. For this model, we looked at the influence of the parameters involved. The most important parameter for bone ingrowth seems to be P_{bs}. A low value gives a slow bone growth process. An increase of the value of D_{gc} gives a high concentration of cartilage growth factors, which slightly enhances cartilage formation. However, bone formation is hardly influenced. An increase of d_{gb} leads to an increased cartilage formation and a delayed bone formation. Changing F_3 hardly has any influence, but an increase of F_1 reduces the growth of cartilage and growth of bone starts a little earlier. An increase of A_{bo} leads to a higher osteoblast density and a lower peak density of cartilage, whereas bone grows faster.

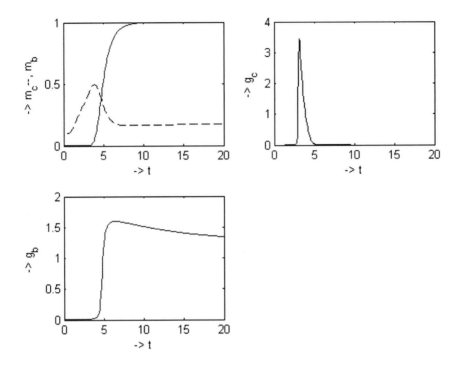

Fig. 9 The evolution of the cartilage and bone densities (top-left), where $m_b \to 1$ as $t \to \infty$, and the evolution of the growth factor concentrations (cartilage top-right; bone bottom) at a point on the middle of the fracture gap. The time is presented in days.

4 The model due to Adam

The model due to Adam is based on the so-called critical size, which is defined as the smallest intra-osseous wound that does not heal. The model is applied to wound healing both on skin tissue and bone. Wound healing, if it takes place, proceeds by various processes: chemotaxis (cell movement up a concentration gradient), neovascularization, synthesis of extracellular matrix proteins and scar remodeling. Growth factors likely play a crucial role in bone regeneration as in the model due to Bailon-Plaza and Van der Meulen. Furthermore, the supply of oxygen is crucially important for the rate and quality of wound healing. Hence, angiogenesis, which is the formation of capillaries in the vicinity of the wound, is crucially important for the healing process. In the model due to Adam, it is assumed that around the wound periphery, there is a thin band of tissue where tissue regeneration takes place. This thin band is referred to as the active layer.

4.1 The model equations

Let Ω_1 be a regular simply connected domain, surrounded by Ω_2, which is surrounded by Ω_3, with $\Omega = \cup_{p=1}^{3}\Omega_p \cup_{p=1}^{2}(\overline{\Omega}_p \cap \overline{\Omega}_{p+1})$, see Figure 10. The model for wound healing due to Adam is governed by the following equations:

$$\frac{\partial c}{\partial t} - D\Delta c + \lambda c = Pf(x,y), \text{ for } (x,y) \in \Omega, \quad (30)$$

$$\frac{\partial c}{\partial n}(x,y,t) = 0, \quad \text{for } (x,y) \in \Gamma, \quad (31)$$

$$c(x,y,0) = 0, \quad \text{for } (x,y) \in \Omega, \quad (32)$$

$$\text{further } f(x,y) = \mathbf{1}_{\Omega_2} = \begin{cases} 1, & \text{for } (x,y) \in \Omega_2 \text{ open domain.} \\ 0, & \text{for } (x,y) \in \Omega_1 \cup \Omega_3. \end{cases} \quad (33)$$

Here c is the concentration of a generic growth factor that stimulates cell division and hence healing of an epidermal or intra-osseous wound, and f can be seen as an indicator function on Ω_2 over Ω. The term with λc represents a decay of the growth factor. The PDE is supplemented with a homogeneous Neumann condition. Further, the outer boundary is denoted by $\Gamma = \overline{\Omega} \setminus \Omega$. In the present study we use the assumption that the wound heals if and only if the growth factor concentration exceeds a threshold concentration $\hat{c}$, at the wound edge $W(t) = \overline{\Omega}_1 \cap \overline{\Omega}_2$. Hence

$$\begin{aligned} &v_n > 0 \text{ if and only if } c(x,y,t) \geq \hat{c} \text{ for } (x,y) \in W(t), \\ &\text{else } v_n = 0. \end{aligned} \quad (34)$$

This implies that in order to determine whether the wound heals at a certain location at W at a certain time t, one needs to know c there. We assume that the healing rate is an affine function of the local curvature, $\hat{\kappa}$, at the wound edge $W(t)$, hence

$$v_n = -\frac{1}{2}(\alpha + \beta\hat{\kappa})w(c(\mathbf{x},t) - \hat{c}), \text{ for } \mathbf{x} \in W, \quad (35)$$

where $\alpha, \beta \geq 0$ ($\alpha + \beta \geq 0$). Here the function $w(s)$ falls within the class of Heaviside functions, that is $w(s) \in H(s)$, where $H(.)$ represents the family of Heaviside functions, for which we have

$$H : s \to \begin{cases} 0, & \text{if } s < 0, \\ [0,1], & \text{if } s = 0, \\ 1, & \text{if } s > 0. \end{cases} \quad (36)$$

Until now, some mathematical analysis on the existence, uniqueness and conditions for (retardated) healing has been performed. Further, a Finite Element solution has been obtained where the level set method has been used to track the wound edge, $W(t)$. To determine the concentration at the wound edge, local mesh refinement is

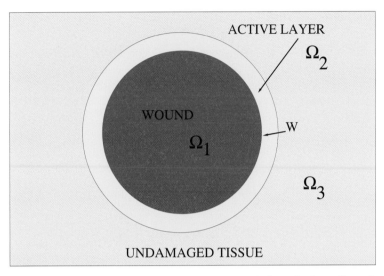

Fig. 10 A schematic of the computational domain: The wound Ω_1, active layer Ω_2, undamaged tissue Ω_3 and interface Γ.

used in the vicinity of the wound edge to enhance the numerical accuracy. Some results are shown Figure 11, 12 and 13. For more details concerning this issue, we refer to Javierre [12]. In Figure 11, we see the initial shape of a hypothetical star-shaped wound. This peculiar geometry is only used to illustrate the potential of the level set method to treat this moving boundary problem. In Figures 12 and 13, we see the gradual breaking up of the wound, which illustrates the topological changes, which are undergone by the healing wound.

5 Conclusions

A model has been developed for bone-ingrowth into a prosthesis. Parameters that were used were obtained from literature and animal experiments. For small forces exerted, bone develops mainly near the interface, between the prosthesis and host bone, and close to the applied force. For large forces, bone develops far away from the interface. For a complete ingrowth, oscillatory forces are to be applied. Linear-linear (displacement-pressure) elements are applicable for this two-dimensional problem.

Further, an accepted model for the healing of a bone fracture has been presented and some results have been shown. A parameter sensitivity analysis has been shown.

Finally, a model for the healing of an intra-osseous wound has been described and some results were shown. The model, being simple in its nature, poses a challenging

Fig. 11 The initial shape of the star-shaped wound and a local mesh refinement around the wound edge.

numerical problem due to the incorporation of a moving boundary to model wound closure.

References

1. J.A. Adam. A simplified model of wound healing (with particular reference to the critical size defect). *Mathematical and Computer Modelling*, 30:23–32, 1999.

Fig. 12 The shape of the star-like wound once the initial wound area has healed for 40 %.

2. G. Aguilar, F. Gaspar, F. Lisbona, and C. Rodrigo. Numerical stabilization of Biot's consolidation model by a perturbation on the flow equation. *International Journal of Numerical Methods in Engineering*, 75:1282–1300, 2008.
3. Ch. Ament and E.P. Hofer. A fuzzy logic model of fracture healing. *Journal of Biomechanics*, 33:961–968, 2000.
4. A. Andreykiv. *Simulation of bone ingrowth*. Thesis at the Delft University, Faculty of Mechanical Engineering, 2006.
5. A. Andreykiv, F. van Keulen, and P.J. Prendergast. Computational mechanobiology to study the effect of surface geometry on peri-implant tissue differentiation. *Journal of Biomechanics*, 130 (5):051015-1–11, 2008.
6. A. Andreykiv, F. van Keulen, and P.J. Prendergast. Simulation of fracture healing incorporating mechanoregulation of tissue differentiation and dispersal/proliferation of cells. *Biomechanical Models in Mechanobiology*, 7:443–461, 2008.

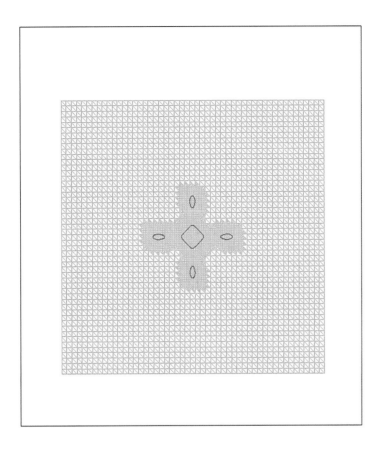

Fig. 13 The shape of the star-like wound once the initial wound area has healed for 90 %. The topology of the wound changes due to breaking up.

7. A. Bailon-Plaza and M. C. H. van der Meulen. A mathematical framework to study the effect of growth factors that influence fracture healing. *Journal of Theoretical Biology*, 212:191–209, 2001.
8. J. Bear. *Dynamics of fluids in porous media*. American Elsevier Publishing Inc., New York, 1972.
9. D. Braess. *Finite elements: theory, fast solvers, and applications in solid mechanics*. Cambridge University Press, Cambridge, 7th edition, 2007.
10. R. Huiskes, W. D. van Driel, P. J. Prendergast, and K. Søballe. A biomechanical regulatory model for periprosthetic fibrous-tissue differentiation. *Journal of Materials Science: Materials in Medicine*, 8:785–788, 1997.
11. W. Hundsdorfer and J. G. Verwer. *Numerical solution of time-dependent advection-diffusion-reaction equations*. Springer Series in Computational Mathematics, Berlin-Heidelberg, 2003.

12. E. Javierre, F.J. Vermolen, C. Vuik, and S. van der Zwaag. A mathematical model approach to epidermal wound closure: model analysis and computer simulations. *Report at DIAM, Delft University of Technology, and to appear in Journal of Mathematical Biology*, 07-14, 2007.
13. D. LaCroix and P.J. Prendergast. A mechano-regulation model for tissue differentiation during fracture healing: analysis of gap size and loading. *Journal of Biomechanics*, 35 (9):1163–1171, 2002.
14. P.J. Prendergast, R. Huiskes, and K. Søballe. Biophysical stimuli on cells during tissue differentiation at implant interfaces. *Journal of Biomechanics*, 30(6):539–548, 1997.

Numerical Modeling of the Electromechanical Interaction in MEMS

S.D.A. Hannot and D.J. Rixen

1 Introduction

Microsystems or Micro–Electro–Mechanical Systems (MEMS) are small (micrometer size) machines usually built by lithographic technologies originally developed for microchips. MEMS are designed to integrate sensing and actuation (and even data processing) on a single chip, therefore they often include moving and deforming parts. Currently microsystem technologies are used for a wide variety of purposes such as: read/write heads in hard-disk drives, ink-jet printheads, Digital Light Processing (DLP) chips in video projection systems and several types of sensors for pressure, flow, acceleration or bio-elements.

Due to their very small size, the dominant driving forces differ from the ones in the macro–world. For instance, the gravitational forces are negligible with respect to the elastic, adhesive and electrostatic forces [43]. These scaling effects may cause a strong coupling between different physical domains. For instance small silicon devices can be heated by an electric current relatively fast and they will cool down rapidly by conduction to the surrounding parts. Devices can be heated and cooled several times per second. Due to the thermal expansion of the material this effect can be used to actuate a device at frequencies that are high compared to macroscopic systems [22, 23, 27].

Another coupling that becomes useful at small scales is the coupling between structural displacement through electrostatic forces. At these small scales electrostatic forces are big enough to actuate devices, which is the type of coupling discussed in this document. Two types of MEMS that utilize the electromechanical

---

S.D.A. Hannot
Delft University of Technology, Faculty of Mechanical Maritime and Materials Engineering, Mekelweg 2, 2628 CD Delft, The Netherlands, e-mail: s.d.a.hannot@tudelft.nl

D.J. Rixen
Delft University of Technology, Faculty of Mechanical Maritime and Materials Engineering, Mekelweg 2, 2628 CD Delft, The Netherlands, e-mail: d.j.rixen@tudelft.nl

Fig. 1 Micro bridge

Fig. 2 Comb drive

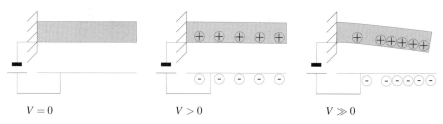

Fig. 3 Electromechanical coupling

coupling are micro switches and comb drives. Micro switches are the oldest type of microsystems. They are similar to magnetic relay switches, but the magnetic actuation is replaced by electrostatic actuation thanks to the very small dimensions [34, 44, 31]. Closely related to these devices is the micro mirror used in optics, which is basically a large switch with a reflecting surface. A picture of a small micro bridge that can be used as micro switch is given in fig. 1. The comb drive is an array of small fingers which can be used to maximize the electrostatic force between a moving and a fixed set of fingers while still allowing a large displacement of the moving set. A picture is shown in fig. 2. Modeling this electro–mechanical coupling is particularly challenging because the resulting equations are non-linear and the physics are strongly coupled.

1.1 Electromechanical coupling

The basic physical principle of electromechanically actuated microsystems is illustrated in fig. 3. Once a potential difference between a moving (or deformable) electrode and a fixed electrode is applied a charge difference is created. Coulomb forces between the charges will make the flexible electrode deform. This effect depends non-linearly on the shape of the electrode and the deformation itself, because the charge will concentrate near the point where the gap between the electrodes is smallest, which causes even higher Coulomb forces at this point [2].

A coupled electromechanical model normally consists of two domains: a structural domain and an electric domain. The structural domain contains all moving electrodes. The electric domain in principle contains the whole model, but if it is assumed that the electrodes are ideal conductors the electrostatic equation has to be solved only in the domain between the conductors. The applied potentials on the electrodes can then be modeled as fixed potential boundary conditions (Dirichlet boundary conditions).

Once the ideal conductor assumption has been made and when it is furthermore assumed that there are no free charges in the electrical domain, the equation that describes the electrostatics in the air gap is the electrostatic equation [18, 25]:

$$\frac{\partial}{\partial x_i}\left(\varepsilon \frac{\partial \phi}{\partial x_i}\right) = 0, \qquad (1)$$

where ε is the electric permittivity in the gap and $\phi = \phi(x,y,z)$ is the electric potential. Einstein's summation convention will be used in this document as it was used in this equation for $i = x, y, z$.

The static mechanical equilibrium equations, when assuming small deformations, can be written as [15]:

$$\frac{\partial}{\partial x_i}\sigma_{ji} + F_j = 0, \qquad (2)$$

where σ is the mechanical stress tensor and $\mathbf{F}$ the external force field which incorporates all applied forces. For simplicity it is assumed that the only applied forces are the Coulomb forces due to the electrostatic field. When there are no moments proportional to a volume, which is the case for most solid materials the stress tensor is symmetric:

$$\sigma_{ij} = \sigma_{ji}. \qquad (3)$$

To compute the displacement caused by an applied load a relation between stress σ and displacement $\mathbf{u}$ is required. This is done in two steps. First the strain tensor ε is determined as a function of $\mathbf{u}$ and then the stress as a function of the strain. These relations are in principle non linear but once it is assumed that the stresses and strains remain relatively small the relations can be assumed to be linear[1], which yields the linear relations:

$$\sigma_{ij} = C_{ijkl}\varepsilon_{kl}, \qquad (4)$$

$$\varepsilon_{ij} = \tfrac{1}{2}\left(\frac{\partial u_j}{\partial x_i} + \frac{\partial u_i}{\partial x_j}\right), \qquad (5)$$

where C is Hooke's tensor which has constant coefficients.

Thus the electric problem is linear and the mechanical problem is assumed to be linear, however the coupled problem is non-linear. The non-linearity stems on one

---

[1] Note that when modeling thin MEMS the non-linear mechanical equations need to be considered. For simplicity linear mechanics is assumed but all the concepts discussed in this text can be extended to non-linear mechanics

hand from the fact that the electrostatic domain is defined by the deformed position of the electrodes, and on the other hand from the fact that the electrostatic forces depend on the accumulated charges on the surfaces of the electrodes. There the distributed force follows from Coulomb's law [18]:

$$F_j = qE_j, \qquad (6)$$

where q is the distributed charge on the surface and **E** the electric field generated by the other electrode. In the vicinity of charges on the surface of a conductor one has to note that half of the electric field is generated by those charges themselves and does therefore not contribute to the force. Hence, if **E** is the total electric field in the domain one must use $F_j = q(E_j/2)$ [18].

According to Gauss' law the surface charge q on a conductor can be written as a function of the electric field at the boundary and the unit normal vector **n** pointing towards the electric domain:

$$q = \varepsilon E_i n_i. \qquad (7)$$

The electric potential is defined such that the electric field is (minus) the gradient of an electric potential ϕ, therefore the force can be defined as a function of this potential [18]:

$$F_j = \tfrac{1}{2}\varepsilon \left(\frac{\partial \phi}{\partial x_i} \frac{\partial \phi}{\partial x_i} \right) n_j, \qquad (8)$$

where it has to be noted that the force is always pointing into the electric domain. Again we stress that the non-linear coupling is caused by the non-linear dependency of **F** on ϕ and the fact that the electric domain is shaped by the structural displacements.

1.2 A one-dimensional example

The consequences of this non-linear relation between structural displacements and electrostatic forces can be explained best by a simple one-dimensional model. Figure 4 gives such a very simple model of a micro actuator. In an air gap, the electrostatic problem is described by Laplace's equation, which reduces to the following equation in 1D:

$$\varepsilon \frac{\partial^2 \phi}{\partial \xi^2} = 0, \qquad (9)$$

where ϕ is the electrostatic potential, ξ is the spatial coordinate and ε the electric permittivity. The domain on which this differential equation has to be solved is $\xi = [0,x]$, x describing the position of the moving electrode. The boundary conditions are: $\phi = V$ at $\xi = x$ and $\phi = 0$ at $\xi = 0$. The simple solution to this boundary value problem is:

$$\phi = \frac{V}{x}\xi. \qquad (10)$$

Numerical Modeling of the Electromechanical Interaction in MEMS

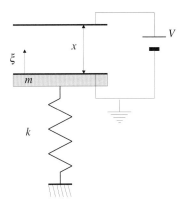

Fig. 4 1D micro resonator

The electric field is then by definition:

$$E = -\frac{\partial \phi}{\partial \xi} = -\frac{V}{x}, \qquad (11)$$

and the electrostatic force on the mass is:

$$F_{es} = \frac{1}{2}\varepsilon E^2 = \frac{1}{2}\varepsilon \frac{V^2}{x^2}. \qquad (12)$$

The mechanical force applied to the electrode arising from the spring is

$$F_{mech} = k(x - x_0), \qquad (13)$$

with k the linear stiffness of the spring and x_0 the position when the spring is at its natural length. The static equilibrium is found by summing all applied forces:

$$k(x - x_0) + \frac{1}{2}\varepsilon \frac{V^2}{x^2} = 0, \qquad (14)$$

Clearly, to a given applied potential V corresponds an equilibrium position x and vice-versa. It can be shown that there exists a voltage and corresponding displacement for which $\frac{\partial V}{\partial x} = 0$, thus for which the voltage as function of displacement reaches a maximum. This voltage equals: $V_{PI} = \sqrt{\frac{8kx_0^3}{27\varepsilon}}$. Using this maximum voltage to normalize equation (14) it can be written as:

$$\left(\frac{V}{V_{PI}}\right)^2 = \frac{27}{4}\left(\frac{x_0 - x}{x_0}\right)\left[\left(\frac{x_0 - x}{x_0}\right) - 1\right]^2. \qquad (15)$$

From the two solutions obtained for this equation it can be shown that the solution closest to the bottom conductor is unstable: at this position a small perturbation of

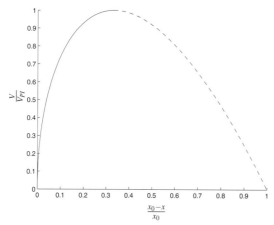

Fig. 5 1D voltage-displacement curve

the moving electrode will make it either return to the equilibrium position close to the undeformed position or smash against the fixed electrode (the electrodes are pulled together). The other solution is stable and can be used to perform a linearized vibrational analysis. There exists a voltage above which there exist no more static solutions, this voltage (equal to V_{PI} mentioned above) is called the *pull-in voltage*. This is graphically indicated in fig. 5. The pull-in voltage is the maximum of the curve. The solid part of the curve gives stable solutions, the dashed part being unstable solutions.

The result in fig. 5 also provides an indication about the real behavior of electrostatic microsystems. As mentioned the pull-in voltage is the maximum of the curve. This means that when a voltage above this pull-in voltage is applied to the system, the moving electrode will snap to the fixed electrode and they will stick together, possibly short-circuiting the device. Since this pull-in phenomenon is a very critical mode of operation, an important goal of MEMS modeling is the prediction of reliable pull-in voltages and displacements [8, 12].

An important characteristic of pull-in is its similarity to limit point buckling. Limit point buckling is defined by a maximum point of the applied force in the force-displacement relationship for a non-linear mechanical problem [6, 41]. A simple version of limit point buckling is presented in fig. 6. At this buckling point the derivative of the applied load with the generalized coordinate is zero:

$$\frac{\partial P}{\partial \alpha} = 0, \qquad (16)$$

which is very similar to the definition of the pull-in point:

$$\frac{\partial V}{\partial x} = 0. \qquad (17)$$

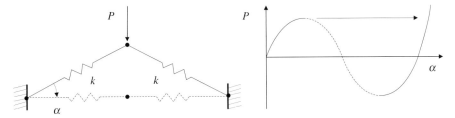

Fig. 6 Limit point buckling of a two spring model actuated by force P [41]

There exist many numerical techniques designed to handle the specific problems of limit point buckling. Several of those techniques can be adapted for pull-in computations as explained in the following sections.

1.3 Numerical modeling

The coupled equilibrium equations (1) and (2) are partial differential equations. To solve these equations numerically they have to be discretized. Different discretization techniques can be applied, but the mechanical equations (2) are almost always handled with the Finite Element Method (FEM) for which abundant literature exists (e.g. [4, 10]). The electrostatic problem (1) is a Laplace equation which also describes other physical fields such as thermics, potential fluid flow or membrane displacements. Therefore scientists from several application fields have investigated the Laplace equation. Structural engineers prefer to use the Finite Element Method for solving this equation, whereas engineers from the fluid mechanics field typically apply the Finite Volume Method (FVM) [42, 14] or Finite Difference schemes. The Boundary Element Method (BEM) is also quite popular because it handles infinitely extending domains in a natural manner [26].

Once such a discretization procedure has been chosen the discretized equations have the following form:

$$\mathsf{K}^m_{uu}\mathbf{u} = \mathbf{f}^{ext}(\mathbf{u},\mathbf{v}), \tag{18}$$

$$\mathsf{K}_{vv}(\mathbf{u})\mathbf{v} = \mathbf{q}(\mathbf{u},V), \tag{19}$$

where K_{vv} is the electric stiffness matrix, K^m_{uu} the mechanical stiffness matrix, $\mathbf{v}$ the set of discretized potentials (nodal potentials for a FEM model), $\mathbf{u}$ the set of discretized displacements, $\mathbf{q}$ are applied charges (on the boundaries or in the electric domain) and V the imposed potentials on Dirichlet boundaries. $\mathbf{f}^{ext}$ are the applied mechanical forces that include Coulomb forces obtained from eq. (8) or, alternatively, by formulations based on variational calculus and total energy approaches. The latter methods to construct the electrostatic forces are more efficient when the structure exhibits sharp corners [37, 21]. From the discretized equations (18) it can be clearly seen that the non-linear electrostatic coupling originates on one hand from

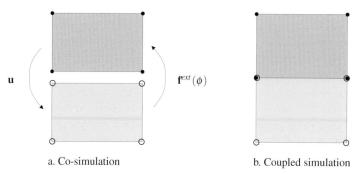

a. Co-simulation   b. Coupled simulation

Fig. 7 Modeling strategies

the fact that the electrostatic forces applied on the structure are non-linear functions of the potential field, and on the other by the fact that the electrostatic domain (hence the operator K_{vv}) depends on the structural response $\mathbf{u}$.

Defining the internal forces and internal charges as

$$\mathbf{f}^{int} = K_{uu}^m \mathbf{u}, \qquad (20)$$
$$\mathbf{q}^{int} = K_{vv}(\mathbf{u})\mathbf{v}, \qquad (21)$$

the coupled equilibrium equations can be written in the following monolithic form [37]:

$$\mathbf{F}^{ext}(\mathbf{U},V) - \mathbf{F}^{int}(\mathbf{U}) = 0, \qquad (22)$$

where

$$\mathbf{U} = \begin{bmatrix} \mathbf{u} \\ \mathbf{v} \end{bmatrix}, \qquad (23)$$

$$\mathbf{F} = \begin{bmatrix} \mathbf{f} \\ \mathbf{q} \end{bmatrix}, \qquad (24)$$

There are two ways to obtain these sets of equilibrium equations: *co-simulation* or *fully coupled simulation*.

In *co-simulation* separate structural and electric models are handled by separate codes to solve the different physical domains, and the results are communicated between the codes at some synchronization moments in the solution procedure: mechanical displacements are sent to the electric model to update its geometry and electric potentials or fields are sent to the structural model to generate electrostatic forces. This is schematically depicted figure 7a. An advantage of co-simulation is that different discretization procedures can be used for the different domains and established specialized software can be used for each physical domain. For instance a structural FEM model and a separate electrostatic FEM model can be co-simulated [24, 45], but it is also common in the literature to combine a structural FEM model with electrostatic BEM approach [2, 32, 17]. Structural FEM models coupled to electrostatic FVM models are rare, which is surprising because structural FEM models coupled with fluid FVM models are very common [13], even structural FVM with fluid FVM does exist [39]. Since FVM handles singularities at corners better than FEM, coupling FEM and FVM in electrostatics is probably an interesting aspect for future research.

In *fully coupled simulation* the entire problem is handled in a single code: only one mesh is used to model both domains as schematically explained in figure 7b. In that case all the properties of the multiphysical problem are known in a single code, which greatly simplifies analyzing global properties such as eigenfrequencies for instance (see section 4). An advantage of the coupled simulation is that it is much easier to obtain a linearization of the coupled equations, which enables the creation of very efficient coupled solvers [37].

2 Solution techniques

Traditionally non-linear solvers rely on successive linear iterations to find a solution to (22). In this chapter the distinction will be made between solvers that solve the coupled problem in a *staggered* manner and solvers that solve physics in *monolithic* way.

A staggered solver basically solves one physical domain at a time. Figure 8 presents a simple diagram for a staggered solver: at iteration k first the electrostatic force is computed with the electric solution from step $k-1$, next the new displacements are computed by the mechanical solver and sent to the electric solver, which then can compute the new electrostatic solution.

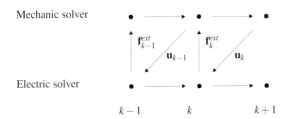

Fig. 8 Staggered solution procedure

This scheme can be interpreted from a mathematical point of view as a block Gauss-Seidel iteration and can be written as [3, 7]:

$$\mathbf{v}_k = \mathsf{K}_{vv}^{-1}(\mathbf{u}_{k-1})\mathbf{q}(\mathbf{u}_{k-1}, V), \qquad (25)$$

$$\mathbf{u}_k = (\mathsf{K}_{uu}^m)^{-1} \mathbf{f}^{ext}(\mathbf{u}_{k-1}, \mathbf{v}_k). \qquad (26)$$

This can be repeated until convergence. The algorithm is converged when the update for the displacements is smaller than a specified tolerance ($\|\mathbf{u}_k - \mathbf{u}_{k-1}\| < \text{tol}$).

It is clear that this algorithm can be implemented when the problem is handled by co-simulation, that is when the structural and electrostatic equations are solved in separate codes. But obviously the staggered approach can also be used when the discretization is done in a fully coupled simulation. An advantage of a staggered solver is that the structural and electrostatic problems taken separately are, for many applications, linear. Also, for each subproblem, efficient specialized solvers can be used.

Monolithic solvers attempt to solve the fully coupled equations as a whole and lead to specific solution procedures [7] (see figure 9). Clearly, monolithic solvers need to have access to the fully coupled model and are not appropriate for co-simulation strategies. So far mainly FEM-FEM coupling can be found in the literature [19, 37]. The advantage is that the convergence rate tends to be faster but, the full equilibrium equation being non-linear, the monolithic approaches are typically using successive linearized problems and thus often require computing the global tangent stiffness matrix $\mathsf{K} = \frac{\partial \mathbf{F}}{\partial \mathbf{U}}$.

Fig. 9 Monolithic solution procedure

Using equations (18) and (19) the monolithic linearized problem can be written as

$$\begin{bmatrix} \mathsf{K}_{uu}(\mathbf{u},\mathbf{v}) & \mathsf{K}_{uv}(\mathbf{u},\mathbf{v}) \\ \mathsf{K}_{uv}^T(\mathbf{u},\mathbf{v}) & \mathsf{K}_{vv}(\mathbf{u}) \end{bmatrix} \begin{bmatrix} \Delta \mathbf{u} \\ \Delta \mathbf{v} \end{bmatrix} = \begin{bmatrix} \Delta \mathbf{f}(\mathbf{u},\mathbf{v}) \\ \Delta \mathbf{q}(\mathbf{u},V) \end{bmatrix}, \qquad (27)$$

where

$$\mathsf{K}_{uu} = \mathsf{K}_{uu}^m + \frac{\partial \mathbf{f}^{ext}(\mathbf{u},\mathbf{v})}{\partial \mathbf{u}}, \qquad (28)$$

$$\mathsf{K}_{uv} = \frac{\partial \mathbf{f}^{ext}(\mathbf{u},\mathbf{v})}{\partial \mathbf{v}}. \qquad (29)$$

It can be shown that the tangent matrix is symmetrical, namely that $\mathsf{K}_{vu} = \mathsf{K}_{uv}^T$ [37]. Obtaining this tangent stiffness matrix from for instance the FEM discretization can be intricate [19, 36, 37] and solving the linearized problem (27) is time consuming. Without going in further specific details it can be seen that all four sub-matrices of K depend on the state vector $\mathbf{U}$ and thus need to be updated regularly in the solution process. Hence generally speaking one can state that monolithic approaches exhibit good convergence rates and are robust for badly conditioned problems, but every iteration incurs significant computational effort.

The first monolithic approach one can think of is the basic Newton-Raphson algorithm. At each step k the monolithic problem is linearized around the estimate $\mathbf{U}_{k-1}$ and a correction $\Delta\mathbf{U}_k$ is computed. Starting from the force imbalance for estimate $\mathbf{U}_{k-1}$

$$\Delta\mathbf{F}_k = \mathbf{F}^{ext}(\mathbf{U}_{k-1},V) - \mathbf{F}^{int}(\mathbf{U}_{k-1}), \qquad (30)$$

an update for the unknowns is computed using the linearized equilibrium equations

$$\Delta\mathbf{U}_k = \mathbf{K}^{-1}\Delta\mathbf{F}_k, \qquad (31)$$

and finally the solution is updated:

$$\mathbf{U}_k = \mathbf{U}_{k-1} + \Delta\mathbf{U}_k. \qquad (32)$$

The global tangent stiffness being updated every step, the Newton-Raphson update is optimal in the sense that quadratic convergence is guaranteed in the vicinity of the solution. For the Gauss-Seidel iterations (i.e. the staggered solution technique) described by (25, 26) the variation of the coupling effects is not accounted for: only the external forces and the pure mechanical and electrostatic operators are updated. For that reason the convergence of staggered schemes is relatively slow. As mentioned, the faster convergence of the Newton-Raphson method comes at the price of updating the stiffness. Therefore, an alternative procedure can be used where the stiffness matrix is approximated or updated only for some iteration steps. These methods are known as the modified Newton and quasi-Newton-Raphson method [10]. For some type of problems these methods are overall computationally more efficient even though the convergence is slower.

3 Finding the pull-in curve

So far only the method for solving the displacement at a single applied voltage V has been discussed. But for a proper characterization of the pull-in point a set of successive solutions has to be found defining the pull-in curve. Therefore it seems very logical to use so-called incremental-iterative procedures. In this section we will shortly discuss the different approaches that can be used to find the pull-in curve.

3.1 Voltage stepping

The simplest incremental-iterative method is the voltage iterative method: the solution is computed at a certain load V_i by an iterative procedure (staggered or monolithic, see previous section). The load is incremented by a value ΔV: $V_{i+1} = (V_i + \Delta V)$. At this new voltage the solution is computed and next V is incremented again, until a part of the pull-in curve has been determined. The problem with simple

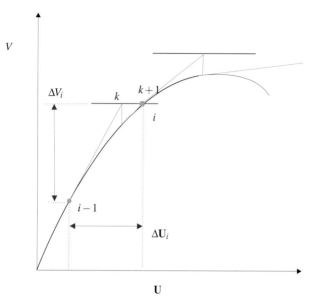

Fig. 10 Several voltage iterations

voltage stepping is illustrated in figure 10: for a voltage above the pull-in limit no solution can be found and thus the iterative solution techniques will not converge.

The voltage stepping method is conceptually similar to the load stepping method, classically applied in non-linear buckling analysis of structures [4]. And also for buckling analysis, the Newton-Raphson scheme is not well suited to handle limit point buckling [28]. Limit point buckling was defined by a maximum point of the applied force in the force-displacement relationship, at which the mechanical tangent stiffness matrix has a zero determinant [6, 41]. This phenomenon is very similar to pull-in where the voltage-displacement relationship has a maximum and the fully coupled tangent stiffness matrix has a zero determinant [37]. Therefore the block Gauss-Seidel and Newton-Raphson algorithms will fail near the pull-in point as well. Moreover they cannot find the solutions in the unstable part of the pull-in curve.

3.2 Path following methods

To overcome the shortcomings of the strategy described above where stepping is performed on the voltage and the Newton-Raphson or Gauss-Seidel iterations used to find the corresponding response, one can combine the stepping and the solution iterations in a unified iterative solution procedure where one constrains the norm of the increment in the hyperspace that includes the driving parameter (the voltage in this case) and the response. These techniques were developed for buckling

Numerical Modeling of the Electromechanical Interaction in MEMS

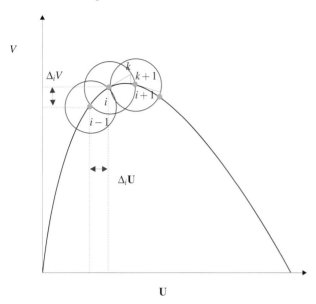

Fig. 11 Several path following iterations

problems and are known as *path-following methods* [28]. Since buckling is a purely mechanical problem, these algorithms were developed as being inherently monolithic. However the concept of path-following can also be used to develop staggered strategies [20].

Path-following methods go back to the early papers by E. Riks [35] and M.A. Crisfield [9]. The basic idea consists in adding to the set of equilibrium equations a constraint on norm of the increment for $\mathbf{U}$ and V. The circular version of such a constraint is shown in figure 11. The algorithm with a circular constraint is known as the Riks-Crisfield algorithm (or the arc-length method).

To apply such a circular constraint to find the next point $i+1$ on the pull-in curve (see figure 11) a separation is made between the total change ($\Delta_i V_k$, $\Delta_i \mathbf{U}_k$), and the update of the increments during the iterative process ($dV_{k+1}, d\mathbf{U}_{k+1}$) such that the new estimate for the increments is written as

$$\Delta_i \mathbf{U}_{k+1} = \Delta_i \mathbf{U}_k + d\mathbf{U}_{k+1}, \tag{33}$$

$$\Delta_i V_{k+1} = \Delta_i V_k + dV_{k+1} . \tag{34}$$

The iterations for the increment ($\Delta_i V, \Delta_i \mathbf{U}$) stop when the point satisfies all equilibrium equations. During an update of the increments the norm of allowed total displacement increment (structural dofs and potential dofs) augmented with the driving parameter increment is kept constant:

$$\Delta_i \mathbf{U}_{k+1}^T \Delta_i \mathbf{U}_{k+1} + \Delta_i V_{k+1}^2 = S^2 , \tag{35}$$

where S is the constraint on the norm of the increment. The equation can also be written as

$$\Delta_i \mathbf{U}_{k+1}^T \Delta_i \mathbf{U}_{k+1} + \Delta_i V_{k+1}^2 = \Delta_i \mathbf{U}_k^T \Delta_i \mathbf{U}_k + \Delta_i V_k^2 \, . \tag{36}$$

Substituting (33,34) in this last relation, one finds a constraint for the updates $(d\mathbf{U}_{k+1}, d\mathbf{U}_{k+1})$. These constraints together with the linearized equilibrium equations around the latest estimate define the equations for the next estimate. For more details on different methods using the path-following concept see e.g. [10].

In a block iterative procedure it proves to be very difficult to use the full circular constraint on all dofs because the full tangent stiffness is not known. But it is possible to assume that the problem is purely mechanical with an intricate external force function of the potential V and displacements $\mathbf{u}$:

$$\mathbf{f}^{ext} = \mathbf{f}^{ext}(\mathbf{u}, \mathbf{v}(V)) \, . \tag{37}$$

In that case it is possible to write a constraint equation on the mechanical displacements ($\mathbf{u}$) and applied external load (V):

$$\Delta_i \mathbf{u}_{k+1}^T \Delta_i \mathbf{u}_{k+1} + \Delta_i V_{k+1}^2 = \Delta_i \mathbf{u}_k^T \Delta_i \mathbf{u}_k + \Delta_i V_k^2 \, . \tag{38}$$

Results obtained with an algorithm that used a linearized version of constraint (38) were presented in [20].

3.3 Displacement stepping

Another way to construct the pull-in curve is to take as driving parameter a displacement of the structure: a displacement dof is prescribed and an iterative solver is used to compute the applied voltage needed to satisfy equilibrium. This approach is schematically illustrated in figure 12.

However for problems with more than one dof this approach is not trivial since the structural displacements (except the one chosen as driving parameter) and the electric potentials need to be found. Therefore a relaxation scheme called DIPIE has been proposed in the literature [5]. At a given iteration of the algorithm, an estimate $\mathbf{u}_k$ is known for the structural displacement. The algorithm uses the quadratic dependency of $\mathbf{f}^{ext}$ on the applied potential V in the 1D problem and assumes that this is still a good assumption in a more complex geometry:

$$\mathbf{f}^{ext}(V) = \left(\mathbf{f}^{ext}(\mathbf{u}_k, V=1)\right) V^2 \, , \tag{39}$$

At equilibrium one must have for the structure

$$\mathbf{f}^{ext} = \mathbf{f}^{int} \, , \tag{40}$$

where:

$$\mathbf{f}^{int} = \mathsf{K}_{uu}^m \mathbf{u} \, . \tag{41}$$

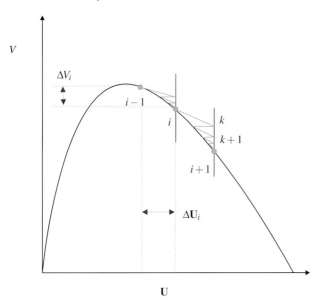

Fig. 12 Several displacement iterations

So an estimate of the voltage needed to satisfy equilibrium in the structure can be constructed from an averaged force ratio

$$V^2 = \frac{1}{n}\sum_{e=1}^{n}\frac{f_e^{int}(\mathbf{u}_k)}{f_e^{ext}(\mathbf{u}_k, V=1)}, \qquad (42)$$

where e indicates those n dofs that experience the electric forces. Once this guess is made an update of the mechanical force can be made:

$$\mathbf{f}^{int}(\mathbf{u}_{k+1}) = V^2 \mathbf{f}^{ext}(\mathbf{u}_k, V=1), \qquad (43)$$

which can be used to update the displacement. This can be repeated until convergence.

3.4 Charge stepping

Another approach to travel along the pull-in curve is inspired by some strategies used to control the motion of MEMS in practice [33, 38]: prescribe the total electric charge instead of the voltage on the moving conductor. This idea is explained in [20]. The voltage can be determined in a post processing step. The advantage of this method is illustrated on the 1D example problem presented in sect. 1.2 and shown in fig. 4.

Coulomb's law for the force on a point charge is:

$$\mathbf{F}_{es} = \frac{1}{2} Q \mathbf{E} , \qquad (44)$$

with $\mathbf{E}$ the field outside the conductor.

Writing Gauss' law for a control volume crossed by a unit conductor surface one finds

$$\int_{surface} \mathbf{E} dA = \frac{1}{\varepsilon} Q , \qquad (45)$$

which means that in 1D

$$E = \frac{1}{\varepsilon} Q . \qquad (46)$$

Thus the force becomes equal to:

$$F_{es} = \frac{1}{2\varepsilon} Q^2 . \qquad (47)$$

Direct substitution of equation (46) into the force function, equation (12), gives the same result. Therefore it is possible to consider the charge on the conductor as the applied electric load and as driving parameter. Thus using (47) to write the mechanical equilibrium equation one finds

$$k(x - x_0) + \frac{1}{2\varepsilon} Q^2 = 0 , \qquad (48)$$

which can be written as

$$Q = \sqrt{2\varepsilon k(x_0 - x)} . \qquad (49)$$

The charge at a zero gap ($x = 0$) is:

$$Q(x = 0) = \sqrt{2\varepsilon k x_0} . \qquad (50)$$

Thus the normalized charge becomes:

$$\frac{Q}{Q(x=0)} = \sqrt{\frac{x_0 - x}{x_0}} , \qquad (51)$$

which is a monotonously increasing function as can be seen in fig. 13. Hence unlike the voltage-displacement curve it does not have an extremum and moreover it shows less non-linear behavior. It will be easier to follow the charge-displacement curve and extract from it the voltage-displacement curve and pull-in characteristics.

For more complicated structures it is not possible to simply apply the charge on the moving structure, because the charge distribution over the boundary of the structure depends on the shape of the electric domain. However due to the assumption that electrodes are perfect conductors it is known that the potential over the conductor is constant, which effectively defines an equipotential constraint. Thus the charge can be applied on one dof of the conductor boundary only in combina-

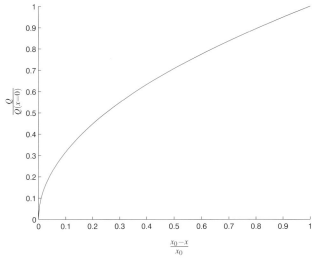

Fig. 13 1D charge-displacement curve

tion with the equipotential constraint. This equipotential constraint can be implemented using several methods, for instance with Lagrange multipliers. The equilibrium equations when the charge is imposed can again be solved using staggered or monolithic approaches (see previous section). So far only a block Gauss-Seidel approach for charge loading can be found in the literature [20].

3.5 General remarks

An overview of all the different methods to compute the pull-in curve that have been presented in literature and discussed here is shown in figure 14.

An aspect of stepping that is common to all the algorithms is the determination of the step size. Once a converged solution for an increment has been found one has to determine the step to use for the next increment. A popular, heuristically derived algorithm has been proposed by Ramm [28], where the new step size ΔV_{i+1} (or Δu_{i+1}, ΔQ_{i+1}, S_{i+1} for displacement stepping, charge stepping or path-following respectively) is:

$$\Delta V_{i+1} = \Delta V_i \sqrt{\frac{k_{desired}}{k_{realized}}}, \qquad (52)$$

where $k_{desired}$ is the desired number of steps per iteration loop and $k_{realized}$ the number of steps needed in the last converged solution step i.

A final remark on pull-in detection. Pull-in is characterized by a singular stiffness matrix K. This can be detected by checking if $\det(K) = 0$. Detection is very simple because before pull-in $\det(K) > 0$ but once pull-in is past and the unstable solutions

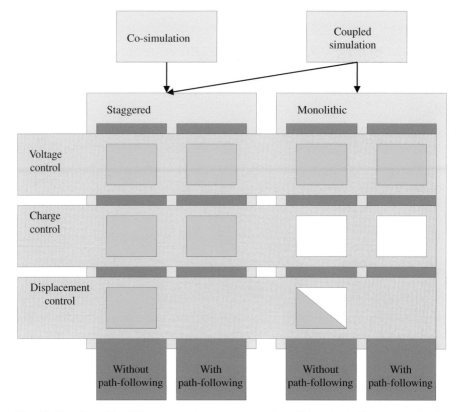

Fig. 14 Overview of the different procedures to compute the pull-in point (white squares give the theoretically possible methods, green squares give methods realized in literature so far, the green and white square gives a method which is explained in literature but never truly implemented in literature)

are computed $\det(K) < 0$. If one includes such a test after each converged iteration, the methods can be easily enhanced with a bisection algorithm that precisely computes the pull-in point. In principle this is only possible if the full coupled stiffness matrix is available, therefore only for the coupled simulation approach. Although hybrid methods are conceivable where a staggered method is used to compute the equilibrium position, after which the monolithic tangent stiffness is computed at equilibrium only for evaluation of $\det(K)$.

3.6 Example

A small comparison between the approaches is done with the model of a simple beam. The geometry of the beam and the air gap below it is shown in fig. 15. The finite element method is used for discretization of the problem. Both domains are

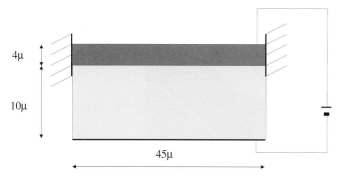

Fig. 15 Linear beam

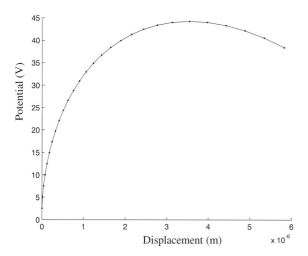

Fig. 16 Voltage displacement curve

meshed with quadrilaterals. The structural beam is modeled with 2D solid elements under plane strain assumption and with linear shape functions to approximate the displacement field [4]. The Young's modulus is $1 \times 10^5\,GPa$ and the Poisson's ratio is 0.3. The air gap is modeled with standard electrostatic elements with linear shape functions. The vacuum electric permittivity of $8.85 \times 10^{-12}\,F/m$ is used to characterize the electric domain.

The geometry of the air gap depends on the deformation of the beam, therefore the mesh of the gap has to be updated or deformed. Here the displacements of the nodes of the electric problem are determined by a pseudo-structure model: an elastic stiffness matrix is assembled on the mesh of the electric problem and the deformations of the beam are applied to the pseudo structure as imposed displacements. These displacements are used to update the coordinates of the electrostatic mesh, influencing the electric stiffness matrix.

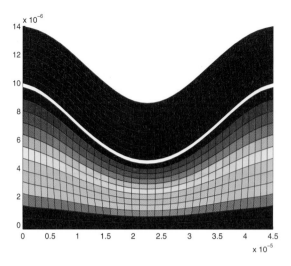

Fig. 17 Deformed shape of the simple beam

Figure 16 presents the voltage displacement curve for this beam up to a displacement of 5.5μm, which is 55% of the initial air gap. The absolute displacement of the middle node is shown. The deformed geometry of the beam and air gap at this displacement is shown in fig. 17. In this figure the beam is the part above the white division line and the gap is the part below the line. The beam has a uniform color and the colors in the gap indicate the potential distribution, the maximum potential near the beam and the minimum near the grounded electrode at the bottom.

The voltage-displacement curve in fig. 16 was computed with the charge loaded block-iterative procedure, but the same results were also computed with the other approaches. Computation times are listed in table 1. The computations were performed in Matlab on a simple desktop computer (3 GHz, 2 GB RAM). The first thing that can be noted is that the relative error tolerance on the residual of the structural force for the DIPIE method is less strict than that for the other methods: this is due to the fact that it was not possible to make it converge to a relative error below 5×10^{-2}. The fastest of the methods that actually makes a reliable estimate of pull-in is the charge loaded block Gauss-Seidel. Also noteworthy is the fact that monolithic path-following with voltage control is faster than the block-iterative version, but not faster than using charge control.

The second comparison between the approaches is presented in fig. 18, which shows the decrease of the relative error during an iteration procedure. These error curves were determined at an increment near the pull-in point. The reason of the fast computation time for the charge algorithm in table 1 can be easily explained because its error curve has the steepest descent of the staggered methods. Only the monolithic arc-length method converges faster, but it requires so much more computation time per iteration step that the overall computation time is higher than the charge loaded method.

Table 1 Computation times.

Method	Computation time (s)	Error tolerance (relative)
Voltage block GS with path following	277	10^{-3}
DIPIE	296	5×10^{-2}
Charge block Gauss-Seidel	76	10^{-3}
Voltage monolithic path following	181	10^{-3}

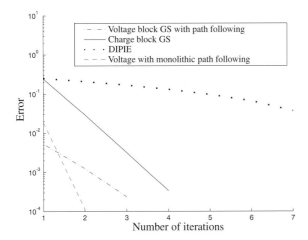

Fig. 18 Error curves for the simple beam

4 Coupled eigenfrequencies

One of the earliest applications of electromechanical MEMS is the microresonator [40]. An advantage of these electromechanical resonators is that the eigenfrequency depends on the bias voltage that is applied [29, 1, 11]. This can simply be explained with the 1D example from section 1.2. In that section static equilibrium was defined. The dynamic equation of motion can be written as [37]:

$$m\ddot{x} + k(x - x_0) + \frac{1}{2}\varepsilon \frac{V^2}{x^2} = 0. \quad (53)$$

It is assumed that the voltage is fixed and the mass moves only with small displacements around the static equilibrium equilibrium position $x = x_{eq} + \Delta x$, and thus $\ddot{x} = \Delta\ddot{x}$. Equation (53) is linearized as follows:

$$m\Delta\ddot{x} + \frac{\partial}{\partial x}\left(k(x - x_0) + \frac{1}{2}\varepsilon \frac{V^2}{x^2}\right)\bigg|_{x=x_{eq}} \Delta x = 0, \quad (54)$$

$$m\Delta\ddot{x} + k\Delta x - \varepsilon \frac{V^2}{x_{eq}^3}\Delta x = 0 , \tag{55}$$

The eigenfrequency of this linear differential equation is then by definition:

$$\omega = \sqrt{\frac{k}{m} - \varepsilon \frac{V^2}{mx_{eq}^3}} , \tag{56}$$

which shows that the eigenfrequency can be changed by giving the device a bias voltage.

4.1 Staggered method

For the mechanical system the eigenfrequency of a multiple dof system is well defined [16]. Once the inertia terms are included in the mechanical equilibrium equation (2) this results in:

$$\frac{\partial}{\partial x_i}\sigma_{ji} + F_j - \rho\ddot{u}_j = 0 , \tag{57}$$

where $\ddot{u}_j$ is the acceleration in direction j and ρ is the density of the material. After discretization one obtains the classical equation of motion for a system of multiple dofs:

$$\mathbf{M}_{uu}\ddot{\mathbf{u}} + \mathbf{K}_{uu}^m \mathbf{u} = \mathbf{f}^{ext} , \tag{58}$$

where $\mathbf{M}_{uu}$ is the mass matrix. The linear eigenfrequencies ω_i and eigenmodes $\mathbf{x}_i$ are then defined as the solutions of the eigenvalue problem:

$$-\omega_i^2 \mathbf{M}_{uu}\mathbf{x}_i + \mathbf{K}_{uu}^m \mathbf{x}_i = 0 . \tag{59}$$

Of these eigenfrequencies only the lowest few are of practical interest. For instance a tunable microresonator normally operates at it lowest eigenfrequency.

However for an electromechanically coupled problem the external force $\mathbf{f}^{ext}$ depends on $\mathbf{u}$:

$$\mathbf{M}_{uu}\ddot{\mathbf{u}} + \mathbf{K}_{uu}^m \mathbf{u} = \mathbf{f}^{ext}(\mathbf{u}, V) . \tag{60}$$

In a staggered model, this effect is not incorporated in the mechanical stiffness matrix, therefore not in the eigenfrequencies. In the literature a method exists to compute the influence of this dependency on the eigenfrequency [30]. The idea starts with projection of the non-linear equation on the lowest mechanical eigenmode. In that case it is assumed that the displacement vector around an equilibrium can be approximated by $\mathbf{u} = \mathbf{u}_{eq} + \mathbf{x}_1 \Delta u$. When the non linear equation of motion is projected on the first eigenmode one finds

$$\mathbf{x}_1^T \mathbf{M}_{uu}\mathbf{x}_1 \Delta\ddot{u} + \mathbf{x}_1^T \mathbf{K}_{uu}^m \mathbf{x}_1 \Delta u + \mathbf{x}_1^T \mathbf{K}_{uu}^m \mathbf{u}_{eq} = \mathbf{x}_1^T \mathbf{f}^{ext}(\mathbf{u}_{eq} + \mathbf{x}_1 \Delta u, V) , \tag{61}$$

Since at equilibrium

$$\mathbf{K}_{uu}^m \mathbf{u}_{eq} = \mathbf{f}^{ext}(\mathbf{u}_{eq}, V) , \tag{62}$$

one can also write

$$\mathbf{x}_1^T \mathbf{M}_{uu} \mathbf{x}_1 \Delta \ddot{u} + \mathbf{x}_1^T \mathbf{K}_{uu}^m \mathbf{x}_1 \Delta u = \mathbf{x}_1^T \left(\mathbf{f}^{ext}(\mathbf{u}_{eq} + \mathbf{x}_1 \Delta u, V) - \mathbf{f}^{ext}(\mathbf{u}_{eq}, V) \right) , \tag{63}$$

which reduces the set of equations to a single differential equation:

$$m_r \Delta \ddot{u} + k_r \Delta u = \Delta f_r(\mathbf{u}_{eq}, \mathbf{x}_1 \Delta u, V) , \tag{64}$$

where m_r is the modal mass, k_r the modal stiffness and f_r the modal participation of the force.

An estimate of the coupled eigenfrequency can then be found as:

$$\omega = \sqrt{\frac{k_r}{m_r} - \frac{1}{m_r}\frac{\partial \Delta f_r}{\partial \Delta u}} , \tag{65}$$

where the term $\frac{\partial \Delta f_r}{\partial \Delta u}$ can be obtained by a finite difference approximation. Of course this method only works properly if the electric forces do not change the modeshape too much.

4.2 Monolithic method

Monolithically the computation is much easier. To obtain the linearized equation of motion only inertia terms are added to the linearized equilibrium equations (27) [37]:

$$\mathbf{M} \begin{bmatrix} \Delta \ddot{u} \\ \Delta \ddot{v} \end{bmatrix} + \mathbf{K} \begin{bmatrix} \Delta u \\ \Delta v \end{bmatrix} = \begin{bmatrix} \Delta f \\ \Delta q \end{bmatrix} , \tag{66}$$

where M is the coupled inertia matrix. Since the electric domain was assumed to be electrostatic, there are no inertia terms for the potentials, hence:

$$\mathbf{M} = \begin{bmatrix} \mathbf{M}_{uu} & 0 \\ 0 & 0 \end{bmatrix} . \tag{67}$$

The full stiffness matrix, as defined by equation (27), is evaluated at the static equilibrium position. Therefore the eigenvalue problem becomes:

$$\omega_i^2 \begin{bmatrix} \mathbf{M}_{uu} & 0 \\ 0 & 0 \end{bmatrix} \begin{bmatrix} \mathbf{x}_i^u \\ \mathbf{x}_i^v \end{bmatrix} - \begin{bmatrix} \mathbf{K}_{uu} & \mathbf{K}_{uv} \\ \mathbf{K}_{uv}^T & \mathbf{K}_{vv} \end{bmatrix} \begin{bmatrix} \mathbf{x}_i^u \\ \mathbf{x}_i^v \end{bmatrix} = 0 , \tag{68}$$

which will give the correct eigenfrequencies and modeshapes since the full electromechanical coupling is accounted for. The only problem is that the inertia matrix is singular, hence there are infinitely large eigenfrequencies. However the static dofs

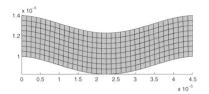 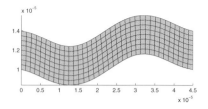

Fig. 19 First mechanical eigenmode of the beam   **Fig. 20** Second mechanical eigenmode of the beam

can be condensed out, in other words the electric dofs can be condensed out. The bottom line of eq. (68) defines the relation between $\mathbf{x}_i^v$ and $\mathbf{x}_i^u$. If this relation is substituted into the upper line of (68) it results in

$$\omega_i^2 \mathbf{M}_{uu} \mathbf{x}_i^u - \left(\mathbf{K}_{uu} - \mathbf{K}_{uv} \mathbf{K}_{vv}^{-1} \mathbf{K}_{uv}^T \right) \mathbf{x}_i^u = 0 , \qquad (69)$$

which defines an eigenvalue problem that can be solved exactly like the purely mechanical eigenvalue problem.

4.3 Example

These algorithms were tested on the model of a simple beam in fig. 15. The density used was $5 \times 10^3 \, kg/m^3$. The first and third purely mechanical eigenmodes of the beam are shown in figs. 19 and 20. After the computation of each static solution shown in fig. 16 the first and second eigenfrequency and eigenmode around this static solution were determined. The first eigenmode is very similar to the shape of the static deflection shown in fig. 17, thus the modeshape will not be changed very much by the electric field. Therefore it is expected that the staggered method will be accurate for the first eigenfrequency. The second mode shape is more complex. The electrostatic load might affect the modeshape, thus the eigenfrequency. This will affect the accuracy of the results obtained with the staggered method, but not with the monolithic method.

The curves for the first eigenfrequency shown in fig. 21 confirm that the staggered method computes the same curve as the monolithic method. It also shows that the frequency goes to zero at the pull-in point, similarly to what is found in limit point buckling. Both methods compute the same curve, however the curve obtained with the staggered method is less smooth. A disadvantage of the staggered method is that the accuracy depends heavily on the size of the generalized displacement step that is used for the finite difference approximation.

The curves for the second eigenfrequency are shown in fig. 22. As expected the curves are very similar, but there is a small discrepancy due to the geometrical complexity of the mode which induces more influence of the electrostatic forces on the

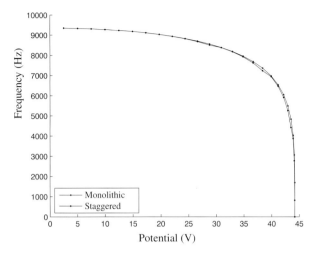

Fig. 21 First eigenfrequency as a function of applied potential

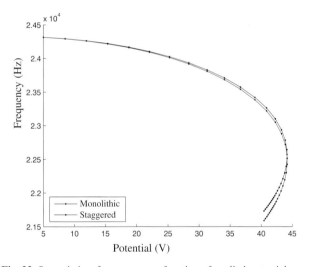

Fig. 22 Second eigenfrequency as a function of applied potential

mode shape. It is interesting to note that the second eigenvalue is not zero at pull-in which is the case for all the higher modes.

5 Summary and conclusions

We have presented an overview of important issues in the numerical modeling of electromechanical coupling as seen with MEMS. In particular the solution techniques have been classified in staggered and monolithic solvers. Several different staggered and monolithic algorithms for finding the static load displacement curve were outlined and illustrated. Finally the computation of linearized eigenfrequencies around the static equilibria was discussed.

From these discussions it can be concluded that staggered methods can be advantageous. Staggered schemes can easily couple different discretization procedures for different problems, such as the often seen combination of structural FEM with electric BEM. When the proper algorithm is used (charge loading) these algorithms can be faster than monolithic methods.

An advantage of monolithic methods is that knowledge of the full tangent stiffness matrix enables eigenvalue and eigenfrequency analysis. Definitely for the higher modes these eigenvalues can only be properly computed with monolithic methods. However these computations can be performed with hybrid methods, methods that compute the pull-in curve staggered, and assemble the full tangent stiffness matrix to do the eigenvalue analysis only when the solution of an increment has converged.

References

1. Adams, S., Bertsch, F., Shaw, K., Hartwell, P., Moon, F., MacDonald, N.: Capacitance based tunable resonators. Jounal of Micromechanics and Microengineering **8**, 15–23 (1998)
2. Aluru, N., White, J.: An efficient numerical technique for electrochemical simulation of complicated microelectromechanical structures. Sensors & Actuators: A. Physical **58**(1), 1–11 (1997)
3. Bailey, C., Taylor, G., Cross, M., Chow, P.: Discretisation procedures for multi-physics phenomena. Journal of Computational and Applied Mathematics **103**(1), 3–17 (1999)
4. Bathe, K.: Finite element procedures. Prentice Hall, Englewood Cliffs, NJ (1996)
5. Bochobza-Degani, O., Elata, D., Nemirovsky, Y.: An efficient DIPIE algorithm for CAD of electrostatically actuated MEMS devices. Microelectromechanical Systems, Journal of **11**(5), 612–620 (2002)
6. Budianski, B.: Theory of buckling and post-buckling behavior of elastic structures. In: C. Yih (ed.) Advances in Applied Mechanics, vol. 14, pp. 1–65. Academic Press (1974)
7. Cervera, M., Codina, R., Galindo, M.: On the computational efficiency and implementation of block-iterative algorithms for nonlinear coupled problems. Engineering Computations **13**(6), 4–30 (1996)
8. Cheng, J., Zhe, J., Wu, X.: Analytical and finite element model pull-in study of rigid and deformable electrostatic microactuators. Journal of Micromechanics and Microengineering **14**(1), 57–68 (2004)
9. Crisfield, M.: A fast incremental/iterative solution procedure that handles snap-through. Computers and Structures **13**(1-3), 55–62 (1981)
10. Crisfield, M.: Non-Linear Finite Element Analysis of Solids and Structures, Volume 1: Essentials. John Wiley & Sons, New York (1991)

11. Dai, C., Yu, W.: A micromachined tunable resonator fabricated by the CMOS post-process of etching silicon dioxide. Microsystem Technologies **12**(8), 766–772 (2006)
12. Degani, O., Socher, E., Lipson, A., Lejtner, T., Setter, D., Kaldor, S., Nemirovsky, Y.: Pull-in study of an electrostatic torsion microactuator. Microelectromechanical Systems, Journal of **7**(4), 373–379 (1998)
13. Farhat, C., Lesoinne, M., Maman, N.: Mixed explicit/implicit time integration of coupled aeroelastic problems: three-field formulation, geometric conservation and distributed solution. International Journal for Numerical Methods in Fluids **21**(10), 807–835 (1995)
14. Ferziger, J., Perić, M.: Computational Methods for Fluid Dynamics. Springer (2002)
15. Fung, Y.: Foundations of Solid Mechanics. Prentice-Hall, Englewood Cliffs, NJ (1968)
16. Géradin, M., Rixen, D.: Mechanical Vibrations. John Wiley & Sons, New York (1997)
17. Gilbert, J., Legtenberg, R., Senturia, S.: 3D coupled electro-mechanics for MEMS: applications of CoSolve-EM. Micro Electro Mechanical Systems, 1995, MEMS'95, Proceedings. IEEE
18. Griffiths, D.: Introduction to Electrodynamics. Prentice-Hall, Englewood Cliffs, NJ (1999)
19. Gyimesi, M., Avdeev, I., Ostergaard, D.: Finite-element simulation of microelectromechanical systems (MEMS) by strongly coupled electromechanical transducers. Magnetics, IEEE Transactions on **40**(2), 557–560 (2004)
20. Hannot, S., Rixen, D.: Determining pull-in curves with electromechanical FEM models. In L.J.Ernst et al. (eds.) Proceedings of the 9th International Conference on Thermal, Mechanical and Multiphysics Simulation and Experiments in Micro-Electronics and Micro-Systems, Freiburg, Germany, April 2008, pp. 528–535 (2008)
21. Hannot, S., Rixen, D., Rochus, V.: Rounding the corners in an electromechanical FEM model. In E. Onate et al. (Eds.) Proceedings of the Second International Conference on Computational Methods for Coupled Problems in Science and Engineering, Ibiza, Spain, May 2008 pp. 507–510 (2007)
22. Henneken, V., Tichem, M., Sarro, P.: In-package MEMS-based thermal actuators for micro-assembly. Journal of Micromechanics and Microengineering **16**(6), S107–S115 (2006)
23. Henneken, V., Tichem, M., Sarro, P.: Improved thermal U-beam actuators for micro-assembly. Sensors & Actuators: A. Physical **142**(1), 298–305 (2008)
24. Hung, E., Senturia, S.: Generating efficient dynamical models for microelectromechanical systems from a few finite-element simulation runs. Microelectromechanical Systems, Journal of **8**(3), 280–289 (1999)
25. Jackson, J.: Classical Electrodynamics. John Wiley & Sons (1975)
26. Katsikadelis, J.: Boundary Elements: Theory and Applications. Elsevier (2002)
27. Kolesar, E., Allen, P., Howard, J., Wilken, J., Boydston, N.: Thermally-actuated cantilever beam for achieving large in-plane mechanical deflections. Thin Solid Films **355**, 295–302 (1999)
28. Kouhia, R., Mikkola, M.: Some aspects of efficient path-following. Computers and Structures **72**(4-5), 509–524 (1999)
29. Lee, K., Cho, Y.: A triangular electrostatic comb array for micromechanical resonant frequency tuning. Sensors & Actuators: A. Physical **70**(1-2), 112–117 (1998)
30. Lee, W., Kwon, K., Kim, B., Cho, J., Youn, S.: Frequency-shifting analysis of electrostatically tunable micro-mechanical actuator. Journal of Modeling and Simulation of Microsystems **2**(1), 83–88 (2001)
31. Majumder, S., McGruer, N., Adams, G., Zavracky, P., Morrison, R., Krim, J.: Study of contacts in an electrostatically actuated microswitch. Sensors & Actuators: A. Physical **93**(1), 19–26 (2001)
32. Mukherjee, S., Bao, Z., Roman, M., Aubry, N.: Nonlinear mechanics of MEMS plates with a total Lagrangian approach. Computers & Structures **83**, 758–768 (2005)
33. Nadal-Guardia, R., Dehe, A., Aigner, R., Castaner, L.: Current drive methods to extend the range of travel of electrostatic microactuators beyond the voltage pull-in point. Microelectromechanical Systems, Journal of **11**(3), 255–263 (2002)
34. Nathanson, H., Newell, W., Wickstrom, R., Davis Jr, J.: The resonant gate transistor. Electron Devices, IEEE Transactions on **14**(3), 117 133 (1967)

35. Riks, E.: An incremental approach to the solution of snapping and buckling problems. International Journal of Solids and Structures **15**(7), 529–551 (1979)
36. Rochus, V.: Finite element modeling of strong electro-mechanical coupling in MEMS. Université de Liège, PhD Thesis, Belgium (2006)
37. Rochus, V., Rixen, D., Golinval, J.: Monolithic modelling of electro-mechanical coupling in micro-structures. International Journal for Numerical Methods in Engineering **65**(4), 461–493 (2006)
38. Seeger, J., Boser, B.: Charge control of parallel-plate, electrostatic actuators and the tip-in instability. Microelectromechanical Systems, Journal of **12**(5), 656–671 (2003)
39. Slone, A., Pericleous, K., Bailey, C., Cross, M.: Dynamic fluid–structure interaction using finite volume unstructured mesh procedures. Computers and Structures **80**(5-6), 371–390 (2002)
40. Tang, W., Nguyen, T., Howe, R.: Laterally driven polysilicon resonant microstructures. Sensors & Actuators: A. Physical **20**(1-2), 25–32 (1989)
41. Thompson, J., Hunt, G.: Elastic instability phenomena. John Wiley & Sons (1984)
42. Versteeg, H., Malalasekera, W.: An Introduction to Computational Fluid Dynamics: The Finite Volume Method. Longman Scientific & Technical, Harlow, Essex (1995)
43. Wautelet, M.: Scaling laws in the macro-, micro-and nanoworlds. European Journal of Physics **22**(6), 601–611 (2001)
44. Zhang, L., Zhao, Y.: Electromechanical model of RF MEMS switches. Microsystem Technologies **9**(6), 420–426 (2003)
45. Zhulin, V., Owen, S., Ostergaard, D.: Finite element based electrostatic-structural coupled analysis with automated mesh morphing. Proceedings of the International Conference on Modeling and Simulation of Microsystems MSM pp. 501–504 (2000)

Simulation of Progressive Failure in Composite Laminates

F.P. van der Meer and L.J. Sluys

Abstract Fiber reinforced polymers are materials with excellent mechanical properties and relatively much design freedom. However, because complex failure mechanisms originating from the microstructure of the material may occur, realistic simulation of the failure process is still a challenge. Two alternative models for the modeling of failure in composite laminates are presented. The first is a continuum damage model that is supposed to cover all ply failure mechanisms. A limitation of the continuum approach with respect to the modeling of matrix failure is illustrated. Therefore, a discontinuous model has been developed for matrix failure specifically.

1 Introduction

Fiber reinforced polymers are promising structural materials, because of their excellent strength-to-weight and stiffness-to-weight ratios. The materials are composed of fibers with high strength and longitudinal stiffness (e.g. glass or carbon) held together with a polymer matrix (e.g. epoxy). An advantage of these materials is that there is design freedom in choosing the fiber direction such that the stiffness and strength are highest in the most critical direction. To exploit this design freedom, proper predictive tools are needed. However, the development of reliable numerical models is a challenging task, because the failure process is complex, involving different processes such as splitting, fiber/matrix debonding, delamination, fiber breaking, and fiber pull out (see e.g. [14, 16, 35]).

This contribution is restricted to composite laminates with unidirectional plies, i.e. each layer of the laminate contains long straight fibers which are all oriented in a single direction, and the strength and stiffness of the laminate as a whole are con-

---

Faculty of Civil Engineering and Geosciences, Section of Structural Mechanics,
Delft University of Technology, P.O. Box 5048, 2600 GA Delft, The Netherlands
e-mail: f.p.vandermeer@tudelft.nl

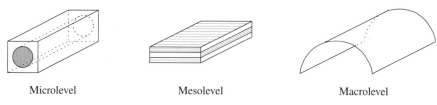

| Microlevel | Mesolevel | Macrolevel |

Fig. 1 Three scales levels of observation of laminated composites.

trolled by stacking layers with different fiber direction. In analyzing these laminates, three levels of observation may be distinguished (see Fig. 1):

- the microlevel, on which the fiber, the matrix and the fiber-matrix interface are explicitly present,
- the mesolevel, on which each ply is considered to be a homogeneous continuum, and
- the macrolevel, in which the laminate is considered as a plate with a single set of properties after through-thickness homogenization.

Fig. 2 Complex failure in a composite laminate (Green et al. [14], used with permission from Elsevier).

In Fig. 2, a complex failure mechanism is shown for a laminate with a circular hole loaded in uni-axial tension. It can be observed that there is much delamination, which allows the different plies with different fiber orientation to fail each in a different failure plane. In the $\pm 45°$-plies, splitting occurs, i.e. the failure plane is parallel to the fiber. Also in the $90°$-plies, the failure plane is located between the fibers. Only in the $0°$-plies, fibers break and an irregular failure pattern is observed.

While Fig. 2 is typical for the mesolevel approach, Fig. 3 shows a schematic representation on the microlevel of the processes that constitute ply failure under tension in fiber direction. Also on this level, there are different processes involved such as, matrix cracking, fiber/matrix debonding, fiber failure and fiber pull-out.

How to represent these different mechanisms in a single computational framework that is generic with respect to geometry, loading conditions and ply layup? That is the central question in the research presented here. Is it possible to build a model with which the complex laminate failure mechanisms can be simulated and

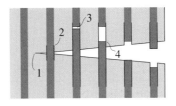

Fig. 3 Pull out process with different micromechanical failure mechanisms.

eventually predicted? If this can be done efficiently, the number of tests required to assess the safety of a composite structure may be reduced drastically [10], while the designer is given more freedom to exploit the structural capacity of the materials, and the manufacturers are provided with tools to optimize their products.

Modeling of laminate failure is often endeavored on the mesolevel, where a laminate is considered to be consisting of homogeneous plies, each with orthotropic properties that depend on the fiber orientation. Although initiation and propagation of failure inside a ply are typically described on the microlevel, the mesolevel approach is adopted because explicit modeling of the entire microstructure is computationally unaffordable. As a consequence, assumptions must be made on how to represent micromechanical failure in the homogenized model. Because the intact ply is modeled as a continuum, a relatively straightforward way to represent failure is with a continuum failure model, i.e. by introducing softening behavior in the homogenized constitutive law.

A popular continuum failure theory is the continuum damage model [25], in which the stiffness of the material is reduced gradually after a certain failure criterion has been violated. In the scope of orthotropic composite materials, several continuum damage formulations have been proposed [7, 12, 21–24, 26–28, 36, 37], with different degrees of complexity in failure criteria and degradation laws.

A well known problem with the application of continuum failure models is that a mesh dependence is introduced, due to ill-posedness of the mathematical problem. Localization of deformations will occur in one row of elements, which causes the amount of dissipated energy to depend on the element size, approaching zero for infinitesimally small elements. In several of the damage descriptions that have been proposed for composites [22, 27, 37], this is mitigated with the crack band method, in which the local stress strain behavior depends on the element size [2]. However, this does not solve the mesh sensitivity problem completely; element shape and orientation still influence the solution. Moreover, in the case of non-localized material degradation, which may occur in laminates as distributed transverse matrix cracking, this approach causes an opposite mesh dependency, i.e. a smaller element size causes an increase in the amount of energy th at is dissipated.

For the development of reliable predictive tools, more advanced localization limiters, which introduce an internal length scale in the model, are to be preferred, such as non-local [3] and gradient models [5]. With these models, the localization zone has a finite width, the mathematical problem is well-posed, and a truly mesh independent solution exists, which will be obtained under the condition that the typical element size in the localization region is smaller than this width. Of these, the implicit gradient model has been applied to anisotropic materials like composites by Germain et al. [12]. Another possibility is to resolve mesh dependence with artificial viscosity [42, 49]. Artificial viscosity is an appealing option, because of its easy implementation and limited extra computational costs, although it entails an unphysical rate dependence and, consequently, difficulties in model calibration and validation. Maimí et al. [27] and Lapczyk and Hurtado [22] already proposed to use artificial viscosity, but only to improve stability of the simulation. In recent work, we have presented a continuum damage model and a softening plasticity in both of which artificial viscosity ensures mesh objectivity of the results [29]. Unfortunately, the continuum models proved to be inadequate for the simulation of failure mechanisms such as illustrated in Fig. 2.

Besides continuum failure theories, there are discontinuous failure models. With these, failure is represented as a discontinuity in the displacement field instead of as localized strain, and a cohesive zone may be modeled by defining tractions that work on the crack surface as a function of the displacement jump. Examples of discontinuous failure methods are interface elements and the eXtended Finite Element Method (XFEM). Interface elements are special elements that are placed between regular continuum finite elements, and that allow the elements to separate after the traction between the elements exceeds a certain threshold. In the context of composite materials, interface elements are regularly applied for the modeling of delamination [6, 31, 41, 45, 53] and, more exceptionally, splitting [20, 52, 53] and fiber-matrix debonding [8, 13]. With XFEM (or the partition of unity method) a discontinuity is introduced inside the finite elements by adding degrees of freedom that are related to discontinuous basis functions. This method has also been used for the simulation of delamination [39, 50] and fiber-matrix debonding [38]. In comparison with continuum models, discontinuous methods demand additional implementation effort, especially XFEM, but they allow for the use of larger elements to capture the kinematics of the problem.

In this contribution, the performance of a continuum model for ply failure is assessed along with that of a discontinuous method for splitting. A limitation of continuum models for the modeling of matrix failure is illustrated, and the discontinuous method is presented as an alternative. Section 2 deals with the continuum damage model and Section 3 with the phantom node method, which is equivalent to the above mentioned XFEM, but differs in implementation.

2 Continuum damage

In this section, a continuum damage model for a homogenized unidirectional ply is presented. In recent years, several comparable models have been introduced (see e.g. [22, 23, 26–28, 37]). Attention is given to three essential ingredients in three subsections. Firstly, the material degradation laws, which form the core of the constitutive model, viz. the procedure with which stress is computed from strain. Secondly, the consistent tangent formulation, which is necessary for robustness of the Newton-Raphson procedure in implicit nonlinear finite element analysis. And thirdly, the regularization method, which is necessary to obtain mesh independent results. In subsections 2.4 and 2.5 the performance of the continuum damage model is assessed, with emphasis on its insufficiency for capturing matrix failure patterns.

2.1 Material degradation

In the continuum damage theory, material degradation is understood as the development of voids, defects or microcracks, which reduce the effective volume of the material. The effective stress $\hat{\sigma}$ is the stress acting on the intact material, computed with linear elasticity

$$\hat{\sigma} = \mathsf{D}^e \varepsilon, \tag{1}$$

in which D^e is the elastic stiffness matrix and ε is the strain in the bulk material.

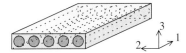

Fig. 4 Local coordinate system for unidirectional ply aligned with fiber direction.

The constitutive law which relates nominal stress σ to strain, using small strain theory, is written in the form

$$\varepsilon = \mathsf{C}\sigma, \tag{2}$$

where C is the secant compliance matrix, and ε and σ contain strain and stress values in the local coordinate frame of the ply (see Fig. 4). Following Matzenmiller et al. [28], damage is taken into account via modification of the values on the diagonal of the compliance matrix (in Voigt notation) as:

$$\mathsf{C} = \begin{bmatrix} \frac{1}{(1-\tilde{d}_\mathrm{f})E_1} & -\frac{v_{21}}{E_2} & -\frac{v_{21}}{E_2} & 0 & 0 & 0 \\ -\frac{v_{21}}{E_2} & \frac{1}{(1-\tilde{d}_\mathrm{m})E_2} & -\frac{v_{23}}{E_2} & 0 & 0 & 0 \\ -\frac{v_{21}}{E_2} & -\frac{v_{23}}{E_2} & \frac{1}{(1-\tilde{d}_\mathrm{m})E_2} & 0 & 0 & 0 \\ 0 & 0 & 0 & \frac{1}{(1-\tilde{d}_\mathrm{m})G_{23}} & 0 & 0 \\ 0 & 0 & 0 & 0 & \frac{1}{(1-\tilde{d}_\mathrm{m})G_{12}} & 0 \\ 0 & 0 & 0 & 0 & 0 & \frac{1}{(1-\tilde{d}_\mathrm{m})G_{12}} \end{bmatrix}, \tag{3}$$

where $\tilde{d}_\mathrm{f}$ and $\tilde{d}_\mathrm{m}$ are the damage variables related to fiber failure and matrix failure, respectively, E_1 and E_2 are the longitudinal and transverse Young's moduli, v_{21}

and v_{23} are the longitudinal and transverse Poisson ratios, and G_{12} and G_{23} are the longitudinal and transverse shear moduli, with $G_{23} = E_2/(2+2v_{23})$. The assumption that the components C_{22}–C_{66} evolve synchronously is debatable, but considered acceptable for the in-plane load cases presented in this paper.

Hashin's failure criteria [18, 19] are applied to evaluate the effective stress state. The criteria for different failure modes are the following:

- Tensile fiber mode:

$$f_{\text{ft}} = \frac{\hat{\sigma}_1}{F_{1t}}, \tag{4}$$

- Compressive fiber mode:

$$f_{\text{fc}} = -\frac{\hat{\sigma}_1}{F_{1c}}, \tag{5}$$

- Tensile matrix mode:

$$f_{\text{mt}} = \sqrt{\frac{(\hat{\sigma}_2+\hat{\sigma}_3)^2}{F_{2t}^2} + \frac{\hat{\tau}_{23}^2-\hat{\sigma}_2\hat{\sigma}_3}{F_{23}^2} + \frac{\hat{\tau}_{31}^2+\hat{\tau}_{12}^2}{F_{12}^2}}, \tag{6}$$

- Compressive matrix mode:

$$f_{\text{mc}} = \sqrt{\left[\left(\frac{F_{2c}}{2F_{23}}\right)^2 - 1\right]\frac{\hat{\sigma}_2+\hat{\sigma}_3}{F_{2c}} + \frac{(\hat{\sigma}_2+\hat{\sigma}_3)^2}{4F_{23}^2} + \frac{\hat{\tau}_{23}^2-\hat{\sigma}_2\hat{\sigma}_3}{F_{23}^2} + \frac{\hat{\tau}_{31}^2+\hat{\tau}_{12}^2}{F_{12}^2}}, \tag{7}$$

where F_{1t} and F_{1c} are the tensile and compressive strength in fiber direction, F_{2t} and F_{2c} are the tensile and compressive transverse strength, and F_{23} and F_{12} are the transverse and longitudinal shear strength. The numeric subscripts in the stress components $\sigma_1 \ldots \tau_{12}$ refer to the local coordinate system in the ply (see Figure 4).

With the failure criteria, the two loading functions ϕ_i, one corresponding to fiber failure and the other to matrix failure, are evaluated:

$$\phi_f = \begin{cases} f_{\text{ft}}, & \hat{\sigma}_1 \geq 0, \\ f_{\text{fc}}, & \hat{\sigma}_1 < 0, \end{cases} \tag{8}$$

$$\phi_m = \begin{cases} f_{\text{mt}}, & \hat{\sigma}_2+\hat{\sigma}_3 \geq 0, \\ f_{\text{mc}}, & \hat{\sigma}_2+\hat{\sigma}_3 < 0. \end{cases} \tag{9}$$

For both fiber and matrix failure, there is a single state variable r_i. For simplicity, there is no distinction between compressive and tensile failure regarding the influence on the state of the material. An artificial viscosity, limiting the rate of state variable r_i, is introduced for regularization, such that the evolution of r_i is defined as (cf. [1, 21, 54])

$$\dot{r}_i = B_i \left\langle 1 - \frac{1}{\phi_i - r_i} \right\rangle, \quad i = \text{f}, \text{m}, \tag{10}$$

where B_i is the maximum rate with which r_i is allowed to increase and the operator $\langle \cdot \rangle = \max\{\cdot, 0\}$ is used to ensure irreversibility of the damage process. In the rate independent limit $B_i \to \infty$, Eq. (10) reduces to $r_i = \sup(\phi_i - 1)$.

Damage variables $\tilde{d}_i$ are computed from r_i according to

$$\tilde{d}_i = \min\left\{1, \frac{A_i r_i}{(r_i + 1)(A_i - 1)}\right\}, \quad i = \mathrm{f}, \mathrm{m}, \tag{11}$$

which corresponds to a bilinear stress-strain relation for uni-axial tests without viscosity. Initially $\tilde{d}_i = r_i = 0$ and when $\tilde{d}_i = 1$ the material has failed in either fiber of matrix mode[1]. The softening parameter A_i is related to the strain level at which complete failure occurs in this bilinear relation (see Fig. 5), and hence to the fracture energy dissipated in a uni-axial test, which is generally considered a material constant. With these, the secant compliance matrix is computed (Eq. (3)), and inverted to obtain the secant stiffness matrix, which is in turn used to compute the nominal stress.

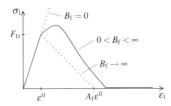

Fig. 5 Schematic representation of uni-axial stress-strain relation for continuum damage with artificial viscosity.

For postprocessing, a variable d_i is introduced, which is defined as

$$d_i = \min\left\{1, \frac{r_i}{A_i - 1}\right\}, \quad i = \mathrm{f}, \mathrm{m}. \tag{12}$$

Note that $\tilde{d}_i = 0 \Leftrightarrow d_i = 0$ and $\tilde{d}_i = 1 \Leftrightarrow d_i = 1$. For minor damage, d_i increases less rapidly than $\tilde{d}_i$.

2.2 Consistent tangent

For robustness of the incremental iterative finite element analysis, it is necessary to employ the consistent tangent stiffness matrix, the derivation of which is given below for the presented continuum damage model.

---

[1] To avoid singularity in the stiffness matrix, the maximum value of $\tilde{d}_i$ is set to $1 - 10^{-5}$ instead of 1.

Expanding the constitutive law, Eq. (2), around a small variation in stress and strain gives

$$\delta\varepsilon = \delta C \sigma + C \delta \sigma. \tag{13}$$

The secant compliance C depends on strain via the damage law, therefore the variation in the compliance may be written as

$$\delta C = \frac{\partial C}{\partial \varepsilon} \delta \varepsilon. \tag{14}$$

Because only the values on the diagonal of C depend on the strain, the first term of the right-hand side of (13) may be rewritten (in index notation) as

$$\begin{aligned}\delta C_{ij}\sigma_j &= \delta_{ij}\frac{\partial C_{ij}}{\partial \varepsilon_k}\delta\varepsilon_k \sigma_j \\ &= \sigma_i \frac{\partial C_{ii}}{\partial \varepsilon_k}\delta\varepsilon_k \\ &= M_{ik}\delta\varepsilon_k,\end{aligned} \tag{15}$$

where δ_{ij} is the Kronecker delta and

$$M_{ik} = \sigma_i \frac{\partial C_{ii}}{\partial \tilde{d}_p}\frac{\partial \tilde{d}_p}{\partial r_p}\frac{\partial r_p}{\partial \phi_p}\frac{\partial \phi_p}{\partial \hat{\sigma}_q}\frac{\partial \hat{\sigma}_q}{\partial \varepsilon_k}, \qquad p = \text{f}, \text{m}. \tag{16}$$

The artificial viscosity is accounted for in the evaluation of $\partial r_p / \partial \phi_p$.

Substitution of (15) into (13) and reordering gives

$$[I - M]\delta\varepsilon = C\delta\sigma. \tag{17}$$

Hence, the consistent tangent for the continuum damage model is defined as

$$D^{\text{con}} = \frac{\partial \sigma}{\partial \varepsilon} = C^{-1}[I - M]. \tag{18}$$

2.3 Regularization

The model described above contains a viscous term. In this section, it is shown that introduction of this term gives rise to mesh independent results. For this purpose, we consider a bone shaped specimen loaded in tension (see Fig. 6); the geometry is such that localization is forced to take place at the middle of the specimen. The material properties are as summarized in Table 1, with the strong direction aligned with the load. 8-Node brick elements are used with only one element in width and thickness direction. The elements are equidistributed in size along the length of the specimen. For all analyses presented with the continuum damage model, the loading rate is $\dot{\varepsilon} = 1\%\,\mathrm{s}^{-1}$, where ε is the averaged strain in load direction.

Table 1 Material parameters for mesh refinement study and notched plate analysis.

Elasticity [53]		Ply strength [11]		Damage	
E_1	140 GPa	F_{1t}	2280 MPa	A_f	15
E_2	10 GPa	F_{1c}	1725 MPa	A_m	2
v_{12}	0.21	F_{2t}	76 MPa	B_f	40 s^{-1}
v_{23}	0.21	F_{2c}	228 MPa	B_m	10 s^{-1}
G_{12}	5 GPa	F_{12}	76 MPa		
		F_{23}	76 MPa		

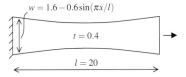

Fig. 6 Geometry of bone shaped specimen, dimensions in mm.

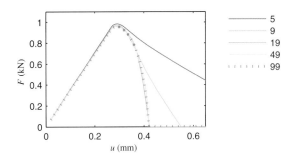

Fig. 7 Response of bone shaped specimen converges to a unique solution upon mesh refinement.

The load displacement diagrams obtained in a mesh refinement study are shown in Fig. 7. It can be observed that the results converge to a unique solution with a finite amount of dissipated energy. The meshes with 5 and 9 elements are too coarse, which can be explained as that, with these meshes, the width of the localized zone is smaller than one finite element. The strain evolution for the solution with 99 elements is shown in Fig. 8, in which the profile of axial strain over the specimen length is depicted for a number of time levels. In both cases, the localization zone is wider than one element, witness the smoothness of the curves, which were obtained by connecting the mid-element strain values.

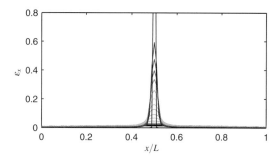

Fig. 8 Localization of strain in analysis with 99 elements.

2.4 Notched plate

An analysis which has many of the components of the typical complex laminate failure, as illustrated in Fig. 2, is the case of a rectangular $[0/90]_s$ laminated plate[2] with an interior notch (see Fig. 9). In experiments it has been observed that well before failure a splitting crack grows from the notch tip in the 0°-ply, which is accommodated by delamination [44]. After that, ultimate failure occurs in both plies in the plane of the notch. This case has been analyzed computationally by Wisnom and Chang [52] and by Yang and Cox [53], who modeled the split with spring or interface elements. Here we examine whether it is also possible to use continuum models. This is motivated by the higher predictive potential of continuum models, since it is not necessary to specify the location of the crack in advance. Moreover, extension of the simulation to ultimate failure of the laminate is possible.

In the numerical model only the upper two plies are taken into account, due to symmetry. Both in-plane axes of symmetry are also used to reduce the size of the model. Each ply is modeled with a single layer of 8-node brick elements. To allow for delamination, the plies are connected with interface elements. These are integrated with a Newton-Cotes scheme to avoid spurious oscillations [40]. Failure in the interface elements, which have an initial elastic dummy stiffness K, is modeled with the damage law developed by Turon et al. [45], containing five material parameters: normal strength t_n, shear strength t_s, mode I fracture energy G_I, mode II fracture energy G_{II} and mode interaction parameter η. Values provided by Daniel and Ishai [11] and Yang and Cox [53] for carbon epoxy laminates are used as lamina and interface properties (see Tables 1 and 2). Interfacial strength parameters are reduced as proposed by Turon et al. [46] in order to allow for a relatively coarse mesh.

Results from two subsequent analyses are presented.

---

[2] That is, the fibers in top and bottom ply are parallel to the x-axis, and those in the two middle plies are perpendicular to the x-axis

Table 2 Material parameters for interface in notched plate analysis [53].

Interface	
K	5×10^5 N/mm
t_n	25 MPa
t_s	15 MPa
G_I	0.35 N/mm
G_{II}	0.7 N/mm
η	1

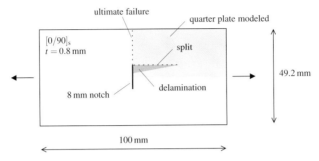

Fig. 9 Geometry and expected failure for notched plate analysis.

1. The 0° ply is analyzed with the damage model, and the 90° ply is kept elastic (just as it is in [53]). The parameters are calibrated to realize approximately the same amount of delamination due to splitting as presented by Yang and Cox [53].
2. Both plies are analyzed with the damage model, with the same data as for the 0° ply in the previous analysis. This is supposed to be the most realistic input, because the plies are made of the same material and differ only in orientation.

Firstly, results are presented from the analysis with the continuum damage model assigned to the 0°-ply and elastic properties assigned to the 90°-ply. In Fig. 10, the deformed mesh is shown for both material models. Note that the strain level is such that tensile failure in the plane of the notch has not yet initiated. The localization zone related to the split is wider than the typical element size, which means that the viscous regularization effectively preserves the mesh objectivity of the results.

Figure 11 shows the damage in the interface for different strain levels. The strain levels for which results are plotted are the same as in [53], and the agreement is satisfactory: there is strong resemblance in shape of the delaminated area, delamination has initiated before $\varepsilon = 0.4\%$, a traction free zone appears between $\varepsilon = 0.4\%$ and $\varepsilon = 0.8\%$ and this zone grows rapidly before $\varepsilon = 0.9\%$. The cohesive zone in the interface, i.e. the zone where $0 < d < 1$, is wider than in the results presented by Yang and Cox. This is the intended effect of the reduction of the interface strength as proposed by Turon et al. [46] in order to allow for the use of relatively coarse meshes.

Secondly, the damage model is assigned to the 90°-ply as well. This is supposed to be the most realistic representation, since both plies are of the same material. In Fig. 12, deformed meshes from both analyses are shown from two different perspec-

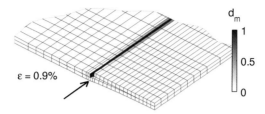

Fig. 10 Deformed mesh from analysis with elastic 90°-ply and continuum damage in the 0°-ply.

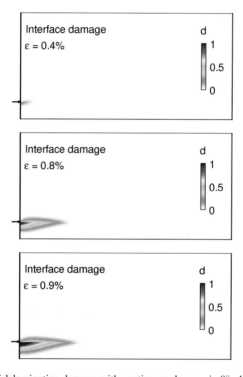

Fig. 11 Evolution of delamination damage with continuum damage in 0°-ply and elastic 90°-ply

tives. Unrealistic failure behavior is obtained. For both material models, localized strains appear in the 90°-ply following the split in the 0°-ply. As a consequence, there is no delamination. Apparently, the damage related to this localized strain in the 90°-ply consumes less energy than what would be needed to cause the expected delamination.

Apparently, something is wrong with the continuum damage model, such that a band with matrix failure that is not aligned with the fibers can be formed more

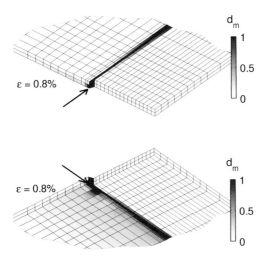

Fig. 12 Subcritical failure when realistic parameters are used for both plies with continuum damage as seen from above (left) and below (right).

easily than in the real material. This we explain with the following micromechanical considerations. A band with matrix shear failure in fiber direction can develop into a macrocrack running between the fibers, which is a relatively brittle mechanism, while a band with matrix shear failure in any other direction is crossed by fibers, and the corresponding failure mechanism is therefore more ductile (see Figure 13). In the presented continuum failure models, however, there is no influence of the orientation of a band with matrix failure on the stress-strain behavior. A matrix crack running between the fibers (Fig. 13 (a)) is represented in the displacement field with localized ε_{21}, and a band with matrix microcracking (Fig. 13 (b)) with localized ε_{12}, both corresponding with the same unique relation between τ_{12} and $\gamma_{12} = (\varepsilon_{12} + \varepsilon_{21})/2$, while in the real material the averaged stress strain response is different for the two mechanisms (Fig. 13 (c)). Notably, no continuum damage models for composite materials exist which provide a solution to this problem.

A positive remark with respect to the last results can also be made, namely that smeared degradation in the 90° ply is observed, which may be understood as representing distributed matrix cracking. In the simulation results, this is clearly visible (see Fig. 12). This is considered an appropriate mesolevel representation of the micromechanical distributed cracking, because this phenomenon is not characterized by individual matrix cracks but rather by the quasi-homogeneous degradation. However, the interply damage that is present around the matrix cracks on the microlevel is not obtained in the mesomodel interface.

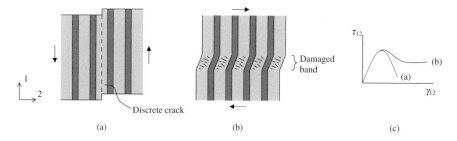

Fig. 13 Micromechanical representation of matrix failure oriented in fiber direction (a) and matrix failure in a band crossed by fibers (b). The difference in averaged stress-strain response is illustrated schematically (c).

2.5 Off-axis tensile test

Next, we consider an off-axis tensile test on a 10° unidirectional laminate. This is a standard test for the determination of the in-plane shear strength [9,47]. Experiments show brittle matrix failure. In a sudden event, the specimen breaks, with the crack running in fiber direction, as shown in Fig. 14. With this relatively simple example we try to further emphasize and clarify the pathology encountered in the previous example.

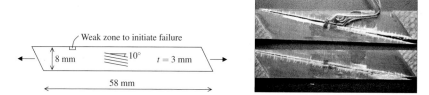

Fig. 14 Setup for off-axis tensile test and experimentally observed failure (Van Paepegem et al. [48], used with permission from Elsevier).

Table 3 Material parameters for off-axis tensile test. Elasticity and tensile strength parameters are taken from Van Paepegem et al. [47] (glass/epoxy).

Elasticity		Ply strength		Damage	
E_1	38.9 GPa	F_{1t}	901 MPa	A_f	100 -
E_2	13.3 GPa	F_{1c}	800 MPa	A_m	3 -
v_{12}	0.26 -	F_{2t}	36.5 MPa	B_f	60 s^{-1}
v_{23}	0.4 -	F_{2c}	70 MPa	B_m	60 s^{-1}
G_{12}	5.13 GPa	F_{12}	52 MPa		
		F_{23}	34.8 MPa		

A horizontal displacement is applied to the right side of the specimen, which is free in vertical direction. Oblique ends with an angle of 54° with the load direction are used, so that for linear elasticity the stress state is homogeneous, thus eliminating edge effects. In order to trigger localization the longitudinal shear strength F_{12} is reduced from 56 to 40 MPa in an area of 1×0.5 mm. Two different meshes are used, with an in-plane element size of 0.5×0.5 mm and 0.25×0.25 mm in the area of interest.

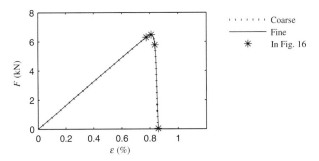

Fig. 15 Load-displacement relation for off-axis tensile test with continuum damage model and two different meshes.

Figure 15 shows the load displacement diagram and final deformed mesh obtained with the continuum damage model. The influence of the element size on the load-displacement behavior is sufficiently small, which is related to the fact that the band with localized strain is wider than the elements, due to the viscous term. From the deformed mesh pictures (Fig. 16), which are solutions related to the coarse mesh, it can be observed that the failure pattern is completely different from that observed in experiments.

Notably, there is a significant displacement perpendicular to the load direction. The deformation is such that the strain in fiber direction ε_1 remains relatively small in the localization area. This is caused by the fact that the stiffness in fiber direction remains almost completely unaffected during the analysis. This behavior stems from the important feature that, locally, a stress state for which matrix failure may be expected, does not give rise to a strain state which implies fiber failure.

However, although the local behavior is correct, the global behavior is not. The fact that ε_1 remains small, is not sufficient to obtain localized deformation aligned with the fibers. In Fig. 16, it can be observed that, immediately when damage is initiated outside the region with reduced strength, the localization occurs at the wrong position. There is also a band with damage which is located approximately in fiber direction, but this remains secondary throughout the analysis.

The cause for this behavior lies in the fact that the direction of failure propagation is governed by the stress concentration rather than by the fiber direction. This is a consequence of the homogenization which is fundamental to the continuum models. In the real material the failure will be contained in the weaker matrix domain, while

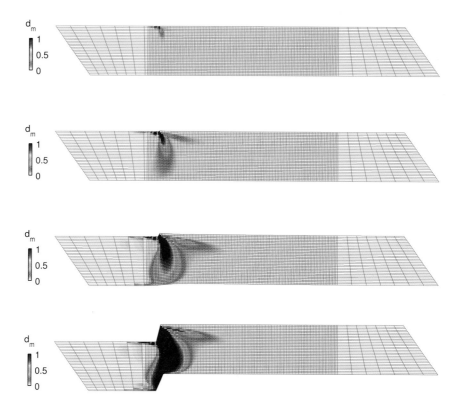

Fig. 16 Evolution of matrix damage in off-axis tensile test with continuum damage ($\Delta x = 0.25$ mm).

the homogenized continuum damage model does not offer this domain separation, as illustrated in Fig. 17. We stress that this is not resolved by distinguishing between matrix and fiber failure in the degradation of the continuum, as is common practice in continuum damage models for composite materials.

With this example, the consequences of the limitation encountered in the previous example are clearly visible. It can be concluded that there is a fundamental problem in the modeling of failure in composites with continuum models. The micromechanical cause for cracks to grow in fiber direction, is not present in continuum models, at least not as long as the model descriptions are purely local. It is unlikely that mechanisms, in which macrocracks in fiber direction play a role in different plies with different fiber orientation, such as the delamination failure in Fig. 2, can be predicted using state-of-the-art continuum models for ply failure, irrespective of the failure criteria and damage evolution laws that are applied. However, for other failure mechanisms, the continuum description might serve well, e.g. when failure in all plies is localized in a single plane [7]. In some cases the matrix crack will emerge correctly, such as the 45°crack presented by Pinho et al. [37], the split near a circular hole as reported by Cox and Yang [10] and the split in the 0°-ply in the notched plate example. But, as far as localized matrix failure in a single ply is concerned, the predictive quality of continuum models should be doubted.

Laš and Zemčik [24] have the same difficulty in correctly representing cracks that are oriented in fiber direction correctly. They solve it by aligning elements with the fiber direction in the critical region in combination with a non-regularized material model. Obviously, this only works when mesh-objectivity is not regarded, which limits the predictive quality of the model.

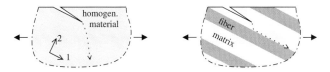

Fig. 17 Crack propagation in a homogeneous orthotropic medium and in a fiber-matrix material.

3 Phantom node method

In this section, a discontinuous representation of matrix cracking is presented. Following the work by Hansbo and Hansbo [17], Mergheim et al. [30] and Song et al. [43], a crack is introduced by addition of an extra element on top of an existing element. It has been shown [43], that this method is equivalent to the extended finite element method (XFEM) in which a discontinuity in the displacement field is introduced by enrichment of the shape functions with the Heaviside step function [32, 51]. With this method, we can force the crack to grow in fiber direction and as such avoid the spurious mechanism encountered with the continuum damage approach, thus incorporating the micromechanical observation that cracks tend to grow between the fibers (Fig. 17) in the mesomodel.

First the kinematic and equilibrium relations of the phantom node method are introduced, then the applied crack propagation criterion, the constitutive relation and the consistent tangent are discussed. The performance of the model is illustrated with the off-axis tensile test introduced in the previous section.

3.1 Kinematical and equilibrium relations

In Fig. 18, it is illustrated how the discontinuous displacement field is composed of the displacements of two overlapping elements, referred to as element A and element B. Four phantom nodes, $\tilde{n}_1, \ldots, \tilde{n}_4$, are introduced on top of the four existing nodes, $n_1,\ldots,n_4$. The element domain is subdivided by a crack segment, Γ, into subdomains Ω_A and Ω_B; Ω_A is corresponding with the active part of element A and Ω_B with the active part of element B. The connectivity of the overlapping elements is

$$\begin{aligned} \text{nodes}_A &= [\tilde{n}_1, \tilde{n}_2, n_3, n_4], \\ \text{nodes}_B &= [n_1, n_2, \tilde{n}_3, \tilde{n}_4]. \end{aligned} \quad (19)$$

The displacement field is defined as

$$\mathbf{u}(\mathbf{x}) = \begin{cases} \mathbf{N}(\mathbf{x})\mathbf{u}_A, & \mathbf{x} \in \Omega_A, \\ \mathbf{N}(\mathbf{x})\mathbf{u}_B, & \mathbf{x} \in \Omega_B, \end{cases} \quad (20)$$

where $\mathbf{N}(\mathbf{x})$ are the standard finite element shape functions and $\mathbf{u}_A$ and $\mathbf{u}_B$ are the nodal displacements of element A and element B, respectively. The displacement jump over the crack is

$$\boldsymbol{\delta}(\mathbf{x}) = \mathbf{N}(\mathbf{x})(\mathbf{u}_A - \mathbf{u}_B), \quad \mathbf{x} \in \Gamma. \quad (21)$$

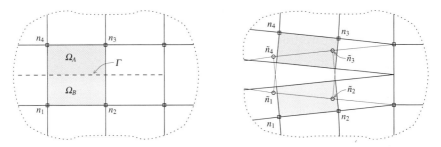

Fig. 18 Connectivity and active parts of two overlapping elements in phantom node method.

Figure 18 also shows that closure of the crack tip is enforced automatically, when no phantom nodes are added on the element boundary that contains the tip.

The contribution to the nodal forces from the bulk material of the old element (there is no conceptual difference between the old and the new element, both are connected to some old and some new nodes and integrated over part of the original domain, but in the code the 'old' element is the one that has the original element number) are computed with

$$\mathbf{f}_A^{\text{bulk}} = \int_{\Omega_A} \mathbf{B}^T \boldsymbol{\sigma}(\boldsymbol{\varepsilon}_A) \, \mathrm{d}\Omega r, \quad (22)$$

with

$$\boldsymbol{\varepsilon}_A = \mathbf{B}\mathbf{u}_A, \quad (23)$$

where Ω_A is the active part of the domain associated with the old element, σ can be any linear or nonlinear law of ε, and $\mathbf{u}_A$ are the nodal displacements associated with the old element and B is the strain nodal displacement matrix. And similarly:

$$\mathbf{f}_B^{\text{bulk}} = \int_{\Omega_B} \mathbf{B}^T \sigma(\varepsilon_B) \, d\Omega, \tag{24}$$

with

$$\varepsilon_B = \mathbf{B}\mathbf{u}_B. \tag{25}$$

The contribution to the nodal forces from the cohesive traction are computed with

$$\mathbf{f}_A^{\text{coh}} = \int_\Gamma \mathbf{N}^T \mathbf{t}(\delta) \, d\Gamma \tag{26}$$

and

$$\mathbf{f}_B^{\text{coh}} = -\int_\Gamma \mathbf{N}^T \mathbf{t}(\delta) \, d\Gamma. \tag{27}$$

The displacement jump is defined on each point $\mathbf{x}$ on Γ as

$$\delta(\mathbf{x}) = \mathbf{u}_A(\mathbf{x}) - \mathbf{u}_B(\mathbf{x}) = \mathbf{N}(\mathbf{x})\mathbf{u}_A - \mathbf{N}(\mathbf{x})\mathbf{u}_B. \tag{28}$$

This jump may be rotated to the local frame with

$$\bar{\delta} = \begin{bmatrix} \delta_n \\ \delta_s \end{bmatrix} = \mathbf{Q}\delta, \tag{29}$$

where (in 2D)

$$\mathbf{Q} = \begin{bmatrix} -\sin\phi & \cos\phi \\ \cos\phi & \sin\phi \end{bmatrix} = \begin{bmatrix} \mathbf{n}^T \\ \mathbf{s}^T \end{bmatrix}. \tag{30}$$

3.2 Crack propagation

For the modeling of crack propagation with the XFEM or the phantom node method, generally two criteria are needed:

- a criterion to decide whether a crack grows or not ,
- a criterion to asses in which direction the crack grows .

In the particular context of splitting in laminates, the second criterion becomes trivial, because the propagation direction is fixed, viz. equal to the fiber direction. To test for propagation, the stress in the tip element is checked with a stress criterion.

$$f(\sigma) = \sqrt{\left(\frac{\langle \sigma_2 \rangle}{F_{2t}}\right)^2 + \left(\frac{\tau_{12}}{F_{12}}\right)^2}, \tag{31}$$

which is the plane stress equivalent of Hashin's criterion for tensile matrix failure Eq. (6). With the operator $\langle \cdot \rangle = \max\{\cdot, 0\}$ compressive matrix stresses are neglected.

3.3 Cohesive law

A cohesive traction is applied on the crack surface. With this, the amount of energy that is dissipated as the crack propagates can be controlled and the singularity in the strain and stress field near the crack tip is avoided. Ideally, the traction **t** would be defined as a function of the displacement jump δ. However, the application of a direct traction separation law for mixed mode cracking leads to computational instability. This stems from the fact that the traction is not uniquely defined for zero crack opening; in a uni-axial case it is obvious that the traction should be equal to the strength, but in a mixed mode formulation it can be either equal to the normal strength with no shear traction, or to the shear strength with no normal traction, or something in between. The traction evaluation itself is always feasible, because after crack extension the crack opening which gives equilibrium will not be equal to zero, but the highly nonlinear nature of the traction separation law does endanger the stability of the Newton Raphson procedure.

However, the traction law may be constrained using equilibrium considerations. When the bulk element that is cut by the crack segment is predominantly loaded in tension, equilibrium demands that the traction in the crack is also predominantly tensile traction, and when the bulk element is loaded in shear, the traction should be shear traction. With this in mind, a new traction law, in which the bulk stress is taken into account, has been developed by Moonen et al. [33]. This law is presented here in adapted form.

Similar to the continuum damage model, an effective traction is computed, which can be interpreted as the effective traction working on the reduced surface of a partially cracked domain. The effective traction is defined in the local frame $\{n, s\}$:

$$\hat{\mathbf{t}} = \begin{bmatrix} \hat{t}_n \\ \hat{t}_s \end{bmatrix} = \mathbf{Q}\sigma\mathbf{n} + T\mathbf{Q}\delta. \tag{32}$$

When T is replaced with the acoustic tensor $\mathbf{n}^T \mathbf{D}^e \mathbf{n}$, this effective traction can be related to the elastic stress in the damaged cross section [33, 34]. In the present model, however, the exact value of T is of little importance, it only serves as a stabilization parameter. Therefore it is considered safe to use a scalar quantity T for the sake of simplicity. For the results presented in this work, we set T equal to the first diagonal entry of the acoustic tensor.

In Voigt notation, $\sigma\mathbf{n}$ is computed as

$$\sigma \mathbf{n} = \mathsf{H}\sigma, \tag{33}$$

with (in 2D)

$$\mathsf{H} = \begin{bmatrix} n_1 & 0 & n_2 \\ 0 & n_2 & n_1 \end{bmatrix}. \tag{34}$$

Notably, since the crack is parallel to the fiber, the vector $\mathsf{H}\sigma$ contains the material stress components σ_2 and τ_{12}.

The bulk stress at Γ, which is used in Eq. (32), is not uniquely defined. Indifference to which element is labeled A and which is labeled B is maintained by computing the stress with the averaged strain:

$$\sigma_\Gamma = \sigma(\varepsilon_\Gamma), \tag{35}$$

with

$$\begin{aligned}\varepsilon_\Gamma &= \frac{1}{2}\left(\varepsilon_{\Gamma_+} + \varepsilon_{\Gamma_-}\right) \\ &= \frac{1}{2}\mathsf{B}\left(\mathbf{u}_A + \mathbf{u}_B\right).\end{aligned} \tag{36}$$

Similar to continuum damage, the effective traction is reduced with a damage variable (denoted ω). The traction in local frame is defined as:

$$\bar{\mathbf{t}} = [\mathsf{I} - \Omega]\hat{\mathbf{t}}, \tag{37}$$

with

$$\Omega = \omega \begin{bmatrix} \frac{\langle \hat{t}_n \rangle}{\hat{t}_n} & 0 \\ 0 & 1 \end{bmatrix}. \tag{38}$$

Introduction of the factor $\langle \hat{t}_n \rangle / \hat{t}_n$, which is equal to 1 when $\hat{t}_n > 0$, and equal to 0 otherwise, suffices to prevent interpenetration of the cracked parts[3], because with this factor, $\hat{t}_n < 0 \Rightarrow t_n = \hat{t}_n$, and, with the assumption of equilibrium ($t_n = (\sigma\mathbf{n})_1$), this gives in turn $t_n = \hat{t}_n \Rightarrow \delta_n = 0$. Because equilibrium is only weakly met, limited interpenetration may occur, but this can be expected to vanish upon mesh refinement.

The traction computed with Eq. (37) is rotated back to the global coordinate frame, such that the cohesive law can be summarized as:

$$\mathbf{t} = \mathsf{Q}^T[\mathsf{I} - \Omega]\mathsf{Q}\{\mathsf{H}\sigma + T\delta\}. \tag{39}$$

During the failure process, the effective traction increases. The damage evolution is driven by a state variable, denoted κ which is the temporal maximum of a scalar

---

[3] Provided that rotations remain small. Extension to large rotations is possible but not presented here.

measure for the magnitude of the effective traction:

$$\kappa(\tau) = \max_{t \leq \tau} \sqrt{\left(\frac{\langle \hat{t}_n \rangle}{F_{2t}}\right)^2 + \left(\frac{\hat{t}_s}{F_{12}}\right)^2}. \tag{40}$$

Damage variable ω evolves with state variable κ with the following law that results in a linear (softening) relation between displacement jump and traction:

$$\omega = \begin{cases} \dfrac{\kappa^f(\kappa - 1)}{\kappa(\kappa^f - 1)}, & \kappa < \kappa^f, \\ 1, & \kappa \geq \kappa^f. \end{cases} \tag{41}$$

The derivation of the expression for κ^f will be outlined in the remainder of this section.

We assume that the material behavior can be described properly with the phenomenological relation proposed for delamination by Benzeggagh and Kenane [4], which was applied in interface elements by Camanho, Turon et al. [6,45]:

$$G_{Tc} = G_{Ic} + (G_{IIc} - G_{Ic})\left(\frac{G_{II}}{G_I + G_{II}}\right)^\eta, \tag{42}$$

in which G_{Ic} is the mode I fracture energy, G_{IIc} is the mode II fracture energy, η is an additional material parameter, and the ratio $G_{II}/(G_I + G_{II})$ indicates the actual mode mixity.

We define the mode ratio β as:

$$\beta = \frac{\hat{t}_s^2}{\hat{t}_s^2 + \langle \hat{t}_n \rangle^2}. \tag{43}$$

This expression can be rewritten as

$$\langle t_n \rangle = |t_s|\sqrt{\frac{1-\beta}{\beta}}. \tag{44}$$

Assuming equilibrium between the cohesive traction acting on the crack surface and the bulk stress directly next to this surface ($\mathbf{H}\sigma = \mathbf{t}$), it follows from Eq. (39) that the ratio between the displacement jump components is necessarily the same:

$$\delta_n = |\delta_s|\sqrt{\frac{1-\beta}{\beta}}. \tag{45}$$

Therefore, for a fixed mode ratio β, the ratio between the energy release of the two modes is fixed:

$$\frac{G_I}{G_{II}} = \frac{\int t_n \, d\delta_n}{\int t_s \, d\delta_s} = \frac{1-\beta}{\beta} \tag{46}$$

and, consequently

$$\frac{G_{II}}{G_I + G_{II}} = \beta. \tag{47}$$

Substitution of Eq. (44) and $\kappa = 1$ in Eq. (40) gives the relation between the initial shear traction, $\hat{t}_s^0$, and β, which may be written as:

$$|\hat{t}_s^0| = F_m \sqrt{\beta/\Phi}, \tag{48}$$

with

$$F_m^2 = F_{2t}^2 + F_{12}^2 \tag{49}$$

and

$$\Phi = \frac{1-\beta}{1-\phi} + \frac{\beta}{\phi}, \tag{50}$$

$$\phi = \frac{F_{12}^2}{F_m^2}. \tag{51}$$

Similarly, substitution of (45) in (40) with $\kappa = \kappa^f$ gives the relation between the shear crack opening that corresponds with complete failure, δ_s^f, and β:

$$|\delta_s^f| = \frac{\kappa^f}{T} F_m \sqrt{\beta/\Phi}. \tag{52}$$

With Eqs. (48) and (52), the energy release in mode II can be computed as a function of β (considering that the sign of t_s^0 will always be equal to that of δ_s^f):

$$G_{II} = \frac{1}{2} \hat{t}_s^0 \delta_s^f = \frac{F_m^2 \kappa^f \beta}{2T\Phi}. \tag{53}$$

Finally, with Eq. (47), the total energy release as a function of β is:

$$G_T = G_I + G_{II} = \frac{F_m^2 \kappa^f}{2T\Phi}. \tag{54}$$

Now, κ^f can be defined as a function of β such that G_T in (54) equals G_{Tc} in (42):

$$\kappa^f = \frac{2T\Phi}{F_m^2} \left(G_{Ic} + (G_{IIc} - G_{Ic}) \beta^\eta \right). \tag{55}$$

Because equilibrium was assumed for this derivation, while equilibrium is only weakly met, the amount of energy dissipated in mixed mode conditions does not satisfy Eq. (42) exactly, but it does approach the correct value upon mesh refinement.

3.4 Consistent tangent

The element matrix for the two overlapping elements together is

$$\begin{bmatrix} K_{AA} & K_{AB} \\ K_{BA} & K_{BB} \end{bmatrix} = \begin{bmatrix} K_A^{bulk} & 0 \\ 0 & K_B^{bulk} \end{bmatrix} + \begin{bmatrix} K_\delta^{coh} & -K_\delta^{coh} \\ -K_\delta^{coh} & K_\delta^{coh} \end{bmatrix} + \begin{bmatrix} K_\sigma^{coh} & K_\sigma^{coh} \\ -K_\sigma^{coh} & -K_\sigma^{coh} \end{bmatrix}, \quad (56)$$

with

$$K_A^{bulk} = \int_{\Omega_A} B^T DB \, d\Omega, \quad (57)$$

$$K_B^{bulk} = \int_{\Omega_B} B^T DB \, d\Omega, \quad (58)$$

$$K_\delta^{coh} = T \int_\Gamma N^T Q^T AQN \, d\Gamma, \quad (59)$$

$$K_\sigma^{coh} = \frac{1}{2} \int_\Gamma N^T Q^T AQHDB \, d\Gamma. \quad (60)$$

The nonlinearity of the system is in matrix **A**, which is defined as

$$A = \frac{\partial \bar{t}}{\partial \hat{t}} = I - \Omega - \bar{t} \otimes \frac{\partial \omega}{\partial \hat{t}}, \quad (61)$$

with

$$\frac{\partial \omega}{\partial \hat{t}} = \frac{\partial \omega}{\partial \kappa} \frac{\partial \kappa}{\partial \hat{t}} + \frac{\partial \omega}{\partial \kappa^f} \frac{\partial \kappa^f}{\partial \beta} \frac{\partial \beta}{\partial \hat{t}}. \quad (62)$$

3.5 Off-axis tensile test

The phantom node method with the introduced constitutive relation is applied to the 10° off-axis tensile test. The same elastic parameters and ply strength parameters as in Section 2.5 are used. Additional parameters are: $G_{Ic} = G_{IIc} = 3$ N/mm and $\eta = 1$. The length of the specimen is 44 mm, apart from which, dimensions are the same as in Section 2.5. There is no weak zone to trigger the initialization, but the location where the crack will initiate is specified in advance. Notably, the geometry is such, that the stress field is homogeneous until nonlinearities occur, which means that the failure criterion is exactly satisfied in the entire specimen when the crack is initiated. As a consequence a sharp snap-back occurs. An arc-length method with energy release control [15] is employed to follow this behavior.

In Fig. 19, the load displacement relation is plotted for two different meshes, one with an average element size 1.1×0.5 mm and the other with average element size 0.8×0.25 mm. It can be observed that the discontinuous approach does not require

Simulation of Progressive Failure in Composite Laminates

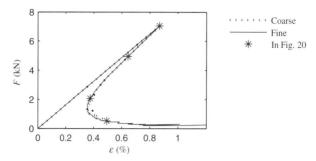

Fig. 19 Load displacement data for off-axis test with phantom node method

any precautions to obtain mesh independent results as opposed to the continuum damage model.

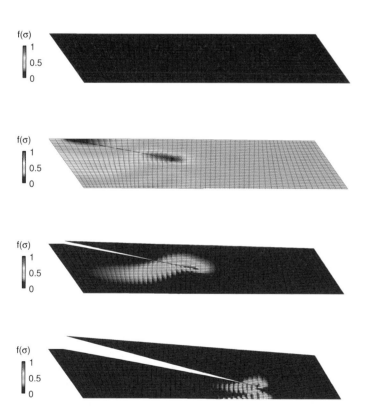

Fig. 20 Off-axis tensile test results with phantom node method (fine mesh, unscaled deformations.

Figure 20 shows the deformed mesh from the fine mesh solution for different stages. The normalized stress quantity that is used for the shading is the failure criterion in Eq. (31). In Fig. 19, the corresponding points on the load displacement curve are indicated. It can be observed that in a sudden event at the peak load level, a cohesive crack of considerable length appears, after which the load drops without much additional crack growth. Then the cohesive zone, i.e. the part of the crack in which tractions are present, becomes smaller and the crack finds its way toward the opposite boundary of the specimen. Contrary to the analysis discussed in Section 2, the crack grows in the correct direction, not surprisingly since the growth direction is fixed. In this example not only the direction of propagation, but also the location of initiation was predefined, which renders it a rather trivial problem that could also be tackled by meshing the crack with interface elements. However, the presented model can also be applied to more complex geometries and cases with multiple cracking, where the advantage that the crack is represented in a mesh-independent fashion becomes significant. The use of interface elements would require complex discretized configurations.

4 Conclusions

A continuum damage model with viscous regularization has been introduced. With this regularization method mesh independent results can be obtained, which is an issue of concern in the computational modeling of failure in materials.

It has been shown that the regularized continuum damage model is not always adequate to predict the correct failure pattern in composite laminates. A relevant microstructural property is lost in the homogenization procedure, namely the preference of matrix damage to propagate in fiber direction. Still bands with matrix failure may be obtained that are oriented correctly, especially because orthotropic elasticity properties still promote this behavior, but bands with different orientation may also arise. This is not only the case for the presented model, but also for other continuum damage models presented in literature in recent years. The predictive quality of continuum models for localized matrix failure in isolated plies should be doubted.

One way to overcome this is to abandon the continuum approach for the specific mechanism of splitting. For this purpose, a phantom node method (equivalent to XFEM) with crack propagation in fixed direction is proposed. This method allows for mesh-independent modeling of splitting cracks, which is of particular importance when location and number of cracks is not known a priori. A new cohesive law has been applied, which can deal with varying mixed mode conditions robustly. In combination with an energy arc-length method, it is possible to simulate cracking in a 10° off-axis tensile test. After combination of this splitting formulation with models for other ply failure mechanisms and delamination, simulation of complex laminate failure mechanisms on the mesolevel may become feasible.

Acknowledgments

This research is supported by the Technology Foundation STW (under grant DCB.6623) and the Ministry of Public Works and Water Management, The Netherlands.

References

1. Allix, O., Feissel, P., Thévenet, P.: A delay damage mesomodel of laminates under dynamic loading: basic aspects and identification issues. Computers and Structures **81**(12), 1177–1191 (2003)
2. Bazant, Z.P., Oh, B.: Crack band theory for fracture of concrete. Materials and Structures **16**(3), 155–177 (1983)
3. Bazant, Z.P., Pijaudier-Cabot, G.: Measurement of characteristic length of non-local continuum. Journal of Engineering Mechanics **115**(4), 755–767 (1989)
4. Benzeggagh, M.L., Kenane, M.: Measurement of mixed-mode delamination fracture toughness of unidirectional glass/epoxy composites with mixed mode bending apparatus. Composites Science and Technology **56**(4), 439–449 (1996)
5. de Borst, R., Mühlhaus, H.B.: Gradient-dependent plasticity: Formulation and algorithmic aspects. International Journal for Numerical Methods in Engineering **35**(3), 521–539 (1992)
6. Camanho, P.P., Dávila, C.G., de Moura, M.F.: Numerical simulation of mixed-mode progressive delamination in composite materials. Journal of Composite Materials **37**(16), 1415–1438 (2003)
7. Camanho, P.P., Maimí, P., Dávila, C.G.: Prediction of size effects in notched laminates using continuum damage mechanics. Composites Science and Technology **67**(13), 2715–2727 (2007)
8. Caporale, A., Luciano, R., Sacco, E.: Micromechanical analysis of interfacial debonding in unidirectional fiber-reinforced composites. Computers and Structures **84**(31), 2200–2211 (2006)
9. Chamis, C.C., Sinclair, J.H.: Ten-deg off-axis test for shear properties in fiber composites. Experimental Mechanics **17**(9), 339–346 (1977)
10. Cox, B.N., Yang, Q.D.: In quest of virtual tests for structural composites. Science **314**(5802), 1102–1107 (2006)
11. Daniel, I.M., Ishai, O.: Engineering Mechanics of Composite Materials, Second edn. Oxford University Press, New York (2006)
12. Germain, N., Besson, J., Feyel, F.: Composite layered materials: Anisotropic nonlocal damage models. Computer Methods in Applied Mechanics and Engineering **196**(41–44), 4272–4282 (2007)
13. González, C., LLorca, J.: Multiscale modeling of fracture in fiber-reinforced composites. Acta Materialia **54**(16), 4171–4181 (2006)
14. Green, B.G., Wisnom, M.R., Hallett, S.R.: An experimental investigation into the tensile strength scaling of notched composites. Composites: Part A **38**(3), 867–878 (2007)
15. Gutiérrez, M.A.: Energy release control for numerical simulations of failure in quasi-brittle solids. Communications in Numerical Methods in Engineering **20**(1), 19–29 (2004)
16. Hallett, S.R., Wisnom, M.R.: Experimental investigation of progressive damage and the effect of layup in notched tensile tests. Journal of Composite Materials **40**(2), 119–141 (2006)
17. Hansbo, A., Hansbo, P.: A finite element method for the simulation of strong and weak discontinuities in solid mechanics. Computer Methods in Applied Mechanics and Engineering **193**(33–35), 3523–3540 (2004)
18. Hashin, Z.: Failure criteria for unidirectional fiber composites. Journal of Applied Mechanics **47**, 329–334 (1980)

19. Hashin, Z., Rotem, A.: A fatigue failure criterion for fiber reinforced materials. Journal of Composite Materials **7**, 448–464 (1973)
20. Jiang, W.G., Hallett, S.R., Green, B.G., Wisnom, M.R.: A concise interface constitutive law for analysis of delamination and splitting in composite materials and its application to scaled notched tensile specimens. International Journal for Numerical Methods in Engineering **69**(9), 1982–1995 (2007)
21. Ladevèze, P.: A damage computational approach for composites: Basic aspects and micromechanical relations. Computational Mechanics **17**(1–2), 142–150 (1995)
22. Lapczyk, I., Hurtado, J.A.: Progressive damage modeling in fiber-reinforced materials. Composites: Part A **38**(11), 2333–2341 (2007)
23. Laurin, F., Carrère, N., Maire, J.F.: A multiscale progressive failure approach for composite laminates based on thermodynamical viscoelastic and damage models. Composites: Part A **38**(1), 198–209 (2007)
24. Laš, V., Zemčik, R.: Progressive damage of unidirectional composite panels. Journal of Composite Materials **42**(1), 25–44 (2008)
25. Lemaitre, J., Chaboche, J.L.: Mechanics of Solid Materials. Cambridge University Press, Cambridge (1990)
26. Maimí, P., Camanho, P.P., Mayugo, J.A., Dávila, C.G.: A continuum damage model for composite laminates: Part I – Constitutive model. Mechanics of Materials **39**(10), 897–908 (2007)
27. Maimí, P., Camanho, P.P., Mayugo, J.A., Dávila, C.G.: A continuum damage model for composite laminates: Part II – Computational implementation and validation. Mechanics of Materials **39**(10), 909–919 (2007)
28. Matzenmiller, A., Lubliner, J., Taylor, R.L.: A constitutive model for anisotropic damage in fiber-composites. Mechanics of Materials **20**(2), 125–152 (1995)
29. van der Meer, F.P., Sluys, L.J.: Continuum models for the analysis of progressive failure in composite laminates. Journal of Composite Materials (*accepted for publication*)
30. Mergheim, J., Kuhl, E., Steinmann, P.: A finite element method for the computational modelling of cohesive cracks. International Journal for Numerical Methods in Engineering **63**(2), 276–289 (2005)
31. Mi, Y., Crisfield, A., Hellweg, H.B., Davies, G.A.O.: Progressive delamination using interface elements. Journal of Composite Materials **32**(14), 1246–1272 (1998)
32. Moës, N., Belytschko, T.: Extended finite element method for cohesive crack growth. Engineering Fracture Mechanics **69**(7), 813–833 (2002)
33. Moonen, P., Sluys, L.J., Carmeliet, J.: Modeling the hygro-mechanical response of quasi-brittle materials. Philosophical Magazine (*submitted*)
34. Oliver, J.: On the discrete constitutive models induced by strong discontinuity kinematics and continuum constitutive equations. International Journal of Solids and Structures **37**(48-50), 7207–7229 (2000)
35. Pierron, F., Green, B., Wisnom, M.R., Hallett, S.R.: Full-field assessment of the damage process of laminated composite open-hole tensile specimens. Part II: Experimental results. Composites: Part A **38**(11), 2321–2332 (2007)
36. Pinho, S.T., Robinson, P., Iannucci, L.: Physically based failure models and criteria for laminated fibre-reinforced composites with emphasis on fibre kinking. Part I: Development. Composites: Part A **37**(1), 63–73 (2006)
37. Pinho, S.T., Robinson, P., Iannucci, L.: Physically based failure models and criteria for laminated fibre-reinforced composites with emphasis on fibre kinking. Part II: FE implementation. Composites: Part A **37**(5), 766–777 (2006)
38. Remmers, J.J.C.: Discontinuities in materials and structures. Ph.D. thesis, Delft University of Technology (2006)
39. Remmers, J.J.C., Wells, G.N., de Borst, R.: A solid-like shell element allowing for arbitrary delaminations. International Journal for Numerical Methods in Engineering **58**, 2013–2040 (2003)
40. Schellekens, J.C.J., de Borst, R.: On the numerical integration of interface elements. International Journal for Numerical Methods in Engineering **36**(1), 43–66 (1993)

41. Schellekens, J.C.J., de Borst, R.: Free edge delamination in carbon-epoxy laminates: a novel numerical/experimental approach. Composite Structures **28**(4), 357–373 (1994)
42. Sluys, L.J.: Wave propagation, localisation and dispersion in softening solids. Ph.D. thesis, Delft University of Technology (1992)
43. Song, J.H., Areias, P.M.A., Belytschko, T.: A method for dynamic crack and shear band propagation with phantom nodes. International Journal for Numerical Methods in Engineering **67**(6), 868–893 (2006)
44. Spearing, S.M., Beaumont, P.W.R.: Fatigue damage mechanics of composite materials. I: Experimental measurement of damage and post-fatigue properties. Composites Science and Technology **44**(2), 159–168 (1992)
45. Turon, A., Camanho, P.P., Costa, J., Dávila, C.G.: A damage model for the simulation of delamination in advanced composites under variable-mode loading. Mechanics of Materials **38**(11), 1072–1089 (2006)
46. Turon, A., Dávila, C.G., Camanho, P.P., Costa, J.: An engineering solution for mesh size effects in the simulation of delamination using cohesive zone models. Engineering Fracture Mechanics **74**(10), 1665–1682 (2007)
47. Van Paepegem, W., De Baere, I., Degrieck, J.: Modelling the nonlinear shear stress–strain response of glass fibre-reinforced composites. Part I: Experimental results. Composites Science and Technology **66**(10), 1455–1464 (2006)
48. Van Paepegem, W., De Baere, I., Degrieck, J.: Modelling the nonlinear shear stress–strain response of glass fibre-reinforced composites. Part II: Model development and finite element simulations. Composites Science and Technology **66**(10), 1465–1478 (2006)
49. Wang, W.M., Sluys, L.J., de Borst, R.: Viscoplasticity for instabilities due to strain softening and strain-rate softening. International Journal for Numerical Methods in Engineering **40**(20), 3839–3864 (1997)
50. Wells, G.N., de Borst, R., Sluys, L.J.: A consistent geometrically non-linear approach for delamination. International Journal for Numerical Methods in Engineering **54**, 1333–1355 (2002)
51. Wells, G.N., Sluys, L.J.: A new method for modelling cohesive cracks using finite elements. International Journal for Numerical Methods in Engineering **50**(12), 2667–2682 (2001)
52. Wisnom, M.R., Chang, F.K.: Modelling of splitting and delamination in notched cross-ply laminates. Composites Science and Technology **60**(15), 2849–2856 (2000)
53. Yang, Q.D., Cox, B.N.: Cohesive models for damage evolution in laminated composites. International Journal of Fracture **133**(2), 107–137 (2005)
54. Zuo, Q.H., Addessio, F.L., Dienes, J.K., Lewis, M.W.: A rate-dependent damage model for brittle materials based on the dominant crack. International Journal of Solids and Structures **43**(11–12), 3350–3380 (2006)

Numerical Modeling of Wave Propagation, Breaking and Run-Up on a Beach

G.S. Stelling and M. Zijlema

Abstract A numerical method for free-surface flow is presented to study water waves in coastal areas. The method builds on the nonlinear shallow water equations and utilizes a non-hydrostatic pressure term to describe short waves. A vertical boundary-fitted grid is used with the water depth divided into a number of layers. A compact finite difference scheme is employed that takes into account the effect of non-hydrostatic pressure with a small number of vertical layers. As a result, the proposed technique is capable of simulating relatively short wave propagation, where both frequency dispersion and nonlinear shoaling play an important role, in an accurate and efficient manner. Mass and momentum are strictly conserved at discrete level while the method only dissipates energy in the case of wave breaking. A simple wet-dry algorithm is applied for a proper calculation of wave run-up on the beach. The computed results show good agreement with analytical and laboratory data for wave propagation, transformation, breaking and run-up within the surf zone.

1 Introduction

Wave transformation in the surf zone plays an important role in coastal engineering since, nearshore waves are the driving forces for many nearshore phenomena, such as longshore currents, water level set-up, sedimentation, erosion, and coastal-structure loading. Surf zones are characterised by the irreversible transformation of organized wave motion of the incident short, wind-generated waves into motions of different types and scales e.g., turbulence and low-frequency motion (well-known

---

G.S. Stelling
Delft University of Technology, Faculty of Civil Engineering and Geosciences, P.O. Box 5048, 2600 GA Delft, The Netherlands, e-mail: g.s.stelling@tudelft.nl

M. Zijlema
Delft University of Technology, Faculty of Civil Engineering and Geosciences, P.O. Box 5048, 2600 GA Delft, The Netherlands, e-mail: m.zijlema@tudelft.nl

as the "surf-beat"). The main features associated with the transformation of coastal waves across a surf zone are illustrated in Fig. 1. (SWL denotes the still water level.) In the pre-breaking region, wave transformation is described by the effects of wave

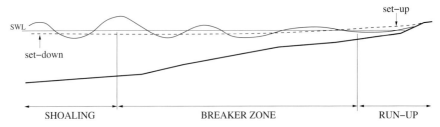

Fig. 1 Wave transformation across a typical surf zone.

steepness due to the nonlinearity (or amplitude dispersion) and wave shortness due to the frequency dispersion. The front face of a wave will steepen continuously until the front becomes vertical. Therefore, the frequency and amplitude dispersion effects must balance each other, so that waves of finite amplitude and permanent form are possible [31]. The bathymetric variations in shallow water distort this balance and cause instabilities and subsequent wave breaking. Once the wave breaks, turbulence is generated and becomes a dominating feature of the flow field. The wavebreaker-generated turbulence balances the steepening of the front and stabilizes the surface profile [31]. The broken waves propagate with a gradual change of form and resemble steady bores. At the end, they become relatively long and run up on the beach. The run-up starts when the bore reaches the shoreline corresponding to a stage of the motion with no water in front of the bore.

The simulation of broken waves and wave run-up amounts to the solution of the nonlinear shallow water (NLSW hereinafter) equations for free-surface flow without viscosity terms in a depth-integrated form [16, 21, 32, 5, 19]. These hyperbolic equations are mathematically equivalent to the Euler equations for compressible flows. Discontinuities are admitted through the weak form of these equations and can take the form of bores which are the hydraulic equivalent of shock waves in aerodynamics. The conservation of energy does not hold across the discontinuities but the conservation of mass and momentum remains valid. By considering the similarity between broken waves and steady bores, energy dissipation due to turbulence generated by wave breaking is inherently accounted for [16, 5].

In the pre-breaking region, however, the NLSW equations do not hold since, they assume a hydrostatic pressure distribution. These equations prohibit a correct calculation of frequency-dispersive or short waves. Moreover, they predict that the front face of any wave or bore will steepen continuously until a vertical front is formed. Only the deviations from hydrostatic pressure can balance the steepening of the front and stabilize the surface profile before it becomes vertical. In this study, we discuss an extension of the NLSW equations to include the effect of vertical acceleration so that the propagation of short, nonlinear waves with finite amplitudes can be simulated.

A main difficulty occurring in the simulation of free-surface flows is the proper handling of a moving free surface since this is part of the solution itself. Many methods for the treatment of the free surface are described in the literature. The most well-known are the Marker-and-Cell (MAC) method [15], the Volume-of-Fluid (VOF) method [18] and the level-set method [30]. An overview of these methods can be found in [25]. Although, these techniques can describe wave overturning in a very accurate manner, they yield more detailed information than necessary for many coastal engineering applications. Moreover, they are too computing intensive when applied to the large-scale wave evolution in the surf zone.

A much simpler approach is the one in which the free-surface motion is tracked using a single-valued function of the horizontal plane as done in the NLSW methodology. Recently, the development of so-called non-hydrostatic models using this approach has been a popular topic of many ocean and coastal modeling activities. Well-known papers on this subject are Casulli and Stelling [9] and Stansby and Zhou [26]. The models in these papers consist of the NLSW equations with the addition of a vertical momentum equation and non-hydrostatic pressure in horizontal momentum equations. As such, the total pressure is decomposed into hydrostatic and non-hydrostatic components. The underlying motivation for this approach is that existing shallow water packages need to be adapted slightly only, since the correction to the hydrostatic pressure is done after the NLSW equations have been solved. As a consequence, this reduces the effort of software extension and maintenance to a minimum. Also, Mahadevan et al. [22] have shown that this technique leads to a more stable and efficient non-hydrostatic calculation than in the case without splitting the pressure into hydrostatic and non-hydrostatic parts. Moreover, the non-hydrostatic models require much less grid cells in the vertical direction than the MAC and VOF methods. These benefits make simulations of wave transformation in coastal waters much more feasible and efficient. There is, however, still ongoing research on this approach. The choice of an appropriate numerical approach for a non-hydrostatic free-surface flow model appears to be non-trivial. Several different solution procedures (fractional step versus pressure-correction approach, modeling of the free-surface boundary condition, Cartesian versus $\sigma-$coordinates, etc.) have been proposed by different authors; for an overview, see [35] and the references therein.

Another issue that remains to be discussed is wave breaking. In principle, non-hydrostatic models represent a good balance between nonlinearity (enables wave shoaling) and frequency dispersion (corrects celerity of shoaling wave) so that initiation of the wave breaking process and the associated energy losses can be described adequately by these models. However, most of the well-established non-hydrostatic models, [22, 9, 26], are by no means momentum-conservative. It is evident that the numerical schemes involved must treat shock propagation adequately in order to model broken waves in the surf zone. Traditionally, the shock-capturing schemes applied to shallow water flows at collocated grids are based on the Godunov-type approach, where a discontinuity in the unknown variables (water depth and discharge) is assumed at the cell edges and a Riemann solver is employed to compute the flux across the cell interface [33]. Applications of this approach to the NLSW equations are given e.g., in Ref. [19, 6, 20]. However, this method suffers of four main shortcomings. First, in the case of variable topography, a numerical imbalance may occur, resulting in an artificial flow, caused by inconsistent approximation of the flux-gradients and source terms due to the bottom slope. Secondly, this technique often uses explicit time integration. As a consequence, time steps may be very small due to the CFL condition related to the shallow depth. Thirdly, although, extension to three dimensions by the common local dimensionally split approach is trivial, it is disputable. Finally, extension of the Godunov-type methods to non-hydrostatic flows is non-trivial.

To our knowledge, no papers have been published on the simulation of surf zone phenomena involving breaking waves and wave run-up employing NLSW equations including non-hydrostatic pressure. The reason for this is probably of a historical nature. Traditionally, the effect of non-hydrostatic pressure is taken into account by a Boussinesq-type approximation through adding higher order derivative terms to the NSLW equations [24]. As such, the Boussinesq-type wave models are based on an efficient depth-integrated formulation and have become very popular for real-life applications involving wave dynamics in coastal regions and harbours. One of the main challenges in the development of the Boussinesq-type models is the accurate simulation of wave propagation from deep water through the surf zone. Because such models are strictly valid only for fairly long waves, there have been attempts to improve frequency dispersion that may complicate the underlying formulation. Since then a continual extension of Boussinesq theory has been ongoing with recent advances in its application to highly nonlinear waves and deeper water. See Ref. [11] for a survey of the field. In addition, because of the approximations involved it may not be guaranteed that the Boussinesq-type wave models can predict the onset of wave breaking and its energy losses correctly. It seems that strict energy dissipation can only be proven by adding a dissipation model to the Boussinesq equations; see [11] and the references quoted there.

The purpose of the present work is to report on the experiences that have been gathered in the development of a non-hydrostatic model for coastal waves in the surf zone. We apply an implicit finite difference method for staggered grids as described in [28], originally developed for modeling subcritical flows in, e.g. coastal seas, lakes and estuaries (see, e.g. [27]). The rationale behind this approach is that a discretized form of the NLSW equations can automatically be shock-capturing if the momentum conservation is retained in the numerical scheme. As a consequence, this simple and efficient scheme is able to track the actual location of wave breaking and compute the associated energy dissipation without the aid of analytical solutions for bore approximation or empirical formulations for energy dissipation. In order to resolve the frequency dispersion up to an acceptable level of accuracy using as few layers as possible, a technique as proposed in [29] is employed that is tailored to wave propagation applications. It is based on a compact difference scheme for the approximation of vertical gradient of the non-hydrostatic pressure. Unlike Boussinesq-type wave models, which rely on higher order derivative terms for better dispersion characteristics, the present model improves its frequency dispersion by increasing the number of vertical layers rather than increasing the order of derivatives of the dependent variables. Hence, it contains at most second order spatial derivatives. Therefore, the second order accurate finite difference approximations are considered to be sufficiently accurate from a numerical point of view. For the calculation of wave run-up on the beach, use of moving boundary conditions is required. Several numerical strategies have been proposed for a proper representation of the interface of water and land. We refer to [1, 7] for overviews on this subject. In the present work, a very simple approach as treated in [28] is adopted. This method tracks the motion of the shoreline very accurately without posing numerical instabilities by ensuring non-negative water depths.

2 Governing equations

We consider a two-dimensional wave motion in the vertical plane. The waves are assumed to approach the beach perpendicularly. The physical domain represented in a Cartesian coordinate system (x,z) is bounded vertically by the free-surface level above the reference plane, $z = \zeta(x,t)$, and the bottom level measured from the reference plane positively downwards, $z = -d(x)$. Furthermore, t is the time. The water depth is $H = \zeta + d$, see Fig. 2.

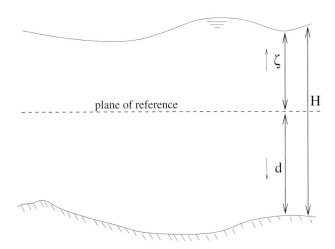

Fig. 2 Water area with free surface and bottom.

The governing equations are the Euler equations for the flow of an incompressible, inviscid fluid with a constant density ρ_0, given by

$$\frac{\partial u}{\partial x} + \frac{\partial w}{\partial z} = 0, \tag{1}$$

$$\frac{\partial u}{\partial t} + \frac{\partial u^2}{\partial x} + \frac{\partial wu}{\partial z} + \frac{g}{\rho_0}\frac{\partial \zeta}{\partial x} + \frac{1}{\rho_0}\frac{\partial q}{\partial x} = 0, \qquad (2)$$

$$\frac{\partial w}{\partial t} + \frac{\partial uw}{\partial x} + \frac{\partial w^2}{\partial z} + \frac{1}{\rho_0}\frac{\partial q}{\partial z} = 0, \qquad (3)$$

where $u(x,z,t)$ and $w(x,z,t)$ are the mean velocity components in the horizontal $x-$ and vertical $z-$direction, respectively, g is the acceleration of gravity, $q(x,z,t)$ is the non-hydrostatic pressure. For convenience, we choose $\rho_0 = 1$. For accuracy reasons, the total pressure has been split into two components, i.e. hydrostatic, $g(\zeta - z)$, and non-hydrostatic, q; for details see, e.g., Ref. [9] and [35]. Next, the kinematic conditions are given by

$$w|_{z=\zeta} = \frac{\partial \zeta}{\partial t} + u\frac{\partial \zeta}{\partial x}, \quad w|_{z=-d} = -u\frac{\partial d}{\partial x}. \qquad (4)$$

To compute the free surface, we integrate Eq. (1) over the water depth H and use the kinematic condition at the free surface (4), giving the following free-surface equation

$$\frac{\partial \zeta}{\partial t} + \frac{\partial Q}{\partial x} = 0, \quad Q \equiv UH = \int_{-d}^{\zeta} u\, dz, \qquad (5)$$

with Q the flow rate and U the depth-averaged horizontal velocity.

To get a unique solution, proper numbers and types of boundary conditions are required at all boundaries of the physical domain considered. We distinguish four types of boundaries: i) free surface, ii) bottom, iii) offshore and iv) onshore. In principle, one normal and one tangential component of the velocity and/or stress are imposed at these boundaries. At the free surface, we assume no wind (tangential stress) and $q|_{z=\zeta} = 0$ (normal stress). At the bottom, we assume no bottom friction (tangential stress) and the normal velocity is imposed through the kinematic condition (4). Because of continuity, the discharge UH must equal $c_g \zeta$, with c_g the group velocity [11]. For arbitrary depths, we have

$$c_g = n\frac{\omega}{k}, \quad n = \frac{1}{2}\left(1 + \frac{2kH}{\sinh 2kH}\right), \qquad (6)$$

where ω and k are the angular frequency and the wave number, respectively, of the first input harmonic. Thus, we impose the depth-averaged velocity $U = n\omega\zeta/kH$ at the offshore boundary. Finally, we may consider two types of the onshore condition. The moving shoreline, in the case of calculation of wave run-up on the beach, requires a numerical treatment which will be outlined in Sect. 3.2. In the pre-breaking zone, an artificial outflow condition is imposed. Usually, the so-called Sommerfeld radiation condition is employed, which allows the (long) waves to cross the outflow boundary without reflections [11]. This condition is given by

$$\frac{\partial f}{\partial t} + c\frac{\partial f}{\partial x} = 0, \qquad (7)$$

where f represents the surface elevation or the tangential velocity and c is the wave phase velocity, which equals $c = \sqrt{gH}$ for long waves.

The set of equations (1)–(3), (5) can be considered as mass- and momentum-conservative NLSW equations with the inclusion of the vertical acceleration. These equations are valid in both pre-breaking and breaking regions describing nonlinear shoaling, breaking, dissipation after breaking and run-up of waves.

3 Numerical framework

The numerical framework is briefly presented and discussed. Distinction is made between space and time discretizations as treated in Sect. 3.1 and 3.2, respectively, after which the solution technique is outlined in Sect. 3.3.

3.1 Space discretization

3.1.1 Grid schematization

The physical domain is discretized by employing a structured grid. A distinction is made between the definition of the grid in the horizontal and vertical direction. In the horizontal planes, we consider a regular grid $\{x_{i+1/2}|x_{i+1/2} = i\Delta x, i = 0,...,I\}$ with Δx the length of the cell. The location of the cell centre is given by $x_i = (x_{i-1/2} + x_{i+1/2})/2$. In the vertical direction, a boundary-fitted grid is employed. The domain is divided into K layers. The interface between two layers is denoted as $z_{k+1/2}(x,t)$ with $k = 0,...,K$. The layer thickness is defined as $h_k = z_{k+1/2} - z_{k-1/2} = f_k H$ with $0 \leq f_k \leq 1$ and $\sum_k f_k = 1$; see Fig. 3. The water level $z_{K+1/2} = \zeta$ and the bottom level $z_{1/2} = -d$ are located at $x = x_i$. As a consequence, $H = \zeta + d$ is given in point i and the water depth at a cell vertex is not uniquely defined. The water depth at $i+1/2$ is denoted as $\hat{H}$ and its approximation depends on the direction of $Q_{i+1/2}$, i.e. the flow rate normal to the face of the water column $i+1/2$, as follows,

$$\hat{H}_{i+1/2} = \begin{cases} H_i, & \text{if } Q_{i+1/2} > 0. \\ H_{i+1}, & \text{if } Q_{i+1/2} < 0. \\ \max(\zeta_i, \zeta_{i+1}) + \min(d_i, d_{i+1}), & \text{if } Q_{i+1/2} = 0. \end{cases} \quad (8)$$

The approximation of $\hat{H}_{i+1/2}$ in case of $Q_{i+1/2} = 0$ is heuristically based and appears to be very robust. For consistency, we have

$$z_{i+1/2,1/2} = \begin{cases} -d_i, & \text{if } Q_{i+1/2} > 0. \\ -d_{i+1}, & \text{if } Q_{i+1/2} < 0. \\ -\min(d_i, d_{i+1}), & \text{if } Q_{i+1/2} = 0. \end{cases} \quad (9)$$

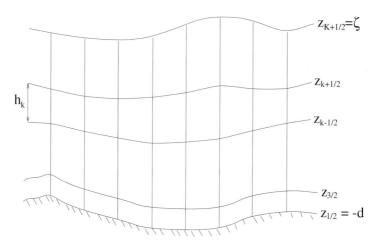

Fig. 3 Vertical grid definition with layer interfaces.

and

$$z_{i+1/2,K+1/2} = \begin{cases} \zeta_i, & \text{if } Q_{i+1/2} > 0. \\ \zeta_{i+1}, & \text{if } Q_{i+1/2} < 0. \\ \max(\zeta_i, \zeta_{i+1}), & \text{if } Q_{i+1/2} = 0. \end{cases} \quad (10)$$

and

$$z_{i+1/2,k+1/2} = z_{i+1/2,k-1/2} + f_k \hat{H}_{i+1/2}, \quad k = 1,\ldots,K-1. \quad (11)$$

This completes our description of the grid.

The vertical grid schematization gives rise to the definition of the vertical velocity with respect to the moving layer interfaces. The vertical velocity relative to layer interface $z_{k+1/2}$, denoted as $\omega_{k+1/2}$, is defined as the difference between the vertical velocity along the streamline and the vertical velocity along the interface, as follows,

$$\omega_{k+1/2} = w(z_{k+1/2}) - \frac{\partial z_{k+1/2}}{\partial t} - u(z_{k+1/2})\frac{\partial z_{k+1/2}}{\partial x}. \quad (12)$$

The kinematic boundary conditions, in terms of relative vertical velocity, are $\omega_{1/2} = \omega_{K+1/2} = 0$.

3.1.2 Location of grid variables

A staggered grid arrangement is used in which the velocity components u and w are located at the centers of the cell faces $(i+1/2,k)$ and $(i,k+1/2)$, respectively. The water level ζ is located at i. Concerning the non-hydrostatic pressure q, two ways to assign this unknown to grid points may be employed. This variable can be given either at the cell center (i,k) or at the face $(i,k+1/2)$. The choice depends on the discretization of the vertical pressure gradient, namely, explicit central differences and an edge-based compact finite difference scheme [17], respectively. Since, the present work deals with the application to wave propagation, only the latter discretization will be considered [29]. The former approximation is particularly meant for applications where vertical structures are important e.g., stratified flows with density currents. Like w, the relative vertical velocity ω is located at the face $(i, k+1/2)$. Fig. 4 shows the staggered grid layout.

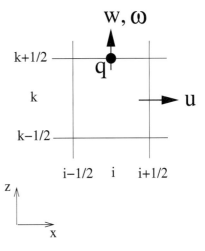

Fig. 4 Arrangement of the unknowns in a staggered grid.

Unknowns not present at points where they are required are computed by interpolation using the fewest number of interpolation points unless stated otherwise. So, $\overline{\varphi}_i^x$ indicates arithmetic averaging of the unknown φ in x-direction over their two points of definition that are nearest to i. The unknown φ not given at layer interface $z_{k+1/2}$ is approximated at this interface as

$$\overline{\varphi}_{k+1/2}^z = \varphi(z_{k+1/2}) \approx \frac{\varphi_k h_{k+1} + \varphi_{k+1} h_k}{h_k + h_{k+1}}. \tag{13}$$

Note that $\overline{\varphi}_k^z = (\varphi_{k+1/2} + \varphi_{k-1/2})/2$ since, arithmetic averaging inside a layer is exact. Finally, $\overline{\varphi}_{i,k}^{xz}$ gives the average value of φ at (i,k) resulting from the two one-dimensional interpolation formulas in each direction.

Space discretization of the governing equations is carried out in a finite volume/finite difference fashion. For each unknown, we define a collection of a finite number of non-overlapping control volumes that covers the whole domain. Each unknown, except the water level, is considered as volume-averaged and is at the centre of its control volume,

$$u_{i+1/2,k} = \frac{1}{\widehat{h}_{i+1/2,k}} \int_{z=z_{k-1/2}}^{z_{k+1/2}} u|_{x=x_{i+1/2}} dz, \quad w_{i,k+1/2} = \frac{1}{h_{i,k+1/2}} \int_{z=z_k}^{z_{k+1}} w|_{x=x_i} dz,$$

$$q_{i,k+1/2} = \frac{1}{h_{i,k+1/2}} \int_{z=z_k}^{z_{k+1}} q|_{x=x_i} dz, \tag{14}$$

with
$$\hat{h}_{i+1/2,k} = f_k \hat{H}_{i+1/2} \tag{15}$$
and
$$h_{i,k+1/2} = \frac{1}{2}(h_{i,k} + h_{i,k+1}). \tag{16}$$

3.1.3 Space discretization of global continuity equation

A global mass conservative approximation of Eq. (5) is given by

$$\frac{d\zeta_i}{dt} + \frac{\hat{H}_{i+1/2}U_{i+1/2} - \hat{H}_{i-1/2}U_{i-1/2}}{\Delta x} = 0, \tag{17}$$

with

$$U_{i+1/2} = \frac{1}{\hat{H}_{i+1/2}} \sum_{k=1}^{K} u_{i+1/2,k} \hat{h}_{i+1/2,k}. \tag{18}$$

3.1.4 Space discretization of local continuity equation

The space discretization of Eq. (1) consists of two steps. First, this equation is integrated vertically over its control volume and thereafter, an appropriate finite difference scheme is applied to each horizontal term of the equation. The layer-averaged continuity equation (1) for layer $1 \leq k \leq K$ is obtained using the Leibniz' rule, as follows,

$$\int_{z_{k-1/2}}^{z_{k+1/2}} \left(\frac{\partial u}{\partial x} + \frac{\partial w}{\partial z}\right) dz = \frac{\partial h_k u_k}{\partial x} - u \frac{\partial z}{\partial x}\Big|_{z_{k-1/2}}^{z_{k+1/2}} + w_{k+1/2} - w_{k-1/2} = 0. \tag{19}$$

By virtue of (12), this equation becomes

$$\frac{\partial h_k}{\partial t} + \frac{\partial h_k u_k}{\partial x} + \omega_{k+1/2} - \omega_{k-1/2} = 0, \tag{20}$$

so that the total amount of water in a moving cell with thickness h_k is conserved. Discretization of Eq. (20) in x–direction gives

$$\frac{dh_{i,k}}{dt} + \frac{\phi_{i+1/2,k} - \phi_{i-1/2,k}}{\Delta x} + \omega_{i,k+1/2} - \omega_{i,k-1/2} = 0, \tag{21}$$

with

$$\phi_{i+1/2,k} = \hat{h}_{i+1/2,k} u_{i+1/2,k}. \tag{22}$$

3.1.5 Space discretization of horizontal momentum equation

Again, the space discretization of Eq. (2) consists of two steps as outlined in Sect. 3.1.4. The derivation of layer-averaging of the terms in Eq. (2) is given in [35] and will not be repeated here. The layer-averaged $u-$momentum equation in conservative form reads

$$\frac{\partial h_k u_k}{\partial t} + \frac{\partial h_k u_k^2}{\partial x} + \overline{u}^z_{k+1/2}\omega_{k+1/2} - \overline{u}^z_{k-1/2}\omega_{k-1/2}$$

$$+ gh_k \frac{\partial \zeta}{\partial x} + \frac{\partial h_k \overline{q}^z_k}{\partial x} - q_{k+1/2}\frac{\partial z_{k+1/2}}{\partial x} + q_{k-1/2}\frac{\partial z_{k-1/2}}{\partial x} = 0. \quad (23)$$

A spatial discretization of Eq. (23) is given by

$$\frac{d\overline{h}^x_{i+1/2,k} u_{i+1/2,k}}{dt} + \frac{\hat{u}_{i+1,k}\overline{\phi}^x_{i+1,k} - \hat{u}_{i,k}\overline{\phi}^x_{i,k}}{\Delta x}$$

$$+ \overline{u}^z_{i+1/2,k+1/2}\overline{\omega}^x_{i+1/2,k+1/2} - \overline{u}^z_{i+1/2,k-1/2}\overline{\omega}^x_{i+1/2,k-1/2}$$

$$+ g\overline{h}^x_{i+1/2,k} \frac{\zeta_{i+1} - \zeta_i}{\Delta x} + \frac{h_{i+1,k}\overline{q}^z_{i+1,k} - h_{i,k}\overline{q}^z_{i,k}}{\Delta x}$$

$$- \overline{q}^x_{i+1/2,k+1/2}\frac{z_{i+1,k+1/2} - z_{i,k+1/2}}{\Delta x} + \overline{q}^x_{i+1/2,k-1/2}\frac{z_{i+1,k-1/2} - z_{i,k-1/2}}{\Delta x} = 0. \quad (24)$$

The one-sided second order upwind scheme is used to approximate $\hat{u}$ at (i,k) [17],

$$\hat{u}_{i,k} = \begin{cases} \frac{3}{2}u_{i-1/2,k} - \frac{1}{2}u_{i-3/2,k}, & \text{if } \overline{\phi}^x_{i,k} \geq 0. \\ \frac{3}{2}u_{i+1/2,k} - \frac{1}{2}u_{i+3/2,k}, & \text{if } \overline{\phi}^x_{i,k} < 0. \end{cases} \quad (25)$$

This scheme generates a limited amount of numerical dissipation which is sufficient to effectively suppress spurious waves with wavelength $2\Delta x$. These undesired wave components are due to nonlinearities.

Since, the velocity component u is the primitive variable and not $\phi = hu$, Eq. (24) is not appropriate for further implementation. For the reformulation, we first consider the discretized form of Eq. (20) in point $(i+1/2,k)$,

$$\frac{d\overline{h}^x_{i+1/2,k}}{dt} + \frac{\overline{\phi}^x_{i+1,k} - \overline{\phi}^x_{i,k}}{\Delta x} + \overline{\omega}^x_{i+1/2,k+1/2} - \overline{\omega}^x_{i+1/2,k-1/2} = 0. \quad (26)$$

Multiplying Eq. (26) with $u_{i+1/2,k}$ and subtracting the result from Eq. (24), after which it is divided by $\overline{h}^x_{i+1/2,k}$, yields

$$\frac{du_{i+1/2,k}}{dt} + \frac{1}{\overline{h}^x_{i+1/2,k}} \left(\frac{\overline{\phi}^x_{i+1,k}(\hat{u}_{i+1,k} - u_{i+1/2,k}) - \overline{\phi}^x_{i,k}(\hat{u}_{i,k} - u_{i+1/2,k})}{\Delta x} \right)$$

$$+ \frac{\overline{\omega}^x_{i+1/2,k+1/2}}{\overline{h}^x_{i+1/2,k}} (\overline{u}^z_{i+1/2,k+1/2} - u_{i+1/2,k}) - \frac{\overline{\omega}^x_{i+1/2,k-1/2}}{\overline{h}^x_{i+1/2,k}} (\overline{u}^z_{i+1/2,k-1/2} - u_{i+1/2,k})$$

$$+ g\frac{\zeta_{i+1} - \zeta_i}{\Delta x} + \frac{1}{\overline{h}^x_{i+1/2,k}} \left(\frac{h_{i+1,k}\overline{q}^z_{i+1,k} - h_{i,k}\overline{q}^z_{i,k}}{\Delta x} \right)$$

$$- \frac{\overline{q}^x_{i+1/2,k+1/2}}{\overline{h}^x_{i+1/2,k}} \frac{z_{i+1,k+1/2} - z_{i,k+1/2}}{\Delta x} + \frac{\overline{q}^x_{i+1/2,k-1/2}}{\overline{h}^x_{i+1/2,k}} \frac{z_{i+1,k-1/2} - z_{i,k-1/2}}{\Delta x} = 0. \quad (27)$$

Eq. (27) guarantees conservation of momentum and is thus valid for simulation of breaking waves. Note that Eq. (27) does not contain a bed slope source term. Hence, transition from sub- to supercritical flows near steep bed slopes can be computed correctly.

3.1.6 Space discretization of vertical momentum equation

The final discretized w−momentum equation can be derived in exactly the same manner as done for the u−momentum equation except for the pressure gradient. The equation is given by

$$\frac{dw_{i,k+1/2}}{dt} + \frac{\overline{\phi}^z_{i+1/2,k+1/2}}{2\overline{h}^z_{i,k+1/2}} \frac{w_{i+1,k+1/2} - w_{i,k+1/2}}{\Delta x}$$

$$+ \frac{\overline{\phi}^z_{i-1/2,k+1/2}}{2\overline{h}^z_{i,k+1/2}} \frac{w_{i,k+1/2} - w_{i-1,k+1/2}}{\Delta x}$$

$$+ \frac{\overline{\omega}^z_{i,k+1}}{2\overline{h}^z_{i,k+1/2}} (w_{i,k+3/2} - w_{i,k+1/2}) + \frac{\overline{\omega}^z_{i,k}}{2\overline{h}^z_{i,k+1/2}} (w_{i,k+1/2} - w_{i,k-1/2})$$

$$+ \frac{1}{\overline{h}^z_{i,k+1/2}} \int_{z_k}^{z_{k+1}} \frac{\partial q}{\partial z}|_i dz = 0. \quad (28)$$

Note that central differences have been used in Eq. (28), i.e.,

$$\hat{w}_{i+1/2,k+1/2} = \frac{1}{2}(w_{i,k+1/2} + w_{i+1,k+1/2}). \quad (29)$$

Since, the accuracy of the frequency dispersion for relatively short waves strongly depends on the discretization of vertical motion, we apply a second order compact scheme for the approximation of the vertical gradient of non-hydrostatic pressure, allowing very few vertical grid points with relatively low numerical dispersion and dissipation [17]. Firstly, we consider the w−momentum equation at $z_{k+1/2}$, Eq. (28), in which the pressure gradient $\partial q/\partial z$ is approximated through backward differencing and subsequent the w−momentum equation at $z_{k-1/2}$ where the approximation of $\partial q/\partial z$ is obtained by means of forward differencing. Thereafter, we take the average of the discretized w−momentum equations at $z_{k-1/2}$ and $z_{k+1/2}$ onto the layer k. Thus, the integral of pressure gradient in Eq. (28) is approximated by means of backward differencing,

$$\frac{1}{\overline{h}^z_{i,k+1/2}}\int_{z_k}^{z_{k+1}}\frac{\partial q}{\partial z}|_i dz = \frac{q(z_{i,k+1})-q(z_{i,k})}{\overline{h}^z_{i,k+1/2}} \approx \frac{q_{i,k+1/2}-q_{i,k-1/2}}{h_{i,k}}. \quad (30)$$

The w−momentum equation at interface $z_{k-1/2}$ is obtained from Eq. (28) by decreasing the index k by 1. However, the integral of pressure gradient is evaluated using forward differencing. This gives

$$\frac{1}{\overline{h}^z_{i,k-1/2}}\int_{z_{k-1}}^{z_k}\frac{\partial q}{\partial z}|_i dz = \frac{q(z_{i,k})-q(z_{i,k-1})}{\overline{h}^z_{i,k-1/2}} \approx \frac{q_{i,k+1/2}-q_{i,k-1/2}}{h_{i,k}}. \quad (31)$$

Finally, we take the average of the w−momentum equations at interfaces $z_{k-1/2}$ and $z_{k+1/2}$, giving

$$\frac{d(w_{i,k+1/2}+w_{i,k-1/2})}{2dt} + \frac{1}{2}\left((L^w w)_{i,k+1/2}+(L^w w)_{i,k-1/2}\right) + \frac{q_{i,k+1/2}-q_{i,k-1/2}}{h_{i,k}} = 0, \quad (32)$$

with L^w the discrete operator representing advection terms as outlined before. Due to the use of the compact scheme, Eq. (32) contains two time derivatives for w.

It must be emphasized that Eq. (32) is solved for layers $2 \leq k \leq K$, i.e. including the free surface, but excluding the bottom. Condition $q|_{z=\zeta} = 0$ can be readily incorporated in Eq. (32) for $k=K$ as $q_{i,K+1/2}=0$. At the bottom ($k=1$), the kinematic condition $w|_{z=-d} = -u\partial d/\partial x$ is imposed.

3.2 Time integration

The spatial discretization, explained in the previous section, yields a system of ordinary differential equations as given by Eqs. (17), (21), (27) and (32). For transparency, we summarize the space-discretized momentum equations:

$$\frac{du_{i+1/2,k}}{dt} + (L^u u)_{i+1/2,k} + (G^1_x \zeta)_{i+1/2,k} + (G^2_x q)_{i+1/2,k} = 0 \quad (33)$$

and

$$\frac{d(w_{i,k+1/2}+w_{i,k-1/2})}{dt} + 2(G_z q)_{i,k} + (L^w w)_{i,k+1/2} + (L^w w)_{i,k-1/2} = 0. \quad (34)$$

In Eqs. (33) and (34), the finite difference operators L^u and L^w are linear and include approximations of the advection terms, whereas G_x^1 and G_x^2 are linear operators representing the gradients in x-direction of the water level and non-hydrostatic pressure, respectively. The linear operator G_z refers to the compact scheme for the vertical gradient of the non-hydrostatic pressure within a layer.

For time discretization we use a linear combination of the explicit and implicit Euler method, the so-called θ-method with θ lying between zero and unity. For brevity, we denote $\varphi^{n+\theta} = \theta \varphi^{n+1} + (1-\theta)\varphi^n$ for some quantity φ with n indicating the time level $t^n = n\Delta t$ where Δt is the time step. For $\theta = \frac{1}{2}$ we obtain the second order Crank-Nicolson scheme and for $\theta = 0$ and $\theta = 1$ the first order explicit and implicit Euler schemes are obtained, respectively. For stability, we take $\theta \geq \frac{1}{2}$.

Integration of Eq. (17) in time in a semi-implicit manner yields

$$\frac{\zeta_i^{n+1} - \zeta_i^n}{\Delta t} + \frac{\hat{H}_{i+1/2}^n U_{i+1/2}^{n+\theta} - \hat{H}_{i-1/2}^n U_{i-1/2}^{n+\theta}}{\Delta x} = 0. \qquad (35)$$

Based on the expressions for $\hat{H}_{i+1/2}$, as given by (8), it can be shown that if the time step is chosen such that $\Delta t |U_{i+1/2}^{n+\theta}|/\Delta x \leq 1$ at every time step, then the water depth H_i^{n+1} is non-negative at every time step [28]. Hence, flooding never happens faster than one grid size per time step, which is physically correct. This implies that the calculation of the dry areas does not need any special feature. For this reason, no complicated drying and flooding procedures as described in [27] and [1] are required. For computational efficiency, the momentum equations are not solved and velocities are set to zero if the water depth $\hat{H}_{i+1/2}$ is below a threshold value. For the examples in this study it equals 10^{-5} m.

Eq. (21) is discretised fully implicitly in time, as follows,

$$\frac{h_{i,k}^{n+1} - h_{i,k}^n}{\Delta t} + \frac{\phi_{i+1/2,k}^{n+1} - \phi_{i-1/2,k}^{n+1}}{\Delta x} + \omega_{i,k+1/2}^{n+1} - \omega_{i,k-1/2}^{n+1} = 0. \qquad (36)$$

Concerning the momentum equations, time discretization takes place by explicit time stepping for advection terms and semi-implicit time stepping using the θ-scheme for both surface level and pressure gradients, as follows,

$$\frac{u_{i+1/2,k}^{n+1} - u_{i+1/2,k}^n}{\Delta t} + (L^u u^n)_{i+1/2,k} + (G_x^1 \zeta^{n+\theta})_{i+1/2,k} + (G_x^2 q^{n+\theta})_{i+1/2,k} = 0 \qquad (37)$$

and

$$\frac{w_{i,k+1/2}^{n+1} - w_{i,k+1/2}^n}{\Delta t} + \frac{w_{i,k-1/2}^{n+1} - w_{i,k-1/2}^n}{\Delta t} + 2(G_z q^{n+\theta})_{i,k}$$

$$+ (L^w w^n)_{i,k+1/2} + (L^w w^n)_{i,k-1/2} = 0. \qquad (38)$$

3.3 Solution method

After the spatial and temporal discretization, the both locally and globally mass conserved solution ($\zeta_i^{n+1}, q_{i,k+1/2}^{n+1}, u_{i+1/2,k}^{n+1}, w_{i,k+1/2}^{n+1}$) of Eqs. (35)–(38) is found in two steps. First, the solution ($\zeta_i^{n+1}, u_{i+1/2,k}^*$) for hydrostatic flows is obtained with conservation of global mass only. Note that $u_{i+1/2,k}^*$ is not the final solution since local mass is not conserved yet. Next, the solution ($q_{i,k+1/2}^{n+1}, u_{i+1/2,k}^{n+1}, w_{i,k+1/2}^{n+1}$) is found such that local mass is conserved. In both steps, a projection method is applied, where correction to the velocity fields for the change in respectively water level and non-hydrostatic pressure is incorporated. The projection method is a well-established predictor-corrector approach for solving the incompressible Navier-Stokes equations and is usually referred to as the pressure correction technique [17].

To find the globally but not necessarily locally mass conserved solution, $U_{i+1/2}^{n+1}$ is replaced by

$$U_{i+1/2}^* = \frac{1}{\hat{H}_{i+1/2}^n} \sum_{k=1}^{K} \hat{h}_{i+1/2,k}^n u_{i+1/2,k}^*, \qquad (39)$$

and instead of Eq. (35), we now have

$$\frac{\zeta_i^{n+1} - \zeta_i^n}{\Delta t} + \frac{\hat{H}_{i+1/2}^n U_{i+1/2}^{n+\theta^*} - \hat{H}_{i-1/2}^n U_{i-1/2}^{n+\theta^*}}{\Delta x} = 0, \qquad (40)$$

with $U^{n+\theta^*} = \theta U^* + (1-\theta) U^n$. Furthermore, $u_{i+1/2,k}^*$ is the solution of the following equation

$$\frac{u_{i+1/2,k}^* - u_{i+1/2,k}^n}{\Delta t} + (L^u u^n)_{i+1/2,k} + (G_x^1 \zeta^{n+\theta})_{i+1/2,k} + (G_x^2 q^n)_{i+1/2,k} = 0. \qquad (41)$$

Note that Eq. (41) contains the non-hydrostatic pressure at the preceding time level so that u^* will not satisfy Eq. (36). Eqs. (40) and (41) are solved using a predictor-corrector procedure as follows. An estimate of the u^*−velocity, denoted as u^{**}, is made that does not satisfy Eq. (40). This is achieved by means of solving Eq. (41) with the best available guess for the water level,

$$\frac{u_{i+1/2,k}^{**} - u_{i+1/2,k}^n}{\Delta t} + (L^u u^n)_{i+1/2,k} + (G_x^1 \zeta^n)_{i+1/2,k} + (G_x^2 q^n)_{i+1/2,k} = 0. \qquad (42)$$

Next, a correction is computed involving the water level as follows. An expression for u^* is obtained by subtracting Eq. (42) from Eq. (41), to give

$$u_{i+1/2,k}^* = u_{i+1/2,k}^{**} - g\theta \Delta t (G_x^1 \Delta \zeta)_{i+1/2,k}, \qquad (43)$$

with $\Delta \zeta \equiv \zeta^{n+1} - \zeta^n$ the surface level correction. The principle of the projection method is that $\Delta \zeta$ must be such that u^* is the solution of Eq. (40) so that mass conservation for each water column is obtained. Multiplying Eq. (43) with $\hat{h}_{i+1/2,k}^n$, summing it from bottom to free surface and substituting into Eq. (40) gives

$$\frac{\Delta \zeta_i}{\Delta t} - \frac{g\theta^2 \Delta t}{\Delta x} \left(\hat{H}^n_{i+1/2} (G^1_x \Delta \zeta)_{i+1/2} - \hat{H}^n_{i-1/2} (G^1_x \Delta \zeta)_{i-1/2} \right) =$$

$$- \frac{\theta}{\Delta x} \left(\sum_{k=1}^{K} \hat{h}^n_{i+1/2,k} u^{**}_{i+1/2,k} - \sum_{k=1}^{K} \hat{h}^n_{i-1/2,k} u^{**}_{i-1/2,k} \right)$$

$$- \frac{1-\theta}{\Delta x} \left(\hat{H}^n_{i+1/2} U^n_{i+1/2} - \hat{H}^n_{i-1/2} U^n_{i-1/2} \right). \tag{44}$$

For each point i, we thus have an equation for $\Delta \zeta_i$, $\Delta \zeta_{i-1}$ and $\Delta \zeta_{i+1}$. The resulting tri-diagonal system of equations is solved directly by the Thomas algorithm [17].

Once the water level ζ^{n+1} and the intermediate velocity component u^* are determined, a prediction for the intermediate vertical velocity w^* is computed by using Eq. (38) with the best known non-hydrostatic pressure q^n,

$$\frac{w^*_{i,k+1/2} - w^n_{i,k+1/2}}{\Delta t} + \frac{w^*_{i,k-1/2} - w^n_{i,k-1/2}}{\Delta t} + 2(G_z q^n)_{i,k}$$

$$+ (L^w w^n)_{i,k+1/2} + (L^w w^n)_{i,k-1/2} = 0. \tag{45}$$

The computed velocities (u^*, w^*) will not accurately fulfil the local continuity equation (36) and the non-hydrostatic pressure must be corrected to achieve this. The velocities can then be modified accordingly. In deriving an equation for the solution of pressure correction, $\Delta q \equiv q^{n+1} - q^n$, Eqs. (41) and (45) are subtracted from Eqs. (37) and (38), respectively, resulting in

$$\frac{u^{n+1}_{i+1/2,k} - u^*_{i+1/2,k}}{\Delta t} + \theta (G^2_x \Delta q)_{i+1/2,k} = 0, \tag{46}$$

$$\frac{w^{n+1}_{i,k+1/2} - w^*_{i,k+1/2}}{\Delta t} + 2\theta (G_z \Delta q)_{i,k} = 0, \tag{47}$$

whereby the difference $w^{n+1}_{i,k-1/2} - w^*_{i,k-1/2}$ is neglected. Based on an analysis, it appears that this neglect does not affect the modeling of linear dispersion [35]. Substitution of Eqs. (46) and (47) into Eq. (36) using expression (12) gives a Poisson equation for Δq,

$$- \frac{\theta \Delta t}{\Delta x} \left(\hat{h}^{n+1}_{i+1/2,k} (G^2_x \Delta q)_{i+1/2,k} - \hat{h}^{n+1}_{i-1/2,k} (G^2_x \Delta q)_{i-1/2,k} \right)$$

$$+ \frac{\theta \Delta t \partial z^{n+1}_{i,k+1/2}/\partial x}{2 \left(h^{n+1}_{i,k} + h^{n+1}_{i,k+1} \right)} \left[h^{n+1}_{i,k+1} \left((G^2_x \Delta q)_{i+1/2,k} + (G^2_x \Delta q)_{i-1/2,k} \right) + \right.$$

$$\left. h^{n+1}_{i,k} \left((G^2_x \Delta q)_{i+1/2,k+1} + (G^2_x \Delta q)_{i-1/2,k+1} \right) \right]$$

$$-\frac{\theta \Delta t \partial z_{i,k-1/2}^{n+1}/\partial x}{2\left(h_{i,k-1}^{n+1}+h_{i,k}^{n+1}\right)}\left[h_{i,k}^{n+1}\left((G_x^2 \Delta q)_{i+1/2,k-1}+(G_x^2 \Delta q)_{i-1/2,k-1}\right)+\right.$$

$$\left. h_{i,k-1}^{n+1}\left((G_x^2 \Delta q)_{i+1/2,k}+(G_x^2 \Delta q)_{i-1/2,k}\right)\right]$$

$$-2\theta \Delta t\left[(G_z \Delta q)_{i,k}-(G_z \Delta q)_{i,k-1}\right]=$$

$$-\frac{1}{\Delta x}\left(\hat{h}_{i+1/2,k}^{n+1}u_{i+1/2,k}^{*}-\hat{h}_{i-1/2,k}^{n+1}u_{i-1/2,k}^{*}\right)$$

$$+\overline{u}_{i,k+1/2}^{*}{}^{xz}\frac{\partial z_{i,k+1/2}^{n+1}}{\partial x}-\overline{u}_{i,k-1/2}^{*}{}^{xz}\frac{\partial z_{i,k-1/2}^{n+1}}{\partial x}$$

$$-\left(w_{i,k+1/2}^{*}-w_{i,k-1/2}^{*}\right). \tag{48}$$

Once Δq is obtained, we can calculate $u_{i+1/2,k}^{n+1}$ and $w_{i,k+1/2}^{n+1}$, respectively, through Eqs. (46) and (47). Local mass is conserved.

The matrix of (48) is a non-symmetric discrete Laplacian and contains 15 non-zero diagonals. For the solution, we adopt the BiCGSTAB method [34] preconditioned with the incomplete LU factorizations: ILU [23] and MILU (Modified ILU) [13]. Based on several numerical experiments, an optimum in the convergence rate is found by taking 55% of MILU and 45% of ILU. It has been observed that the pressure correction is slowly time varying. This suggests that there is no need for the system of equations (48) to be preconditioned at every time step. Since preconditioning is relative expensive with respect to amount of work, much CPU-time can be saved by preconditioning the system every ten to twenty time steps, as suggested by our experiments.

The overall solution for a time step can be summarized as follows:

1. Start the sequence by taking the unknowns ζ^n, u^n, w^n, q^n, either initially or from the previous time level.
2. Solve Eq. (42) to obtain u^{**}.
3. Solve Eq. (44) to obtain the correction $\Delta \zeta$ for water level.
4. Correct the water level and horizontal velocity by means of $\zeta^{n+1} = \zeta^n + \Delta \zeta$, Eq. (43) for u^*.
5. Solve Eq. (45) to obtain w^*.
6. Solve the Poisson equation (48) to obtain the correction Δq.
7. Update the non-hydrostatic pressure and velocities using $q^{n+1} = q^n + \Delta q$, Eq. (46) for u^{n+1} and Eq. (47) for w^{n+1}.
8. Update the relative vertical velocity $\omega_{k+1/2}$ from Eq. (12).

4 Numerical experiments

Our main interest concerns the simulation of transformation of non-linear waves over rapidly varying bathymetry in coastal zones. The present method using the compact scheme is validated by applying it to a number of test cases for which experimental data exist. Concerning the range of applicability of the model to values of kH, indicating the relative importance of linear wave dispersion, results of our numerical analysis, as depicted in Fig. 5, suggest that two layers are sufficient to

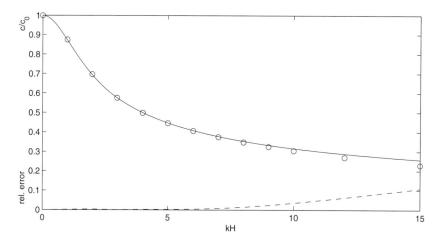

Fig. 5 Normalized wave celerity vs relative depth for linear dispersion. Non-hydrostatic model with two layers (circles), exact (solid line), relative error (dashed line). The quantities $c_0 = \sqrt{gH}$ and c are the long wave celerity and the wave phase velocity, respectively, and the relative error is $|c_{\text{computed}} - c_{\text{exact}}|/c_{\text{exact}}$.

compute linear dispersive waves up to $kH \leq 7$ (typical for coastal areas) with a relative error of at most 1%. Hence, only two equidistant layers are therefore taken in the present numerical experiments.

Simulations of breaking waves and wave run-up are presented in this section. Not only regular waves on a plane sloping bed that are well documented in the literature will be validated but also irregular waves over a barred cross-shore profile. In the test cases discussed, different types of wave breakers for given offshore wave characteristics and beach slope are given, notably, spilling (predominant on flat slopes of beaches) and plunging (predominant on steep slopes) breakers. Details may be found in [12].

While, the cross-shore motion is the main issue in this study, calculation of wave shoaling, refraction and diffraction around a shoal in two horizontal dimensions is also discussed in this section. This relatively computing-intensive application aims among other things at assessing the computational cost per grid point per time step.

The numerical results presented below have been published in previous work; see [29, 35, 36].

4.1 Regular wave breaking on a slope

A number of regular wave experiments on plane slopes were performed by Hansen and Svendsen [14]. The experiments were conducted in a wave flume with a plain slope of 1:34.26. The waves were generated at a depth of 0.36 m. A second order Stokes wave at the toe of the slope is imposed. The wave height is 3.6 cm and the period is 2.0 s. In this case the breaker type is spilling. Time series of the surface elevation were taken at a number of locations along the flume. The simulation period of 120 s has been carried out with a time step of 0.05 s. The first order implicit Euler scheme for time integration is applied ($\theta = 1$). The 15 m flume is covered with 600 grid cells with a grid size of 0.025 m.

Fig. 6 shows the comparison between the measured and calculated wave height and mean free surface (the slope starts at $x=0$m). The agreement for wave height

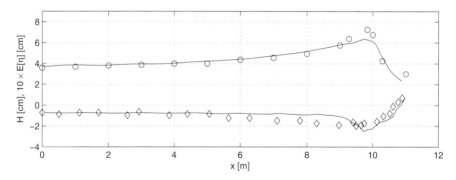

Fig. 6 Computed wave height (upper trend) and set-up (lower trend) compared to data from Hansen and Svendsen [14] for regular spilling breaker. Present method (solid line), experiment (circles, diamonds).

is quite good. Also, the model correctly predicts both shoaling and the position of the breaking point. The set-up tends to be underpredicted shoreward of the breaking point. Furthermore, the model could not reproduce the shoreward shift of the set-up relative to the breaking point. These observed deficiencies are believed to be attributed to a relative inaccurate vertical distribution of the horizontal velocity in the breaking zone, since only two layers are adopted here. This may be improved by adding more layers, possibly combined with a turbulence model. Still, with the present model using two layers, the trend of both wave height and set-up is consistently fairly well predicted.

4.2 Periodic wave run-up on a planar beach

An analytical solution for periodic wave run-up on a plane slope by Carrier and Greenspan [8] is used to verify the accuracy of the shoreline movement calculation. This classical test has been used frequently for assessing the quality of various shoreline boundary condition techniques used in the NLSW equations; see e.g., [16, 20].

A sinusoidal wave with height of 0.006 m and period of 10 s is propagating over a beach with slope 1:25. The maximum still water depth is 0.5 m. In the numerical experiment, a grid spacing of Δx=0.04 m and and a time step of Δt=0.05 s is employed. This time step has been chosen such that the water depth is non-negative everywhere. Furthermore, $\theta = 1$ is chosen. The computational flume has a length of 2 incident wavelengths. Only one layer is adopted here. Since the dispersive effects are relatively small, the non-hydrostatic pressure is not included in the depth-averaged calculation. No wave breaking occurs.

Comparison between the computed free surface envelope and the analytical solution is plotted in Fig. 7. Good agreement is obtained between the computed and

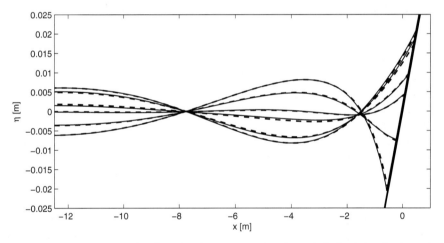

Fig. 7 Computed envelope of surface elevations compared to the analytical solution for the periodic wave run-up on a planar beach. Present method (solid line), theory (dashed line).

theoretical values. This also holds for the horizontal movement of the shoreline as demonstrated in Fig. 8.

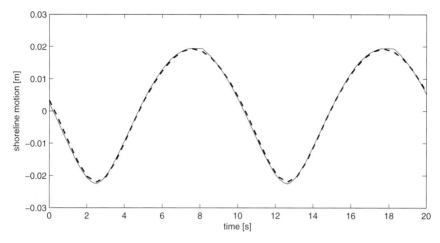

Fig. 8 Computed horizontal shoreline movement compared to the analytical solution for the periodic wave run-up on a planar beach. Present method (solid line), theory (dashed line).

4.3 Regular breaking waves over a submerged bar

In Ref. [10], an overview is given of the physical tests of regular waves over a submerged trapezoidal bar in a wave flume. The description of the experimental set-up for the bar tests can also be found in Ref. [2]. These tests have been used frequently for the evaluation of the performance of various Boussinesq-type wave models. In particular, the bound higher harmonics generated on the upward slope of the bar become free behind the bar, resulting in an irregular wave pattern. This puts heavy demands on the accuracy of the computed dispersion relation. Moreover, contrary to breaking on a slope, the position of incipient wave breaking on the horizontal part of the bar is more difficult to be detected by breaking initiation criteria usually employed in Boussinesq-type models [11].

The computational flume has a length of 30 m. The still water depth is 0.4 m, which is reduced to 0.1 m at the bar. The offshore slope is 1:20 and the shoreward slope is 1:10. The geometry is depicted in Fig. 9 where the regular wave enters from the left (x=0m). Three measurement conditions have been considered in [10] of which one of them is discussed here, namely fairly long wave with a wave period of 2.525 s and a wave height of 2.9 cm. Spilling breakers have been observed in the region between 13.3 m (station 6) and 15.3 m (station 8); see also the snapshot of surface elevation shown in Fig. 9.

In the numerical experiment, a grid spacing of Δx=0.05 m and a time step of Δt=0.01 s is employed. The duration of the simulation is set to 40 s ($\theta = 1$), so that the higher harmonics will reach the farthest station at 21 m before the end of the computation. At the outgoing boundary, the depth at the beach with a slope of 1:25 (starting at x=25 m) has been limited to 0.2 m, so that Sommerfeld radiation condition (7) for long waves can be applied.

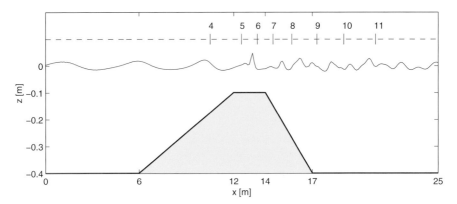

Fig. 9 A snapshot of the free surface and bottom geometry with location of wave gauges for the test of submerged bar.

Comparisons between the measurements and the results of the computations at different locations are plotted in Fig. 10. Good agreement, both in magnitude and phase, is obtained between computed surface elevations and the observed values. Further, it can be seen that the nonlinear shoaling process is well described by the proposed model. Also, the breaking zone between station 6 and 8, in which wave heights on top of the bar are decreased significantly, is represented well. Finally, the dispersion of the free waves behind the bar is predicted quite well.

4.4 Irregular wave breaking in a laboratory barred surf zone

The laboratory flume test of Boers [4] is considered, in which random, uni-directional waves propagate towards a bar-trough beach profile that was adopted from an actual barred sandy beach (see Fig. 11). The origin of the x-axis is at the beginning of the slope. During the experiments, physical parameters in the surf zone such as wave heights and periods have been collected based on the measured free surface elevations at 70 locations. In Ref. [4] a number of wave conditions with different significant wave heights and peak periods for generated incident waves are considered. In this study, a case with a relatively low wave steepness where waves break in the shallow region only, is discussed. The breaker type appears to be weakly plunging.

At the offshore boundary, an irregular wave is imposed with the significant wave height of 0.103 m and the peak period of 3.33 s. The grid size is set to 0.025 m and the time step is taken as 0.025 s. The simulation time is set to 1700 s. Since only permanent waves occur, $\theta = 1$ is chosen for time discretization.

In Fig. 12, spectral comparisons with the numerical and laboratory data are made. The spatial evolution of the wave spectra is characterized by an amplification of spectral levels at both sub- and super-harmonic ranges, consistent with three-wave interaction rules, followed by a transformation toward a broad spectral shape in the

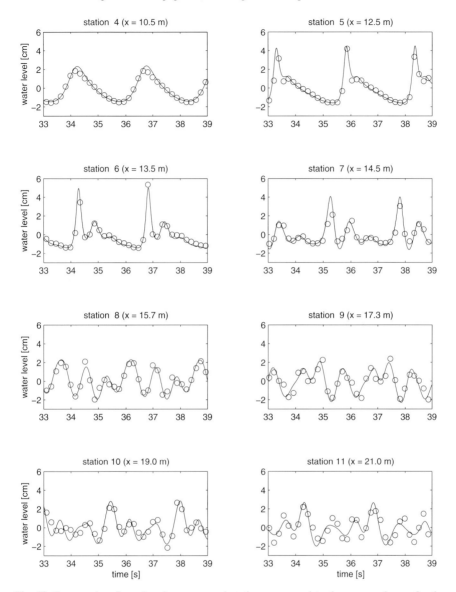

Fig. 10 Computed surface elevations at several stations compared to the measured ones for the wave over submerged bar. Present method (solid line), experiment (circles).

surf zone, attributed to the nonlinear couplings and dissipation. The present numerical method captures the dominant features of the attendant spectral evolution, both in the shoaling region and the surf zone. Nevertheless, from the breaker bar and further, the wave energy is slightly overestimated, in particular the high-frequency part. Apart from this small defect, the numerical model predicts the transforma-

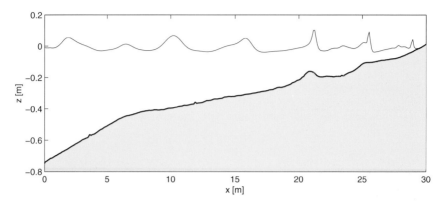

Fig. 11 A snapshot of the free surface and bathymetry of the laboratory flume experiment of Boers [4].

tion of wave energy through the flume where the amount of energy in the short waves reduces, whereas the amount of energy in the long waves increases. Note the slight overestimation of the energy density in the low-frequency part at $x = 26$ m and $x = 28$ m, which might be due to the reflection of infragravity waves against the offshore boundary.

4.5 Deformation of waves by an elliptic shoal on sloped bottom

Deformation of waves by a shoal on plane sloping bed is very interesting because of practical importance in the context of surf zone dynamics. From a physical point of view, this wave transformation is challenging, because the waves are undergoing shoaling, refraction, diffraction and nonlinear dispersion. The experiment conducted by Berkhoff et al. [3] has served as a standard test case for verifying several numerical wave models [11].

The simulations are considered in a rectangular basin $[(x,y): -10 \leq x \leq 10, -10 \leq y \leq 20]$ with a plane slope of 1/50 on which an elliptic shoal is rested; see Fig. 13. Let (x', y') be the slope-oriented coordinates which are related to the (x, y) coordinate system by means of a rotation over -20°. The still water depth without shoal is given in meters by

$$H = \begin{cases} 0.45, & \text{for } y' < -5.484. \\ max(0.10, 0.45 - (5.484 + y')/50), & \text{for } y' \geq -5.484. \end{cases} \quad (49)$$

Instead of shoreline boundary, a minimum depth of 10 cm is employed to prevent breaking waves. The boundary of the shoal is given by

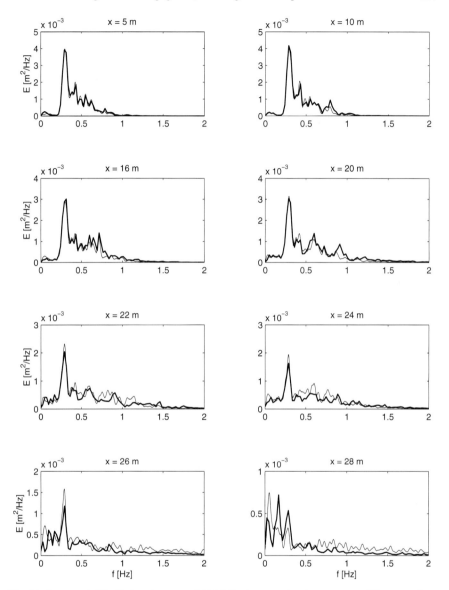

Fig. 12 Computed (thin line) and measured (thick line) energy density spectra at different stations for the irregular wave over bar-trough profile. All spectra use equally spaced frequency intervals and are filtered.

$$\left(\frac{x'}{4}\right) + \left(\frac{y'}{3}\right) = 1, \tag{50}$$

whereas the thickness of the shoal is

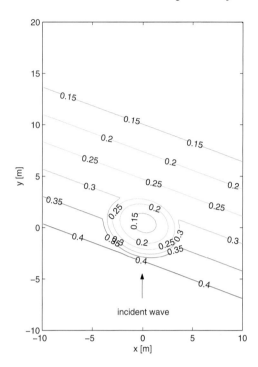

Fig. 13 Bathymetry corresponding to the experiment of Berkhoff et al. [3].

$$d = -0.3 + 0.5\sqrt{1 - \left(\frac{x'}{5}\right)^2 - \left(\frac{y'}{3.75}\right)^2}. \tag{51}$$

Monochromatic waves with wave height of 4.64 cm and wave period of 1.0 s are generated at lower boundary $y = -10$ m. The upper boundary, $y = 20$ m, is of the outflow type where Sommerfeld radiation condition (7) is applied. The left and right boundaries are insulated and the free-slip conditions are imposed.

For the present model, the grid size in both directions is set to 0.05 m. The time step is taken as 0.01 s and the simulation period is set to 30 s, so that a steady-state is reached ($\theta = 1$ is taken). Since $kH \approx 1.9$ in front of the domain, which is relatively large, only computation with two equidistant layers is carried out.

To get an impression, the computation was carried out on a 64-bit AMD processor (1.8 GHz, 4MB L2 cache) with 4GByte internal memory. Code compilation is achieved using Intel Fortran90 compiler 9.1 with the default optimization. The total CPU time per grid point per time step required was about 20 μs.

Profiles of the computed normalized wave height along four transects, which are the most compelling ones, are given in Fig. 14 and compared with the experimental data. The variation of the waves in cross direction representing the effects of combined refraction and diffraction is predicted fairly well as shown by the comparison of the computed and measured profiles along sections 2 and 5. The comparison

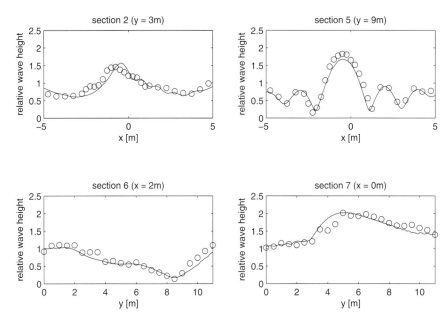

Fig. 14 Computed and measured relative wave heights along different transects for the wave over elliptic shoal. Present method (solid line), experiment (circles).

along sections 6 and 7 indicates that both shoaling and focussing of waves are very well predicted by the present model.

5 Conclusions

A computational method for calculating the conventional nonlinear shallow water equations, including non-hydrostatic pressure has been presented. For accuracy reasons, the pressure is split-up into hydrostatic and non-hydrostatic parts. In the model presented, the water depth is divided into a number of terrain-following layers and the governing equations are integrated in each layer. Next, the second order compact scheme is applied that enables to approximate short wave dynamics with a very limited number of vertical grid points. Simple (semi-)implicit second order finite differences are employed and are based upon a classical staggered grid. In addition, advection terms in the momentum equations are approximated such as to fulfil a proper momentum conservation, which is crucial for accurate computation of energy losses in a wave breaking process. Initiation and cessation of breaking waves can be described adequately by this method. This model does not require any sort of tunable or empirical parameters. Semi-implicit time stepping is done in combination with projection methods, where correction to the velocity fields for the change in both surface elevation and non-hydrostatic pressure is incorporated.

Finally, the algorithm utilizes a simple and numerically stable procedure yielding non-negative water depths with which an accurate representation of the shoreline motion is obtained.

The present method has been employed to model the main features of surf zone dynamics, such as nonlinear shoaling, breaking of waves and wave run-up with good agreement between predictions and observations. The model can be applied in practical applications that comprise areas with spatial dimensions of the order of 10 to 100 wave lengths, particularly in the vicinity of the coast. In the near future, the model will be coupled to a spectral wave model that can be applied on a scale of the order of 100−1000 wave lengths.

References

1. A. Balzano. Evaluation of methods for numerical simulation of wetting and drying in shallow water flow models. *Coast. Eng.*, 34: 83–107, 1998.
2. S. Beji and J.A. Battjes. Experimental investigation of wave propagation over a bar. *Coast. Eng.*, 19: 151–162, 1993.
3. J.C.W. Berkhoff, N. Booy, and A.C. Radder. Verification of numerical wave propagation models for simple harmonic linear water waves. *Coast. Eng.*, 6: 255–279, 1982.
4. M. Boers. Simulation of a surf zone with a barred beach; Part 1: wave heights and wave breaking. Report no. 96-5, Communication on Hydraulic and Geotechnical Engineering, Delft University of Technology, Delft, 1996.
5. M. Brocchini and D.H. Peregrine. Integral flow properties of the swash zone and averaging. *J. Fluid Mech.*, 317: 241–273, 1996.
6. M. Brocchini, R. Bernetti, A. Mancinelli, and G. Albertini. An efficient solver for nearshore flows based on the WAF method. *Coast. Eng.*, 43: 105–129, 2001.
7. M. Brocchini, I.A. Svendsen, R.S. Prasad, and G. Bellotti. A comparison of two different types of shoreline boundary conditions. *Comput. Meth. Appl. Mech. Engng.*, 191: 4475–4496, 2002.
8. G.F. Carrier and H.P. Greenspan. Water waves of finite amplitude on a sloping beach. *J. Fluid Mech.*, 4: 97–109, 1958.
9. V. Casulli and G.S. Stelling. Numerical simulation of 3D quasi-hydrostatic, free-surface flows. *J. Hydr. Eng. ASCE*, 124: 678–686, 1998.
10. M.W. Dingemans. Comparison of computations with Boussinesq-like models and laboratory measurements. MAST G8-M, Report H1684.12, Delft Hydraulics, Delft, 1994.
11. M.W. Dingemans. *Water wave propagation over uneven bottoms*. World Scientific, Singapore, 1997.
12. C.J. Galvin. Breaker type classification on three laboratory beaches. *J. Geophys. Res.*, 73: 3651–3659, 1968.
13. I. Gustafsson. A class of first order factorization methods. *BIT*, 18: 142–156, 1978.
14. J.B. Hansen and I.A. Svendsen. Regular waves in shoaling water: experimental data. Technical report, ISVA series paper 21, Technical University of Denmark, Denmark, 1979.
15. F.H. Harlow and J.E. Welch. Numerical calculation of time-dependent viscous incompressible flow of fluid with a free surface. *Phys. of Fluids*, 8: 2182–2189, 1965.
16. S. Hibberd and D.H. Peregrine. Surf and run-up on a beach: a uniform bore. *J. Fluid Mech.*, 95: 323–345, 1979.
17. C. Hirsch. *Numerical computation of internal and external flows*. John Wiley and Sons, Chichester, 1990.
18. C.W. Hirt and B.D. Nichols. Volume of fluid (VOF) method for the dynamics of free boundaries. *J. Comput. Phys.*, 39: 201–225, 1981.

19. K. Hu, C.G. Mingham, and D.M. Causon. Numerical simulation of wave overtopping of coastal structures using the non-linear shallow water equations. *Coast. Eng.*, 41: 433–465, 2000.
20. M.E. Hubbard and N. Dodd. A 2D numerical model of wave run-up and overtopping. *Coast. Eng.*, 47: 1–26, 2002.
21. N. Kobayashi, A.K. Otta, and I. Roy. Wave reflection and run-up on rough slopes. *J. Waterw. Port Coast. Ocean Eng.*, 113: 282–298, 1987.
22. A. Mahadevan, J. Oliger, and R. Street. A nonhydrostatic mesoscale ocean model. Part II: numerical implementation. *J. Phys. Oceanogr.*, 26: 1881–1900, 1996.
23. J.A. Meijerink and H.A. Van der Vorst. An iterative solution method for linear systems of which the coefficient matrix is a symmetric M-matrix. *Math. Comput.*, 31: 148–162, 1977.
24. D.H. Peregrine. Long waves on a beach. *J. Fluid Mech.*, 27: 815–827, 1967.
25. R. Scardovelli and S. Zaleski. Direct numerical simulation of free-surface and interfacial flow. *Annu. Rev. Fluid Mech.*, 31: 567–603, 1999.
26. P.K. Stansby and J.G. Zhou. Shallow-water flow solver with non-hydrostatic pressure: 2D vertical plane problems. *Int. J. Numer. Meth. Fluids*, 28: 514–563, 1998.
27. G.S. Stelling. *On the construction of computational methods for shallow water flow problems.* PhD thesis, Delft University of Technology, Delft, 1983.
28. G.S. Stelling and S.P.A. Duinmeijer. A staggered conservative scheme for every Froude number in rapidly varied shallow water flows. *Int. J. Numer. Meth. Fluids*, 43: 1329–1354, 2003.
29. G. Stelling and M. Zijlema. An accurate and efficient finite-difference algorithm for non-hydrostatic free-surface flow with application to wave propagation. *Int. J. Numer. Meth. Fluids*, 43: 1–23, 2003.
30. M. Sussman, P. Smereka and S. Osher. A level set approach for computing solutions to incompressible two-phase flow. *J. Comput. Phys.*, 114: 146–159, 1994.
31. I.A. Svendsen and P.A. Madsen. A turbulent bore on a beach. *J. Fluid Mech.*, 148: 73–96, 1984.
32. V.V. Titov and C.E. Synolakis. Modeling of breaking and nonbreaking long-wave evolution and runup using VTCS-2. *J. Waterw. Port Coast. Ocean Eng.*, 121: 308–316, 1995.
33. E.F. Toro. *Shock-capturing methods for free-surface shallow flows.* John Wiley, New York, 2001.
34. H.A. Van der Vorst. Bi-CGSTAB: a fast and smoothly converging variant of Bi-CG for the solution of nonsymmetric linear systems. *SIAM J. Sci. Stat. Comput.*, 13: 631–644, 1992.
35. M. Zijlema and G.S. Stelling. Further experiences with computing non-hydrostatic free-surface flows involving water waves. *Int. J. Numer. Meth. Fluids*, 48: 169–197, 2005.
36. M. Zijlema and G.S. Stelling. Efficient computation of surf zone waves using the nonlinear shallow water equations with non-hydrostatic pressure *Coast. Eng.*, doi:10.1016/j.coastaleng.2008.02.020, 2008.

Hybrid Navier-Stokes/DSMC Simulations of Gas Flows with Rarefied-Continuum Transitions

G. Abbate, B.J. Thijsse, and C.R. Kleijn

1 Introduction

Numerical simulations are an important tool for the design and optimization of gas flow equipment in many areas of science and technology. Most gas flows can be simulated using the continuum transport equations (Navier-Stokes), which describe the transport of mass, momentum and energy. These equations are based on the hypothesis that the mean free path length λ of the gas molecules is very small in comparison to a characteristic dimension L of the flow. This dimension can be either a physical dimension, e.g. a pipe diameter, or a flow dimension, e.g. the gradient length scale $\frac{1}{\phi}\frac{\partial \phi}{\partial x}$ on which some flow property ϕ changes significantly. The dimensionless Knudsen number Kn can be used to describe this situation:

$$Kn = \frac{\lambda}{L}. \qquad (1)$$

When $Kn < 0.01$, gas molecules travel only a small distance (compared to the geometry and flow dimensions) between collisions. For internal flows this means that molecules only very rarely collide with walls, and the flow is dominated by the characteristics of the inter-molecular collisions. As a result, the gas will be in local equilibrium and the velocity distribution of its molecules will be Maxwellian. However, there are situations in which Kn is not so small, e.g. gas flows at low

---

G. Abbate
Department of Multi-Scale Physics & J.M.Burgers Centre for Fluid Mechanics, Delft University of Technology, Prins Bernhardlaan 6, 2628 BW Delft, The Netherlands e-mail: g.abbate@tudelft.nl

B.J. Thijsse
Department of Materials Science and Engineering, Delft University of Technology, Mekelweg 2, 2628 CD Delft, The Netherlands e-mail: b.j.thijsse@tudelft.nl

C.R. Kleijn
Department of Multi-Scale Physics & J.M.Burgers Centre for Fluid Mechanics, Delft University of Technology, Prins Bernhardlaan 6, 2628 BW Delft, The Netherlands e-mail: c.r.kleijn@tudelft.nl

pressure, where λ becomes large, or flows with very small dimension L, e.g. microfluidics. The high Kn numbers in these flows indicate that a molecule travels a significant distance (compared to L) between collisions. For an internal flow, this implies that wall interactions occur more frequently and become important in describing the flow. In this regime, the flow can no longer be described as a continuum and the well known transport equations, or more precisely, the relations for the shear tensor and the heat flux, can no longer be used. In this regime, the particulate nature of the gas becomes important and a different simulation method must be used. The mathematical model at this level is the Boltzmann equation [1]. It provides information on the position, velocity and state of every molecule at all times. As a consequence of its complexity, the Boltzmann equation is not amenable to analytical solution for non-trivial problems.

In the continuum regime, numerical simulations can be done using (commercially available) Computational Fluid Dynamics (CFD) codes based on partial differential equations describing the transport phenomena, e.g. the Navier-Stokes equations. Especially for laminar flows, these codes can produce accurate results for Knudsen numbers up to 0.01, but start deviating from reality for higher Kn. It is generally accepted that the range of applicability of these continuum codes can be extended into the rarefied regime up to $Kn \approx 0.1$ by using special boundary conditions to take into account the possibility of a velocity slip or temperature jump at a surface [2]. In this method, however, the precise formulation of the slip velocity and temperature jump boundary conditions is strongly geometry dependent [3, 4, 5].

Gas flows with $Kn > 10$ are called "free molecular flows". In this regime, intermolecular collisions rarely occur and the flow is completely dominated by the interaction between the gas and the walls. Gas flows in the free molecular regime can be simulated using Molecular Dynamics (MD) or ballistic models.

In the intermediate ($0.01 < Kn < 10$) or rarefied regime, both collisions with solid surfaces and with other gas molecules are important, and therefore have to be included in the simulation to obtain an accurate result. The Direct Simulation Monte Carlo (DSMC) method as developed by Bird [6] is the only practical engineering method that can be used in the rarefied regime. The DSMC method is also valid in free molecular and continuum regimes, although the computational expenses become very large in the latter case. Its computational expenses, in fact, scale with Kn^{-4} and become prohibitively large when Kn becomes lower than ~ 0.05.

In summary, one can simulate gas flows with $Kn < 0.01$ (or, with modifications of boundary conditions < 0.1) using continuum based CFD models, and gas flows with $Kn > 0.05$ with particle based DSMC methods. In many practical applications, however, gas flows undergo spatial and/or temporal transitions from low (< 0.05) to high (> 0.05) Kn numbers, e.g. due to varying pressure or dimensions. Examples include: flow around vehicles at high altitudes, particularly re-entry of vehicles in a planetary atmosphere [7], flow through microfluidic gas devices [8], small cold gas thruster nozzle and plume flows [9], and low pressure thin film deposition processes from expanding plasma or gas jets [10].

Different solutions have been proposed to compute flows undergoing such transitions. The most widely studied method is to use hybrid models coupling continuum solvers and molecular methods, for instance: Molecular Dynamics (MD) and Navier-Stokes (N-S) equations [11], Boltzmann and N-S equations [12], Direct Simulation Monte Carlo (DSMC) and Stokes equations [8], DSMC and incompressible N-S equations [13], DSMC and Euler equations [14] and DSMC and N-S equations [15, 16, 17, 18, 19].

With respect to coupling N-S to DSMC, Garcia et al. [15] constructed a hybrid particle/continuum algorithm with an adaptive mesh and algorithm refinement, in which DSMC was used as a particle method embedded within a Gudonov-type compressible Navier-Stokes solver. There was no overlapping between the continuum and DSMC regions, which were coupled through Neumann-Neumann type boundary conditions. Glass and Gnoffo [16] proposed a 'one-shot' coupled CFD-DSMC method. The interfacial location between the CFD and DSMC zones was identified manually after a 'one-shot' CFD simulation. Results of the CFD simulation at this interface were then used as Dirichlet inflow boundary condition for the DSMC in the rarefied regions. In contrast with that, Wu et al. [17] and Schwartzentruber et al. [18, 19] proposed an 'iterative' coupled CFD-DSMC method, where the coupling is achieved through an overlapped Schwarz method with Dirichlet-Dirichlet type boundary conditions.

In the present chapter, we present a hybrid modeling approach for steady-state and transient continuum/rarefied gas flow, which employs the compressible N-S equations in the continuum regime and DSMC in the rarefied regime. The coupling of the two models is reached through an overlapped Schwarz method [17] with Dirichlet-Dirichlet boundary conditions. It is an adaptive method in which, during the computations, the Kn number with respect to the local gradients is computed to determine the domain interface and divide the CFD domain from the DSMC domain. Our method differs from that by Garcia et al. [15] in that we employ Dirichlet-Dirichlet boundary conditions coupling, as opposed to Neumann-Neumann. The advantage is that Dirichlet boundary conditions are much less sensitive to statistical scatter in the DSMC simulations. The difference with the method proposed by Glass and Gnoffo [16] is that we employ adaptive domain splitting based on dynamic evaluation of local Kn numbers, rather than a pre-determined static interface. In contrast to Wu et al. [17] and Schwartzentruber et al. [18, 19], who demonstrated methods for steady-state flows only, our present method can be applied to both steady-state and transient flows.

In section 2 we demonstrate how and under which conditions the Navier-Stokes equations can approximate the Boltzmann equation. The developed hybrid numerical method is described in section 3. In particular, we first describe the continuum solver in section 3.2 and the DSMC solver in section 3.3. A description of the coupling algorithm in both steady and unsteady formulations is given respectively in sections 3.4.2 and 3.4.3. Results of applying the method to an unsteady 1-D shock tube problem with a sensitivity analysis of the method to various parameters are presented in section 4.1. Finally a validation of the method is performed by applying it respectively to a steady-state rarefied Poiseuille flow, and a steady-state 2-D expanding jet in a low pressure chamber, is presented in sections 4.2 and 4.3.

2 From Boltzmann to Navier-Stokes

The Navier-Stokes equations can be derived from the Boltzmann equation. The derivation can be found in most texts on kinetic theory, e.g. Bird [6], Chapman and Cowling [20], Grad [21], Patterson [22]. For the sake of completeness, the discussion is briefly repeated here.

Considering an ideal mono-atomic gas in the absence of external forces and assuming the gas sufficiently dilute for binary collision to dominate, the Boltzmann equation [1] reads

$$\frac{\partial (nf)}{\partial t} + c_k \frac{\partial (nf)}{\partial x_k} = \left[\frac{\partial (nf)}{\partial t}\right]_{coll}, \quad (2)$$

where n is the number density, f is the velocity distribution function, c_k the molecular velocity in an inertial frame, the repeated index k denotes a sum, and the right-hand side represents the collision integral. Multiplying the Boltzmann equation by any function of molecular velocity $Q(c_i)$ and integrating over velocity space, the moment equations are obtained

$$\frac{\partial (n<Q>)}{\partial t} + \frac{\partial}{\partial x_k}(n<c_k Q>) = \Delta[Q]. \quad (3)$$

In equation (3), the operators $<Q>$ and $\Delta[Q]$ are defined by

$$<Q> = \int_{-\infty}^{\infty}\int_{-\infty}^{\infty}\int_{-\infty}^{\infty} Qf dc_1 dc_2 dc_3 \quad (4)$$

and

$$\Delta[Q] = \int_{-\infty}^{\infty}\int_{-\infty}^{\infty}\int_{-\infty}^{\infty} Q\left[\frac{\partial (nf)}{\partial t}\right]_{coll} dc_1 dc_2 dc_3. \quad (5)$$

Choosing one of the five collisional invariants $Q^{INV} = m\{1, c_i, c^2/2\}$, with m the molecular mass and c^2 the square of the velocity magnitude, as the arbitrary function of molecular velocity $Q(c_i)$, then the corresponding moment of the collision integral is identically zero, i.e. $\Delta[Q] = 0$. This general result is valid for any distribution function f and for any molecular interaction law and it leads to the conservation laws for gas dynamics

$$\frac{\partial}{\partial t}(n<Q^{INV}>) + \frac{\partial}{\partial x_k}(n<c_k Q^{INV}>) = 0. \quad (6)$$

Considering the collisional invariants in turn, the following set of equations can be written

$$\frac{\partial}{\partial t}(\rho) + \frac{\partial}{\partial x_k}(\rho <c_k>) = 0, \quad (7)$$

$$\frac{\partial}{\partial t}(\rho <c_i>) + \frac{\partial}{\partial x_k}(\rho <c_k c_i>) = 0, \quad (8)$$

$$\frac{\partial}{\partial t}(\rho <c^2/2>) + \frac{\partial}{\partial x_k}(\rho <c_k c^2/2>) = 0, \qquad (9)$$

where $\rho = mn$ is the mass density.
In terms of the thermal velocity components $C_i = (c_i - u_i)$, where the mean or fluid velocity is $u_i = <c_i>$, the central moments can be defined

$$P_{ij} = \rho <C_i C_j>, \qquad (10)$$
$$p = P_{kk}/3, \qquad (11)$$
$$\tau_{ij} = -P_{ij} + p\delta_{ij}, \qquad (12)$$
$$e = <C^2/2>, \qquad (13)$$
$$q_i = \rho <C_i C^2/2>, \qquad (14)$$

where P_{ij} is the stress tensor, p is the pressure, τ_{ij} is the viscous stress tensor, e is the internal energy (translational) for a mono-atomic gas, and q_i is the heat flux vector for a mono-atomic gas. Substituting equations (10)–(14) into the equations (7)–(9), the conservation laws for gas dynamics can then be written in the form

$$\frac{\partial}{\partial t}(\rho) + \frac{\partial}{\partial x_k}(\rho u_k) = 0, \qquad (15)$$

$$\frac{\partial}{\partial t}(\rho u_i) + \frac{\partial}{\partial x_k}(\rho u_k u_i - \tau_{ki} + p\delta_{ki}) = 0, \qquad (16)$$

$$\frac{\partial}{\partial t}\left[\rho\left(e + \frac{u^2}{2}\right)\right] + \frac{\partial}{\partial x_k}\left[\rho u_k\left(e + \frac{u^2}{2}\right) - \tau_{ki} u_i + p\delta_{ki} u_i + q_k\right] = 0. \qquad (17)$$

For poly-atomic gases the above procedure is not valid anymore and, therefore, must be modified. The problem is rather difficult, because it includes the question of which equation replaces equation (2), and it becomes necessary to make use of a suitable approximation.

The energy $mc^2/2$ does not properly account for the amount of energy that is carried by a particle with internal structure, and it must be replaced by $(mc^2/2 + \varepsilon)$, where ε is the additional internal energy per particle. Therefore, the collisional invariants become

$$Q^{INV} = \{m, mc_i, (mc^2/2 + \varepsilon)\}. \qquad (18)$$

Assuming that equation (2) continues to hold for the extended distribution function $f(c_i, \varepsilon)$, when applying to both equations (4) and (5), an additional integral over ε is required. The quantities in equation (18) must continue to be conserved in a collision, and consequently, equation (5) still evaluates to zero; thus, equation (6) remains unchanged.

Since integration over ε can be taken first and independently from the c_i integration, evaluating the lefthand side of equation (6), identical results to those obtained for the mono-atomic gas will be found for all quantities that contain polynomials in c_i only.

Therefore equations (7) and (8) and, consequently, (15) and (16) are fully recovered. The same conclusion also applies to the first term in the quantity $(mc^2/2 + \varepsilon)$ and so, equation (9) is replaced by

$$\frac{\partial}{\partial t}(\rho <c^2/2> + n<\varepsilon>) + \frac{\partial}{\partial x_k}(\rho <c_k c^2/2> + n<c_k \varepsilon>) = 0. \tag{19}$$

The unknown term $m e_{int} = <\varepsilon>$ is the additional internal energy. A simple approach is to assume that all internal molecular energy modes are in equilibrium, both internally and with the translational degrees of freedom. Thus, e_{int} can be expressed in terms of the translational temperature T by the equilibrium relation

$$e_{int} = \frac{1}{2}\left(\frac{5-3\gamma}{\gamma-1}\right)RT, \tag{20}$$

with R the gas constant and where the additional internal energy is accounted for through the introduction of the ratio of specific heats γ (for mono-atomic gas, $\gamma = 5/3$). Clearly, in the case of a mono-atomic gas the additional internal energy evaluates to zero, i.e. $e_{int} = 0$.

If we substitute equations (10)–(14) into equation (19) we will get once again equations (15) and (16), but instead of (17) it will lead to

$$\frac{\partial}{\partial t}\left[\rho\left(e + e_{int} + \frac{u^2}{2}\right)\right] +$$
$$+ \frac{\partial}{\partial x_k}\left[\rho u_k\left(e + e_{int} + \frac{u^2}{2}\right) + P_{ki} u_i + q_k + (n<C_k \varepsilon>)\right] = 0. \tag{21}$$

However, if we replace equation (13) by

$$e = (<C^2/2> + e_{int}) \tag{22}$$

and (14) by

$$q_i = \rho <C_i C^2/2> + n <C_i \varepsilon>, \tag{23}$$

we will recover equation (17) as well.

We conclude that, if definitions (22) and (23) are employed in the case where the gas possesses internal structure and a state of equilibrium exists between the internal modes and the translational degrees of freedom, equations (15)–(17) can be used.

The set of conservation equations (15)–(17) can be developed for any general fluid through the use of phenomenological arguments only and, therefore, is more general than the kinetic theory derivation would indicate. Since we are only interested in treating an ideal gas flow, however, the kinetic theory approach is necessary because it shows that the obtained set of equations is valid for any degree of translational nonequilibrium, that is, for any translational velocity distribution function one cares to consider. In case of the equilibrium distribution, namely the Maxwellian distribution f^{Max} [2], then the set becomes the Euler equations, because viscous stress and heat flux are identically zero in this case. On the contrary if one chooses a Chapman-Enskog (CE) distribution f^{CE} [24], then the set becomes the Navier-Stokes equations, because stress and heat flux are then given by the corresponding Chapman-Enskog expressions.

In summary, the conservation equations (15)–(17) are not the N-S equations until one introduces f^{CE}. In fact, one is free to choose any translational velocity distribution function in the equation (6) or in the sets of equations (7)–(9) and (15)–(17), and in this way the set becomes closed, as long as f is fully specified. Otherwise, if f remains general, one is faced with a closure problem, because τ_{ij} and q_i are unknown quantities.

Since f^{CE} is an $O(Kn)$ expansion of the exact solution of f, the resulting N-S equations are an accurate approximation for $Kn \ll 1$ only. For large Kn, solutions to the N-S equations no longer accurately describe the real behaviour of the gas. This is most clearly visible through the occurrence of wall-slip and wall-temperature jumps at high Kn, neither of which is found through N-S.

3 The hybrid numerical method

3.1 Introduction

As described in the previous section, equations (15)–(17) reduce to the compressible Navier-Stokes equations when f^{CE} is an accurate approximation of f, i.e. for $Kn \ll 1$. For large Kn, equations (15)–(17) no longer form a closed set of equations, because of the closure problem for τ_{ij} and q_i. For those cases, one should solve the full Boltzmann equation.

In this section we will describe a finite volume based scheme for explicit time integration of the compressible Navier-Stokes equations in low Kn flows (section 3.2), a discrete, particle based Monte Carlo approach for solving the Boltzmann equations in high Kn flows (section 3.3), and a hybrid approach to dynamically couple these two approaches (section 3.4).

3.2 Finite volume scheme for compressible Navier-Stokes equations

3.2.1 Finite volume discretization

Each of the five separate moment equations represented by either sets of equations (7)–(9) or (15)–(17) can be expressed through the form

$$\frac{\partial U}{\partial t} + \frac{\partial F_k}{\partial x_k} = 0. \qquad (24)$$

Using the notation of equation (6),

$$U = n < Q_{INV} > \qquad (25)$$

is the state vector, and

$$F_n = n < c_n Q_{INV} > \qquad (26)$$

the total flux vector, with c_n the component of the molecular velocity normal to the planar surface.

Finite volume integration of the above model equation over an arbitrary control volume V, and using Gauss' divergence theorem leads to

$$\frac{\partial}{\partial t} \int_V U \, dV + \int_S F_n \, dS = 0, \qquad (27)$$

where S encloses the volume V and F_n is the projection of F_i onto the unit outward pointing normal for the surface element dS.

Considering a Cartesian grid, equation (27) can be written as

$$\Delta V \frac{\partial \bar{U}_{ijk}}{\partial t} = (-\bar{F}\Delta A) \Big|_{i-1/2,j,k}^{i+1/2,j,k} + (-\bar{F}\Delta A) \Big|_{i,j-1/2,k}^{i,j+1/2,k} + (-\bar{F}\Delta A) \Big|_{i,j,k-1/2}^{i,j,k+1/2}, \qquad (28)$$

where e.g. $(.)\Big|_{i-1/2,j,k}^{i+1/2,j,k} = (.)_{i+1/2,j,k} - (.)_{i-1/2,j,k}$ and x_1, x_2 and x_3 are the spatial coordinates in the directions i, j and k respectively, ΔV and $\bar{U}$ are the volume and state variables averaged in the ith cell, ΔA and $\bar{F}$ are the area and the averaged total flux on the relevant cell face.

3.2.2 Time discretization

The finite volume integration of equation (28) over a control volume V must be augmented with a further integration over a finite time step Δt. Considering a first order accurate forward Euler time integration, the 1-D version of equation (28) reads

$$\Delta V (\bar{U}_i^{n+1} - \bar{U}_i^n) = \int_t^{t+\Delta t} (-\bar{F}\Delta A) \Big|_{i-1/2}^{i+1/2} dt, \qquad (29)$$

where n refers at time t and $n+1$ at time $t + \Delta t$. To evaluate the righthand side of the above equation we need to make an assumption. We could use the values at time t or at time $t + \Delta t$ to calculate the integral or, alternatively, a combination of both. Here we considered a first-order explicit scheme in time which uses values at time t, and so equation (29) becomes

$$\bar{U}_i^{n+1} = \bar{U}_i^n - \frac{\Delta t}{\Delta V}\Delta A(\bar{F}_{i+1/2} - \bar{F}_{i-1/2})^n. \tag{30}$$

3.2.3 The MUSCL discretization scheme

To solve equation (30) we must now evaluate the total flux on the relevant cell face $\bar{F}_{i+1/2}$ (and similar for other cell faces), that will be a function of the state variables U at the same cell interface

$$\bar{F}_{i+1/2} = F(U_{i+1/2}). \tag{31}$$

This means that we should estimate $F(U_{i+1/2})$ starting from state variables averaged in the grid cells $\bar{U}$. The flux spitting method consists in splitting the total flux in its positive and negative parts

$$\bar{F}_{i+1/2} = \bar{F}^+_{i+1/2} + \bar{F}^-_{i+1/2}, \tag{32}$$

where the positive and negative parts of the total flux will be functions of the state variables respectively at the left or right of the cell interface

$$\bar{F}^+_{i+1/2} = F^+(U^L_{i+1/2}), \tag{33}$$

$$\bar{F}^-_{i+1/2} = F^-(U^R_{i+1/2}). \tag{34}$$

A second-order spatially accurate MUSCL (Monotone Upstream-centered Scheme for Conservation Laws) [25] scheme is used to approximate the state variables left and right of the cell interface.
The creation of local extrema during the higher order linear reconstruction of fluxes is eliminated by the application of a *minmod* type limiter [26].

3.2.4 Chapman-Enskog split fluxes

We need now to evaluate an expression for the one-side fluxes based on a fixed interface. This is done in a way proposed by Chou and Baganoff [23] and by Lou et al. [27]. The approach is briefly repeated here for the sake of completeness.
From equation (26), introducing a Cartesian coordinate system $(n, t1, t2)$ located in an arbitrary fixed planar surface and replacing both the temperature gradient and the velocity-gradient tensor by the Chapman-Enskog expression for stress and heat flux [28]

$$q_i^{CE} = -K^{(1)}\frac{\partial T}{\partial x_i}, \tag{35}$$

$$\tau_{ij}^{CE} = \mu^{(1)}\left(\frac{\partial u_i}{\partial x_j} + \frac{\partial u_j}{\partial x_i}\right) - \frac{2}{3}\mu^{(1)}\left(\frac{\partial u_k}{\partial x_k}\right)\delta_{ij}, \tag{36}$$

we obtain

$$F^{\pm}_{mass} = \rho\sqrt{RT/2}[(1\pm\alpha_1)S_n \pm \alpha_2(1-\chi_1)], \qquad (37)$$

$$F^{\pm}_{n-mom} = p[(1\pm\alpha_1)(S_n^2+\frac{1}{2}(1-\hat{\tau}^{CE}_{nn})) \pm \alpha_2(S_n+\hat{q}^{CE}_n)], \qquad (38)$$

$$F^{\pm}_{t1-mom} = \sqrt{2RT}[S_{t1}F^{\pm}_{mass}] + \frac{1}{2}p[-(1\pm\alpha_1)\hat{\tau}^{CE}_{nt1} \pm \alpha_2\hat{q}^{CE}_{t1}], \qquad (39)$$

$$F^{\pm}_{tr-energy} = p\sqrt{RT/2}[(1\pm\alpha_1)(S_n(\frac{5}{2}+S^2)+\chi_2) \pm$$
$$\pm \alpha_2(2+S^2+\chi_3)]), \qquad (40)$$

$$F^{\pm}_{int-energy} = (\Delta q^{\pm}_{Eucken} + \rho u_n e^{\pm}_{int}) = \frac{1}{2}(\frac{5-3\gamma}{\gamma-1})RT F^{\pm}_{mass}, \qquad (41)$$

$$F^{\pm}_{energy} = F^{\pm}_{tr-energy} + F^{\pm}_{int-energy}, \qquad (42)$$

where

$$\alpha_1 = erf(S_n), \qquad (43)$$

$$\alpha_2 = \frac{1}{\sqrt{\pi}}e^{-S_n^2}, \qquad (44)$$

$$\chi_1 = S_n\hat{q}^{CE}_n + \frac{1}{2}\hat{\tau}^{CE}_{nn}, \qquad (45)$$

$$\chi_2 = \frac{5}{2}\hat{q}^{CE}_n - (S_n\hat{\tau}^{CE}_{nn} + S_{t1}\hat{\tau}^{CE}_{nt1} + S_{t2}\hat{\tau}^{CE}_{nt2}), \qquad (46)$$

$$\chi_3 = S_{t1}\hat{q}^{CE}_{t1} + S_{t2}\hat{q}^{CE}_{t2} - \chi_1(1+S_{t1}^2+S_{t2}^2) - \hat{\tau}^{CE}_{nn}, \qquad (47)$$

$$S_n = u_n/\sqrt{2RT}, \qquad (48)$$

$$S^2 = S_n^2 + S_{t1}^2 + S_{t2}^2, \qquad (49)$$

$$\hat{\tau}^{CE}_{nn} = \tau^{CE}_{nn}/p, \qquad (50)$$

$$\hat{q}^{CE}_n = \frac{2}{5}q^{CE}_n/(p\sqrt{2RT}). \qquad (51)$$

Since the individual components of S_i, $\hat{\tau}_{ij}$ and $\hat{q}_i$ are all nondimensionalized the same way, they are not listed all. It is interesting to note that we refer to the speed ratio $S = u/\sqrt{2RT}$ instead of the Mach number as frequently in use in kinetic theory. It is simple to check that if we sum the positive and negative parts, we will get once again the total fluxes

$$F_{mass} = \rho u_n, \qquad (52)$$

$$F_{n-mom} = \rho u_n^2 + p - \tau^{CE}_{nn}, \qquad (53)$$

$$F_{t1-mom} = \rho u_n u_{t1} - \tau^{CE}_{nt1}, \qquad (54)$$

$$F_{tr-energy} = \rho u_n \left(\frac{3}{2}RT + \frac{u^2}{2}\right) + pu_n - $$
$$-(\tau_{nn}^{CE} u_n + \tau_{nt1}^{CE} u_{t1} + \tau_{nt2}^{CE} u_{t2}) + q_n^{CE}, \tag{55}$$
$$F_{int-energy} = (\Delta q_{Eucken} + \rho u_n e_{int}) = \frac{1}{2}\left(\frac{5-3\gamma}{\gamma-1}\right) pu_n. \tag{56}$$

3.3 Direct Simulation Monte Carlo scheme for rarefied gas flow simulations

The Boltzmann equation can be solved analytically for some simple problems only. Numerically, solutions can be obtained for a somewhat broader range of problems. For engineering problems, however, it is next to impossible to solve the Boltzmann equation, even numerically. Another disadvantage of the Boltzmann equation is the fact that its definition does not include the possibility for chemical reactions.

The Direct Simulation Monte Carlo (DSMC) method [6], which is closely related to the Boltzmann equation, does not suffer from these shortcomings, and it is therefore the preferred method for simulations of engineering type rarefied gas flows. Rather than solving continuum based partial differential equations like the Navier-Stokes equations, the DSMC method aims at modeling gas flows by calculating the movements and collisions of computational particles which represent molecules in the real flow.

Like the Boltzmann equation [1], the DSMC method assumes a dilute gas and molecular chaos. In a dilute gas, the molecules occupy only a small fraction of the total gas volume. Consequently, the position and velocity distributions of two colliding particles are uncorrelated, which is the definition of molecular chaos. The DSMC method is inherently transient, and steady state solutions are obtained by letting a transient simulation evolve into the long-time, steady state. During the transient calculations, the position, velocity and internal energy of the computational particles are stored and updated each time step. It has been shown [30] that solutions obtained with the DSMC method converge to solutions of the Boltzmann equation in the limit of infinitely small cell size and time step, and infinite number of computational particles.

In addition to the two assumptions mentioned above, the DSMC method involves two more main assumptions:

- It is not necessary to calculate the path of every real molecule, but a relatively small statistical sample of N particles suffices. Typically, $N = 10^5 - 10^7$, which may be compared to e.g. 10^{15} molecules in 1 mm^3 of atmospheric air. The ratio F_{num}, which is defined as the ratio between the number of molecules in the real flow and the number N of simulation particles, can be a very large number (e.g.

$10^{10} - 10^{20}$).

- The translation of the computational particles can be decoupled from their collisions with other computational particles. This implies that each simulation time step can be split into two steps:
  - A translation step in which all particles are displaced and interactions with boundaries are computed;
  - A collision step in which inter-particle collisions are modelled.

  The translation step is purely deterministic, whereas the collision step involves a Monte Carlo type approach, hence the name Direct Simulation Monte Carlo.

The steps for a typical DSMC calculation are:

- Initialization,
- Particle movement,
- Particle collisions,
- Sampling.

They will be discussed in the following sections.

3.3.1 Initialization

At the start of a computation, particles are generated in the flow domain according to the prescribed initial conditions. These include the geometry of the flow domain, the initial temperature T, the initial number density n, and the initial mass-average velocity $\bar{V}_{ma}$. From the prescribed value of F_{num} and the initial flow density the number of computational particles is calculated. Each of these particles is then assigned a location and a velocity. For the most common case of a uniform initial density, the location of the particles is chosen such that they are evenly distributed in the entire domain. The velocities of individual particles are usually sampled from the Maxwellian distribution f^{Max} [2] belonging to the initial temperature. Alternatively, a Chapman-Enskog distribution can be used f^{CE} [24].

3.3.2 Particle movement

In each DSMC time step, the translation of all particles is calculated in a fully deterministic way from the old location $\bar{X}_t$ and velocity $\bar{V}$ of the particle:

$$\bar{X}_{t+\Delta t} = \bar{X}_t + \bar{V}\Delta t. \tag{57}$$

For 1-dimensional flows, only one location variable is required to define the position of a computational particle, and equation (57) reduces to a scalar equation. A similar reasoning holds for 2-dimensional flows. However, because of the requirements of the inter-particle collision treatment, the particle velocities are always treated in 3 dimensions. Of these, only the relevant components are used in equation (57).

If, during its displacement, the path of a computational particle intersects with a solid surface, the interaction with this surface is calculated as fully diffuse, fully specular or a mixture of these two [2]. A symmetry plane is treated identical to a specular surface. When a particle crosses an open boundary, it is removed from the simulation.

During the movement phase of the calculation, new particles also enter the domain through open boundaries. For the implementation of these boundaries, a "buffer zone" or "particle reservoir" approach is used [32]. Some "buffer cells" are considered across the open boundary outside the simulation domain. Every time step, a number of particles, according to the density ρ_{BC} at the boundary, are generated inside these buffer cells with an average temperature T_{BC} and velocity $\bar{V}_{BC}$. A Maxwellian [2] or Chapmann-Enskog [24] distribution is used to create these particles. The created particles are then moved for one time step. Particles that remain in the buffer cells are deleted. The molecules that move into the simulation domain are inserted in the simulation (fig.1).

For a pressure inlet, the temperature and pressure, and therefore the density, are fixed while the velocity is unknown. For a pressure outlet, only the pressure is fixed and the temperature and velocity are unknown. For each "buffer cell" the unknown variables are interpolated from the first cell in the flow nearest to the "buffer cell".

In coupled simulations, the variables ρ_{BC}, T_{BC} and $\bar{V}_{BC}$ are evaluated from the continuum domain as described in the section 3.4.

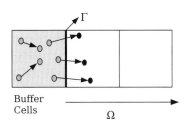

Fig. 1 Buffer zone approach for inlet (outlet) boundary conditions.

3.3.3 Particle collisions

For the purpose of calculating the collisions between computational particles through a Monte Carlo type of approach, the simulation domain is divided into cells with maximum dimensions $\Delta x \cdot \Delta y \cdot \Delta z = (\lambda/3) \cdot (\lambda/3) \cdot (\lambda/3)$, or an equivalent 1D or 2D representation. Here, λ is the particles' mean free path length. Typically, the total number of computational particles is chosen such that the average number of particles in each cell is larger than ~ 30.

In each time step, the collisions between the N computational particles in a cell can be calculated using the number of pairs and the collision probability P for each pair:

$$\#pairs = \frac{N(N-1)}{2}, \tag{58}$$

$$P = F_{num}\frac{\Delta t \sigma_T c_r}{V}. \tag{59}$$

The fraction $\frac{\Delta t \sigma_T c_r}{V}$ is the probability that the computational particles will collide in a time step, with σ_T the total collision cross-section of the two particles, c_r their relative speed, Δt the time step and V the cell volume. By multiplying this probability with the ratio F_{num} between the number of real molecules and the number of computational particles, the correct collision frequency for the real gas is obtained. The probability P is evaluated for each pair and a collision is accepted or rejected by comparing P to a random number.

This method of calculating collisions is not very efficient as the value of P is usually very small. DSMC calculations therefore use an adapted method in which the number of pairs is reduced such that the collision probability for a pair can be increased:

$$\#pairs = \frac{1}{2}N^2 F_{num}\frac{\Delta t (\sigma_T c_r)_{max}}{V}, \tag{60}$$

$$P = \frac{\sigma_T c_r}{(\sigma_T c_r)_{max}}. \tag{61}$$

The value of $(\sigma_T c_r)_{max}$ is estimated at the start of a calculation, and is adjusted if a higher value is found during the calculations. For $F_{num}N$ large compared to unity, the second method of calculating the collisions, equations (60) and (61), approaches the first, equations (58) and (59), and is therefore physically correct, while computationally much more efficient.

For each of the total number of pairs, a pair of computational particles is selected from the cell at random. The colliding particles do not have to be close in physical space (as long as they are within the same computational cell), nor do their paths need to intersect. As long as the cell dimensions are smaller than $\lambda/3$, this does not have a significant effect on the results. To further decrease the effect of separation, a cell may be divided into sub-cells, and a pair is selected from the same sub-cell if possible.

Various collision models can be used to determine the collision cross-section σ_T and the post-collision velocity and internal energy of the computational particles. The parameters of the collision model determine the collision frequency of particles, and the transfer of momentum and energy during a collision. Macroscopically, these parameters determine the diffusion coefficient, thermal diffusivity and viscosity of the gas.

Two collision models frequently used in DSMC are the 1-parameter so-called Variable Hard Sphere (VHS) [33] model, and the 2-parameter so-called Variable Soft Sphere (VSS) [34, 35] model. With the first it is possible to accurately reproduce the temperature dependence of the viscosity, whereas with a VSS model also the temperature dependence of (thermal) diffusivities can be accounted for.

3.3.4 Sampling

Due to the relatively low number of computational particles (compared to the number of molecules in a physical system), DSMC results suffer from statistical noise. The amount of noise is reduced by sampling the molecular properties during many time steps (for a steady problem) or many ensembles (for an unsteady problem).
For steady state flow problems, sampling of the flow properties is performed inside the time step loop and over many time steps once steady state has been reached. Because two consecutive samples are usually highly correlated, sampling is usually done once every $\sim$ 4 times steps. Flow properties are averaged over the same cells as used for the collision routines. Within one cell and at one sampling time, the following particle properties are accumulated:

- number of particles N,
- the sum of their velocities $\sum \bar{V}_i$,
- the sum of the square of their velocities $\sum (\bar{V} \cdot \bar{V})_i$.

All relevant flow data such as the mass-average velocity $\bar{V}_{ma}$, the temperature T and the density ρ can be calculated from these data. The density is calculated as:

$$\rho = F_{num} \frac{Nm}{sV}. \qquad (62)$$

The equation for the mass-average velocity is:

$$\bar{V}_{ma} = \frac{\sum \bar{V}_i}{N}. \qquad (63)$$

Finally, the temperature is determined as:

$$T = \frac{m[\sum (\bar{V} \cdot \bar{V})_i - \bar{V}_{ma} \cdot \bar{V}_{ma}]}{3k_B}. \qquad (64)$$

For unsteady flows, sampling during many time steps is not possible. In this case, many ensembles are calculated, and the flow properties are derived by averaging over all ensembles the samples taken at a specific time. This can be time and memory consuming due to the large number of ensembles which are needed and the necessity of storing sample data also as a function of time.

3.4 Dynamic coupling of Navier-Stokes and DSMC solvers

3.4.1 Breakdown parameter

The first issue in developing a coupled N-S/DSMC method is how to determine the appropriate computational domains for the DSMC and N-S solvers, and the proper interface boundary between these two domains. The continuum breakdown parameter Kn_{max} [36] is employed in the present study as a criterion for selecting the proper solver

$$Kn_{max} = max[Kn_\rho, Kn_V, Kn_T], \qquad (65)$$

where Kn_ρ, Kn_V and Kn_T are the local Knudsen numbers based on density, velocity and temperature length scales, according to

$$Kn_Q = \frac{\lambda}{Q_{ref}} |\nabla Q|. \qquad (66)$$

Here, Q is a flow property (density, velocity or temperature) and λ is the local mean free path length. Q_{ref} is a reference value for Q, which can be either its local value (for temperature or pressure), or a typical value (for the velocity). If the calculated value of the continuum breakdown parameter Kn_{max} in a region is larger than a limiting value Kn_{split}, then that region cannot be accurately modelled using the N-S equations, and DSMC has to be used.

Two different strategies have been implemented for coupling the Navier-Stokes based CFD code and the DSMC code: one for steady state flow simulations, the other for unsteady flow simulations. Both will be described below.

3.4.2 Steady-state formulation

The proposed coupling method for steady flows is based on the Schwarz method [17] and it consists of two stages (fig.2).

The first stage is a prediction stage, where the unsteady N-S equations are integrated in time on the entire domain Ω until a steady state is reached. From this steady state solution, the continuum breakdown parameter Kn_{max} is computed and its values are used to split Ω in the subdomains Ω_{DSMC} ($Kn_{max} > Kn_{split}$), where the flow field will be evaluated using DSMC, and Ω_{CFD} ($Kn_{max} < Kn_{split}$), where the N-S equation will be solved. For Kn_{split} a value of 0.05 was used. Between the DSMC and CFD regions an overlap region is considered, where the flow is computed with both the DSMC and the CFD solver (fig.3).

In the second stage, DSMC and CFD are run in their respective subdomains with their own time steps (Δt_{DSMC} and Δt_{CFD}, respectively), until a steady state is reached.

First DSMC is applied; molecules are allocated in the DSMC subdomain, created

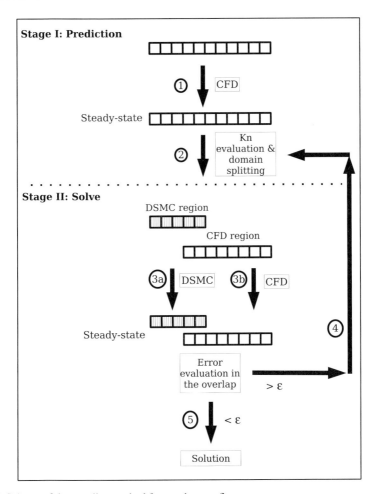

Fig. 2 Scheme of the coupling method for steady-state flows.

from a Chapman-Enskog velocity distribution, according to the density, velocity and temperature obtained from the initial CFD solution. The grid is automatically refined in the DSMC region in order to respect the DSMC requirements ($\Delta x, \Delta y, \Delta z < \frac{\lambda}{3}$). The boundary conditions to the DSMC region come from the solution in the CFD region. As described in the previous section 3.3.2 "particle reservoir cells" are considered outside the overlapping region. In these cells molecules are created according to the density, velocity, temperature and their gradients in the CFD solution with a Chapmann-Enskog velocity distribution.

After running the DSMC, the N-S equations are solved in the CFD region. The boundary conditions come from the solution in the DSMC region, averaged over the CFD cells.

Once a steady state solution has been obtained in both the DSMC and N-S regions,

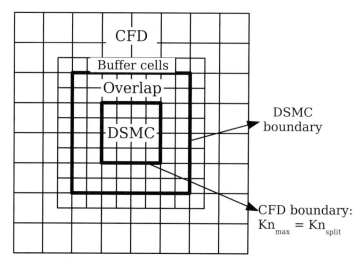

Fig. 3 Illustration of the Schwarz coupling method in a 2-D geometry.

the continuum breakdown parameter Kn_{max} is re-evaluated and a new boundary between the two regions is computed. This second stage is iterated until in the overlapping region the relative difference between the DSMC and CFD solutions

$$\left|\frac{\Delta Q}{Q_{DSMC}}\right| = \left|\frac{Q_{CFD} - Q_{DSMC}}{Q_{DSMC}}\right|, \qquad (67)$$

with Q a flow property (e.g. pressure or temperature), is less than a prescribed value ε (typically, $\varepsilon \approx 0.001$ [8]).

The advantage of using a Schwarz method with Dirichlet-Dirichlet boundary conditions, instead of the more common Neumann-Neumann boundary conditions coupling technique [19], is that the latter requires a much higher number of samples than the Schwarz method [19]. In fact, the DSMC statistical scatter involved in determining the fluxes is much higher than that associated with the macroscopic state variables.

3.4.3 Unsteady formulation

In the unsteady formulation, the coupling method described above is re-iterated every coupling time step $\Delta t_{coupling} \gg \Delta t_{DSMC}, \Delta t_{CFD}$, starting from the solution at the previous time step.

During every coupling time step, the predicted DSMC region is compared to the one of the previous time step. In the cells that still belong to the DSMC region, we consider the same molecules of the previous time step, whose properties were recorded.

In these cells, it is important to consider the same molecules of the previous time step rather than sampling them from continuum variables (temperature, density and velocity) with a Maxwellian or a Chapman-Enskog velocity distribution. The use of a Maxwellian or a Chapman-Enskog velocity distribution, in fact, presumes either equilibrium or near-equilibrium conditions, which is not necessarily true in these cells.

Molecules that are in the cells that no longer belong to the DSMC region are deleted. In cells that have changed from being a CFD cell into a DSMC cell, new molecules are created with a Chapmann-Enskog velocity distribution, according to the density, velocity and temperature of the CFD solution at the previous time step.

At the end of every coupling step, molecule properties are recorded to set the initial conditions in the DSMC region for the next coupling step.

4 Results and discussion

In this section we will demonstrate the use of our hybrid, dynamically coupled, N-S/DSMC solver to various 1D and 2D, transient and steady-state flows with temporal or spatial transitions from high Kn to low Kn.

4.1 Unsteady shock-tube problem

The unsteady coupling method was applied to an unsteady shock tube test case (fig.4).

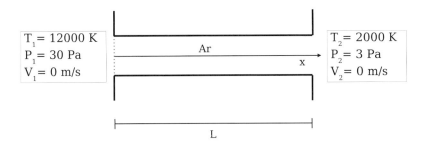

Fig. 4 Shock tube test case.

We simulated the flow field inside a 0.5 m long tube, connecting two infinitely large tanks filled with Argon at different thermodynamic conditions. A membrane at the interface between the first tank and the tube divides the two regions where the fluid is in different conditions: in the left tank it is at a pressure $P_1 = 30$ Pa and at a temperature $T_1 = 12000$ K, while in the right tank and in the tube it is at a pressure $P_2 = 3$ Pa and at a temperature $T_2 = 2000$ K. At $t = 0$ the membrane breaks and the fluid can flow from one region to the other. Two different waves will start travelling from the left to the right with two different velocities: a shock wave and a contact discontinuity. The shock wave produces a rapid increase of the temperature and pressure of the gas passing through it, while through the contact discontinuity, the flow undergoes only a temperature, and not a pressure, variation [25, 26].

The thermodynamic conditions inside the infinitely large tanks remain constant. For this reason the two tanks can be modeled with an inlet and an outlet boundary condition.

Inside the tube, we suppose that the flow is one-dimensional. Upstream (left) from the shock, the gas has a high temperature and relatively high pressure, and gradient length scales are small. Downstream (right) from the shock, both temperature and pressure are much lower, and gradient length scales are large. As a result, the continuum breakdown parameter Kn_{max} (using local values of Q_{ref}) is high upstream from the shock, and low downstream of it. In the hybrid DSMC-CFD approach, DSMC is therefore applied upstream, and CFD is applied downstream. The continuum grid is composed of 100 cells in the x-direction and 1 cell in the y-direction, while the code automatically refines the mesh in the DSMC region to fulfil its requirements. The coupling time step is $\Delta t_{coupling} = 4.0 \times 10^{-6}$ $sec.$ and ensemble averages of the DSMC solution were made on 30 repeated runs.

In figs. 5 and 6 the pressure (a), temperature (b), and velocity (c) inside the tube after 1.5×10^{-5} $sec.$ and 3.0×10^{-5} $sec.$ respectively, evaluated with the coupled DSMC-CFD method, are compared to the results of a full DSMC simulation. The latter was feasible because of the 1-D nature of the problem. Results obtained with a full CFD simulation are shown as well. The full DSMC solution is considered to be the most accurate of the three. In figs. 5(d) and 6(d) also the continuum breakdown parameter, computed using the coupled method, is compared to that same parameter computed with the full CFD simulation.

From the results shown in figs. 5 and 6, it is clear that the full CFD approach fails due to the high values of the local Kn number caused by the presence of the shock. It predicts a shock thickness of ≈ 2 cm, which is unrealistic since even in continuum conditions the shock thickness is one order of magnitude greater than the mean free path ($\lambda \approx 1$ cm) [37]. In the full DSMC approach, therefore, the shock is smeared over almost 10 cm. The results obtained with the hybrid approach are virtually identical to those obtained with the full DSMC solver (maximum difference $< 1.5\%$)..., but were obtained in less than one fifth of the CPU time.

Comparing figs. 5 and 6 it is also possible to see how the DSMC and CFD regions adapt in time to the flow field evolution.

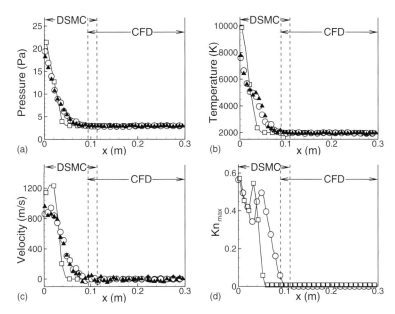

Fig. 5 Pressure (a), temperature (b), velocity (c) and continuum breakdown parameter Kn_{max} (d) in the tube after 1.5×10^{-5} sec. CFD (□), DSMC (▲), Hybrid (O).

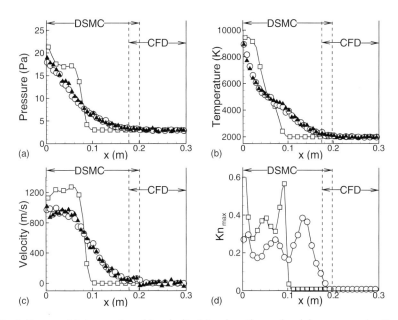

Fig. 6 Pressure (a), temperature (b), velocity (c) and continuum breakdown parameter Kn_{max} (d) in the tube after 3.0×10^{-5} sec. CFD (□), DSMC (▲), Hybrid (O).

4.1.1 Sensitivity to numerical parameters

In this section, the sensitivity of the coupled approach to various numerical parameters is addressed. In particular, the influence of the size of the overlap region, the DSMC noise, and the Courant number, based on the time interval at which DSMC and CFD are coupled, are analysed.

Overlap region: Both DSMC and N-S equations are solved in the overlap region (fig.3). The dependence of the results on the size of the overlap region is investigated by considering various overlap sizes: $\lambda/3$, 3λ, 6λ, 12λ, where λ is the mean free path length.

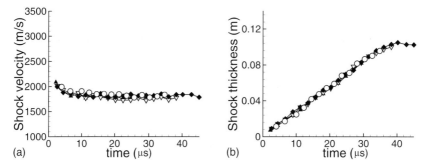

Fig. 7 Computed shock velocity (a) and shock thickness (b) as a function of time, for different sizes of the overlap region. $\lambda/3$ (▲), 3λ (▽), 6λ (◆), 12λ (O).

Fig.7 shows the evolution in time of respectively the shock velocity (a) and its thickness (b), evaluated using the different overlap sizes. From this picture it is clear that the overlap size does not strongly influence the results of the simulation.

If we fix the transition between the CFD and DSMC regions at the location where $Kn = 0.05$ and the overlap region is large, it is important to use an asymmetric overlap that is bounded on one side by the location where $Kn = 0.1$. Otherwise, if the overlap region would extend into regions where $Kn > 0.1$, the program would solve the N-S equations in a region where the continuum hypotheses are no longer valid. As a result, instability problems appear.

Number of repeated runs for the ensemble average: To analyze the effect of the noise in the DSMC solution on the coupling method we considered different numbers of repeated runs for the ensemble average: 5, 30 and 50 runs.

From a comparison (not shown) of the evolution of the shock velocity and thickness similar to the one in fig.7, since the maximum differences in the solutions were all below 5%, it became clear that also the number of repeated runs, over

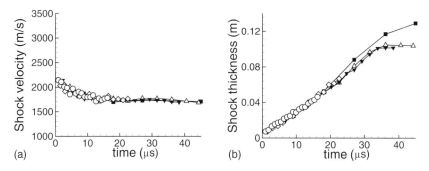

Fig. 8 Computed shock velocity (a) and shock thickness (b) for different coupling Courant numbers. $C = 1.46$ (■), $C = 0.73$ (△), $C = 0.36$ (▼), $C = 0.24$ (◇), $C = 0.15$ (O).

which we average, does not strongly influence the results of the method.
The limited sensitivity of our method to both the size of the overlap region and the reduction of noise through ensemble averaging demonstrates the clear advantage of our Dirichlet-Dirichlet coupling method as compared to Neumann-Neumann coupling schemes [15, 16, 18, 19], which show a strong sensitivity to noise.

Courant number based on the coupling time step: In this section we study the effect of varying the coupling Courant number defined as:

$$C = C_r \frac{\Delta t_{coupling}}{\Delta x_{CFD}}, \qquad (68)$$

where Δx_{CFD} is the size of CFD cells and C_r the molecules most probable velocity.

In fig.8 we present the evolution in time of both the shock velocity (a) and its thickness (b) for different coupling Courant numbers: 0.15, 0.24, 0.36, 0.73 and 1.46. In order to vary the Courant number we fixed $\Delta x_{CFD} = 0.005\ m$ and $C_r = 912\ m/s$ and we considered different values of the coupling time step between 8.0×10^{-7} sec. and 8.0×10^{-6} sec. In terms of multiples of the mean collision time, which is approximately $\Delta t_c = 6.0 \times 10^{-6}$ sec., this corresponds respectively to $0.13\Delta t_c - 1.3\Delta t_c$. Only in the case where the Courant number $C = 1.46 > 1$, the solution is found to deviate from the other solutions. In this case in fact the shock thickness is higher than for the other cases and the error is due to the appearance of instability effects (fig.9).

In order to be sure about the Courant number effects, we also varied the Courant number by varying Δx_{CFD} at fixed $\Delta t_{coupling}$ and fixed C_r, and by varying C_r at fixed Δx_{CFD} and $\Delta t_{coupling}$.

In all cases, instabilities were found to arise when $C > 1$, as expected. It is therefore necessary to keep the Courant number, based on the coupling time step, the

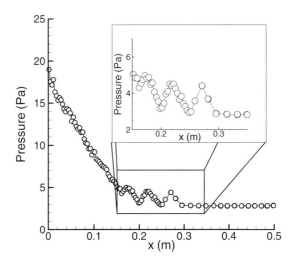

Fig. 9 Instability problems for Courant number $C = 1.46$.

CFD cell size and the molecules most probable velocity, smaller than 1.

4.2 Rarefied Poiseuille flow

The steady-state coupling method in two dimensions was applied to a plane Poiseuille flow (fig.10).

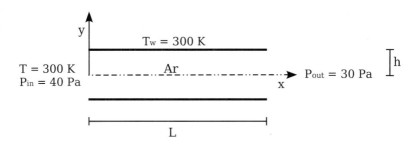

Fig. 10 Rarefied Poiseuille flow.

We consider a flow of Argon at a temperature $T = 300\,K$ in a small channel of height

$2h = 10$ mm and length $L = 50$ mm. The pressure at the inlet is $P_{in} = 40$ Pa and at the outlet $P_{out} = 30$ Pa. The wall temperature is $T_w = 300$ K. The typical mean free path is $\lambda \approx 0.25$ mm under these conditions. At a sufficient distance from the inlet the flow is a Poiseuille flow: its solution under continuum conditions is known and characterized by a linear pressure decay in the x-direction and a parabolic velocity profile in the y-direction. However, because λ/h is not small, deviations from the continuum solution are expected. Because of symmetry we limit the simulation domain to the upper half of the channel. The continuum grid is composed of 50 cells in the x-direction and 10 cells in the y-direction, while the code automatically refines the mesh in the DSMC region to fulfil its requirements ($\Delta x, \Delta y < \frac{\lambda}{3}$).

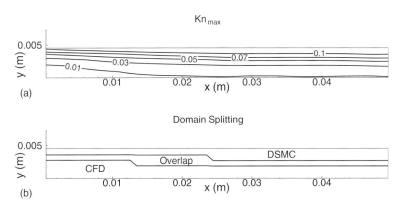

Fig. 11 Rarefied Poiseuille flow: Continuum breakdown parameter (a) and CFD/DSMC domains splitting (b)

In fig.11(a) the continuum breakdown parameter is shown. Because temperature, pressure and density are almost constant in the whole domain, the continuum breakdown parameter identifies with the Knudsen number based on the local velocity gradient length scale ($Kn_{max} = Kn_V$). As a reference velocity, we used V_0 at the symmetry plane under continuum conditions, rather then the local velocity, since the latter approaches zero near the wall

$$Kn_V = \frac{\lambda}{V_0} |\nabla V|, \qquad (69)$$

with

$$V_0 = \frac{h^2}{2\mu} \frac{\Delta p}{L}. \qquad (70)$$

The velocity gradient is small near the axis and large near the wall. This means that the continuum breakdown parameter is high near the wall and low near the axis. In fig.11(b), the resulting division between the DSMC, CFD and overlapping regions is shown.

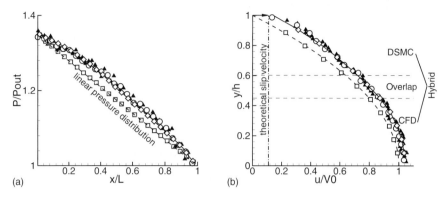

Fig. 12 Poiseuille flow pressure drop in x-direction (a) and velocity profile at $x = 40$ mm (b). Theoretical continuum Poiseuille solution with no-slip BC (– –), CFD with no-slip BC (□), CFD with velocity slip BC (–◊–), DSMC (▲), Hybrid (O).

In fig.12(a) we compare the pressure drop along the x-direction predicted by the hybrid method to the pressure drop obtained from a full DSMC simulation. Results obtained with a full CFD simulation with no-slip, as well as with the theoretically expected velocity slip boundary conditions [2]

$$V_{slip} = \lambda \left| \frac{du}{dy} \right| \tag{71}$$

and the theoretical linear pressure drop for continuum conditions are also shown.
From fig.12(a), it is clear that while the full CFD approach with no-slip boundary conditions predicts a linear pressure drop, the hybrid approach, the DSMC method and the full continuum simulation with velocity slip boundary conditions show the non-linear pressure drop which is known to prevail for rarefied Poiseuille flow [39, 40].
In fig.12(b) the velocity profile at $x = 40$ mm from the hybrid DSMC-CFD method is compared to the results of a full DSMC simulation, results obtained with a full CFD simulation both with no-slip and with velocity slip boundary conditions, and the theoretical solution for continuum Poiseuille flow.
The coupled CFD-DSMC method, the full DSMC simulation and the results of the full CFD simulations with slip boundary conditions give very similar results for the velocity profile, describing a slip velocity at the wall. The computed slip velocity agrees very well with the theoretical prediction [39]. It is also obvious that the slip effect increases the average velocity in the channel at given total pressure drop and reduces the curvature of the velocity profile with respect to the continuum solution [39].

4.3 Steady-state jet expanding in a low pressure chamber

The steady-state coupling method in two dimensional cylindrical coordinates was further applied to a steady state expanding neutral gas jet in a low pressure chamber (fig.13).

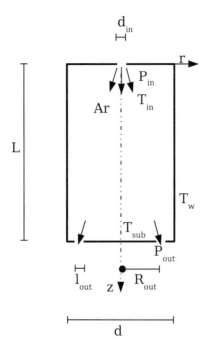

Fig. 13 Scheme of the low pressure chamber (not to scale).

The computational domain is a $d = 32\ cm$ diameter cylinder of length $L = 50\ cm$. From a circular hole of diameter $d_{in} = 8\ mm$, on its top, a flow of 56 standard cubic centimeters per second ($sccs$) of argon is injected at a temperature $T_{in} = 8000\ K$. The top and lateral walls are at a temperature $T_w = 400\ K$, while the bottom wall is at a temperature $T_{sub} = 600\ K$. In our 2-D axi-symmetric model, the pumping exit, which is a circular hole in reality, has been represented as a $l_{out} = 2\ cm$ wide ring on the bottom of the cylinder at a distance of $R_{out} = 12\ cm$ from the axis where the pressure is $P_{out} = 20\ Pa$.

Inside the chamber we assume a 2-D axi-symmetric flow. The continuum grid is composed of 100 cells in the radial direction and 200 cells in the axial direction. The cells are slightly stretched in the radial direction with a ratio of 1.65 between the last and the first cells. In the DSMC regions, the grid is further refined to meet the $\Delta x, \Delta y < \frac{\lambda}{3}$ criterion. Grid independence has been tested by doubling the continuum grid in each of the two directions, leading to variations in the solution below

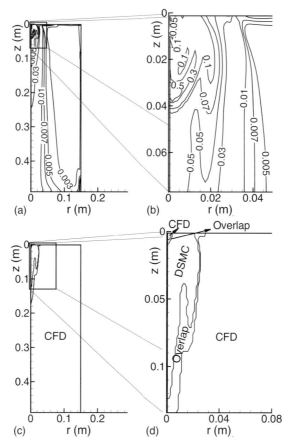

Fig. 14 Contours of the continuum breakdown parameter Kn_{max} in the entire chamber (a), and zoomed in to the expansion-shock region (b), CFD/DSMC domains splitting in the entire chamber (c), and zoomed in to the expansion-shock region (d).

3%.

In fig. 14 we show the continuum breakdown parameter Kn_{max} in the chamber and the consequent division between the DSMC, continuum and overlapping regions in our hybrid method.

There are various counteracting effects influencing the value of Kn_{max}: as a result of the decrease in pressure, the mean free path increases from the inlet to the exit of the chamber. As a result of the cooling of the gas, the temperature decreases from the inlet to the exit of the chamber and the opposite effect occurs. And finally, smaller local gradient length scales are present near the inlet and in the shock, than in the rest of the chamber. The overall effect is that the continuum breakdown parameter is small near the inlet, then it increases becoming high in the expansion-shock region, and finally it becomes low again in the rest of the chamber. Also near the bottom

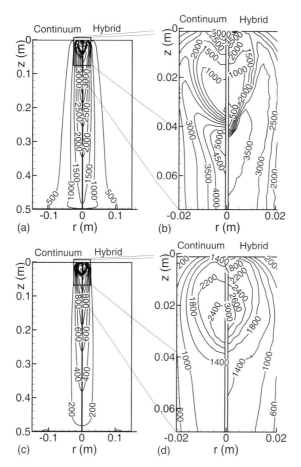

Fig. 15 Temperature field (in K) in the entire chamber (a), and zoomed in to the expansion-shock region (b) and velocity field (in m/s) in the entire chamber (c), and zoomed in to the expansion-shock region (d).

wall the continuum breakdown parameter increases, due to steep velocity and temperature gradients, but not to values exceeding Kn_{split}. This means that the flow first undergoes a continuum-rarefied transition in the near-inlet region, and then a rarefied-continuum transition downstream of the shock (fig. 14(b)).

Temperature and velocity fields obtained with the coupled CFD/DSMC method are compared to results from a full continuum CFD simulation in fig. 15. It should be noted that DSMC simulations intrinsically contain statistical scatter, explaining why the contours in the hybrid simulations are less smooth than in the continuum simulations. From an analysis of the figures, it is evident that far away from the expansion-shock region, the two methods give very similar results (figs. 15(a) and (c)). The use of DSMC in the hybrid method influences only the region where rarefaction effects are present.

The hybrid method predicts a stronger expansion compared to the continuum method, reaching a lower temperature (fig. 15(b)) and higher velocity (fig. 15(d)). Also, compared to the full continuum simulation, the shock is slightly moved downstream along the z-axis in the hybrid simulation. Finally, after the shock the temperature predicted by the hybrid method is significantly $(500 - 1500\ K)$ lower than the one calculated by the continuum approach.

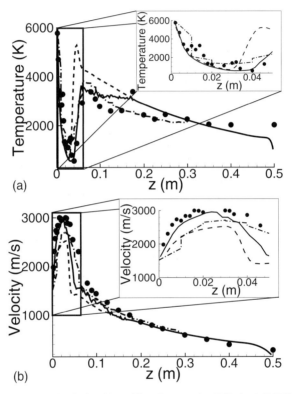

Fig. 16 Temperature (a) and velocity (b) profiles along z-axis. CFD (- -), DSMC data from [41] (- · -), Hybrid (—), Experimental data from [42] (•).

In order to further clarify the effects of rarefaction on the flow field, fig. 16 shows, for the temperature (a) and the velocity (b) along the z-axis, a comparison between the present hybrid method, the present continuum simulation, results from full DSMC simulations performed by Selezneva et al. [41], and experimental data from [42]. Although there is quite some scattering in the experimental data, it is clear that the hybrid method predicts the experimental data better than the other approaches. It is important to explain that the reason why the hybrid approach predicts experimental data even better than the full DSMC simulations, in the current case, is that, as already highlighted by Selezneva et al. [41], in the full DSMC simulations it was not possible to respect the $\Delta x, \Delta y < \frac{\lambda}{3}$ grid requirements of DSMC in the near inlet region, where λ is very small and a too coarse mesh had to be used.

If we first compare the results of the hybrid CFD/DSMC approach to those of the full CFD approach, fig. 16(a) shows that the hybrid approach follows much better than the continuum approach the experimental data in the shock and after-shock region. In the continuum approach the shock wave appears too early and the temperature after the shock is too high, whereas in the hybrid approach the shock moves forward due to rarefaction and the temperature after the shock is lower.

Figure 16(b) demonstrates that the continuum approach is unable to quantitatively predict the velocity profile and maximum velocity in the expansion-shock, and is quantitatively correct only downstream of the expansion-shock region as already shown in [41]. Because of rarefaction, upstream of the shock the expansion is stronger, reaching higher velocity values as predicted by the hybrid solution in agreement with the experimental data.

If we compare the hybrid method to the full DSMC simulations by Selezneva et al. [41], we can notice that the results of the full DSMC simulations and the hybrid method are almost equivalent in the shock and after shock regions. However, in the near inlet and expansion regions the hybrid approach matches the experimental data better than the full DSMC approach.

We can notice that the temperature profiles predicted by the DSMC alone and by the hybrid approach are very similar and they both accurately match the experimental data (fig. 16(a)). However, from fig. 16(b) we can conclude that, because it was not possible to respect DSMC requirements in the near-inlet region, the DSMC method predicts a wrong velocity in this region that influences its solution also in the expansion region. As a result, DSMC predicted a too low value of the maximum velocity reached in the expansion, whereas the hybrid approach accurately predicts these maxima.

5 Conclusion

A hybrid continuum-rarefied flow simulation method has been presented to couple a Navier-Stokes description of a continuum gas flow with a DSMC description of a rarefied gas flow in both steady-state and unsteady conditions. As opposed to other published methods, the presented approach is suitable for both steady-state and transient compressible flows and uses a Schwarz method based Dirichlet-Dirichlet coupling boundary conditions.

The method has been applied to a 1D transient shock tube flow, to a 2D steady-state low Mach number rarefied plane Poiseuille flow, and to a 2D steady-state high Mach number jet flow expanding in a low pressure chamber. These examples illustrate the possibility of efficiently coupling CFD and DSMC simulation approaches and show the potentials of the method for steady and transient, compressible and incompressible, one-dimensional and multi-dimensional flows.

The results of the method were found to be independent on the size of the overlap region for an asymmetric overlap, while instability effects can appear using a symmetric overlap. The method also has limited sensibility to noise, as demonstrated by its insensitivity to the number of DSMC runs for ensemble average to reduce the scattering. However, the coupling time step and the CFD cell size should be chosen such, that the Courant number based on these quantities and the most probable velocity of the molecules is less than one in order to avoid instability effects.

References

1. Boltzmann L (1872) Weitere Studien ueber das Waermegleichgewicht under Gasmolekuelen. In: Sitzungsberichte Akademie der Wissenschaften Wien, 66:275–370
2. Maxwell J C (1879) On stresses in rarefied gases arising from inequalities of temperature. Philos. Trans. R. Soc. London, Ser. B 170:231
3. Alder B (1997) Highly discretized dynamics. Physica A 240:193–195
4. Hadjiconstantinou N G (2003) Comment on Cercignani's second-order slip coefficient. Physics of Fluids 15:2352–2354
5. Hadjiconstantinou N G (2004) Validation of a second-order slip model for transition-regime, gaseous flows. In: 2nd Int. Conf. on Microchannels and Minichannels, Rochester, New York, USA.
6. Bird G A (1998) Molecular gas dynamics and Direct Simulation Monte Carlo. Clarendon Press, Oxford.
7. Sharipov F (2003) Hypersonic flow of rarefied gas near the Brazilian satellite during its re-entry into atmosphere. Brazilian Journal of Physics, vol.33, no.2
8. Aktas O, Aluru N R (2002) A combined Continuum/DSMC technique for multiscale analysis of microfluidic filters. Journal of Computational Physics 178:342–372
9. Cai C, Boyd I D (2005) 3D simulation of Plume flows from a cluster of plasma thrusters. In: 36th AIAA Plasmadynamics and Laser Conference, Toronto, Ontario, Canada, AIAA-2005-4662
10. van de Sanden M C M, Severens R J, Gielen J W A M, Paffen R M J, Schram D C (1996) Deposition of a-Si:H and a-C:H using an expanding thermal arc plasma. Plasma sources Science and Technology 5:268–274
11. Hadjiconstantinou N G (1999) Hybrid Atomistic-Continuum formulations and moving contact-line problem. Journal of Computational Physics 154:245–265
12. Le Tallec P, Mallinger F (1997) Coupling Boltzmann and Navier-Stokes equations by half fluxes. Journal of Computational Physics 136:51–67
13. Wijesinghe H S, Hadijconstantinou N G (2004) Discussion of hybrid Atomistic-Continuum scheme method for multiscale hydrodynamics. International Journal for Multiscale Computational Engineering 2 no.2:189–202
14. Roveda R, Goldstein D B, Varghese P L (1998) Hybrid Euler/particle approach for continuum/rarefied flows. J. Spacecraft Rockets 35:258
15. Garcia A L, Bell J B, Crutchfield W Y, Alder B J (1999) Adaptive mesh and algorithm refinement using Direct Simulation Monte Carlo. Journal of Computational Physics 154:134–155
16. Glass C E, Gnoffo P A (2002) A 3-D coupled CFD-DSMC solution method for simulating hypersonic interacting flow. In: 8th AIAA/ASME Joint Thermophysics and Heat Transfer Conference, AIAA Paper 2002-3099,
17. Wu J S, Lian Y Y, Cheng G, Koomullil R P, Tseng K C (2006) Development and verification of a coupled DSMC-NS scheme using unstructured mesh. Journal of Computational Physics 219:579–607

18. Schwartzentruber T E, Scalabrin L C, Boyd I D (2006) Hybrid Particle-Continuum Simulations of Non-Equilibrium Hypersonic Blunt Body Flows. AIAA Paper, San Francisco, CA, AIAA-2006-3602
19. Schwartzentruber T E, Boyd I D (2006) A Hybrid particle-continuum method applied to shock waves. Journal of Computational Physics 215 No.2:402–416
20. Chapman S, Cowling T G (1960) *The Mathematical Theory of Non-Uniform Gases*. Cambridge University Press, Cambridge.
21. Grad H (1949) On the kinetic theory of rarefied gases. Commun. Pure Appl. Math. 2:331
22. Patterson G N (1956), Molecular flow of gases. Wiley, New York.
23. Chou S Y, Baganoff D (1997) Kinetic flux-vector splitting for the Navier-Stokes equations. Journal of Computational Physics 130:217–230
24. Garcia A L, Alder B J (1998) Generation of the Chapman-Enskog distribution. Journal of Computational Physics 140:66–70
25. van Leer B (1979) Towards the ultimate conservative difference scheme V. a second order sequel to Godunov's method. Journal of Computational Physics 32:101–136
26. Roe P L (1981) Approximate Riemann solvers, parameter vectors and difference schemes. Journal of Computational Physics 43:357–372
27. Lou T, Dahlby D C, Baganoff D (1998) A numerical study comparing kinetic flux-vector splitting for the Navier-Stokes equations with a particle method. Journal of Computational Physics 145:489–510
28. Hirschfelder J O, Curtis C F, Bird R B (1954) *Molecular Theory of Gasses and Liquids*, Wiley, New York
29. Dorsman R (2007) *Numerical Simulations of Rarefied Gas Flows in Thin Film Processes*. PhD Thesis, Delft University of Technology, The Netherlands
30. Wagner W (1992) A convergence proof for Bird's direct simulation method for the Boltzmann equation. Journal of Statistical Physics 66:1011–1044
31. Bhatnagar P L, Gross E P, Krook M (1954) A model for collision processes in gases. Physical Review 94:511–525
32. Nance R P, Hash D B, Hassan H A (1998) Role of boundary condition in Monte Carlo simulation of microelectromechanical system. J.Thermophys.Heat Trans. 12
33. Bird G A (1981) Monte Carlo simulation in an engineering context. In: Fischer S S (ed) 12th International Symposium Rarefied Gas Dynamics - part 1:239–255
34. Koura K, Matsumoto H (1991) Variable soft sphere molecular model for air species. Physics of Fluids 3:2459–2465
35. Koura K, Matsumoto H (1992) Variable soft sphere molecular model for inverse-power-law or Lennard-Jones potential. Physics of Fluids 4:1083–1085
36. Wang W L, Boyd I D (2002) Continuum breakdown in hypersonic viscous flows. In: 40th AIAA Aerospace Sciences Meeting and Exhibit, Reno, NV
37. Mott-Smith H M (1951) The solution of the Boltzmann equation for a shock wave. Phys. Rev. 82 (6) 885–892
38. Hash D, Hassan H (1996) A decoupled DSMC/Navier-Stokes analysis of a transitional flow experiment. AIAA Paper 96-0353
39. Yohung S, Chan W K (2004) Analytical modelling of Rarefied Poiseuille flow in microchannels. J.Vac.Sci.Technol. A 22(2)
40. Cai C, Boyd I D, Fan J (2000) Direct simulation method for low-speed microchannel flows. J.Thermophysics and Heat Transfer 14 No.3
41. Selezneva S E, Boulos M I, van de Sanden M C M, Engeln R, Schram D C (2002) Stationary supersonic plasma expansion: continuum fluid mechanics versus Direct Simulation Monte Carlo method. J. Phys. D: Appl. Phys. 35:1362–1372
42. Engeln R, Mazouffre S, Vankan P, Schram D C, Sadeghi N (2001) Flow dynamics and invasion by background gas of a supersonically expanding thermal plasma. Plasma Sources Sci. Technol. 10:595

Multi-Scale PDE-Based Design of Hierarchically Structured Porous Catalysts

Gang Wang, Chris R. Kleijn, and Marc-Olivier Coppens

Abstract Optimization problems involving the solution of partial differential equations (PDEs) often arise in the context of optimal design, optimal control and parameter estimation. Based on the reduced gradient method, a general strategy is proposed to solve these problems by using existing optimization packages and PDE solvers. For illustration purposes, this strategy was employed to solve a PDE-based optimization problem that arises from the optimal design of hierarchically structured porous catalysts. A Fortran program was developed that combines a gradient-based optimization package, NLPQL, a multigrid PDE solver, MGD9V, and a limited amount of in-house coding. The PDEs were discretized using a finite volume method, applied to a matrix of computational cells. The number of cells ranged from 129×129 to 513×513, and the number of the optimization variables from 101 to 201. For the problems studied here, the optimization typically converged within a limited number (about 9-12) of iterations, requiring a CPU time of only 2.5 to 212 seconds on a 2.16 GHz Intel Core2 Duo processor. The PDE solver was called 36-50 times in each of the numerical tests.

Gang Wang
Howard P. Isermann Department of Chemical and Biological Engineering, Rensselaer Polytechnic Institute, Troy, NY 12180, USA.
e-mail: wangg2@rpi.edu

Chris R. Kleijn
Department of Multi Scale Physics, Delft University of Technology, Prins Bernhardlaan 6, 2628 BW, Delft, The Netherlands.
e-mail: C.R.Kleijn@tudelft.nl

Marc-Olivier Coppens
Howard P. Isermann Department of Chemical and Biological Engineering, Rensselaer Polytechnic Institute, Troy, NY 12180, USA; Physical Chemistry and Molecular Thermodynamics, DelftChemTech, Delft University of Technology, Julianalaan 136, 2628 BL Delft, The Netherlands.
e-mail: coppens@rpi.edu

1 Introduction

PDEs (Partial Differential Equations) emerge as governing equations in diverse disciplines of science and engineering. PDE-based optimization problems are often encountered in the context of optimal design, optimal control and parameter estimation, and are computationally challenging because of the size and complexity of the discretized PDEs [1]. The solution of the PDEs is often the most time-consuming procedure for PDE-based optimizations. Therefore, rather than gradient-free optimization methods (e.g., genetic algorithms), gradient-based optimization methods are typically selected. This minimizes the need to evaluate the objective function and, consequently, the need to actually solve the PDEs. For the solution of PDE-based optimization problems, one can make use of existing, sophisticated PDE solvers and optimization packages. Biegler et al. [1] presented an excellent overview on the issues that need to be addressed for PDE-based optimizations.

The reduced gradient method is an ideal candidate for PDE-based optimization problems, since it employs gradient information to reduce the need for the evaluation of the objective function, and could be implemented with existing software. More importantly, the implementation could be tailored to different PDE-based optimization problems by an appropriate choice of existing software. However, this advantage of the reduced gradient method has not been addressed well, even though the reduced gradient method has been used to solve PDE-based optimization problems that arise from the optimal operation of a chemical vapor deposition (CVD) reactor [2], dynamic optimization of dissipative PDE systems [3], inverse parameter estimation [4] and optimal control of steady incompressible Navier-Stokes flows [5].

The objective of this paper is to discuss the implementation of the reduced gradient method for a PDE-based optimization problem on the basis of existing software. As a showcase, the reduced gradient method is implemented using a multigrid PDE solver, MGD9V [6], and a gradient-based optimization package, NLPQL [7], to solve a model problem that arises from the optimal design of hierarchically structured porous catalysts. The method discussed in this book chapter was presented in [8], whereas results of the application to structured porous catalysts were presented in more details in [9].

2 Implementation framework

To introduce the implementation framework, consider the following generic formulation of nonlinear programming

$$\min \; f_1(\mathbf{x}), \tag{1a}$$
$$\text{s.t.} \; \mathbf{g}_1(\mathbf{x}) \leq \mathbf{0}. \tag{1b}$$

where $f_1 : \mathbf{R}^N \to \mathbf{R}$ and $g_1 : \mathbf{R}^N \to \mathbf{R}^M$ are continuously differentiable functions. Equation (1a) is the objective function, and Eq. (1b) the constraint. $\mathbf{x} \in \mathbf{R}^N$ is the optimization variable vector.

The problem consisting of Eqs. (1a-b) could be readily solved using a number of gradient-based optimization packages, e.g., SNOPT [10], MINOS [11], LANCELOT [12], DONLP2 [13] and GRG2 [14]. To do so, all a user needs to do is to provide the information required by the optimization package. The standard input information includes the value and gradient of the objective function, Eq. (1a), and the Jacobian matrix and value of the constraints, Eq. (1b). To illustrate how to obtain this input information, consider the following PDE-based optimization problem

$$\min \ f_2(\mathbf{x_s}, \mathbf{x_d}) f_2[\mathbf{q}(\mathbf{x_d}), \mathbf{x_d}], \tag{2a}$$

$$\text{s.t.} \ \mathbf{g_2}(\mathbf{x_d}) \leq \mathbf{0}, \tag{2b}$$

where $f_2 : \mathbf{R}^n \to \mathbf{R}$ and $\mathbf{g_2} : \mathbf{R}^n \to \mathbf{R}^l$ are continuously differentiable functions. Equation (2a) is the objective function. $\mathbf{x_s} = \mathbf{q}(\mathbf{x_d})$ is mathematically equivalent to the discretized equations $\mathbf{h}(\mathbf{x_s}, \mathbf{x_d}) = \mathbf{0}$, which result from the discretization of a PDE in terms of finite element or finite difference methods. $\mathbf{x_s} \in \mathbf{R}^m$ and $\mathbf{x_d} \in \mathbf{R}^n$ are state variable and design variable vectors, respectively. State variables are those that could be determined by solving $\mathbf{h}(\mathbf{x_s}, \mathbf{x_d}) = \mathbf{0}$, with given data, i.e., geometry, coefficients, boundary conditions and initial conditions. These data are the design variables.

For the problem consisting of Eqs. (2a-b), the objective function could be evaluated by solving the discretized equations $\mathbf{h}(\mathbf{x_s}, \mathbf{x_d}) = \mathbf{0}$. A numerical solver could be used to carry out these calculations. The Jacobian matrix and value of the non-PDE constraint, Eq. (2b), are easily obtained as follows. To compute the gradient of the objective function, differentiate Eq. (2a):

$$\frac{d f_2}{d \mathbf{x_d}} = \frac{\partial f_2}{\partial \mathbf{x_d}} + \frac{\partial \mathbf{x_s}}{\partial \mathbf{x_d}} \frac{\partial f_2}{\partial \mathbf{x_s}}, \tag{3}$$

$$\frac{\partial f_2}{\partial \mathbf{x_d}} = \begin{bmatrix} \frac{\partial f_2}{\partial x_{d1}} & \frac{\partial f_2}{\partial x_{d2}} & \cdots & \frac{\partial f_2}{\partial x_{dn}} \end{bmatrix}^T, \tag{4a}$$

$$\frac{\partial f_2}{\partial \mathbf{x_s}} = \begin{bmatrix} \frac{\partial f_2}{\partial x_{s1}} & \frac{\partial f_2}{\partial x_{s2}} & \cdots & \frac{\partial f_2}{\partial x_{sm}} \end{bmatrix}^T, \tag{4b}$$

$$\frac{\partial \mathbf{x_s}}{\partial \mathbf{x_d}} = \begin{bmatrix} \frac{\partial x_{s1}}{\partial x_{d1}} & \frac{\partial x_{s2}}{\partial x_{d1}} & \cdots & \frac{\partial x_{sm}}{\partial x_{d1}} \\ \frac{\partial x_{s1}}{\partial x_{d2}} & \frac{\partial x_{s2}}{\partial x_{d2}} & \cdots & \frac{\partial x_{sm}}{\partial x_{d2}} \\ \vdots & \vdots & \ddots & \vdots \\ \frac{\partial x_{s1}}{\partial x_{dn}} & \frac{\partial x_{s2}}{\partial x_{dn}} & \cdots & \frac{\partial x_{sm}}{\partial x_{dn}} \end{bmatrix}. \tag{4c}$$

Typically, Eqs. (4a-b) are easily evaluated either analytically or by numerical differentiation. To compute Eq. (4c), differentiate both sides of the discretized equations, $\mathbf{h}(\mathbf{x_s}, \mathbf{x_d}) = \mathbf{0}$:

$$\frac{d\mathbf{h}}{d\mathbf{x_d}} = \frac{\partial \mathbf{h}}{\partial \mathbf{x_d}} + \frac{\partial \mathbf{x_s}}{\partial \mathbf{x_d}} \frac{\partial \mathbf{h}}{\partial \mathbf{x_s}} = 0, \tag{5}$$

where

$$\frac{\partial \mathbf{h}}{\partial \mathbf{x_d}} = \begin{bmatrix} \frac{\partial h_1}{\partial x_{d1}} & \frac{\partial h_2}{\partial x_{d1}} & \cdots & \frac{\partial h_m}{\partial x_{d1}} \\ \frac{\partial h_1}{\partial x_{d2}} & \frac{\partial h_2}{\partial x_{d2}} & \cdots & \frac{\partial h_m}{\partial x_{d2}} \\ \vdots & \vdots & \ddots & \vdots \\ \frac{\partial h_1}{\partial x_{dn}} & \frac{\partial h_2}{\partial x_{dn}} & \cdots & \frac{\partial h_m}{\partial x_{dn}} \end{bmatrix}, \tag{6a}$$

$$\frac{\partial \mathbf{h}}{\partial \mathbf{x_s}} = \begin{bmatrix} \frac{\partial h_1}{\partial x_{s1}} & \frac{\partial h_2}{\partial x_{s1}} & \cdots & \frac{\partial h_m}{\partial x_{s1}} \\ \frac{\partial h_1}{\partial x_{s2}} & \frac{\partial h_2}{\partial x_{s2}} & \cdots & \frac{\partial h_m}{\partial x_{s2}} \\ \vdots & \vdots & \ddots & \vdots \\ \frac{\partial h_1}{\partial x_{sm}} & \frac{\partial h_2}{\partial x_{sm}} & \cdots & \frac{\partial h_m}{\partial x_{sm}} \end{bmatrix}. \tag{6b}$$

Equations. (6a-b) can be evaluated either analytically or by numerical differentiation. From Eq. (5),

$$\frac{\partial \mathbf{x_s}}{\partial \mathbf{x_d}} = -\frac{\partial \mathbf{h}}{\partial \mathbf{x_d}} \left(\frac{\partial \mathbf{h}}{\partial \mathbf{x_s}} \right)^{-1}. \tag{7}$$

Substitute Eq. (7) into Eq. (3):

$$\frac{d f_2}{d \mathbf{x_d}} = \frac{\partial f_2}{\partial \mathbf{x_d}} - \frac{\partial \mathbf{h}}{\partial \mathbf{x_d}} \left(\frac{\partial \mathbf{h}}{\partial \mathbf{x_s}} \right)^{-1} \frac{\partial f_2}{\partial \mathbf{x_s}}. \tag{8}$$

For numerical efficiency, the inverse of the matrix, Eq. (6b), is not evaluated directly, but calculated by solving the linear equation

$$\left(\frac{\partial \mathbf{h}}{\partial \mathbf{x_s}} \right) \mathbf{z} = \frac{\partial f_2}{\partial \mathbf{x_s}}. \tag{9}$$

Note that the coefficient matrix of the linear equation (9) is a Jacobian matrix of the discretized equations $\mathbf{h}(\mathbf{x_s}, \mathbf{x_d}) = 0$. Substituting $\mathbf{z} \in \mathbf{R}^m$ into Eq. (8), we get:

$$\frac{d f_2}{d \mathbf{x_d}} = \frac{\partial f_2}{\partial \mathbf{x_d}} - \frac{\partial \mathbf{h}}{\partial \mathbf{x_d}} \mathbf{z}. \tag{10}$$

At this point, one could provide all the information that a gradient-based optimization package needs to solve the problem consisting of Eqs. (2a-b). The gradient, Eq. (10), is the so-called reduced gradient, therefore the implementation strategy constructed here is called the reduced gradient method [15]. According to the above discussion, the reduced gradient method could be readily implemented on the basis of existing software to solve a PDE-based optimization problem. Furthermore, in this implementation framework, the software can be freely chosen among those that can perform the aforementioned calculations. This feature can be employed to tailor the implementation to different problems. It should be mentioned that the reduced gradient can be calculated either by Eqs. (9-10) or by numerical differentiation. The fact that both methods should yield the same value for the reduced gradient can be employed in debugging the code.

3 PDE-based design of hierarchically structured porous catalysts

In this section, the above strategy is applied to solve the model problem that arises from the optimal design of hierarchically structured porous catalysts. Catalysts are essential for the fast and selective chemical transformation of raw materials to products, for instance, crude oil to gasoline, diesel and plastics. Discovery of more efficient catalysts therefore has a large economical impact. Note that, among the top ten in the Fortune Global 500, six are chemical companies [16]. Discovery of more efficient catalysts also helps to protect the environment, since catalysts are used in emission control and waste water treatment. It also contributes to saving energy and resources and, consequently, building a sustainable future.

One way to design more efficient catalysts is to structure the catalysts in a rational way. The idea is the following: catalysts are usually nanoporous materials, or they are formed by the dispersion of nanoparticles over the internal surface of nanoporous materials. These nanoporous materials often have an extremely large internal surface area, which is beneficial because catalytic reactions occur on the surface. Note that the internal surface area of one gram of nanoporous catalyst could be as large as the area of five tennis courts! However, this huge internal surface could be inaccessible, limiting the efficient use of the catalytic materials, since molecular transport in the nanopores is substantially slower than that in the bulk, and the nanopores are easy to block. This indicates that, apart from the nanopores where reactions actually occur, a "distribution" network of large pores is needed for the catalyst, just like a road network is needed for a city. One important question is how to design this distribution network for optimal catalyst performance. PDE-based optimizations were used to study this question [9,17-19]. Note that the PDE-based optimization is a multi-scale problem because of two reasons: (1) there is a network of narrow nanopores that are part of a porous matrix (treated as a continuum in the paper, as is justified by earlier research); (2) the size of the large pores is allowed to span a few orders of magnitude in the optimizations, even if it could eventually turn out that a broad distribution of large pore diameters is not the optimal solution.

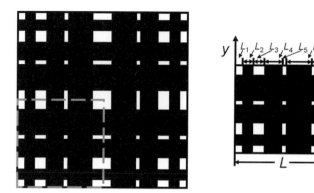

Fig. 1 Hierarchically structured porous catalyst with $2N \times 2N$ large pores (left) and its bottom-left quarter (right). $N = 3$ for illustration only. The nanoporous, catalytically active material (where the nanopores are not shown explicitly) is indicated in white, the large pores in black.

Gheorghiu and Coppens [17] optimized a model for the studied structured porous catalyst, shown in Figure 1. For reasons of computational simplicity, a two-dimensional square geometry was used, without affecting the generality of the conclusions [9, 17-19]. The square hierarchical pore structure is a nanoporous catalytically active material, in which $2N \times 2N$ large pores have been introduced. N is an arbitrary positive integer, typically $1 - 100$. It was assumed that the large pores are straight, perpendicular to the square sides, and have a constant diameter as they go through the entire square. Four-fold symmetry was imposed, as the optimum was assumed to have this symmetry. Therefore, only the bottom-left quarter of the entire catalyst (as shown in the right part of Figure 1) was modelled. The aim was to maximize the normalized catalytic yield η with respect to the design variables $L_i (i = 1, 2, \cdots, 2N+1)$ as labelled in the right part of Figure 1:

$$\max \quad \eta = \iint_\Omega k c \, \mathrm{d}x \mathrm{d}y / (k c_0 L^2), \tag{11a}$$

$$\text{s. t.} \quad \sum_{i=1}^{2N+1} L_i = L, \tag{11b}$$

$$L_i > 0 \quad i = 1, 2, \cdots, 2N+1, \tag{11c}$$

where Ω is the area occupied by the nanoporous material, k the intrinsic rate constant, c the concentration and c_0 the (fixed) concentration on the external surface of the catalyst; L is half of the catalyst size as labelled in the right part of Figure 1. Equations (11b-c) are algebraic (non-PDE) constraints.

The equations governing steady-state diffusion and first-order reaction in the bottom-left quarter of the catalyst are

$$\frac{\partial}{\partial x}\left(D_m \frac{\partial c}{\partial x}\right) + \frac{\partial}{\partial y}\left(D_m \frac{\partial c}{\partial y}\right) = 0 \text{ in the large pores (black),} \quad (12a)$$

$$\frac{\partial}{\partial x}\left(D_e \frac{\partial c}{\partial x}\right) + \frac{\partial}{\partial y}\left(D_e \frac{\partial c}{\partial y}\right) = kc \text{ in the nanoporous material (white),} \quad (12b)$$

with the boundary conditions

$$x = 0: \quad c = c_0, \quad (13a)$$
$$y = 0: \quad c = c_0, \quad (13b)$$
$$x = L: \quad \frac{\partial c}{\partial x} = 0, \quad (13c)$$
$$y = L: \quad \frac{\partial c}{\partial y} = 0, \quad (13d)$$

where D_m is the molecular diffusion coefficient in the large pores, and D_e is the effective diffusion coefficient in the nanoporous material. D_m/D_e is typically $10^2 - 10^4$.

Equations (11-13) can be made dimensionless, as follows:

$$\max \quad \eta = \iint_\Omega \bar{c}\, d\bar{x}\, d\bar{y}, \quad (14a)$$

$$\text{s. t.} \quad \sum_{i=1}^{2N+1} \bar{L}_i = 1, \quad (14b)$$

$$\bar{L}_i > 0 \quad i = 1, 2, \cdots, 2N+1, \quad (14c)$$

and

$$\frac{\partial^2 \bar{c}}{\partial \bar{x}^2} + \frac{\partial^2 \bar{c}}{\partial \bar{y}^2} = 0, \quad \text{in the large pores,} \quad (15a)$$

$$\frac{\partial}{\partial \bar{x}}\left(\beta \frac{\partial \bar{c}}{\partial \bar{x}}\right) + \frac{\partial}{\partial \bar{y}}\left(\beta \frac{\partial \bar{c}}{\partial \bar{y}}\right) = \Phi_0^2\, \bar{c}, \text{ in the nanoporous material,} \quad (15b)$$

with the boundary conditions

$$\bar{x} = 0: \quad \bar{c} = 1, \quad (16a)$$
$$\bar{y} = 0: \quad \bar{c} = 1, \quad (16b)$$
$$\bar{x} = 1: \quad \frac{\partial \bar{c}}{\partial \bar{x}} = 0, \quad (16c)$$
$$\bar{y} = 1: \quad \frac{\partial \bar{c}}{\partial \bar{y}} = 0, \quad (16d)$$

where

$$\Phi_0 = L\sqrt{\frac{k}{D_m}}, \tag{17a}$$

$$\beta = \frac{D_e}{D_m}, \tag{17b}$$

$$\bar{c} = \frac{c}{c_0} \tag{17c}$$

$$\bar{x} = \frac{x}{L}, \tag{17d}$$

$$\bar{y} = \frac{y}{L}, \tag{17e}$$

$$\bar{L}_i = \frac{L_i}{L}, \qquad i = 1, 2, \cdots, 2N+1. \tag{17f}$$

Gheorghiu and Coppens [17] employed a genetic algorithm to solve the above PDE-based optimization problem. The solution of the PDE needed by the genetic algorithm was obtained using FEMLAB. This numerical scheme is easy to implement, but suffers from low efficiency, because the genetic algorithm requires many more solutions of the PDE than a gradient-based method. To avoid these problems, the reduced gradient method is implemented on the basis of the specialized PDE solver, MGD9V [6], and the gradient-based optimization package, NLPQL [7]. Note that the use of FEMLAB does not imply the use of a derivative-free optimization algorithm, e.g, a genetic algorithm [20].

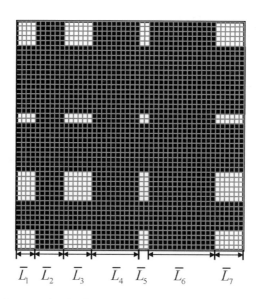

Fig. 2 Illustration of computational cells. The nanoporous, catalytically active material (where the nanopores are not shown explicitly) is indicated in white, the large pores in black.

As shown in Figure 2, the diffusion-reaction equation, Eqs. (15a-b), is discretized in terms of the finite volume method on a matrix of computational cells [21]. The computational cells have nearly the same size, and none of the cells crosses the interfaces between the large pores and the nanoporous material. For the cell (i,j) (the cell in the ith column counting from the left and the jth row counting from the bottom), the discretized equation is

$$H_{ij} = \left(F_{ij,right} - F_{ij,left}\right) + \left(F_{ij,top} - F_{ij,botten}\right) - R_{ij} = 0, \qquad (18)$$

where R_{ij} is the dimensionless reaction rate in the cell (i,j). $F_{ij,top}$, $F_{ij,bottom}$, $F_{ij,right}$ and $F_{ij,left}$ are the dimensionless fluxes through the four sides of the cell (i,j):

$$R_{ij} = \begin{cases} \Phi_0^2 \bar{c}_{ij} \Delta \bar{x}_i \Delta \bar{x}_j, & \text{when the cell } (i,j) \text{ is inside the nanoporous material,} \\ 0, & \text{when the cell } (i,j) \text{ is inside the large pores,} \end{cases} \qquad (19a)$$

$$F_{ij,top} = \begin{cases} \alpha \dfrac{\bar{c}_{i,j+1} - \bar{c}_{i,j}}{(\Delta \bar{x}_j + \Delta \bar{x}_{j+1})/2} \Delta \bar{x}_i, & j = 1, 2, \cdots, n_t - 1, \\ 0, & j = n_t \quad \text{(boundary on the top),} \end{cases} \qquad (19b)$$

$$F_{ij,bottom} = \begin{cases} \alpha \dfrac{\bar{c}_{i,j} - \bar{c}_{i,j-1}}{(\Delta \bar{x}_j + \Delta \bar{x}_{j-1})/2} \Delta \bar{x}_i, & j = 2, 3, \cdots, n_t, \\ \alpha \dfrac{\bar{c}_{ij} - 1}{\Delta \bar{x}_j/2} \Delta \bar{x}_i, & j = 1 \quad \text{(boundary on the bottom),} \end{cases} \qquad (19c)$$

$$F_{ij,right} = \begin{cases} \alpha \dfrac{\bar{c}_{i+1,j} - \bar{c}_{i,j}}{(\Delta \bar{x}_{i+1} + \Delta \bar{x}_i)/2} \Delta \bar{x}_j, & i = 1, 2, \cdots, n_t - 1, \\ 0, & j = n_t \quad \text{(boundary on the right),} \end{cases} \qquad (19d)$$

$$F_{ij,left} = \begin{cases} \alpha \dfrac{\bar{c}_{i,j} - \bar{c}_{i-1,j}}{(\Delta \bar{x}_i + \Delta \bar{x}_{i-1})/2} \Delta \bar{x}_j, & i = 2, 3, \cdots, n_t, \\ \alpha \dfrac{\bar{c}_{ij} - 1}{\Delta \bar{x}_i/2} \Delta \bar{x}_j, & i = 1 \quad \text{(boundary on the left),} \end{cases} \qquad (19e)$$

where $\bar{c}_{ij}$ is the concentration in the center of the cell (i,j). $\Delta \bar{x}_i$ is the length of the cell (i,j) in the horizontal direction. Note that, as a result of the four-fold symmetry, $\Delta \bar{x}_i$ is also the length of the cell (j,i) in the vertical direction. n_t is the number of the cells in a row or a column.

When the flux passes the interface between the large pores and the nanoporous material, the dimensionless diffusivity, α, is determined from [22]

$$\alpha = \frac{2\beta}{\beta + 1}. \qquad (20a)$$

Otherwise, α is determined from

$$\alpha = \begin{cases} \beta, & \text{when the cell } (i,j) \text{ is inside the nanoporous material.} \\ 1, & \text{when the cell } (i,j) \text{ is inside the large pores.} \end{cases} \quad (20b)$$

A system of discretized equations, which is equivalent to $\mathbf{h}(\mathbf{x_s}, \mathbf{x_d}) = \mathbf{0}$ in Section 2, is obtained by assembly of Eq. (18) for all the cells. There are many existing solvers for the discretized equations. The multigrid solver, MGD9V, is selected here because of two reasons. (1) A multigrid algorithm has proven to be the most efficient method for the equations that result from the discretization of elliptic PDEs [23]. (2) This multigrid solver is specially designed to deal with numerical difficulties that arise from the discontinuous diffusion coefficient across the interface. However, MGD9V might break down for some reason. To handle this exception, an ICCG (Incomplete Cholesky Conjugate Gradient) was implemented to solve the linear equations when MGD9V fails [24]. Note that an algebraic multigrid method might be employed to avoid the need to use the slow ICCG as a safeguard [25]. However, our implementation is sufficiently efficient for the application considered here. Furthermore, ICCG is robust and it is readily implemented. While MGD9V as a black-box solver is easy to use, its source code can be acquired free. This is desirable for the integration of all the codes.

The discretized equations are linear with respect to the state variables (concentrations) and, therefore, the Jacobian matrix of the discretized equations, which is equivalent to $\frac{\partial \mathbf{h}}{\partial \mathbf{x_s}}$ in Section 2, is the same as the coefficient matrix of the discretized equations. As a result, ICCG-safeguarded MGD9V is also used for the solution of the linear equations that are equivalent to Eq. (9) in Section 2. $\frac{\partial H_{ij}}{\partial \bar{L}_k}$ is evaluated using the chain rule to obtain the matrix that is equivalent to $\frac{\partial \mathbf{h}}{\partial \mathbf{x_d}}$ in Section 2:

$$\frac{\partial H_{ij}}{\partial \bar{L}_k} = \sum_{l=1}^{n_t} \frac{\partial H_{ij}}{\partial \Delta \bar{x}_l} \frac{\partial \Delta \bar{x}_l}{\partial \bar{L}_k}, \quad (21)$$

where $\frac{\partial H_{ij}}{\partial \Delta \bar{x}_l}$ is easy to obtain from Eqs. (19-20).

From straightforward geometrical consideration, $\bar{L}_k$ is the sum of a subset of $\Delta \bar{x}_l (l = 1, \cdots, n_t)$, as shown in Figure 2. Therefore, $\frac{\partial \Delta \bar{x}_l}{\partial \bar{L}_k}$ is equal to unity if $\Delta \bar{x}_l$ belongs to that subset and is equal to zero if not. The discretized objective function can be expressed either as the sum of the reaction rates over all the cells or as the sum of the fluxes through the boundary. The two expressions are equivalent, which could be shown by adding Eq. (18) over all the cells. The discretized objective function, expressed as the sum of the fluxes through the boundary, is the following:

$$\eta = \sum_{j=1}^{n_t} F_{n_t j, right} + \sum_{i=1}^{n_t} F_{in_t, top} - \sum_{j=1}^{n_t} F_{1j, left} - \sum_{i=1}^{n_t} F_{i1, bottom}. \quad (22)$$

Since there is no flux through the top and right boundaries, so that, combined with Eqs. (19c) and (19e):

$$\eta = \frac{4}{\Delta \bar{x}_l} \sum_{j=1}^{n_t} \alpha(1 - \bar{c}_{1j}) \Delta \bar{x}_j. \tag{23}$$

Therefore,

$$\frac{\partial \eta}{\partial \bar{c}_{ij}} = \begin{cases} 0 & i \neq 1. \\ -4\alpha \dfrac{\Delta \bar{x}_j}{\Delta \bar{x}_1} & i = 1. \end{cases} \tag{24}$$

Note that $\dfrac{\partial \eta}{\partial \bar{c}_{ij}}$ is equivalent to $\dfrac{\partial f_2}{\partial \mathbf{x_s}}$ in Section 2. $\dfrac{\partial \eta}{\partial \bar{L}_i}$, which is equivalent to $\dfrac{\partial f_2}{\partial \mathbf{x_d}}$ in Section 2, is calculated (using the chain rule) from

$$\frac{\partial \eta}{\partial \bar{L}_i} = \sum_{j=1}^{n_t} \frac{\partial \eta}{\partial \Delta \bar{x}_j} \frac{\partial \Delta \bar{x}_j}{\partial \bar{L}_i}, \tag{25}$$

where $\dfrac{\partial \Delta \bar{x}_j}{\partial \bar{L}_i}$ is computed in the same way as discussed above and

$$\frac{\partial \eta}{\partial \Delta \bar{x}_j} = \begin{cases} -\dfrac{4}{(\Delta \bar{x}_1)^2} \sum_{k=2}^{n_t} \alpha(1 - \bar{c}_{1k}) \Delta \bar{x}_k, & j = 1. \\ \dfrac{4\alpha(1 - \bar{c}_{1j})}{\Delta \bar{x}_1}, & j \neq 1. \end{cases} \tag{26}$$

Table 1 Summary of the numerical experiments ($\Phi_0 = 1, \beta = 0.01$): (a) number of computational cells. (b) number of optimization variables. (c) number of evaluations of objective function. (d) number of evaluations of reduced gradient. (e) number of calls for linear solvers. (f) number of calls for ICCG (g) number of iterations (h) CPU time (seconds) (i) optimal value.

	Case 1	Case 2	Case 3	Case 4	Case 5	Case 6	Case 7
(a)	129 × 129	257× 257	257× 257	257 × 257	513× 513	513× 513	513× 513
(b)	101	101	161	201	101	161	201
(c)	26	22	26	21	25	22	18
(d)	12	11	12	10	12	11	9
(e)	50	44	50	41	49	44	36
(f)	6	4	8	7	8	6	6
(g)	12	11	12	10	12	11	9
(h)	2.52	13.86	21.39	19.59	211.52	174.28	166.86
(i)	0.52010	0.52030	0.52078	0.52097	0.52040	0.52090	0.52101

At this point, all the ingredients that are needed to implement the framework established in Section 2 are available for this particular PDE-based optimization problem. A Fortran program was developed by combining NLPQL, MGD9V and a limited amount of in-house coding, in the way described in the framework. All the optimization codes listed earlier might be used here. We chose NLPQL, because it is sufficiently efficient for the application discussed here, and it is accessible free of charge. To test the Fortran implementation, several numerical experiments were performed with different combinations of the numbers of computational cells and optimization variables. The number of computational cells varied from 129×129 to 513×513, and the number of optimization variables from 101 to 201. However, the parameters, Φ_0 and β, remain constant in all cases, in order to study the effects of the numbers of computational cells and optimization variables on the performance of the implementation. All the numerical tests were carried out on a Dell laptop with a 2.16 GHz Intel Core2 Duo processor. NLPQL allows users to control the termination condition by setting the value of a parameter, ACC [7]. The smaller this value is, the stricter the condition is. The optimization does not converge if the value is too large, while it takes many iterations without significant change in the value of the objective function if the value is too small. This parameter was set to be 5×10^{-3} in all the numerical tests.

Table 1 summarizes the results of these numerical tests. The objective function was evaluated $18 - 26$ times in each of these optimizations, and the reduced gradient was evaluated $9 - 12$ times. Note that the evaluation of the objective function requires only the solution of the discretized equations, while the evaluation of the reduced gradient requires both the solution of the discretized equations and of the linear equations, which are equivalent to Eq. (10) in Section 2. Therefore, the total number of calls for the linear solver, ICCG-safeguarded MGD9V, varied between 36 and 50. Among those calls, ICCG was active $4 - 8$ times in each optimization to handle the exceptions caused by the failure of MGD9V. In other words, MGD9V solved more than 80% of the equations. This is desirable since MGD9V is much more efficient than ICCG.

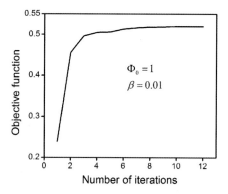

Fig. 3 Objective function value as a function of the number of iterations. Optimization was performed on 129×129 computational cells, with 101 optimization variables.

The CPU time of the optimizations mainly depends on the number of the computational cells, because the solution of the discretized equations is the most time-consuming calculation. The CPU time for cases on 513×513 computational cells is much longer than that for cases on 129×129 or 257×257 computational cells. This observation is attributed to poor scaling of ICCG for the solution of the discretized equations. The number of iterations varied from 9 to 12 for each of the optimizations. The optimal values are independent of the number of computational cells, showing satisfactory convergence. When the structure contains a sufficient number of large pores, the optimal values are also independent of the number of optimization variables, which is a generic feature of this particular optimal design problem. For a more extensive discussion of the results, the reader is referred to [9, 17-19]. As shown in Figure 3, the value of the objective function typically converges within a limited number of iterations to a value of 0.52 for this case. It is interesting to note that for the same case in the absence of large pores, the value of the objective function is only 0.18.

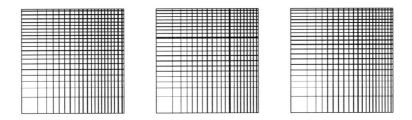

Fig. 4 Optimal structure of the lower-left quarter of a hierarchically structured catalytic square, computed with 41 optimization variables and using different numbers of computational cells: 129×129 (left), 257×257 (center) and 513×513 (right). The optimization results shown correspond to $\Phi_0 = 1$ and $\beta = 0.01$.

Figure 4 presents the optimal structures using different numbers of computational cells, varying from 129×129 to 513×513. Although the optimal structures calculated using different numbers of cells slightly differ from one another, almost no difference in normalized catalytic yield (i.e., the objective function) is found between them (see Table 1). The rationale is that the normalized yield is insensitive to changes in structure when all the nanoporous islands (indicated as white rectangles in Figure 4) are sufficiently small so that diffusion limitations vanish locally. For more discussion on how the optimal catalytic yield and the structure depend on Φ_0 and β, the reader is referred to [9, 18, 19]. As shown in Figure 5, the optimal structures remain essentially the same when β changes from 0.01 to 0.001. The figure also shows that the large pores (indicated in black) in the optimal structures become wider when Φ_0 increases from 1 to 5, corresponding to situations with more significant diffusion limitations, requiring broader pores to access the catalytic material, in order to achieve high yields.

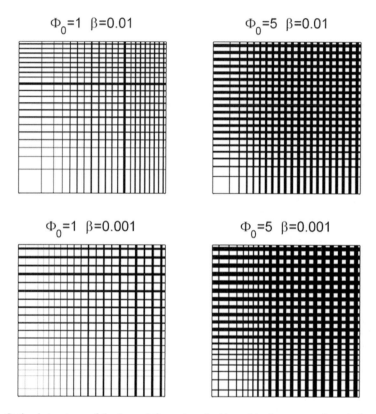

Fig. 5 Optimal structures of the lower-left quarter of a hierarchically structured catalytic square, computed using different values of Φ_0 and β. Optimizations were performed with 41 optimization variables and using 257×257 computational cells.

4 Conclusions

The reduced gradient method was employed to solve the PDE-based optimization problem that arises from the optimal design of hierarchically structured porous catalysts. A Fortran program was developed by combining the gradient-based optimization package NLPQL, the multigrid PDE solver, MGD9V and a limited amount of in-house coding, in the way described. Numerical tests show that this implementation is robust and highly efficient.

Acknowledgements Start-up funds for MOC from Rensselaer Polytechnic Institute, as well as funds from the Delft Research Centre for Computational Science and Engineering to initiate this work, are gratefully acknowledged. GW thanks Prof. Schittkowski (from the University of Bayreuth) for the permission of the use of the optimization package NLPQL.

References

1. Biegler, L.T., Ghattas, O., Heinkenschloss, M., and van Bloemen Waanders, B., Eds. Large-scale PDE-constrained optimization, Springer, Berlin/Heidelberg (2003)
2. Itle, G.C., Salinger, A.G., Pawlowski, R.P., Shadid, J.N., and Biegler, L.T., A tailored optimization strategy for PDE-based design: application to a CVD reactor, Comp. Chem. Eng.**28**,291-302 (2004)
3. Armaou, A., and Christofides, P.D., Dynamic optimization of dissipative PDE systems using nonlinear order reduction, Chem. Eng. Sci. **57**,5083-5114 (2002)
4. Kunisch, K., and Sachs, E.W., Reduced SQP methods for parameter-identification problems, SIAM J. Numer. Anal. **29**,1793-1820 (1992)
5. Ghattas, O., and Bark, J.H., Optimal control of two- and three-dimensional incompressible Navier-Stokes flows, J. Comp. Phys. **136**,231-244 (1997)
6. de Zeeuw, P.M., Matrix-dependent prolongations and restrictions in a blackbox multigrid solver, J. Comp. Appl. Math. **33**,1-27 (1990)
7. Schittkowski, K., NLPQL: a Fortran subroutine for solving constrained nonlinear programming problems, Ann. Oper. Res. **5**,485-500 (1985/86)
8. Wang, G., Kleijn, C.R., and Coppens, M.-O., A tailored strategy for PDE-based design of hierarchically structured porous catalysts, Inter. J. Multiscale Comp. Eng. **5**,179-190 (2008)
9. Wang, G., Johannessen, E., Kleijn, C.R., de Leeuw, S.W., and Coppens, M.-O., Optimizing transport in nanostructured catalysts: a computational study, Chem. Eng. Sci. **62**,5110-5116 (2007)
10. Gill, P.E., Murray, W., and Saunders, M.A., SNOPT: an SQP algorithm for large-scale constrained optimization, SIAM J. Optimi. **12**,979-1006 (2002)
11. Murtagh, B.A., and Saunders, M.A., MINOS 5.4 User's Guide, Report SOL 83-20R, Systems Optimization Laboratory, Stanford University, December 1983 (revised February 1995).
12. Conn, A.R., Gould, N.I.M., and Toint, Ph.L., LANCELOT: A Fortran Package for Large-Scale Nonlinear Optimization (Release A), Springer Series in Computational Mathematics, vol.**17**, Springer-Verlag, Berlin(1992)
13. Spellucci, P., An SQP method for general nonlinear programs using only equality constrained subproblems, Math. Prog. **82**,413-448 (1998)
14. Lasdon, L.S., Warren, A.D., Jain, A., and Ratner, M., Design and testing of a generalized reduced gradient code for nonlinear programming, ACM Trans. Math. Softw. **4**,34-50 (1978)
15. Nocedal, J., and Wright, S., Numerical Optimization, Springer, New York (1999). . http://money.cnn.com/magazines/fortune/global500/2007/
16. Gheorghiu, S., and Coppens, M.-O., Optimal bimodal pore networks for heterogeneous catalysis, AIChE J. **50**,812-820 (2004)
17. Johannessen, E., Wang, G., and Coppens, M.-O. Optimal distributor networks in porous catalyst pellets. I. molecular diffusion, Ind. Eng. Chem. Res.**46**,4245-4256 (2007)
18. Wang, G., and Coppens, M.-O., Calculation of the optimal macropore size in nanoporous catalysts, and its application to DeNOx catalysis. Ind. Eng. Chem. Res. **47**,3847-3855 (2008)
19. Olesen, L.H., Okkels, F., and Bruus, H., A high-level programming-language implementation of topology optimization applied to steady-state Navier-Stokes flow. Inter. J. Numer. Meth. Eng. **65**,975-1001 (2006)
20. Versteeg, H.K., and Malalasekera, W., An Introduction to Computational Fluid Dynamics: The Finite Volume Method, Longman Scientific and Technical, Harlow, Essex, U.K.(1995)

21. Crumpton, P.I., Shaw, G.J., and Ware, A.F., Discretization and multigrid solution of elliptic-equations with mixed derivative terms and strongly discontinuous coefficients, J. Comp. Phys. **116**,343-358 (1995)
22. Wesseling, P., An Introduction to Multigrid Methods, R. T. Edwards, Inc., Flourtown, Pennsylvania (2004)
23. Meijerink, J.A., and van der Vorst, H.A., An iterative solution method for linear systems of which the coefficient matrix is a symmetric M-matrix, Math Comp. **31**,148-162 (1997)
24. Trottenberg, U., Oosterlee, C., and Schüller, A., Multigrid, Academic Press, San Diego (2001)
25. Stüben, K., An introduction to algebraic multigrid. In [24], pages 413-532

From Molecular Dynamics and Particle Simulations towards Constitutive Relations for Continuum Theory

Stefan Luding

Abstract A challenge of today's research is the realistic simulation of disordered atomistic systems or particulate and granular materials like sand, powders, ceramics or composites, which consist of many millions of atoms/particles. The inhomogeneous fine-structure of such materials makes it very difficult to treat these with continuum methods, which typically assume homogeneity and scale separation. As an alternative, particle based methods can be straightforwardly applied, since they intrinsically take the fine-structure into account. The ultimate challenge is to find constitutive relations for continuum theory from these particle-based simulations.
In this chapter, a particle simulation approach, the so-called discrete element method (DEM), as related to molecular dynamics (MD) methods, is introduced and applied to the simulation of many-particle systems. The examples (clustering in granular gases, and bi-axial as well as cylindrical shearing of dense packings) illustrate the micro-macro transition towards continuum theory.
There exist two basically different approaches, the so-called soft particle molecular dynamics and the hard sphere, event-driven method. The former is straightforward, easy to generalize, and has many applications, while the latter is optimized for rigid interactions and is mainly used for collisional, dissipative granular gases. The connection between the two methods will be elaborated on. Models for the forces between the atoms/particles are the basis of both MD and DEM. A set of the most basic contact force models for particles is presented involving elasto-plasticity, adhesion, viscosity, static and dynamic friction as well as rolling- and torsion-resistance. Besides some words about Van der Waals forces, we will not detail on electro-magnetic interactions, dipole moments, H-bonding, and other effects which become important when the objects become smaller and smaller.

Key words: molecular dynamics (MD), discrete element methods (DEM), event driven MD, equation of state, clustering, shear band formation, micro-macro

---

Multi Scale Mechanics, TS, CTW, UTwente, P.O.Box 217, 7500 AE Enschede, Netherlands
e-mail: s.luding@utwente.nl -- www2.msm.ctw.utwente.nl/sluding

1 Introduction

Materials with inhomogeneous fine-structures are the subject of this chapter. As example, we mostly discuss particulate, granular systems where the fine-structures are spherical, polydisperse, plastic, adhesive, and frictional objects.

One approach towards the microscopic understanding of such macroscopic particulate material behavior [19, 25, 20] is the modeling of particles using so-called discrete element methods (DEM). Even though millions of particles can be simulated, the possible length of such a particle system is in general too small in order to regard it as macroscopic. Therefore, methods and tools to perform a so-called micro-macro transition [68, 58, 24] are discussed, starting from the DEM simulations. These "microscopic" simulations of a small sample (representative volume element) can be used to derive macroscopic constitutive relations needed to describe the material within the framework of a macroscopic continuum theory.

For granular materials, as an example, the particle properties and interaction laws are inserted into DEM, which is also often referred to as molecular dynamics (MD), and lead to the collective behavior of the dissipative many-particle system. From a particle simulation, one can extract, e.g., the pressure of the system as a function of density. This equation of state allows a macroscopic description of the material, which can be viewed as a compressible, non-Newtonian complex fluid [48], including a fluid-solid phase transition.

In the following, two versions of the molecular dynamics simulation method are introduced. The first is the so-called soft sphere molecular dynamics (MD=DEM), as described in section 2. It is a straightforward implementation to solve the equations of motion for a system of many interacting particles [5, 59]. For DEM, both normal and tangential interactions, like friction, are discussed for spherical particles. The second method is the so-called event-driven (ED) simulation, as discussed in section 3, which is conceptually different from DEM, since collisions are dealt with via a collision matrix that determines the momentum change on physical grounds. For the sake of brevity, the ED method is only discussed for smooth spherical particles. A comparison and a way to relate the soft and hard particle methods is provided in section 4.

As one ingredient of a micro-macro transition, the stress is defined for a dynamic system of hard spheres, in section 5, by means of kinetic-theory arguments [58], and for a quasi-static system by means of volume averages [26]. Examples are presented in the following sections 6 and 7, where the above-described methods are applied.

2 The soft particle molecular dynamics method

One possibility to obtain information about the behavior of granular media is to perform experiments. An alternative are simulations with the molecular dynamics (MD) or discrete element model (DEM) [68, 9, 8, 6, 19, 63, 64, 65, 27]. Note that both methods are identical in spirit, however, different groups of researchers use these (and also other) names.

Conceptually, the DEM method has to be separated from the hard sphere event-driven (ED) molecular dynamics, see section 3, and also from the so-called Contact Dynamics (CD). Like alternative (stochastic) methods, as there are cell- or lattice-gas-methods these are just named as keywords – not discussed here further.

2.1 Discrete particle model

The elementary units of granular materials are mesoscopic grains which deform under stress. Since the realistic modeling of the deformations of the particles is much too complicated, we relate the interaction force to the overlap δ of two particles, see Fig. 1. Note that the evaluation of the inter-particle forces based on the overlap may not be sufficient to account for the inhomogeneous stress distribution inside the particles. Consequently, our results presented below are of the same quality as the simple assumptions about the force-overlap relation, see Fig. 1.

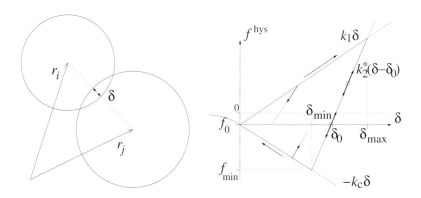

Fig. 1 (Left) Two particle contact with overlap δ. (Right) Schematic graph of the piecewise linear, hysteretic, adhesive force-displacement model used below.

2.2 Equations of motion

If all forces f_i acting on the particle i, either from other particles, from boundaries or from external forces, are known, the problem is reduced to the integration of Newton's equations of motion for the translational and rotational degrees of freedom:

$$m_i \frac{d^2}{dt^2} r_i = f_i + m_i g, \quad \text{and} \quad I_i \frac{d^2}{dt^2} \varphi_i = t_i, \qquad (1)$$

with the mass m_i of particle i, its position r_i the total force $f_i = \sum_c f_i^c$ acting on it due to contacts with other particles or with the walls, the acceleration due to volume forces like gravity g, the spherical particles moment of inertia I_i, its angular velocity $\omega_i = d\varphi_i/dt$ and the total torque $t_i = \sum_c (l_i^c \times f_i^c + q_i^c)$, where q_i^c are torques/couples at contacts other than due to a tangential force, e.g., due to rolling and torsion.

The equations of motion are thus a system of $\mathscr{D} + \mathscr{D}(\mathscr{D}-1)/2$ coupled ordinary differential equations to be solved in $\mathscr{D}$ dimensions. With tools from numerical integration, as nicely described in textbooks as [5, 59], this is straightforward. The typically short-ranged interactions in granular media, allow for a further optimization by using linked-cell or alternative methods [5, 59] in order to make the neighborhood search more efficient. In the case of long-range interactions, (e.g. charged particles with Coulomb interaction, or objects in space with self-gravity) this is not possible anymore, so that more advanced methods for optimization have to be applied – for the sake of brevity, we restrict ourselves to short range interactions here.

2.3 Normal contact force laws

2.3.1 Linear normal contact model

Two spherical particles i and j, with radii a_i and a_j, respectively, interact only if they are in contact so that their overlap

$$\delta = (a_i + a_j) - (r_i - r_j) \cdot n \qquad (2)$$

is positive, $\delta > 0$, with the unit vector $n = n_{ij} = (r_i - r_j)/|r_i - r_j|$ pointing from j to i. The force on particle i, from particle j, at contact c, can be decomposed into a normal and a tangential part as $f^c := f_i^c = f^n n + f^t t$, where f^n is discussed first.

The simplest normal contact force model, which takes into account excluded volume and dissipation, involves a linear repulsive and a linear dissipative force

$$f^n = k\delta + \gamma_0 v_n, \qquad (3)$$

with a spring stiffness k, a viscous damping γ_0, and the relative velocity in normal direction $v_n = -v_{ij} \cdot n = -(v_i - v_j) \cdot n = \dot{\delta}$.

This so-called linear spring dashpot model allows to view the particle contact as a damped harmonic oscillator, for which the half-period of a vibration around an equilibrium position, see Fig. 1, can be computed, and one obtains a typical response time on the contact level,

$$t_c = \frac{\pi}{\omega}, \quad \text{with} \quad \omega = \sqrt{(k/m_{12}) - \eta_0^2}, \qquad (4)$$

with the eigenfrequency of the contact ω, the rescaled damping coefficient $\eta_0 = \gamma_0/(2m_{ij})$, and the reduced mass $m_{ij} = m_i m_j/(m_i + m_j)$. From the solution of the equation of a half period of the oscillation, one also obtains the coefficient of restitution

$$r = v'_n/v_n = \exp(-\pi \eta_0/\omega) = \exp(-\eta_0 t_c), \quad (5)$$

which quantifies the ratio of relative velocities after (primed) and before (unprimed) the collision.

The contact duration in Eq. (4) is also of practical technical importance, since the integration of the equations of motion is stable only if the integration time-step Δt_{DEM} is much smaller than t_c. Furthermore, it depends on the magnitude of dissipation. In the extreme case of an overdamped spring, t_c can become very large. Therefore, the use of neither too weak nor too strong dissipation is recommended.

2.3.2 Adhesive, elasto-plastic normal contact model

Here we apply a variant of the linear hysteretic spring model [69, 31, 67, 39], as an alternative to the frequently applied spring-dashpot models. This model is the simplest version of some more complicated nonlinear-hysteretic force laws [69, 70, 60], which reflect the fact that at the contact point, plastic deformations may take place. The repulsive (hysteretic) force can be written as

$$f^{hys} = \begin{cases} k_1 \delta & \text{for loading,} & \text{if } k_2^*(\delta - \delta_0) \geq k_1 \delta \\ k_2^*(\delta - \delta_0) & \text{for un/reloading,} & \text{if } k_1 \delta > k_2^*(\delta - \delta_0) > -k_c \delta \\ -k_c \delta & \text{for unloading,} & \text{if } -k_c \delta \geq k_2^*(\delta - \delta_0) \end{cases} \quad (6)$$

with $k_1 \leq k_2^*$, see Fig. 1, and Eq. (7) below for the definition of the (variable) k_2^* as function of the constant model parameter k_2.

During the initial loading the force increases linearly with the overlap δ, until the maximum overlap δ_{max} is reached (which has to be kept in memory as a history parameter). The line with slope k_1 thus defines the maximum force possible for a given δ. During unloading the force drops from its value at δ_{max} down to zero at overlap $\delta_0 = (1 - k_1/k_2^*)\delta_{max}$, on the line with slope k_2^*. Reloading at any instant leads to an increase of the force along this line, until the maximum force is reached; for still increasing δ, the force follows again the line with slope k_1 and δ_{max} has to be adjusted accordingly.

Unloading below δ_0 leads to negative, attractive forces until the minimum force $-k_c \delta_{min}$ is reached at the overlap $\delta_{min} = (k_2^* - k_1)\delta_{max}/(k_2^* + k_c)$. This minimum force, i.e. the maximum attractive force, is obtained as a function of the model parameters k_1, k_2, k_c, and the history parameter δ_{max}. Further unloading leads to attractive forces $f^{hys} = -k_c \delta$ on the adhesive branch with slope $-k_c$. The highest possible attractive force, for given k_1 and k_2, is reached for $k_c \to \infty$, so that $f^{hys}_{max} = -(k_2 - k_1)\delta_{max}$. Since this would lead to a discontinuity at $\delta = 0$, it is avoided by using finite k_c.

The lines with slope k_1 and $-k_c$ define the range of possible force values and departure from these lines takes place in the case of unloading and reloading, respectively. Between these two extremes, unloading and reloading follow the same line with slope k_2. Possible equilibrium states are indicated as circles in Fig. 1, where the upper and lower circle correspond to a pre-stressed and stress-free state, respectively. Small perturbations lead, in general, to small deviations along the line with slope k_2 as indicated by the arrows.

A non-linear un/reloading behavior would be more realistic, however, due to a lack of detailed experimental information, we use the piece-wise linear model as a compromise. One refinement is a k_2^* value dependent on the maximum overlap that implies small and large plastic deformations for weak and strong contact forces, respectively. One model, as implemented recently [50, 39], requires an additional model parameter, δ_{max}^*, so that $k_2^*(\delta_{max})$ is increasing from k_1 to k_2 (linear interpolation is used below, however, this is another choice to be made and will depend on the material under consideration) with the maximum overlap, until δ_{max}^* is reached[1]:

$$k_2^*(\delta_{max}) = \begin{cases} k_2 & \text{if } \delta_{max} \geq \delta_{max}^* \\ k_1 + (k_2 - k_1)\delta_{max}/\delta_{max}^* & \text{if } \delta_{max} < \delta_{max}^* \end{cases}. \quad (7)$$

While in the case of collisions of particles with large deformations, dissipation takes place due to the hysteretic nature of the force-law, stronger dissipation of small amplitude deformations is achieved by adding the viscous, velocity dependent dissipative force from Eq. (3) to the hysteretic force, such that $f^n = f^{hys} + \gamma_0 v_n$. The hysteretic model contains the linear contact model as special case $k_1 = k_2 = k$.

2.3.3 Long range normal forces

Medium range Van der Waals forces can be taken into account in addition to the hysteretic force such that $f^n = f_i^{hys} + f_i^{vdW}$ with, for example, the attractive part of a Lennard-Jones Potential

$$f^{vdW} = -6(\varepsilon/r_0)[(r_0/r_{ij})^7 - (r_0/r_c)^7] \quad \text{for } r_{ij} \leq r_c . \quad (8)$$

The new parameters necessary for this force are an energy scale ε, a typical length scale r_0 and a cut-off length r_c. As long as r_c is not much larger than the particle diameter, the methods for short range interactions still can be applied to such a medium range interaction model – only the linked cells have to be larger than twice the cut-off radius, and no force is active for $r > r_c$.

---

[1] A limit to the slope k_2 is needed for practical reasons. If k_2 would not be limited, the contact duration could become very small so that the time step would have to be reduced below reasonable values.

2.4 Tangential forces and torques in general

For the tangential degrees of freedom, there are three different force- and torque-laws to be implemented: (i) friction, (ii) rolling resistance, and (iii) torsion resistance.

2.4.1 Sliding

For dynamic (sliding) and static *friction*, the relative tangential velocity of the contact points,

$$v_t = v_{ij} - n(n \cdot v_{ij}) , \qquad (9)$$

is to be considered for the force and torque computations in subsection 2.5, with the total relative velocity of the particle surfaces at the contact

$$v_{ij} = v_i - v_j + a'_i n \times \omega_i + a'_j n \times \omega_j , \qquad (10)$$

with the corrected radius relative to the contact point $a'_\alpha = a_\alpha - \delta/2$, for $\alpha = i, j$. Tangential forces acting on the contacting particles are computed from the accumulated sliding of the contact points along each other, as described in detail in subsection 2.5.1.

2.4.2 Objectivity

In general, two particles can rotate together, due to both a rotation of the reference frame or a non-central "collision". The angular velocity $\omega_0 = \omega_0^n + \omega_0^t$, of the rotating reference has the tangential-plane component

$$\omega_0^t = \frac{n \times (v_i - v_j)}{a'_i + a'_j} , \qquad (11)$$

which is related to the relative velocity, while the normal component, ω_0^n, is not. Inserting $\omega_i = \omega_j = \omega_0^t$, from Eq. (11), into Eq. (10) leads to zero sliding velocity, proving that the above relations are objective. Tangential forces and torques due to sliding can become active only when the particles are rotating with respect to the common rotating reference frame. [2]

Since action should be equal to reaction, the tangential forces are equally strong, but opposite, i.e., $f^t_j = -f^t_i$, while the corresponding torques are parallel but not necessarily equal in magnitude: $q_i^{\text{friction}} = -a'_i n \times f_i$, and $q_j^{\text{friction}} = (a'_j/a'_i) q_i^{\text{friction}}$. Note that tangential forces and torques *together* conserve the total angular momen-

---

[2] For rolling and torsion, there is no similar relation between rotational and tangential degrees of freedom: for any rotating reference frame, torques due to rolling and torsion can become active only due to rotation relative to the common reference frame, see below.

tum about the pair center of mass

$$L_{ij} = L_i + L_j + m_i r_{icm}^2 \omega_0^t + m_j r_{jcm}^2 \omega_0^t , \qquad (12)$$

with the rotational contributions $L_\alpha = I_\alpha \omega_\alpha$, for $\alpha = i, j$, and the distances $r_{\alpha cm} = |r_\alpha - r_{cm}|$ from the particle centers to the center of mass $r_{cm} = (m_i r_i + m_j r_j)/(m_i + m_j)$, see Ref. [31]. The change of angular momentum consists of the change of particle spins (first term) and of the change of the angular momentum of the two masses rotating about their common center of mass (second term):

$$\frac{dL_{ij}}{dt} = q_i^{\text{friction}} \left(1 + \frac{a_j'}{a_i'}\right) + \left(m_i r_{icm}^2 + m_j r_{jcm}^2\right) \frac{d\omega_0^t}{dt} , \qquad (13)$$

which both contribute, but exactly cancel each other, since

$$q_i^{\text{friction}} \left(1 + \frac{a_j'}{a_i'}\right) = -(a_i' + a_j') n \times f_i \qquad (14)$$

$$= -\left(m_i r_{icm}^2 + m_j r_{jcm}^2\right) \frac{d\omega_0^t}{dt} ,$$

see [37] for more details.

2.4.3 Rolling

A *rolling* velocity $v_r^0 = -a_i' n \times \omega_i + a_j' n \times \omega_j$, defined in analogy to the sliding velocity, is not objective in general [14, 37] – only in the special cases of (i) equal-sized particles or (ii) for a particle rolling on a fixed flat surface.

The rolling velocity should quantify the distance the two surfaces roll over each other (without sliding). Therefore, it is equal for both particles by definition. An *objective rolling velocity* is obtained by using the reduced radius, $a_{ij}' = a_i' a_j'/(a_i' + a_j')$, so that

$$v_r = -a_{ij}' (n \times \omega_i - n \times \omega_j) . \qquad (15)$$

This definition is objective since any common rotation of the two particles vanishes by construction. A more detailed discussion of this issue is beyond the scope of this paper, rather see [14, 37] and the references therein.

A rolling velocity will activate torques, acting against the rolling motion, e.g., when two particles are rotating anti-parallel with spins in the tangential plane. These torques are then equal in magnitude and opposite in direction, i.e., $q_i^{\text{rolling}} = -q_j^{\text{rolling}} = a_{ij} n \times f_r$, with the quasi-force f_r, computed in analogy to the friction force, as function of the rolling velocity v_r in subsection 2.5.2; the quasi-forces for both particles are equal and do not act on the centers of mass. Therefore, the total momenta (translational and angular) are conserved.

2.4.4 Torsion

For *torsion resistance*, the relative spin along the normal direction

$$v_o = a_{ij} (n \cdot \omega_i - n \cdot \omega_j) n , \quad (16)$$

is to be considered, which activates torques when two particles are rotating antiparallel with spins parallel to the normal direction. Torsion is not activated by a common rotation of the particles around the normal direction $n \cdot \omega_0 = n \cdot (\omega_i + \omega_j)/2$, which makes the torsion resistance objective.

The torsion torques are equal in magnitude and directed in opposite directions, i.e., $q_i^{torsion} = -q_j^{torsion} = a_{ij} f_o$, with the quasi-force f_o, computed from the torsion velocity in subsection 2.5.3, and also not changing the translational momentum. Like for rolling, the torsion torques conserve the total angular momentum.

2.4.5 Summary

The implementation of the tangential force computations for f_t, f_r, and f_o as based on v_t, v_r, and v_o, respectively, is assumed to be *identical*, i.e., even the same subroutine is used, but with different parameters as specified below. The difference is that friction leads to a force in the tangential plane (changing both translational and angular momentum), while rolling- and torsion-resistance lead to quasi-forces in the tangential plane and the normal direction, respectively, changing the particles' angular momentum only. For more details on tangential contact models, friction, rolling and torsion, see Refs. [7, 13, 38, 37, 14].

2.5 The tangential force- and torque-models

The tangential contact model presented now is a single procedure (subroutine) that can be used to compute either sliding, rolling, or torsion resistance. The subroutine needs a relative velocity as input and returns the respective force or quasi-force as function of the accumulated deformation. The sliding/sticking friction model will be introduced in detail, while rolling and torsion resistance are discussed where different.

2.5.1 Sliding/sticking friction model

The tangential force is coupled to the normal force via Coulomb's law, $f^t \leq f_C^s := \mu^s f^n$, where for the sliding case one has dynamic friction with $f^t = f_C^t := \mu^d f^n$. The dynamic and the static friction coefficients follow, in general, the relation $\mu^d \leq \mu^s$. The static situation requires an elastic spring in order to allow for a restoring force, i.e., a non-zero remaining tangential force in static equilibrium due to activated Coulomb friction.

If a purely repulsive contact is established, $f^n > 0$, and the tangential force is active. For an adhesive contact, Coulomb's law has to be modified in so far that f^n is replaced by $f^n + k_c \delta$. In this model, the reference for a contact is no longer the zero force level, but it is the adhesive, attractive force level along $-k_c \delta$.

If a contact is active, one has to project (or better rotate) the tangential spring into the actual tangential plane, since the frame of reference of the contact may have rotated since the last time-step. The tangential spring,

$$\xi = \xi' - n(n \cdot \xi'), \qquad (17)$$

is used for the actual computation, where ξ' is the old spring from the last iteration, with $|\xi| = |\xi'|$ enforced by appropriate scaling/rotation. If the spring is new, the tangential spring-length is zero, but its change is well defined after the first, initiation step. In order to compute the changes of the tangential spring, a tangential test-force is first computed as the sum of the tangential spring force and a tangential viscous force (in analogy to the normal viscous force)

$$f_0^t = -k_t \xi - \gamma_t v_t, \qquad (18)$$

with the tangential spring stiffness k_t, the tangential dissipation parameter γ_t, and v_t from Eq. (9). As long as $|f_0^t| \leq f_C^s$, with $f_C^s = \mu^s (f^n + k_c \delta)$, one has static friction and, on the other hand, for $|f_0^t| > f_C^s$, sliding friction becomes active. As soon as $|f_0^t|$ gets smaller than f_C^d, static friction becomes active again.

In the *static friction* case, below the Coulomb limit, the tangential spring is incremented,

$$\xi' = \xi + v_t \Delta t_{MD}, \qquad (19)$$

to be used in the next iteration in Eq. (17), and the tangential force $f^t = f_0^t$ from Eq. (18) is used. In the *sliding friction* case, the tangential spring is adjusted to a length consistent with Coulomb's condition, so that

$$\xi' = -\frac{1}{k_t}\left(f_C^d t + \gamma_t v_t \right), \qquad (20)$$

with the tangential unit vector, $t = f_0^t / |f_0^t|$, defined by Eq. (18), and thus the magnitude of the Coulomb force is used. Inserting ξ' from Eq. (20) into Eq. (18) during the next iteration will lead to $f_0^t \approx f_C^d t$. Note that f_0^t and v_t are not necessarily parallel in three dimensions. However, the mapping in Eq. (20) works always, rotating the new spring such that the direction of the frictional force is unchanged and, at the same time, limiting the spring in length according to Coulomb's law. In short notation the tangential contact law reads

$$f^t = f^t t = +\min\left(f_C, |f_0^t| \right) t, \qquad (21)$$

where f_C follows the static/dynamic selection rules described above. The torque on a particle due to frictional forces at this contact is $q^{\text{friction}} = l_i^c \times f_i^c$, where l_i^c is the branch vector, connecting the center of the particle with the contact point. Note that the torque on the contact partner is generally different in magnitude, since l_i^c can be different, but points in the same direction; see subsection 2.4.2 for details on this.

The four parameters for the friction law are k_t, μ_s, $\phi_d = \mu_d/\mu_s$, and γ_t, accounting for tangential stiffness, the static friction coefficient, the dynamic friction ratio, and the tangential viscosity, respectively. Note that the tangential force described above is identical to the classical Cundall-Strack spring only in the limits $\mu = \mu^s = \mu^d$, i.e., $\phi_d = 1$, and $\gamma_t = 0$. The sequence of computations and the definitions and mappings into the tangential direction can be used in 3D as well as in 2D.

2.5.2 Rolling resistance model

The three new parameters for rolling resistance are k_r, μ_r, and γ_r, while $\phi_r = \phi_d$ is used from the friction law. The new parameters account for rolling stiffness, a static rolling "friction" coefficient, and rolling viscosity, respectively. In the subroutine called, the rolling velocity v_r is used instead of v_t and the computed quasi-force f_r is used to compute the torques, q^{rolling}, on the particles.

2.5.3 Torsion resistance model

The three new parameters for rolling resistance are k_o, μ_o, and γ_o, while $\phi_o = \phi_d$ is used from the friction law. The new parameters account for torsion stiffness, a static torsion "friction" coefficient, and torsion viscosity, respectively. In the subroutine, the torsion velocity v_o is used instead of v_t and the projection is a projection along the normal unit-vector, not into the tangential plane as for the other two models. The computed quasi-force f_o is then used to compute the torques, q^{torsion}, on the particles.

2.6 Background friction

Note that the viscous dissipation takes place in a two-particle contact. In the bulk material, where many particles are in contact with each other, this dissipation mode is very inefficient for long-wavelength cooperative modes of motion [42, 41]. Therefore, an additional damping with the background can be introduced, so that the total force on particle i is

$$f_i = \sum_j \left(f^n n + f^t t \right) - \gamma_b v_i , \qquad (22)$$

and the total torque

$$q_i = \sum_j \left(q^{\text{friction}} + q^{\text{rolling}} + q^{\text{torsion}} \right) - \gamma_{br} a_i^2 \omega_i , \qquad (23)$$

with the damping artificially enhanced in the spirit of a rapid relaxation and equilibration. The sum in Eqs. (22) and (23) takes into account all contact partners j of particle i, but the background dissipation can be attributed to the medium between the particles. Note that the effect of γ_b and γ_{br} should be checked for each set of parameters: it should be small in order to exclude artificial over-damping. The set of parameters is summarized in table 1. Note that only a few parameters are specified with dimensions, while the other parameters are expressed as ratios.

Property	Symbol
Time unit	t_u
Length unit	x_u
Mass unit	m_u
Particle radius	a_0
Material density	ρ
Elastic stiffness (variable)	k_2
Maximal elastic stiffness	$k = k_2$
Plastic stiffness	k_1/k
Adhesion "stiffness"	k_c/k
Friction stiffness	k_t/k
Rolling stiffness	k_r/k
Torsion stiffness	k_o/k
Plasticity depth	ϕ_f
Coulomb friction coefficient	$\mu = \mu_d = \mu_s$
Dynamic to static Friction ratio	$\phi_d = \mu_d/\mu_s$
Rolling "friction" coefficient	μ_r
Torsion "friction" coefficient	μ_o
Normal viscosity	$\gamma = \gamma_n$
Friction viscosity	γ_t/γ
Rolling viscosity	γ_r/γ
Torsion viscosity	γ_o/γ
Background viscosity	γ_b/γ
Background viscous torque	γ_{br}/γ

Table 1 Summary of the microscopic contact model parameters. The longer ranged forces and their parameters, ε, r_0, and r_c are not included here.

2.7 Example: tension test simulation results

In order to illustrate the power of the contact model (especially the adhesive normal model), in this section, uni-axial tension and compression tests are presented. Note that the contact model parameters are chosen once and then one can simulate loose particles, pressure-sintering, and agglomerates with one set of parameters.

With slight extensions, the same model was already applied to temperature-sintering [50] or self-healing [53, 52].

The tests consists of three stages: (i) pressure sintering, (ii) stress-relaxation, and (iii) the compression- or tension-test itself. The contact parameters, as introduced in the previous section, are summarized in table 1 and typical values are given in table 2. These parameters are used for particle-particle contacts, *the same for all tests*, unless explicitly specified.

First, for *pressure sintering*, a very loose assembly of particles is compressed with isotropic stress $p_s 2a/k_2 \approx 0.02$ in a cuboidal volume so that the adhesive contact forces are activated this way. The stress- and strain-controlled wall motion modes are described below in subsection 6.2.2.

Two of the six walls are adhesive, with $k_c^{\text{wall}}/k_2 = 20$, so that the sample sticks to them later, while all other walls are adhesionless, so that they can be easily removed in the next step. Note that during compression and sintering, the walls could all be without adhesion, since the high pressure used keeps the sample together anyway – only later for relaxation, adhesion must switched on. If not, the sample does not remain a solid, and it could also lose contact with the walls, which are later used to apply the tensile strain.

All walls should be frictionless during sintering, while the particles can be slightly adhesive and frictional. If the walls would be frictional, the pressure from a certain wall would not be transferred completely to the respective opposite wall, since frictional forces carry part of the load – an effect that is known since the early work of Janssen [21, 62, 66].

Pressure-sintering is stopped when the kinetic energy of the sample is many orders of magnitude smaller than the potential energy – typically 10 orders of magnitude.

During *stress-relaxation* all wall stresses are slowly released to $p_r/p_s \ll 1$ and the sample is relaxed again until the kinetic energy is much smaller than the potential energy. After this, the sample is ready for the *tension or compression tests*. The non-adhesive side walls still feel a very small external stress that is not big enough to affect the dynamics of the tension test, it is just convenient to keep the walls close to the sample. (This is a numerical and not a physical requirement, since our code uses linked-cells and those are connected to the system size. If the walls would move too far away, either the linked cells would grow, or their number would increase. Both cases are numerically inefficient.)

For the *tension test* wall friction is typically active, but some variation does not show a big effect. One of the sticky walls is slowly and smoothly moved outwards like described and applied in earlier studies [46, 35, 38, 53, 39, 52], following a prescribed cosine-function with time.

2.7.1 Model parameters for tension

The system presented in this subsection contains $N = 1728$ particles with radii a_i drawn from a Gaussian distribution around $a = 0.005$ mm [11, 10]. The contact model parameters are summarized in tables 1 and 2. The volume fraction, $v =$

$\sum_i V(a_i)/V$, with the particle volume $V(a_i) = (4/3)\pi a_i^3$, reached during pressure sintering with $2ap_s/k_2 = 0.01$ is $v_s = 0.6754$. The coordination number is $\mathscr{C} \approx 7.16$ in this state. After stress-relaxation, these values have changed to $v \approx 0.629$ and $\mathscr{C} \approx 6.19$. A different preparation procedure (with adhesion $k_c/k_2 = 0$ during sintering) does not lead to a difference in density after sintering. However, one observes $v \approx 0.630$ and $\mathscr{C} \approx 6.23$ after relaxation. For both preparation procedures the tension test results are virtually identical, so that only the first procedure is used in the following.

Symbol	Value	rescaled units	SI-units
t_u	1	1 μs	10^{-6} s
x_u	1	1 mm	10^{-3} m
m_u	1	1 mg	10^{-6} kg
a_0	0.005	5 μm	$5 \cdot 10^{-6}$ m
ρ	2	2 mg/mm^3	2000 kg/m^3
$k = k_2$	5	5 mg/μs^2	$5 \cdot 10^6$ kg/s^2
k_1/k	0.5		
k_c/k	0.5		
k_t/k	0.2		
$k_r/k = k_o/k$	0.1		
ϕ_f	0.05		
$\mu = \mu_d = \mu_s$	1		
$\phi_d = \mu_d/\mu_s$	1		
$\mu_r = \mu_o$	0.1		
$\gamma = \gamma_n$	$5 \cdot 10^{-5}$	$5 \cdot 10^{-5}$ mg/μs	$5 \cdot 10^1$ kg/s
γ_t/γ	0.2		
$\gamma_r/\gamma = \gamma_o/\gamma$	0.05		
γ_b/γ	4.0		
γ_{br}/γ	1.0		

Table 2 Microscopic material parameters used (second column), if not explicitly specified. The third column contains these values in the appropriate units, i.e., when the time-, length-, and mass-unit are μs, mm, and mg, respectively. Column four contains the parameters in SI-units. Energy, force, acceleration, and stress have to be scaled with factors of 1, 10^3, 10^9, and 10^9, respectively, for a transition from reduced to SI-units.

The material parameters used for the particle contacts are given in table 2. The particle-wall contact parameters are the same, except for cohesion and friction, for which $k_c^{\text{wall}}/k_2 = 20$ and $\mu^{\text{wall}} = 10$ are used – the former during all stages, the latter only during tensile testing.

The choice of numbers and units is such that the particles correspond to spheres with several microns in radius. The magnitude of stiffness k cannot be compared directly with the material bulk modulus C, since it is a contact property. However, there are relations from micro-macro transition analysis, which allow to relate k and $C \sim k\mathscr{C} a^2/V$ [35, 39].

Using the parameter $k = k_2$ in Eq. (4) leads to a typical contact duration (half-period) $t_c \approx 6.5 \cdot 10^{-4}$ μs, for a normal collision of a large and a small particle with

$\gamma = 0$. Accordingly, an integration time-step of $t_{MD} = 5.10^{-6} \mu s$ is used, in order to allow for a "safe" integration of the equations of motion. Note that not only the normal "eigenfrequency" but also the eigenfrequencies in tangential and rotational direction have to be considered as well as the viscous response times $t_\gamma \approx m/\gamma$. All of the physical time-scales should be considerably larger than t_{MD}, whereas the viscous response times should be even larger, so that $t_\gamma > t_c > t_{MD}$. A more detailed discussion of all the effects due to the interplay between the model parameters and the related times is, however, far from the scope of this paper.

2.7.2 Compressive and tensile strength

The tensile (compressive) test is performed uni-axially in x-direction by increasing (reducing) slowly and smoothly the distance between the two sticky walls. (The same initial sample, prepared with $k_c/k_2 = 1/2$, is used for all tests reported here.)

The stress-strain curves for different cohesion are plotted in Fig. 2, for both tension and compression. Note that the shape of the curves and the apparent material behavior (ductile, quasi-brittle, and brittle) depend not only on the contact parameters, but also on the rate at which the deformation is performed (due to the viscous forces introduced above). The present data are for moderate to slow deformation. Faster deformation leads to even smoother curves with larger apparent strength, while considerably slower deformation leads to more brittle behavior (with sharper drops of stress) and somewhat smaller strength.

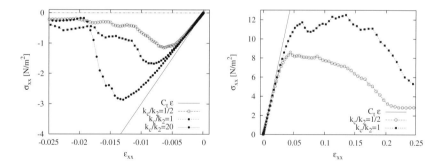

Fig. 2 (Left) Axial tensile stress plotted against tensile strain for simulations with weak, moderate and strong particle contact adhesion; the k_c/k_2 values are given in the inset. The line gives a fit to the linear elastic regime with $C_t = 3.10^{11}$ N/m². (Right) Axial compressive stress plotted against compressive strain for two of the parameter sets from the top panel. The initial slope is the same as in the top panel, indicating that the linear elastic regime is identical for tension and compression.

The axial tensile stress initially increases linearly with strain, practically independent from the contact adhesion strength. With increasing strain, a considerable number of contacts are opened due to tension – contacts open more easily for smaller adhesion (data not shown). This leads to a decrease of the stress-strain slope, then the stress reaches a maximum and, for larger strain, turns into a softening failure mode. As expected, the maximal stress is increasing with contact adhesion k_c/k_2. The compressive strength is $6-7$ times larger than the tensile strength, and a larger adhesion force also allows for larger deformation before failure. The sample with weakest adhesion, $k_c/k_2 = 1/2$, shows tensile and compressive failure at strains $\varepsilon_{xx} \approx -0.006$ and $\varepsilon_{xx} \approx 0.045$, respectively.

Note that for tension, the post-peak behavior for the test with $k_c/k_2 = 20$ is different from the other two cases, due to the strong particle-particle contact adhesion. In this case, the tensile fracture occurs at the wall (except for a few particles that remain in contact with the wall). This is in contrast to the other cases with smaller bulk-adhesion, where the fracture occurs in the bulk, see Fig. 3.

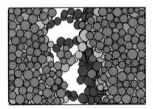 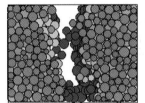

Fig. 3 Snapshots from tensile tests with $k_c/k_2 = 1/5$ and 1, at horizontal strain of $\varepsilon_{xx} \approx -0.8$. The color code denotes the distance from the viewer: blue, green, and red correspond to large, moderate, and short distance.

3 Hard sphere molecular dynamics

In this section, the hard sphere model is introduced together with the event-driven algorithm. A generalized model takes into account the finite contact duration of realistic particles and, besides providing a physical parameter, saves computing time because it avoids the "inelastic collapse".

In the framework of the hard sphere model, particles are assumed to be perfectly rigid and they follow an undisturbed motion until a collision occurs as described below. Due to the rigidity of the interaction, the collisions occur instantaneously, so that an event-driven simulation method [28, 51, 57, 56, 55] can be used. Note that the ED method was only recently implemented in parallel [29, 57]; however, we avoid to discuss this issue in detail.

The instantaneous nature of hard sphere collisions is artificial, however, it is a valid limit in many circumstances. Even though details of the contact- or collision behavior of two particles are ignored, the hard sphere model is valid when binary collisions dominate and multi-particle contacts are rare [44]. The lack of physical information in the model allows a much simpler treatment of collisions than described in section 2 by just using a collision matrix based on momentum conservation and energy loss rules. For the sake of simplicity, we restrict ourselves to smooth hard spheres here. Collision rules for rough spheres are extensively discussed elsewhere, see e.g. [47, 18], and references therein.

3.1 Smooth hard sphere collision model

Between collisions, hard spheres fly independently from each other. A change in velocity – and thus a change in energy – can occur only at a collision. The standard interaction model for instantaneous collisions of identical particles with radius a, and mass m, is used in the following. The post-collisional velocities v' of two collision partners in their center of mass reference frame are given, in terms of the pre-collisional velocities v, by

$$v'_{1,2} = v_{1,2} \mp (1+r)v_n/2, \qquad (24)$$

with $v_n \equiv [(v_1 - v_2) \cdot n]n$, the normal component of the relative velocity $v_1 - v_2$, parallel to n, the unit vector pointing along the line connecting the centers of the colliding particles. If two particles collide, their velocities are changed according to Eq. (24), with the change of the translational energy at a collision $\Delta E = -m_{12}(1-r^2)v_n^2/2$, with dissipation for restitution coefficients $r < 1$.

3.2 Event-driven algorithm

Since we are interested in the behavior of granular particles, possibly evolving over several decades in time, we use an event-driven (ED) method which discretizes the sequence of events with a variable time step adapted to the problem. This is different from classical DEM simulations, where the time step is usually fixed.

In the ED simulations, the particles follow an undisturbed translational motion until an event occurs. An event is either the collision of two particles or the collision of one particle with a boundary of a cell (in the linked-cell structure) [5]. The cells have no effect on the particle motion here; they were solely introduced to accelerate the search for future collision partners in the algorithm.

Simple ED algorithms update the whole system after each event, a method which is straightforward, but inefficient for large numbers of particles. In Ref. [28] an ED algorithm was introduced which updates only those two particles involved in the last collision. Because this algorithm is "asynchronous" in so far that an event, i.e. the *next* event, can occur anywhere in the system, it is so complicated to parallelize it [57]. For the serial algorithm, a double buffering data structure is implemented, which contains the 'old' status and the 'new' status, each consisting of: time of event, positions, velocities, and event partners. When a collision occurs, the 'old' and 'new' status of the participating particles are exchanged. Thus, the former 'new' status becomes the actual 'old' one, while the former 'old' status becomes the 'new' one and is then free for the calculation and storage of possible future events. This seemingly complicated exchange of information is carried out extremely simply and fast by only exchanging the pointers to the 'new' and 'old' status respectively. Note that the 'old' status of particle i has to be kept in memory, in order to update the time of the next contact, t_{ij}, of particle i with any other object j if the latter, independently, changed its status due to a collision with yet another particle. During the simulation such updates may be necessary several times so that the predicted 'new' status has to be modified.

The minimum of all t_{ij} is stored in the 'new' status of particle i, together with the corresponding partner j. Depending on the implementation, positions and velocities after the collision can also be calculated. This would be a waste of computer time, since before the time t_{ij}, the predicted partners i and j might be involved in several collisions with other particles, so that we apply a delayed update scheme [28]. The minimum times of event, i.e. the times, which indicate the next event for a certain particle, are stored in an ordered heap tree, such that the next event is found at the top of the heap with a computational effort of $O(1)$; changing the position of one particle in the tree from the top to a new position needs $O(\log N)$ operations. The search for possible collision partners is accelerated by the use of a standard linked-cell data structure and consumes $O(1)$ of numerical resources per particle. In total, this results in a numerical effort of $O(N \log N)$ for N particles. For a detailed description of the algorithm see Ref. [28]. Using all these algorithmic tricks, we are able to simulate about 10^5 particles within reasonable time on a low-end PC [45], where the particle number is more limited by memory than by CPU power. Parallelization, however, is a means to overcome the limits of one processor [57].

As a final remark concerning ED, one should note that the disadvantages connected to the assumptions made that allow to use an event driven algorithm limit the applicability of this method. Within their range of applicability, ED simulations are typically much faster than DEM simulations, since the former account for a collision in one basic operation (collision matrix), whereas the latter require about one hundred basic steps (integration time steps). Note that this statement is also true in the dense regime. In the dilute regime, both methods give equivalent results, because collisions are mostly binary [41]. When the system becomes denser, multi-particle collisions can occur and the rigidity assumption within the ED hard sphere approach becomes invalid.

The most striking difference between hard and soft spheres is the fact that soft particles dissipate less energy when they are in contact with many others of their kind. In the following chapter, the so called TC model is discussed as a means to account for the contact duration t_c in the hard sphere model.

4 The link between ED and DEM via the TC model

In the ED method the contact duration is implicitly zero, matching well the corresponding assumption of instantaneous contacts used for the kinetic theory [17, 22]. Due to this artificial simplification (which disregards the fact that a real contact takes always finite time) ED algorithms run into problems when the time between events t_n gets too small: In dense systems with strong dissipation, t_n may even tend towards zero. As a consequence the so-called "inelastic collapse" can occur, i.e. the divergence of the number of events per unit time. The problem of the inelastic collapse [54] can be avoided using restitution coefficients dependent on the time elapsed since the last event [51, 44]. For the contact that occurs at time t_{ij} between particles i and j, one uses $r = 1$ if at least one of the partners involved had a collision with another particle later than $t_{ij} - t_c$. The time t_c can be seen as a typical duration of a contact, and allows for the definition of the dimensionless ratio

$$\tau_c = t_c/t_n . \tag{25}$$

The effect of t_c on the simulation results is negligible for large r and small t_c; for a more detailed discussion see [51, 45, 44].

In assemblies of soft particles, multi-particle contacts are possible and the inelastic collapse is avoided. The TC model can be seen as a means to allow for multi-particle collisions in dense systems [43, 30, 51]. In the case of a homogeneous cooling system (HCS), one can explicitly compute the corrected cooling rate (r.h.s.) in the energy balance equation

$$\frac{d}{d\tau}E = -2I(E,t_c) , \tag{26}$$

with the dimensionless time $\tau = (2/3)At/t_E(0)$ for 3D systems, scaled by $A = (1 - r^2)/4$, and the collision rate $t_E^{-1} = (12/a)vg(v)\sqrt{T/(\pi m)}$, with $T = 2K/(3N)$. In these units, the energy dissipation rate I is a function of the dimensionless energy $E = K/K(0)$ with the kinetic energy K, and the cut-off time t_c. In this representation, the restitution coefficient is hidden in the rescaled time via $A = A(r)$, so that inelastic hard sphere simulations with different r scale on the same master-curve. When the classical dissipation rate $E^{3/2}$ [17] is extracted from I, so that $I(E,t_c) = J(E,t_c)E^{3/2}$, one has the correction-function $J \to 1$ for $t_c \to 0$. The deviation from the classical HCS is [44]:

$$J(E,t_c) = \exp(\Psi(x)) , \tag{27}$$

with the series expansion $\Psi(x) = -1.268x + 0.01682x^2 - 0.0005783x^3 + \mathcal{O}(x^4)$ in the collision integral, with $x = \sqrt{\pi}t_c t_E^{-1}(0)\sqrt{E} = \sqrt{\pi}\tau_c(0)\sqrt{E} = \sqrt{\pi}\tau_c$ [44]. This is close to the result $\Psi_{LM} = -2x/\sqrt{\pi}$, proposed by Luding and McNamara, based on probabilistic mean-field arguments [51] [3].

Given the differential equation (26) and the correction due to multi-particle contacts from Eq. (27), it is possible to obtain the solution numerically, and to compare

---
[3] Ψ_{LM} thus neglects non-linear terms and underestimates the linear part

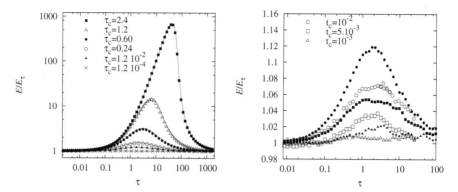

Fig. 4 (Left) Deviation from the HCS, i.e. rescaled energy E/E_τ, where E_τ is the classical solution $E_\tau = (1+\tau)^{-2}$. The data are plotted against τ for simulations with different $\tau_c(0) = t_c/t_E(0)$ as given in the inset, with $r = 0.99$, and $N = 8000$. Symbols are ED simulation results, the solid line results from the third order correction. (Right) E/E_τ plotted against τ for simulations with $r = 0.99$, and $N = 2197$. Solid symbols are ED simulations, open symbols are DEM (soft particle simulations) with three different t_c as given in the inset.

it to the classical $E_\tau = (1+\tau)^{-2}$ solution. Simulation results are compared to the theory in Fig. 4 (left). The agreement between simulations and theory is almost perfect in the examined range of t_c-values, only when deviations from homogeneity are evidenced one expects disagreement between simulation and theory. The fixed cut-off time t_c has no effect when the time between collisions is very large $t_E \gg t_c$, but strongly reduces dissipation when the collisions occur with high frequency $t_E^{-1} \gtrsim t_c^{-1}$. Thus, in the homogeneous cooling state, there is a strong effect initially, and if t_c is large, but the long time behavior tends towards the classical decay $E \to E_\tau \propto \tau^{-2}$.

The final check if the ED results obtained using the TC model are reasonable is to compare them to DEM simulations, see Fig. 4 (right). Open and solid symbols correspond to soft and hard sphere simulations respectively. The qualitative behavior (the deviation from the classical HCS solution) is identical: The energy decay is delayed due to multi-particle collisions, but later the classical solution is recovered. A quantitative comparison shows that the deviation of E from E_τ is larger for ED than for DEM, given that the same t_c is used. This weaker dissipation can be understood from the strict rule used for ED: Dissipation is inactive if any particle had a contact already. The disagreement between ED and DEM is systematic and should disappear if an about 30% smaller t_c value is used for ED. The disagreement is also plausible, since the TC model disregards all dissipation for multi-particle contacts, while the soft particles still dissipate energy - even though much less - in the case of multi-particle contacts.

The above simulations show that the TC model is in fact a "trick" to make hard particles soft and thus connecting between the two types of simulation models: soft and hard. The only change made to traditional ED involves a reduced dissipation for (rapid) multi-particle contacts.

5 The stress in particle simulations

The stress tensor is a macroscopic quantity that can be obtained by measurement of forces per area, or via a so-called micro-macro homogenization procedure. Both methods will be discussed below. During derivation, it also turns out that stress has two contributions, the first is the "static stress" due to particle contacts, a *potential energy density*, the second is the "dynamics stress" due to momentum flux, like in the ideal gas, a *kinetic energy density*. For the sake of simplicity, we restrict ourselves to the case of smooth spheres here.

5.1 Dynamic stress

For dynamic systems, one has momentum transport via flux of the particles. This simplest contribution to the stress tensor is the standard stress in an ideal gas, where the atoms (mass points) move with a certain fluctuation velocity v_i. The kinetic energy $E = \sum_{i=1}^{N} m v_i^2/2$ due to the fluctuation velocity v_i can be used to define the temperature of the gas $k_B T = 2E/(\mathscr{D} N)$, with the dimension $\mathscr{D}$ and the particle number N. Given a number density $n = N/V$, the stress in the ideal gas is then isotropic and thus quantified by the pressure $p = n k_B T$; note that we will disregard k_B in the following. In the general case, the dynamic stress is $\sigma = (1/V) \sum_i m_i v_i \otimes v_i$, with the dyadic tensor product denoted by '$\otimes$', and the pressure $p = \mathrm{tr}\sigma/\mathscr{D} = nT$ is the kinetic energy density.

The additional contribution to the stress is due to collisions and contacts and will be derived from the principle of virtual displacement for soft interaction potentials below, and then be modified for hard sphere systems.

5.2 Static stress from virtual displacements

From the centers of mass r_1 and r_2 of two particles, we define the so-called branch vector $l = r_1 - r_2$, with the reference distance $l = |l| = 2a$ at contact, and the corresponding unit vector $n = l/l$. The deformation in the normal direction, relative to the reference configuration, is defined as $\delta = 2an - l$. A virtual change of the deformation is then

$$\partial \delta = \delta' - \delta \approx \partial l = \varepsilon \cdot l \,, \tag{28}$$

where the prime denotes the deformation after the virtual displacement described by the tensor ε. The corresponding potential energy density due to the contacts of one pair of particles is $u = k\delta^2/(2V)$, expanded to second order in δ, leading to the virtual change

$$\partial u = \frac{k}{V} \left(\delta \cdot \partial \delta + \frac{1}{2}(\partial \delta)^2 \right) \approx \frac{k}{V} \delta \cdot \partial l^n \,, \tag{29}$$

where k is the spring stiffness (the prefactor of the quadratic term in the series expansion of the interaction potential), V is the averaging volume, and $\partial l^n = n(n \cdot \varepsilon \cdot l)$ is the normal component of ∂l. Note that ∂u depends only on the normal component of $\partial \delta$ due to the scalar product with δ, which is parallel to n.

From the potential energy density, we obtain the stress from a virtual deformation by differentiation with respect to the deformation tensor components

$$\sigma = \frac{\partial u}{\partial \varepsilon} = \frac{k}{V} \delta \otimes l = \frac{1}{V} f \otimes l, \tag{30}$$

where $f = k\delta$ is the force acting at the contact, and the dyadic product $\otimes$ of two vectors leads to a tensor of rank two.

5.3 Stress for soft and hard spheres

Combining the dynamic and the static contributions to the stress tensor [49], one has for smooth, soft spheres:

$$\sigma = \frac{1}{V} \left[\sum_i m_i v_i \otimes v_i - \sum_{c \in V} f_c \otimes l_c \right], \tag{31}$$

where the right sum runs over all contacts c in the averaging volume V. Replacing the force vector by momentum change per unit time, one obtains for hard spheres:

$$\sigma = \frac{1}{V} \left[\sum_i m_i v_i \otimes v_i - \frac{1}{\Delta t} \sum_n \sum_j p_j \otimes l_j \right], \tag{32}$$

where p_j and l_j are the momentum change and the center-contact vector of particle j at collision n, respectively. The sum in the left term runs over all particles i, the first sum in the right term runs over all collisions n occurring in the averaging time Δt, and the second sum in the right term concerns the collision partners of collision n [51].

Exemplary stress computations from DEM and ED simulations are presented in the following section.

6 2D simulation results

Stress computations from two dimensional DEM and ED simulations are presented in the following subsections. First, a global equation of state, valid for all densities, is proposed based on ED simulations, and second, the stress tensor from a slow, quasi-static deformation is computed from DEM simulations with frictional particles.

6.1 The equation of state from ED

The mean pressure in two dimensions is $p = (\sigma_1 + \sigma_2)/2$, with the eigenvalues σ_1 and σ_2 of the stress tensor [48, 49, 32]. The 2D dimensionless, reduced pressure $P = p/(nT) - 1 = pV/E - 1$ contains only the collisional contribution and the simulations agree nicely with the theoretical prediction $P_2 = 2vg_2(v)$ for elastic systems, with the pair-correlation function $g_2(v) = (1 - 7v/16)/(1 - v)^2$, and the volume fraction $v = N\pi a^2/V$, see Fig. 5. A better pair-correlation function is

$$g_4(v) = \frac{1 - 7v/16}{(1-v)^2} - \frac{v^3/16}{8(1-v)^4}, \qquad (33)$$

which defines the non-dimensional collisional stress $P_4 = 2vg_4(v)$. For a system with homogeneous temperature, as a remark, the collision rate is proportional to the dimensionless pressure $t_n^{-1} \propto P$.

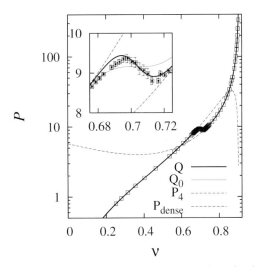

Fig. 5 The dashed lines are P_4 and P_{dense} as functions of the volume fraction v, and the symbols are simulation data, with standard deviations as given by the error bars in the inset. The thick solid line is Q, the corrected global equation of state from Eq. (34), and the thin solid line is Q_0 without empirical corrections.

When plotting P against v with a logarithmic vertical axis, in Fig. 5, the simulation results can almost not be distinguished from P_2 for $v < 0.65$, but P_4 leads to better agreement up to $v = 0.67$. Crystallization is evidenced at the point of the liquid-solid transition $v_c \approx 0.7$, and the data clearly deviate from P_4. The pressure is strongly reduced due to the increase of free volume caused by ordering. Eventually, the data diverge at the maximum packing fraction $v_{max} = \pi/(2\sqrt{3})$ for a perfect triangular array.

For high densities, one can compute from free-volume models, the reduced pressure $P_{fv} = 2v_{max}/(v_{max} - v)$. Slightly different functional forms do not lead to much better agreement [32]. Based on the numerical data, we propose the corrected high density pressure $P_{dense} = P_{fv}h(v_{max} - v) - 1$, with the empirical fit function $h(x) = 1 + c_1 x + c_3 x^3$, and $c_1 = -0.04$ and $c_3 = 3.25$, in perfect agreement with the simulation results for $v \geq 0.73$.

Since, to our knowledge, there is no conclusive theory available to combine the disordered and the ordered regime [23], we propose a global equation of state

$$Q = P_4 + m(v)[P_{dense} - P_4], \qquad (34)$$

with an empirical merging function $m(v) = [1 + \exp(-(v - v_c)/m_0)]^{-1}$, which selects P_4 for $v \ll v_c$ and P_{dense} for $v \gg v_c$, with the transition density v_c and the width of the transition m_0. In Fig. 5, the fit parameters $v_c = 0.702$ and $m_0 \approx 0.0062$ lead to qualitative and quantitative agreement between Q (thick line) and the simulation results (symbols). However, a simpler version $Q_0 = P_2 + m(v)[P_{fv} - P_2]$, (thin line) without empirical corrections leads already to reasonable agreement when $v_c = 0.698$ and $m_0 = 0.0125$ are used. In the transition region, this function Q_0 has no negative slope but is continuous and differentiable, so that it allows for an easy and compact numerical integration of P. We selected the parameters for Q_0 as a compromise between the quality of the fit on the one hand and the simplicity and treatability of the function on the other hand.

As an application of the global equation of state, the density profile of a dense granular gas in the gravitational field has been computed for monodisperse [49] and bidisperse situations [48, 32]. In the latter case, however, segregation was observed and the mixture theory could not be applied. The equation of state and also other transport properties are extensively discussed in Refs. [4, 1, 3, 2] for 2D, bi-disperse systems.

6.2 Quasi-static DEM simulations

In contrast to the dynamic, collisional situation discussed in the previous section, a quasi-static situation, with all particles almost at rest most of the time, is discussed in the following.

6.2.1 Model parameters

The systems examined in the following contain $N = 1950$ particles with radii a_i randomly drawn from a homogeneous distribution with minimum $a_{\min} = 0.5\,10^{-3}$ m and maximum $a_{\max} = 1.5\,10^{-3}$ m. The masses $m_i = (4/3)\rho\pi a_i^3$, with the density $\rho = 2.0\,10^3$ kg m^{-3}, are computed as if the particles were spheres. This is an artificial choice and introduces some dispersity in mass in addition to the dispersity in size. Since we are mainly concerned about slow deformation and equilibrium situations, the choice for the calculation of mass should not matter. The total mass of the particles in the system is thus $M \approx 0.02$ kg with the typical reduced mass of a pair of particles with mean radius, $m_{12} \approx 0.42\,10^{-5}$ kg. If not explicitly mentioned, the material parameters are $k_2 = 10^5$ N m^{-1} and $\gamma_0 = 0.1$ kg s^{-1}. The other spring-constants k_1 and k_c will be defined in units of k_2. In order to switch on adhesion, $k_1 < k_2$ and $k_c > 0$ is used; if not mentioned explicitly, $k_1 = k_2/2$ is used, and k_2 is constant, independent of the maximum overlap previously achieved.

Using the parameters $k_1 = k_2$ and $k_c = 0$ in Eq. (4) leads to a typical contact duration (half-period): $t_c \approx 2.03\,10^{-5}$ s for $\gamma_0 = 0$, $t_c \approx 2.04\,10^{-5}$ s for $\gamma_0 = 0.1$ kg s^{-1}, and $t_c \approx 2.21\,10^{-5}$ s for $\gamma_0 = 0.5$ kg s^{-1} for a collision. Accordingly, an integration time-step of $t_{\mathrm{DEM}} = 5\,10^{-7}$ s is used, in order to allow for a 'safe' integration of contacts involving smaller particles. Large values of k_c lead to strong adhesive forces, so that also more energy can be dissipated in one collision. The typical response time of the particle pairs, however, is not affected so that the numerical integration works well from a stability and accuracy point of view.

6.2.2 Boundary conditions

The experiment chosen is the bi-axial box set-up, see Fig. 6, where the left and bottom walls are fixed, and stress- or strain-controlled deformation is applied. In the first case a wall is subject to a predefined pressure, in the second case, the wall is subject to a pre-defined strain. In a typical 'experiment', the top wall is strain controlled and slowly shifted downwards while the right wall moves stress controlled, dependent on the forces exerted on it by the material in the box. The strain-controlled position of the top wall as function of time t is here

$$z(t) = z_{\mathrm{f}} + \frac{z_0 - z_{\mathrm{f}}}{2}(1 + \cos\omega t), \quad \text{with} \quad \varepsilon_{zz} = 1 - \frac{z}{z_0}, \tag{35}$$

where the initial and the final positions z_0 and z_{f} can be specified together with the rate of deformation $\omega = 2\pi f$ so that after a half-period $T/2 = 1/(2f)$ the extremal deformation is reached. With other words, the cosine is active for $0 \le \omega t \le \pi$. For larger times, the top-wall is fixed and the system can relax indefinitely. The cosine function is chosen in order to allow for a smooth start-up and finish of the motion so that shocks and inertia effects are reduced, however, the shape of the function is arbitrary as long as it is smooth.

The stress-controlled motion of the side-wall is described by

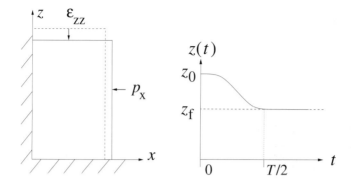

Fig. 6 (Left) Schematic drawing of the model system. (Right) Position of the top-wall as function of time for the strain-controlled situation.

$$m_w \ddot{x}(t) = F_x(t) - p_x z(t) - \gamma_w \dot{x}(t) , \qquad (36)$$

where m_w is the mass of the right side wall. Large values of m_w lead to slow adaption, small values allow for a rapid adaption to the actual situation. Three forces are active: (i) the force $F_x(t)$ due to the bulk material, (ii) the force $-p_x z(t)$ due to the external pressure, and (iii) a strong frictional force which damps the motion of the wall so that oscillations are reduced.

6.2.3 Initial configuration and compression

Initially, the particles are randomly distributed in a huge box, with rather low overall density. Then the box is compressed, either by moving the walls to their desired position, or by defining an external pressure $p = p_x = p_z$, in order to achieve an isotropic initial condition. Starting from a relaxed, isotropic initial configuration, the strain is applied to the top wall and the response of the system is examined. In Fig. 7, snapshots from a typical simulation are shown during compression.

In the following, simulations are presented with different side pressures $p = 20$, 40, 100, 200, 400, and 500. The behavior of the averaged scalar and tensor variables during the simulations is examined in more detail for situations with small and large confining pressure. The averages are performed such that 10 to 20% of the total volume is disregarded in the vicinity of each wall in order to avoid boundary effects. A particle contact is taken into account for the average if the contact point lies within the averaging volume V.

6.2.4 Compression and dilation

The first quantity of interest is the density (volume fraction) v and, related to it, the volumetric strain $\varepsilon_V = \Delta V/V$. From the averaged data, we evidence compression

From Particles to Continuum Theory

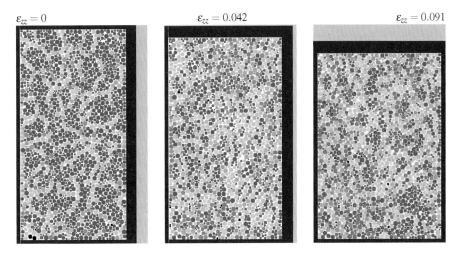

Fig. 7 Snapshots of the simulation at different ε_{zz} for constant side pressure p. The color code corresponds to the potential energy of each particle, decaying from red over green to blue and black. The latter black particles are so-called rattlers that do not contribute to the static contact network.

for small deformation and large side pressure. This initial regime follows strong dilation, for all pressures, until a quasi-steady-state is reached, where the density is almost constant besides a weak tendency towards further dilation.

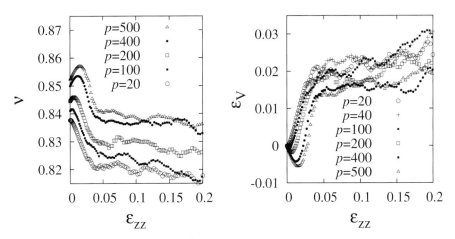

Fig. 8 (Left) Volume fraction $v = \sum_i \pi a_i^2 / V$ for different confining pressure p. (Right) Volumetric strain – negative values mean compression, whereas positive values correspond to dilation.

An initially dilute granular medium (weak confining pressure) thus shows dilation from the beginning, whereas a denser granular material (strong confining pressure) can be compressed even further by the relatively strong external forces until dilation starts. The range of density changes is about 0.02 in volume fraction and spans up to 3 % changes in volumetric strain.

From the initial slope, one can obtain the Poisson ratio of the bulk material, and from the slope in the dilatant regime, one obtains the so-called dilatancy angle, a measure of the magnitude of dilatancy required before shear is possible [46, 33].

The anisotropy of the granular packing is quantified by the deviatoric fabric (data not shown). The anisotropy is initially of the order of a few percent at most – thus the initial configurations are already not perfectly isotropic - even though isotropically prepared. With increasing deviatoric deformation, the anisotropy grows, reaches a maximum and then saturates on a lower level in the critical state flow regime. The scaled fabric grows faster for smaller side pressure and is also relatively larger for smaller p. The non-scaled fabric deviator, astonishingly, grows to values around $f_D^{\max} \text{tr} F \approx 0.56 \pm 0.03$, independently of the side pressures used here (data not shown, see [33, 34] for details). Using the definition $f_D := \text{dev} F / \text{tr} F$, the functional behavior,

$$\frac{\partial f_D}{\partial \varepsilon_D} = \beta_f (f_D^{\max} - f_D) , \qquad (37)$$

was evidenced from simulations in Ref. [33], with $f_D^{\max} \text{tr} F \approx$ const., and the deviatoric rate of approach $\beta_f = \beta_f(p)$, decreasing with increasing side pressure. The differential equation is solved by an exponential function that describes the approach of the anisotropy f_D to its maximal value, $1 - (f_D / f_D^{\max}) = \exp(-\beta_f \varepsilon_D)$, but not beyond.

6.2.5 Stress tensor

The sums of the normal and the tangential stress-contributions are displayed in Fig. 9 for two side-pressures $p = 20$ and $p = 200$. The lines show the stress measured on the walls, and the symbols correspond to the stress measured via the micro-macro average in Eq. (31), proving the reasonable quality of the micro-macro transition as compared to the wall stress "measurement".

There is also other macroscopic information hidden in the stress-strain curves in Fig. 9. From the initial, rapid increase in stress, one can determine moduli of the bulk-material, i.e, the stiffness under confinement p. Later, the stress reaches a peak at approximately $2.6p$ and then saturates at about $2p$. From both peak- and saturation stress, one obtains the yield stresses at peak and in critical state flow, respectively [61].

Note that for the parameters used here, both the dynamic stress and the tangential contributions to the stress tensor are more than one order of magnitude smaller than the normal contributions. As a cautionary note, we remark also that the artificial stress induced by the background viscous force is negligible here (about 2%), when $\gamma_b = 10^{-3} \text{kg s}^{-1}$ and a compression frequency $f = 0.1 \text{s}^{-1}$ are used. For faster

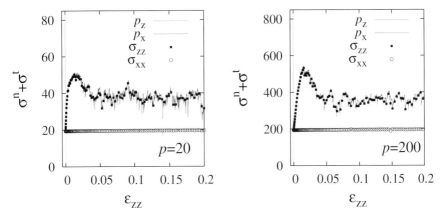

Fig. 9 Total stress tensor $\sigma = \sigma^n + \sigma^t$ for small (Left) and high (Right) pressure – the agreement between the wall pressure and the averaged stress is almost perfect.

compression with $f = 0.5\,\mathrm{s}^{-1}$, one obtains about 10% contribution to stress from the artificial background force.

The behavior of the stress is displayed in Fig. 10, where the isotropic stress $\frac{1}{2}\mathrm{tr}\,\sigma$ is plotted in units of p, and the deviatoric fraction is plotted in units of the isotropic stress. Note that the tangential forces do not contribute to the isotropic stress here since the corresponding entries in the averaging procedure compensate. From Fig. 10, we evidence that both normal contributions, the non-dimensional trace and the non-dimensional deviator behave similarly, independent of the side pressure: Starting from an initial value, a maximum is approached, where the maximum is only weakly dependent on p.

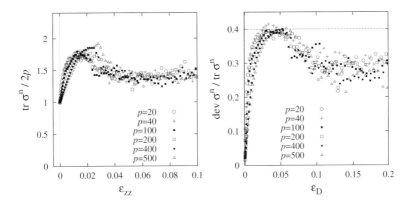

Fig. 10 Non-dimensional stress tensor contributions for different p. The isotropic (Left) and the deviatoric fractions (Right) are displayed as functions of the vertical and deviatoric strain, respectively.

The increase of stress is faster for lower p. After the maximum is reached, the stresses decay and approach a smaller value in the critical state flow regime. Using the definitions $s_V := \operatorname{tr} \sigma /(2p) - 1$ and $s_D := \operatorname{dev} \sigma / \operatorname{tr} \sigma$, the maximal (non-dimensional) isotropic and deviatoric stresses are $s_V^{\max} \approx 0.8 \pm 0.1$ and $s_D^{\max} \approx 0.4 \pm 0.02$, respectively, with a rather large error margin. The corresponding values at critical state flow are $s_V^c \approx 0.4 \pm 0.1$ and $s_D^c \approx 0.29 \pm 0.04$.

The evolution of the *deviatoric stress* fraction, s_D, as function of ε_D, is displayed in Fig. 10. Like the fabric, also the deviatoric stress exponentially approaches its maximum. This is described by the differential equation

$$\frac{\partial s_D}{\partial \varepsilon_D} = \beta_s \left(s_D^{\max} - s_D \right), \tag{38}$$

where $\beta_s = \beta_s(p)$ is decaying with increasing p (roughly as $\beta_s \approx p^{-1/2}$). For more details on the deviatoric stress and also on the tangential contribution to the stress, see [33, 34, 36, 35].

7 Larger computational examples

In this section, several examples of rather large particle numbers simulated with DEM and ED are presented. The ED algorithm is first used to simulate a freely cooling dissipative gas in two and three dimensions [45, 56]. Then, a peculiar three dimensional ring-shear experiment is modeled with soft sphere DEM.

7.1 Free cooling and cluster growth (ED)

In the following, a two-dimensional system of length $L = l/d = 560$ with $N = 99856$ dissipative particles of diameter $d = 2a$ is examined [51, 45], with volume fraction $v = 0.25$ and restitution coefficient $r = 0.9$. This 2D system is compared to a three-dimensional system of length $L = l/d = 129$ with $N = 512000$ dissipative spheres of diameter d and volume fraction $v = 0.25$ with $r = 0.3$ [56].

7.1.1 Initial configuration

Initially the particles are arranged on a square lattice with random velocities drawn from an interval with constant probability for each coordinate. The mean total velocity, i.e. the random momentum due to the fluctuations, is eliminated in order to have a system with its center of mass at rest. The system is allowed to evolve for some time, until the arbitrary initial condition is forgotten, i.e. the density is homogeneous, and the velocity distribution is a Gaussian in each coordinate. Then dissipation is switched on and the evolution of the system is reported for the selected r. In order to avoid the inelastic collapse, the TC model is used, which reduces dissipation if the time between collisions drops below a value of $t_c = 10^{-5}$ s.

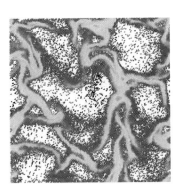 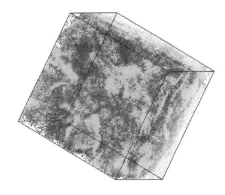

Fig. 11 (Left) Collision frequency of individual particles from a 2D simulation, after about 5200 collisions per particle. (Right) Cluster visualization from a 3D simulation. The colors in both panels indicate large (red), medium (green), and small (blue) collision rates.

7.1.2 System evolution

For the values of r used here, the system becomes inhomogeneous quite rapidly [45, 56]. Clusters, and thus also dilute regions, build up and have the tendency to grow. Since the system is finite, their extension will reach system size at a finite time. Thus we distinguish between three regimes of system evolution: (i) the initially (almost) homogeneous state, (ii) the cluster growth regime, and (iii) the system size dependent final stage where the clusters have reached system size. We note that a cluster does not behave like a solid body, but has internal motion and can eventually break into pieces after some time. These pieces (small clusters) collide and can merge to larger ones.

In Fig. 11, snapshots are presented and the collision rate is color-coded. The collision rate and the pressure are higher inside the clusters than at their surface. Note that most of the computational effort is spent in predicting collisions and to compute the velocities after the collisions. Therefore, the regions with the largest collision frequencies require the major part of the computational resources. Due to the TC model, this effort stays limited and the simulations can easily continue for many thousand collisions per particle.

7.1.3 Discussion

Note that an event driven simulation can be 10-100 times faster than a soft-particle DEM code applied to model the same particle number. However, ED is rather limited to special, simple interactions between the particles.

7.2 3D (ring) shear cell simulation

The simulation in this section models a ring-shear cell experiment, as recently proposed [15, 16]. The interesting observation in the experiment is a universal shear zone, initiated at the bottom of the cell and becoming wider and moving inwards while propagating upwards in the system.

In the following, the shear-band will be examined, and the micro-macro transition from particle quantities, like forces, to continuum quantities, like stresses, will be performed, leading to a yield stress (or flow function) based on a single simulation. This is in contrast to the two-dimensional example from the previous chapter, where the yield stress had to be determined from different simulations with different side stress p. In the ring shear cell, space- and time-averaging is possible, so that - at different radial and vertical positions, one obtains data for different density, stress, velocity gradient, etc.

7.2.1 Model system

The numerical model chosen here is DEM with smooth particles in three dimensions. In order to save computing time, only a quarter of the ring-shaped geometry is simulated. The walls are cylindrical, and are rough on the particle scale due to some attached particles. The outer cylinder wall with radius R_o, and part of the bottom $r > R_s$ are rotating around the symmetry axis, while the inner wall with radius R_i, and the attached bottom-disk $r < R_s$ remain at rest. In order to resemble the experiment, the geometry data are $R_i = 0.0147$ m, $R_s = 0.085$ m, and $R_o = 0.110$ m. Note that the small R_i value is artificial, but it does not affect the results for small and intermediate filling heights.

The slit in the bottom wall at $r = R_s$ triggers a shear band. In order to examine the behavior of the shear band as function of the filling height H, this system is filled with 6000 to 64000 spherical particles with mean radius 1.0 mm and radii range 0.5 mm $< a <$ 1.5 mm, which interact here via repulsive and dissipative forces only. The particles are forced towards the bottom by the gravity force $f_g = mg$ here and are kept inside the system by the cylindrical walls. In order to provide some wall roughness, a fraction of the particles (about 3%) that are originally in contact with the walls are glued to the walls and move with them.

7.2.2 Material and system parameters

The material parameters for the particle-particle and particle-wall interactions are $k = 10^2$ N/m and $\gamma_0 = 2.10^{-3}$ kg/s. Assuming a collision of the largest and the smallest particle used, the reduced mass $m_{12} = 2.94\,10^{-6}$ kg, leads to a typical contact duration $t_c = 5.4\,10^{-4}$ s and a restitution coefficient of $r = 0.83$. The integration time step is $t_{\text{DEM}} = 5.10^{-6}$ s, i.e. two orders of magnitude smaller than the contact duration.

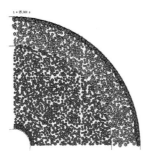 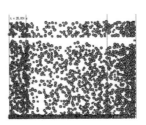

Fig. 12 Snapshots from the quarter-cylinder geometry. Visible are here only those particles glued to the wall; the cylinder and slit positions are indicated by the lines. (Left) Top-view and (Right) front-view. The colors blue and red correspond to static and moving wall particles.

The simulations run for 25 s with a rotation rate $f_o = 0.01$ s^{-1} of the outer cylinder, with angular velocity $\Omega_o = 2\pi f_o$. For the average of the displacement, only times $t > 10$ s are taken into account. Within the averaging accuracy, the system has seemingly reached a quasi-steady state after about 8 s. The empty cell is shown in Fig. 12, while three realizations with different filling height are displayed in Fig. 13, both as top- and front-view.

7.2.3 Shear deformation results

From the top-view, it is evident that the shear band moves inwards with increasing filling height, and it also becomes wider. From the front-view, the same information can be evidenced and, in addition, the shape of the shear band inside the bulk is visible: The inwards displacement happens deep in the bulk and the position of the shear band is not changing a lot closer to the surface.

In order to allow for a more quantitative analysis of the shear band, both on the top and as function of depth, we perform fits with the universal shape function proposed in [15]:

$$\frac{v_\varphi(r)}{r\Omega_o} = A\left(1 + \text{erf}\left(\frac{r - R_c}{W}\right)\right), \quad (39)$$

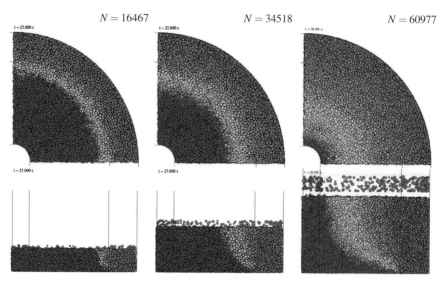

Fig. 13 Snapshots from simulations with different filling heights seen from the top and from the front, and the particle number N is given in the inset. The colors blue, green, orange and red denote particles with $rd\phi \leq 0.5$ mm, $rd\phi \leq 2$ mm, $rd\phi \leq 4$ mm, and $rd\phi > 4$ mm, i.e. the displacement in tangential direction per second, respectively. The filling heights in these simulations are $H = 0.018$ m, 0.037 m, and 0.061 m (from left to right).

where A is a dimensionless amplitude $A = 0.50 \pm 0.02$, R_c is the center of the shearband, and W its width.

The fits to the simulations confirm qualitatively the experimental findings in so far that the center of the shear band, as observed on top of the material, see Fig. 14, moves inwards with a $R_c \propto H^{5/2}$ behavior, and that the width of the shear band increases almost linearly with H. For filling heights larger than $H \approx 0.05$ m, deviations from this behavior are observed, because the inner cylinder is reached and thus sensed by the shearband. Slower shearing does not affect the center, but reduces slightly the width - as checked by one simulation.

Like in the experiments, the behavior of the shearband within the bulk, see Fig. 15, deviates qualitatively from the behavior seen from the top. Instead of a slow motion of the shear band center inwards, the shear band rapidly moves inwards at small heights h, and reaches a saturation distance with small change closer to the surface. Again, a slower rotation does not affect the center but reduces the width.

From the velocity field in the bulk it is straightforward to compute the velocity gradient tensor and, from this extracting the (symmetric) strain rate:

$$\dot{\gamma} = \sqrt{d_1^2 + d_2^2} = \frac{1}{2}\sqrt{\left(\frac{\partial v_\phi}{\partial r} - \frac{v_\phi}{r}\right)^2 + \left(\frac{\partial v_\phi}{\partial z}\right)^2}, \tag{40}$$

i.e., the shear intensity in the shear plane [40]. Note that the solid-body rotation term v_ϕ/r comes from the cylindrical coordinate system used. The shear planes are

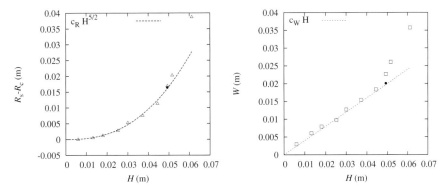

Fig. 14 (Left) Distance of the top-layer shearband center from the slit, both plotted against the filling height H. The open symbols are simulation results, the solid symbol is a simulation with slower rotation $f_o = 0.005\,\text{s}^{-1}$, and the line is a fit with constant $c_R = 30$. (Right) Width of the shearband from the same simulations; the line is a fit with $c_W = 2/5$.

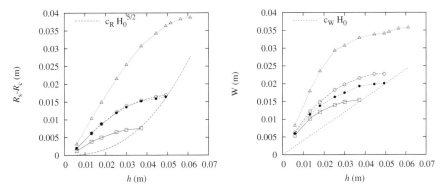

Fig. 15 (Left) Distance of the bulk shearband center from the slit and, (Right) width of the shearband, both plotted against the height h. The open symbols are simulation results obtained with $f_o = 0.01\,\text{s}^{-1}$, the solid symbols are obtained with slower rotation $f_o = 0.005\,\text{s}^{-1}$. Squares, circles and triangles correspond to the filling heights $H = 0.037\,\text{m}$, $0.049\,\text{m}$, and $0.061\,\text{m}$, respectively. The curves are identical to those plotted in Fig. 14.

in fact described by a normal unit vector $\hat{\gamma} = (\cos\theta, 0, \sin\theta)$, with $\theta = \theta(r,z) = \arccos(d_1/\dot{\gamma})$, as predicted [12]. The center of the shear band indicates the direction of the unit-vector $\hat{\gamma}$. In the system with friction, we observe that the average particles spin is also normal to the shear-plane, i.e., parallel to $\hat{\gamma}$, within the rather strong fluctuations (data not shown).

From the stress, as computed according to Eq. (31), the shear stress is extracted (in analogy to the strain rate) as proposed in [12]:

$$|\tau| = \sqrt{\sigma_{r\phi}^2 + \sigma_{z\phi}^2}\,. \tag{41}$$

Remarkably, the shear stress intensity $|\tau|/p \approx \mu$ is almost constant for practically all averaging volumina with strain rates larger than some threshold value, i.e., $\dot{\gamma} > \dot{\gamma}_c$, with $\dot{\gamma}_c \approx 0.02\,\text{s}^{-1}$. Whether the threshold has a physical meaning or is only an artefact due to the statistical fluctuations in the average data has to be examined further by much longer runs with better statistics.

From the constant shear stress intensity in the shear zone, one can determine the Mohr-Coulomb-type friction angle of the equivalent macroscopic constitutive law, see Fig. 16, as $\psi \approx \arcsin\mu$. Interestingly, without friction, ψ is rather large, i.e., much larger than expected from a frictionless material, whereas it is astonishingly small with friction (data not shown), i.e., smaller than the microscopic contact friction $\mu = 0.4$ used, see Ref. [40].

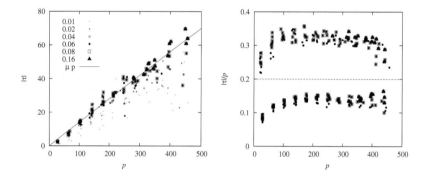

Fig. 16 (Left) Shear stress $|\tau|$ and (Right) shear stress intensity $|\tau|/p$ plotted against pressure. The size of the points is proportional to the shear rate, and the dashed line (right panel) separates the data from simulations without (Bottom) and with (Top) friction, see [40].

7.2.4 Discussion

In summary, the example of a ring shear cell simulation in 3D has shown, that even without the more complicated details of fancy interaction laws, experiments can be reproduced at least qualitatively. A more detailed study of quantitative agreement has been performed in 2D [27], and is in progress for the 3D case.

A challenge for the future remains the micro-macro transition, for which a first result has been shown, i.e. the yield stress can be extracted from a single 3D DEM simulation for various pressures and shear rates. Open remains an objective continuum theory formulation of the shear band problem.

8 Conclusion

The present study is a summary of the most important details about soft particle molecular dynamics (MD), widely referred to as discrete element methods (DEM) in engineering, and hard particle event driven (ED) simulations, together with an attempt to link the two approaches in the dense limit where multi-particle contacts become important.

As an example for a micro-macro transition, the stress tensor was defined and computed for dynamic and quasi-static systems. This led, for example, to a global equation of state, valid for all attainable densities, and also to the partial stresses due to normal and tangential (frictional) contacts. For the latter situation, the micro-macro average is compared to the macroscopic stress (=force/area) measurement (with reasonable agreement) and, at least in 3D, a yield stress function can be extracted from a single ring shear cell simulation.

In conclusion, discrete element methods have proven a helpful tool for the understanding of many granular systems, while MD is the standard tool for atomistic and molecular systems. The methods presented in this paper can be applied to both DEM and MD simulation results with the goal to obtain micro- and particle-based constitutive relations for continuum theory.

The qualitative approach on DEM of the early years has now developed into the attempt of a quantitative predictive modeling tool for the diverse modes of complex behavior in granular media. To achieve this goal will be a research challenge for the next decades, involving enhanced kinetic theories for dense collisional flows and elaborate constitutive models for quasi-static, dense systems with shear band localisation.

In the future this tool will allow to impose a desired behavior by control or design, with particular application in mind as, e.g., modern sintered materials, reactors involving catalysts, and many others.

Acknowledgements We acknowledge the financial support of several funding institutions that supported the reviewed research, and also the helpful discussions with, and contributions from the many persons that contributed to these results.

References

1. Alam, M., Luding, S.: How good is the equipartition assumption for transport properties of a granular mixture. Granular Matter **4**(3), 139–142 (2002)
2. Alam, M., Luding, S.: First normal stress difference and crystallization in a dense sheared granular fluid. Phys. Fluids **15**(8), 2298–2312 (2003)
3. Alam, M., Luding, S.: Rheology of bidisperse granular mixtures via event driven simulations. J. Fluid Mech. **476**, 69–103 (2003)
4. Alam, M., Willits, J.T., Arnarson, B.O., Luding, S.: Kinetic theory of a binary mixture of nearly elastic disks with size and mass-disparity. Physics of Fluids **14**(11), 4085–4087 (2002)
5. Allen, M.P., Tildesley, D.J.: Computer Simulation of Liquids. Oxford University Press, Oxford (1987)

6. van Baars, S.: Discrete element analysis of granular materials. Ph.D. thesis, Technische Universiteit Delft, Delft, Nederlands (1996)
7. Bartels, G., Unger, T., Kadau, D., Wolf, D.E., Kertesz, J.: The effect of contact torques on porosity of cohesive powders. Granular Matter **7**, 139 (2005)
8. Bashir, Y.M., Goddard, J.D.: A novel simulation method for the quasi-static mechanics of granular assemblages. J. Rheol. **35**(5), 849–885 (1991)
9. Cundall, P.A., Strack, O.D.L.: A discrete numerical model for granular assemblies. Géotechnique **29**(1), 47–65 (1979)
10. David, C.T., Garcia-Rojo, R., Herrmann, H.J., Luding, S.: Powder flow testing with 2d and 3d biaxial and triaxial simulations. Particle and Particle Systems Characterization **24**(1), 29–33 (2007)
11. David, C.T., Rojo, R.G., Herrmann, H.J., Luding, S.: Hysteresis and creep in powders and grains. In: R. Garcia-Rojo, H.J. Herrmann, S. McNamara (eds.) Powders and Grains 2005, pp. 291–294. Balkema, Leiden, Netherlands (2005)
12. Depken, M., van Saarloos, W., van Hecke, M.: Continuum approach to wide shear zones in quasistatic granular matter. Phys. Rev. E **73**, 031,302 (2006)
13. Dintwa, E., van Zeebroeck, M., Tijskens, E., Ramon, H.: Torsion of viscoelastic spheres in contact. Granular Matter **7**, 169 (2005)
14. Els, D.: Definition of roll velocity for spherical particles (2006). Submitted
15. Fenistein, D., van Hecke, M.: Kinematics – wide shear zones in granular bulk flow. Nature **425**(6955), 256 (2003)
16. Fenistein, D., van de Meent, J.W., van Hecke, M.: Universal and wide shear zones in granular bulk flow. Phys. Rev. Lett. **92**, 094,301 (2004). E-print cond-mat/0310409
17. Haff, P.K.: Grain flow as a fluid-mechanical phenomenon. J. Fluid Mech. **134**, 401–430 (1983)
18. Herbst, O., Müller, P., Otto, M., Zippelius, A.: Local equation of state and velocity distributions of a driven granular gas. Phys. Rev. E **70**, 051,313-1–14 (2004). E-print cond-mat/0402104
19. Herrmann, H.J., Hovi, J.P., Luding, S. (eds.): Physics of dry granular media - NATO ASI Series E 350. Kluwer Academic Publishers, Dordrecht (1998)
20. Hinrichsen, H., Wolf, D.E.: The Physics of Granular Media. Wiley VCH, Weinheim, Germany (2004)
21. Janssen, H.A.: Versuche über Getreidedruck in Silozellen. Zeitschr. d. Vereines Deutscher Ingenieure **39**(35), 1045–1049 (1895)
22. Jenkins, J.T., Richman, M.W.: Kinetic theory for plane shear flows of a dense gas of identical, rough, inelastic, circular disks. Phys. of Fluids **28**, 3485–3494 (1985)
23. Kawamura, H.: A simple theory of hard disk transition. Prog. Theor. Physics **61**, 1584–1596 (1979)
24. Kirkwood, J.G., Buff, F.P., Green, M.S.: The statistical mechanical theory of transport processes. J. Chem. Phys. **17**(10), 988 (1949)
25. Kishino, Y. (ed.): Powders & Grains 2001. Balkema, Rotterdam (2001)
26. Lätzel, M., Luding, S., Herrmann, H.J.: Macroscopic material properties from quasi-static, microscopic simulations of a two-dimensional shear-cell. Granular Matter **2**(3), 123–135 (2000). E-print cond-mat/0003180
27. Lätzel, M., Luding, S., Herrmann, H.J., Howell, D.W., Behringer, R.P.: Comparing simulation and experiment of a 2d granular Couette shear device. Eur. Phys. J. E **11**(4), 325–333 (2003)
28. Lubachevsky, B.D.: How to simulate billards and similar systems. J. Comp. Phys. **94**(2), 255 (1991)
29. Lubachevsky, B.D.: Simulating billiards: Serially and in parallel. Int.J. in Computer Simulation **2**, 373–411 (1992)
30. Luding, S.: Surface waves and pattern formation in vibrated granular media. In: Powders & Grains 97, pp. 373–376. Balkema, Amsterdam (1997)
31. Luding, S.: Collisions & contacts between two particles. In: H.J. Herrmann, J.P. Hovi, S. Luding (eds.) Physics of dry granular media - NATO ASI Series E350, p. 285. Kluwer Academic Publishers, Dordrecht (1998)

32. Luding, S.: Liquid-solid transition in bi-disperse granulates. Advances in Complex Systems **4**(4), 379–388 (2002)
33. Luding, S.: Micro-macro transition for anisotropic, frictional granular packings. Int. J. Sol. Struct. **41**, 5821–5836 (2004)
34. Luding, S.: Molecular dynamics simulations of granular materials. In: H. Hinrichsen, D.E. Wolf (eds.) The Physics of Granular Media, pp. 299–324. Wiley VCH, Weinheim, Germany (2004)
35. Luding, S.: Anisotropy in cohesive, frictional granular media. J. Phys.: Condens. Matter **17**, S2623–S2640 (2005)
36. Luding, S.: Shear flow modeling of cohesive and frictional fine powder. Powder Technology **158**, 45–50 (2005)
37. Luding, S.: About contact force-laws for cohesive frictional materials in 2d and 3d. In: P. Walzel, S. Linz, C. Krülle, R. Grochowski (eds.) Behavior of Granular Media, pp. 137–147. Shaker Verlag (2006). Band 9, Schriftenreihe Mechanische Verfahrenstechnik, ISBN 3-8322-5524-9
38. Luding, S.: Contact models for very loose granular materials. In: P. Eberhard (ed.) Symposium on Multiscale Problems in Multibody System Contacts, pp. 135–150. Springer (2007). ISBN 978-1-4020-5980-3
39. Luding, S.: Cohesive frictional powders: Contact models for tension. Granular Matter **10**, 235–246 (2008)
40. Luding, S.: The effect of friction on wide shear bands. Part. Science and Technology **26**(1), 33–42 (2008)
41. Luding, S., Clément, E., Blumen, A., Rajchenbach, J., Duran, J.: Anomalous energy dissipation in molecular dynamics simulations of grains: The "detachment effect". Phys. Rev. E **50**, 4113 (1994)
42. Luding, S., Clément, E., Blumen, A., Rajchenbach, J., Duran, J.: The onset of convection in molecular dynamics simulations of grains. Phys. Rev. E **50**, R1762 (1994)
43. Luding, S., Clément, E., Rajchenbach, J., Duran, J.: Simulations of pattern formation in vibrated granular media. Europhys. Lett. **36**(4), 247–252 (1996)
44. Luding, S., Goldshtein, A.: Collisional cooling with multi-particle interactions. Granular Matter **5**(3), 159–163 (2003)
45. Luding, S., Herrmann, H.J.: Cluster growth in freely cooling granular media. Chaos **9**(3), 673–681 (1999)
46. Luding, S., Herrmann, H.J.: Micro-macro transition for cohesive granular media (2001). In: Bericht Nr. II-7, Inst. für Mechanik, Universität Stuttgart, S. Diebels (Ed.)
47. Luding, S., Huthmann, M., McNamara, S., Zippelius, A.: Homogeneous cooling of rough dissipative particles: Theory and simulations. Phys. Rev. E **58**, 3416–3425 (1998)
48. Luding, S., Lätzel, M., Herrmann, H.J.: From discrete element simulations towards a continuum description of particulate solids. In: A. Levy, H. Kalman (eds.) Handbook of Conveying and Handling of Particulate Solids, pp. 39–44. Elsevier, Amsterdam, The Netherlands (2001)
49. Luding, S., Lätzel, M., Volk, W., Diebels, S., Herrmann, H.J.: From discrete element simulations to a continuum model. Comp. Meth. Appl. Mech. Engng. **191**, 21–28 (2001)
50. Luding, S., Manetsberger, K., Muellers, J.: A discrete model for long time sintering. Journal of the Mechanics and Physics of Solids **53**(2), 455–491 (2005)
51. Luding, S., McNamara, S.: How to handle the inelastic collapse of a dissipative hard-sphere gas with the TC model. Granular Matter **1**(3), 113–128 (1998). E-print cond-mat/9810009
52. Luding, S., Suiker, A.: Self-healing of damaged particulate materials through sintering (2008). Submitted to Philosophical Magazine
53. Luding, S., Suiker, A., Kadashevich, I.: Discrete element modeling of self-healing processes in damaged particulate materials. In: A.J.M. Schmets, S. van der Zwaag (eds.) Proceedings of the 1st International Conference on Self Healing Materials. Springer series in Material Science, Berlin, Germany (2007). ISBN 978-1-4020-6249-0
54. McNamara, S., Young, W.R.: Inelastic collapse in two dimensions. Phys. Rev. E **50**(1), R28–R31 (1994)

55. Miller, S.: Clusterbildung in granularen Gasen. Ph.D. thesis, Universität Stuttgart (2004)
56. Miller, S., Luding, S.: Cluster growth in two- and three-dimensional granular gases. Phys. Rev. E **69**, 031,305 (2004)
57. Miller, S., Luding, S.: Event driven simulations in parallel. J. Comp. Phys. **193**(1), 306–316 (2004)
58. Pöschel, T., Luding, S. (eds.): Granular Gases. Springer, Berlin (2001). Lecture Notes in Physics 564
59. Rapaport, D.C.: The Art of Molecular Dynamics Simulation. Cambridge University Press, Cambridge (1995)
60. Sadd, M.H., Tai, Q.M., Shukla, A.: Contact law effects on wave propagation in particulate materials using distinct element modeling. Int. J. Non-Linear Mechanics **28**(2), 251 (1993)
61. Schwedes, J.: Review on testers for measuring flow properties of bulk solids. Granular Matter **5**(1), 1–45 (2003)
62. Sperl, M.: Experiments on corn pressure in silo cells. translation and comment of Janssen's paper from 1895. Granular Matter **8**(2), 59–65 (2006)
63. Thornton, C.: Numerical simulations of deviatoric shear deformation of granular media. Géotechnique **50**(1), 43–53 (2000)
64. Thornton, C., Antony, S.J.: Quasi-static deformation of a soft particle system. Powder Technology **109**(1-3), 179–191 (2000)
65. Thornton, C., Zhang, L.: A DEM comparison of different shear testing devices. In: Y. Kishino (ed.) Powders & Grains 2001, pp. 183–190. Balkema, Rotterdam (2001)
66. Tighe, B.P., Sperl, M.: Pressure and motion of dry sand: translation of Hagen's paper from 1852. Granular Matter **9**(3/4), 141–144 (2007)
67. Tomas, J.: Particle adhesion fundamentals and bulk powder consolidation. KONA **18**, 157–169 (2000)
68. Vermeer, P.A., Diebels, S., Ehlers, W., Herrmann, H.J., Luding, S., Ramm, E. (eds.): Continuous and Discontinuous Modelling of Cohesive Frictional Materials. Springer, Berlin (2001). Lecture Notes in Physics 568
69. Walton, O.R., Braun, R.L.: Viscosity, granular-temperature, and stress calculations for shearing assemblies of inelastic, frictional disks. J. Rheol. **30**(5), 949–980 (1986)
70. Zhu, C.Y., Shukla, A., Sadd, M.H.: Prediction of dynamic contact loads in granular assemblies. J. of Applied Mechanics **58**, 341 (1991)

Editorial Policy

1. Volumes in the following three categories will be published in LNCSE:

i) Research monographs
ii) Lecture and seminar notes
iii) Conference proceedings

Those considering a book which might be suitable for the series are strongly advised to contact the publisher or the series editors at an early stage.

2. Categories i) and ii). These categories will be emphasized by Lecture Notes in Computational Science and Engineering. **Submissions by interdisciplinary teams of authors are encouraged**. The goal is to report new developments – quickly, informally, and in a way that will make them accessible to non-specialists. In the evaluation of submissions timeliness of the work is an important criterion. Texts should be well-rounded, well-written and reasonably self-contained. In most cases the work will contain results of others as well as those of the author(s). In each case the author(s) should provide sufficient motivation, examples, and applications. In this respect, Ph.D. theses will usually be deemed unsuitable for the Lecture Notes series. Proposals for volumes in these categories should be submitted either to one of the series editors or to Springer-Verlag, Heidelberg, and will be refereed. A provisional judgment on the acceptability of a project can be based on partial information about the work: a detailed outline describing the contents of each chapter, the estimated length, a bibliography, and one or two sample chapters – or a first draft. A final decision whether to accept will rest on an evaluation of the completed work which should include

– at least 100 pages of text;
– a table of contents;
– an informative introduction perhaps with some historical remarks which should be accessible to readers unfamiliar with the topic treated;
– a subject index.

3. Category iii). Conference proceedings will be considered for publication provided that they are both of exceptional interest and devoted to a single topic. One (or more) expert participants will act as the scientific editor(s) of the volume. They select the papers which are suitable for inclusion and have them individually refereed as for a journal. Papers not closely related to the central topic are to be excluded. Organizers should contact Lecture Notes in Computational Science and Engineering at the planning stage.

In exceptional cases some other multi-author-volumes may be considered in this category.

4. Format. Only works in English are considered. They should be submitted in camera-ready form according to Springer-Verlag's specifications.
Electronic material can be included if appropriate. Please contact the publisher.
Technical instructions and/or LaTeX macros are available via http://www.springer.com/authors/book+authors?SGWID=0-154102-12-417900-0. The macros can also be sent on request.

General Remarks

Lecture Notes are printed by photo-offset from the master-copy delivered in camera-ready form by the authors. For this purpose Springer-Verlag provides technical instructions for the preparation of manuscripts. See also *Editorial Policy*.

Careful preparation of manuscripts will help keep production time short and ensure a satisfactory appearance of the finished book.

The following terms and conditions hold:

Categories i), ii), and iii):
Authors receive 50 free copies of their book. No royalty is paid. Commitment to publish is made by letter of intent rather than by signing a formal contract. Springer- Verlag secures the copyright for each volume.

For conference proceedings, editors receive a total of 50 free copies of their volume for distribution to the contributing authors.

All categories:
Authors are entitled to purchase further copies of their book and other Springer mathematics books for their personal use, at a discount of 33.3% directly from Springer-Verlag.

Addresses:

Timothy J. Barth
NASA Ames Research Center
NAS Division
Moffett Field, CA 94035, USA
e-mail: barth@nas.nasa.gov

Michael Griebel
Institut für Numerische Simulation
der Universität Bonn
Wegelerstr. 6
53115 Bonn, Germany
e-mail: griebel@ins.uni-bonn.de

David E. Keyes
Department of Applied Physics
and Applied Mathematics
Columbia University
200 S. W. Mudd Building
500 W. 120th Street
New York, NY 10027, USA
e-mail: david.keyes@columbia.edu

Risto M. Nieminen
Laboratory of Physics
Helsinki University of Technology
02150 Espoo, Finland
e-mail: rni@fyslab.hut.fi

Dirk Roose
Department of Computer Science
Katholieke Universiteit Leuven
Celestijnenlaan 200A
3001 Leuven-Heverlee, Belgium
e-mail: dirk.roose@cs.kuleuven.ac.be

Tamar Schlick
Department of Chemistry
Courant Institute of Mathematical
Sciences
New York University
and Howard Hughes Medical Institute
251 Mercer Street
New York, NY 10012, USA
e-mail: schlick@nyu.edu

Mathematics Editor at Springer:
Martin Peters
Springer-Verlag
Mathematics Editorial IV
Tiergartenstrasse 17
D-69121 Heidelberg, Germany
Tel.: *49 (6221) 487-8185
Fax: *49 (6221) 487-8355
e-mail: martin.peters@springer.com

Lecture Notes
in Computational Science
and Engineering

1. D. Funaro, *Spectral Elements for Transport-Dominated Equations.*

2. H. P. Langtangen, *Computational Partial Differential Equations.* Numerical Methods and Diffpack Programming.

3. W. Hackbusch, G. Wittum (eds.), *Multigrid Methods V.*

4. P. Deuflhard, J. Hermans, B. Leimkuhler, A. E. Mark, S. Reich, R. D. Skeel (eds.), *Computational Molecular Dynamics: Challenges, Methods, Ideas.*

5. D. Kröner, M. Ohlberger, C. Rohde (eds.), *An Introduction to Recent Developments in Theory and Numerics for Conservation Laws.*

6. S. Turek, *Efficient Solvers for Incompressible Flow Problems.* An Algorithmic and Computational Approach.

7. R. von Schwerin, ***M**ulti **B**ody **S**ystem **SIM**ulation.* Numerical Methods, Algorithms, and Software.

8. H.-J. Bungartz, F. Durst, C. Zenger (eds.), *High Performance Scientific and Engineering Computing.*

9. T. J. Barth, H. Deconinck (eds.), *High-Order Methods for Computational Physics.*

10. H. P. Langtangen, A. M. Bruaset, E. Quak (eds.), *Advances in Software Tools for Scientific Computing.*

11. B. Cockburn, G. E. Karniadakis, C.-W. Shu (eds.), *Discontinuous Galerkin Methods.* Theory, Computation and Applications.

12. U. van Rienen, *Numerical Methods in Computational Electrodynamics.* Linear Systems in Practical Applications.

13. B. Engquist, L. Johnsson, M. Hammill, F. Short (eds.), *Simulation and Visualization on the Grid.*

14. E. Dick, K. Riemslagh, J. Vierendeels (eds.), *Multigrid Methods VI.*

15. A. Frommer, T. Lippert, B. Medeke, K. Schilling (eds.), *Numerical Challenges in Lattice Quantum Chromodynamics.*

16. J. Lang, *Adaptive Multilevel Solution of Nonlinear Parabolic PDE Systems.* Theory, Algorithm, and Applications.

17. B. I. Wohlmuth, *Discretization Methods and Iterative Solvers Based on Domain Decomposition.*

18. U. van Rienen, M. Günther, D. Hecht (eds.), *Scientific Computing in Electrical Engineering.*

19. I. Babuška, P. G. Ciarlet, T. Miyoshi (eds.), *Mathematical Modeling and Numerical Simulation in Continuum Mechanics.*

20. T. J. Barth, T. Chan, R. Haimes (eds.), *Multiscale and Multiresolution Methods.* Theory and Applications.

21. M. Breuer, F. Durst, C. Zenger (eds.), *High Performance Scientific and Engineering Computing.*

22. K. Urban, *Wavelets in Numerical Simulation.* Problem Adapted Construction and Applications.

23. L. F. Pavarino, A. Toselli (eds.), *Recent Developments in Domain Decomposition Methods.*

24. T. Schlick, H. H. Gan (eds.), *Computational Methods for Macromolecules: Challenges and Applications.*

25. T. J. Barth, H. Deconinck (eds.), *Error Estimation and Adaptive Discretization Methods in Computational Fluid Dynamics.*

26. M. Griebel, M. A. Schweitzer (eds.), *Meshfree Methods for Partial Differential Equations.*

27. S. Müller, *Adaptive Multiscale Schemes for Conservation Laws.*

28. C. Carstensen, S. Funken, W. Hackbusch, R. H. W. Hoppe, P. Monk (eds.), *Computational Electromagnetics.*

29. M. A. Schweitzer, *A Parallel Multilevel Partition of Unity Method for Elliptic Partial Differential Equations.*

30. T. Biegler, O. Ghattas, M. Heinkenschloss, B. van Bloemen Waanders (eds.), *Large-Scale PDE-Constrained Optimization.*

31. M. Ainsworth, P. Davies, D. Duncan, P. Martin, B. Rynne (eds.), *Topics in Computational Wave Propagation.* Direct and Inverse Problems.

32. H. Emmerich, B. Nestler, M. Schreckenberg (eds.), *Interface and Transport Dynamics.* Computational Modelling.

33. H. P. Langtangen, A. Tveito (eds.), *Advanced Topics in Computational Partial Differential Equations.* Numerical Methods and Diffpack Programming.

34. V. John, *Large Eddy Simulation of Turbulent Incompressible Flows.* Analytical and Numerical Results for a Class of LES Models.

35. E. Bänsch (ed.), *Challenges in Scientific Computing - CISC 2002.*

36. B. N. Khoromskij, G. Wittum, *Numerical Solution of Elliptic Differential Equations by Reduction to the Interface.*

37. A. Iske, *Multiresolution Methods in Scattered Data Modelling.*

38. S.-I. Niculescu, K. Gu (eds.), *Advances in Time-Delay Systems.*

39. S. Attinger, P. Koumoutsakos (eds.), *Multiscale Modelling and Simulation.*

40. R. Kornhuber, R. Hoppe, J. Périaux, O. Pironneau, O. Wildlund, J. Xu (eds.), *Domain Decomposition Methods in Science and Engineering.*

41. T. Plewa, T. Linde, V.G. Weirs (eds.), *Adaptive Mesh Refinement – Theory and Applications.*

42. A. Schmidt, K.G. Siebert, *Design of Adaptive Finite Element Software.* The Finite Element Toolbox ALBERTA.

43. M. Griebel, M.A. Schweitzer (eds.), *Meshfree Methods for Partial Differential Equations II.*

44. B. Engquist, P. Lötstedt, O. Runborg (eds.), *Multiscale Methods in Science and Engineering.*

45. P. Benner, V. Mehrmann, D.C. Sorensen (eds.), *Dimension Reduction of Large-Scale Systems.*

46. D. Kressner, *Numerical Methods for General and Structured Eigenvalue Problems.*

47. A. Boriçi, A. Frommer, B. Joó, A. Kennedy, B. Pendleton (eds.), *QCD and Numerical Analysis III.*

48. F. Graziani (ed.), *Computational Methods in Transport.*

49. B. Leimkuhler, C. Chipot, R. Elber, A. Laaksonen, A. Mark, T. Schlick, C. Schütte, R. Skeel (eds.), *New Algorithms for Macromolecular Simulation.*

50. M. Bücker, G. Corliss, P. Hovland, U. Naumann, B. Norris (eds.), *Automatic Differentiation: Applications, Theory, and Implementations.*

51. A.M. Bruaset, A. Tveito (eds.), *Numerical Solution of Partial Differential Equations on Parallel Computers.*

52. K.H. Hoffmann, A. Meyer (eds.), *Parallel Algorithms and Cluster Computing.*

53. H.-J. Bungartz, M. Schäfer (eds.), *Fluid-Structure Interaction.*

54. J. Behrens, *Adaptive Atmospheric Modeling.*

55. O. Widlund, D. Keyes (eds.), *Domain Decomposition Methods in Science and Engineering XVI.*

56. S. Kassinos, C. Langer, G. Iaccarino, P. Moin (eds.), *Complex Effects in Large Eddy Simulations.*

57. M. Griebel, M.A Schweitzer (eds.), *Meshfree Methods for Partial Differential Equations III.*

58. A.N. Gorban, B. Kégl, D.C. Wunsch, A. Zinovyev (eds.), *Principal Manifolds for Data Visualization and Dimension Reduction.*

59. H. Ammari (ed.), *Modeling and Computations in Electromagnetics: A Volume Dedicated to Jean-Claude Nédélec.*

60. U. Langer, M. Discacciati, D. Keyes, O. Widlund, W. Zulehner (eds.), *Domain Decomposition Methods in Science and Engineering XVII.*

61. T. Mathew, *Domain Decomposition Methods for the Numerical Solution of Partial Differential Equations.*

62. F. Graziani (ed.), *Computational Methods in Transport: Verification and Validation.*

63. M. Bebendorf, *Hierarchical Matrices. A Means to Efficiently Solve Elliptic Boundary Value Problems.*

64. C.H. Bischof, H.M. Bücker, P. Hovland, U. Naumann, J. Utke (eds.), *Advances in Automatic Differentiation.*

65. M. Griebel, M.A. Schweitzer (eds.), *Meshfree Methods for Partial Differential Equations IV.*

66. B. Engquist, P. Lötstedt, O. Runborg (eds.), *Multiscale Modeling and Simulation in Science.*

67. I.H. Tuncer, Ü. Gülcat, D.R. Emerson, K. Matsuno (eds.), *Parallel Computational Fluid Dynamics.*

68. S. Yip, T. Diaz de la Rubia (eds.), *Scientific Modeling and Simulations.*

69. A. Hegarty, N. Kopteva, E. O'Riordan, M. Stynes (eds.), *BAIL 2008 – Boundary and Interior Layres.*

70. M. Bercovier, M.J. Gander, R. Kornhuber, O. Widlund (eds.), *Domain Decomposition Methods in Science and Engineering XVIII.*

71. B. Koren, C. Vuik (eds.), *Advanced Computational Methods in Science and Engineering*

For further information on these books please have a look at our mathematics catalogue at the following URL: www.springer.com/series/3527

Monographs in Computational Science and Engineering

1. J. Sundnes, G.T. Lines, X. Cai, B.F. Nielsen, K.-A. Mardal, A. Tveito, *Computing the Electrical Activity in the Heart.*

For further information on this book, please have a look at our mathematics catalogue at the following URL: www.springer.com/series/7417

Texts in Computational Science
and Engineering

1. H. P. Langtangen, *Computational Partial Differential Equations.* Numerical Methods and Diffpack Programming. 2nd Edition

2. A. Quarteroni, F. Saleri, *Scientific Computing with MATLAB and Octave.* 2nd Edition

3. H. P. Langtangen, *Python Scripting for Computational Science.* 3rd Edition

4. H. Gardner, G. Manduchi, *Design Patterns for e-Science.*

5. M. Griebel, S. Knapek, G. Zumbusch, *Numerical Simulation in Molecular Dynamics.*

6. H.P. Langtangen, *A Primer on Scientific Programming with Python.*

For further information on these books please have a look at our mathematics catalogue at the following URL: www.springer.com/series/5151